The CHEMISTRY *of* CEMENTS

VOLUME I

The CHEMISTRY *of* CEMENTS

Edited by

H. F. W. TAYLOR

Department of Chemistry
University of Aberdeen, Scotland

VOLUME I

1964
ACADEMIC PRESS · LONDON and NEW YORK

ACADEMIC PRESS INC. (LONDON) LTD
Berkeley Square House
Berkeley Square
London, W.1

U.S. Edition published by
ACADEMIC PRESS INC.
111 Fifth Avenue
New York 3, New York

Library of Congress Catalog Card Number: 64-14225

Printed in Great Britain by
Spottiswoode, Ballantyne and Company Limited
London and Colchester

List of Contributors

S. Brunauer, *Portland Cement Association Research and Development Laboratories, Skokie, Illinois, U.S.A.* (p. 287)

L. E. Copeland, *Portland Cement Association Research and Development Laboratories, Skokie, Illinois, U.S.A.* (p. 313)

Å. Grudemo, *Swedish Cement and Concrete Research Institute, Royal Institute of Technology, Stockholm, Sweden* (p. 371)

J. W. Jeffery, *Birkbeck College Crystallography Laboratory, Torrington Square, London, England* (p. 131)

D. L. Kantro, *Portland Cement Association Research and Development Laboratories, Skokie, Illinois, U.S.A.* (pp. 287, 313)

H. G. Midgley, *Building Research Station, Garston, Watford, Hertfordshire, England* (p. 89)

H. W. W. Pollitt, *Associated Portland Cement Manufacturers Limited, Research Department, Greenhithe, Kent, England* (p. 27)

T. C. Powers, *Portland Cement Association Research and Development Laboratories, Skokie, Illinois, U.S.A.* (p. 391)

H. F. W. Taylor, *Department of Chemistry, University of Aberdeen, Scotland* (pp. 1, 167, 417)

R. Turriziani, *Instituto di Chimica Applicata e Metallurgia, Facoltà di Ingegneria, University of Cagliari, Italy* (p. 233)

J. H. Welch, *Late of the Building Research Station, Garston, Watford, Hertfordshire, England* (p. 49)

Preface

Cement chemistry is largely the chemistry of the calcium silicates and aluminates, both anhydrous and hydrated. This book describes the chemistry of these and related compounds, and deals with the more important chemical aspects of the manufacture and use of cements. All the main cements used in building are considered, including Portland, aluminous, slag, and expansive cements, pozzolanas, and calcium silicate products. Crystallographic and other data for the principal compounds are given in an appendix.

Much emphasis is placed on the underlying chemistry; my object has been not so much to produce a technical handbook, as to set out the basic chemistry and to show how this can be applied. There are several reasons for this approach. The more technical aspects of cement chemistry are adequately treated in other books, but there is no recent and full summary of the basic chemistry. The intention, moreover, is not only to explain the chemistry of existing materials and processes, but also to provide a book which will be useful to all who are concerned with the development of new ones. For this purpose, a wide chemical background is essential.

I have tried to make this book useful and comprehensible to those entering the field from other branches of chemistry. Chapters have therefore been included that deal with the more specialized experimental techniques used, such as high-temperature phase equilibria, electron microscopy and X-ray diffraction.

This is an edited book. Cement chemistry is a subject in which intensive research is taking place, and it is inevitable that differences of opinion or approach will exist between different authors. Cross references have been inserted to draw the reader's attention to conflicting views.

I am grateful to many friends and colleagues, and especially to all the contributors for their willing cooperation in reducing gaps and overlaps. I should like to thank Dr. R. W. Nurse of the Building Research Station for much helpful advice, and to Mrs. S. Kelly for editorial assistance.

H. F. W. Taylor

Aberdeen
January 1964

Contents

Contents of Volume 2

Introduction

H. F. W. TAYLOR

Department of Chemistry,
University of Aberdeen, Scotland

CONTENTS

I. General Points

In its broadest sense, the word cement denotes any kind of adhesive. In building and civil engineering, it denotes a substance which can be used to bind together sand and broken stone, or other forms of aggregate, into a solid mass. In this way are produced such materials as concrete, mortars, and various kinds of asbestos-cement products. A cement may be a single chemical compound, but more often it is a mixture.

A primitive type of cement is represented by hydrated lime ($Ca(OH)_2$), which is still sometimes used together with sand to make mortar. A paste made by mixing hydrated lime with water and sand gradually hardens as a result of drying out and reaction with atmospheric carbon dioxide to give $CaCO_3$. The sand takes no part in the reaction. The development of strength is slow and uneven, and the paste does not harden at all if it is immersed in water.

More important today are hydraulic cements, of which Portland cement is the most familiar example. When made into a paste with water and aggregate, these set and harden as a result of chemical reactions between the water and the compounds present in the cement. Setting and hardening do not depened on drying out, or on adventitious processes

such as reaction with atmospheric carbon dioxide. They occur not only if the paste is left to stand in air but also if it is placed in water. With a good hydraulic cement, development of strength is predictable, uniform and relatively rapid. The product is of low permeability to water, and is nearly insoluble in water which, therefore, under normal conditions, does not destroy the hardened material.

The more familiar hydraulic cements, including of course Portland cement, set and harden on mixing with water at room temperature. Some can also be used at higher temperatures, and other materials exist which act as hydraulic cements only at high temperatures. An example of the latter class is a mixture of lime with finely divided quartz, which acts as a hydraulic cement if it is first made into a paste with water and then treated with saturated steam at high pressure in an autoclave.

Certain terms require definition. The thick slurry formed by mixing a hydraulic cement with aggregate and water in such proportions that setting will occur is called a *paste*. This term is also used to denote the resultant material at all stages of setting and hardening, even after it has become quite rigid. *Setting* is the initial stiffening, which usually occurs within a few hours; *hardening* is the development of strength, which is a slower process. With Portland cement at ordinary temperature, the strength increases detectably over at least two years. The reactions that cause setting and hardening are collectively described as *hydration reactions*, though the use of the word is very loose, as the processes are usually more complicated than the formation of a hydrate from an anhydrous salt. The term *gauging* is sometimes used to denote the initial mixing of the cement with water; *curing* means leaving the paste while setting and hardening proceed, and can be carried out under various conditions, as in air, under water, or in saturated steam.

It can usually be assumed that the aggregate undergoes, at the most, only surface reactions with the cement. Chemical studies on cement hydration can therefore legitimately be made in the absence of aggregate, and it is almost always more convenient to study them under these conditions. Pastes made from a cement with water in the absence of aggregate are called *neat cement* pastes. Mechanical or physical determinations, such as strength tests, are usually made with an aggregate present, as determinations of this type on neat cement pastes can give misleading results.

Reactive materials which are not in themselves cements are sometimes deliberately added to a mix. An example of this is the use of finely divided quartz in mixtures containing also Portland cement and aggregate, which are cured in the autoclave. Such materials are often described as forming part of the aggregate, but it is more logical to regard them as an integral

part of the cement, and their use is considered in this book. Cases have also been reported in which unintended reactions between cement and aggregate have occurred. For a discussion of these, and of the chemistry of aggregates in general, other works must be consulted [*1*, *2*].

Nearly all of the cements commonly used in building owe their action mainly to the formation of hydrated calcium silicates, aluminates, or aluminate sulphates, or to compounds of two or more of these groups. This book is concerned almost entirely with cements of these types. A few cements owe their action to the formation of compounds of other types, such as calcium sulphate hydrates, basic magnesium chlorides, or magnesium silicate hydrates. These are briefly considered later in this introduction.

Chemical formulae in cement chemistry are often expressed as a sum of oxides; thus tricalcium silicate, Ca_3SiO_5, can be written as $3CaO.SiO_2$. This way of writing the formula does not, of course, imply that the constituent oxides have any separate existence within the structure of the compound. It is usual to use abbreviations for the formulae of the oxides most often encountered, such as C for CaO and S for SiO_2; Ca_3SiO_5 thus becomes C_3S. This system is often used in conjunction with orthodox chemical notation, within a single chemical equation, e.g.

$$3CaO + SiO_2 = C_3S \tag{1}$$

or even within a single chemical formula; thus $C_3A.3CaSO_4.32H_2O$ denotes $6CaO.Al_2O_3.3SO_3.32H_2O$. The full list of abbreviations in general use is as follows.

C = CaO	F = Fe_2O_3	N = Na_2O	P = P_2O_5
A = Al_2O_3	M = MgO	K = K_2O	f = FeO
S = SiO_2	H = H_2O	L = Li_2O	T = TiO_2

Others are sometimes seen, such as $\bar{S} = SO_3$ and $\bar{C} = CO_2$. The formulae of the simple oxides are generally written out in full when the compounds themselves are indicated, as in the equation given above, though they are sometimes abbreviated in this context also; thus one may also represent equation (1) as $3C + S = C_3S$.

Many of the more technical aspects of cement chemistry are not treated in detail in this book; the physical and mechanical properties of cement pastes, technical specifications, and the reactions of cement pastes with aggressive substances are among topics which are only considered incidentally. The history of cements has also been omitted. Much useful information on these subjects will be found in other text books [*1*, *3*] and in the proceedings of the three most recent international symposia on cement chemistry [*4–6*].

II. Types of Cement

A. Portland Cements

Portland cement is by far the most important cement in terms of the quantity produced. It is made by heating a mixture of limestone and clay, or other material of similar bulk composition, to a temperature at which partial fusion occurs. The product, which is called clinker, is ground and mixed with a few per cent of gypsum. The clinker contains four main phases: tricalcium silicate (C_3S), β-dicalcium silicate (β-C_2S), tricalcium aluminate (C_3A), and ferrite solid solution (composition between C_2F and C_6A_2F approximately, often approximating to C_4AF).

In most countries where Portland cement is manufactured, several types are recognized having different characteristics. The most important variables are the rate of hardening, the rate and total extent of heat evolution during hydration, and the resistance of the hardened cement to attack by sulphate solutions. These characteristics are influenced by the relative proportions of the four phases mentioned above, and by physical factors such as the fineness of grinding. Specifications for particular types of Portland (or other) cements can be based on tests of three basic types. These are: (i) tests made on the unhydrated cement; (ii) tests made on the behaviour of the cement during hydration; and (iii) tests made on the hardened paste. Especially with groups (ii) and (iii), the conditions of the test must be rigidly defined. Group (i) includes such determinations as bulk chemical analysis and particle size distribution. Group (ii) includes, for example, determinations of setting time or of heat of hydration. Group (iii) includes determinations of such properties as the compressive strength after various times of curing or the resistance of the hardened paste to sulphate attack. In general, combinations of tests belonging to different groups are used. A compilation of specifications exists [*7*].

The classification of Portland cements which is used in the U.S.A. is frequently referred to. It is based partly on the *potential phase composition*, i.e. the proportions of the four main phases calculated from the bulk chemical analysis by a standard method known as the Bogue calculation, which is described in Chapter 2. The potential phase composition is probably in fair agreement with the true phase composition for most Portland cements. Table I gives the names and the average potential phase compositions of the five types of Portland cement manufactured

TABLE I

Potential phase compositions for cements of different types made in the U.S.A. (average values quoted in 1955 by Bogue [3])

Cement		Potential phase composition (%)					Free CaO	$CaSO_4$	Total[1]
Type	Description	C_3S	C_2S	C_3A	C_4AF	MgO	(%)	(%)	(%)
I	General Use	45	27	11	8	2·9	0·5	3·1	98
II	Moderate Heat of Hardening	44	31	5	13	2·5	0·4	2·8	99
III	High Early Strength	53	19	11	9	2·0	0·7	4·0	99
IV	Low Heat	28	49	4	12	1·8	0·2	3·2	98
V	Sulphate Resisting	38	43	4	9	1·9	0·5	2·7	99

(1) The remaining 1–2% consists mainly of moisture, insoluble residue, and alkali oxides combined in various ways.

in the U.S.A. Of these, Type 1 is by far the most important; the others are only made for special purposes. Typical compositions of Type 1 cements made in the U.S.A., and of comparable Portland cements intended for general use made in other countries, have not changed greatly since about 1935, but most cements made there or in other countries before that time contained more C_2S and less C_3S.

The term Rapid Hardening Portland Cement, used in Britain, denotes a product corresponding approximately in properties to the American High Early Strength Portland Cement. The names Browmillerite Portland Cement, Ferrari Cement and Erz Cement (the last of these is no longer manufactured) denote Portland cements high in ferrite and containing little or no C_3A. In contrast, Kühl Cement and Bauxitland Cement are Portland cements high in both C_3A and ferrite.

To a very rough approximation, the four main compounds in the cement clinker hydrate independently of each other. On the basis of this assumption, the relationship between phase composition and properties can be understood.

1. *Compressive Strength*

Figure 1 shows the development of compressive strength in pastes of each of the four compounds, and Fig. 2 shows comparable results for two typical cements of differing compositions. The strength at early ages is due mainly to the C_3S, but as hydration proceeds the β-C_2S becomes increasingly important. High early strength can therefore be obtained by increasing the content of C_3S. It can also be obtained by grinding the

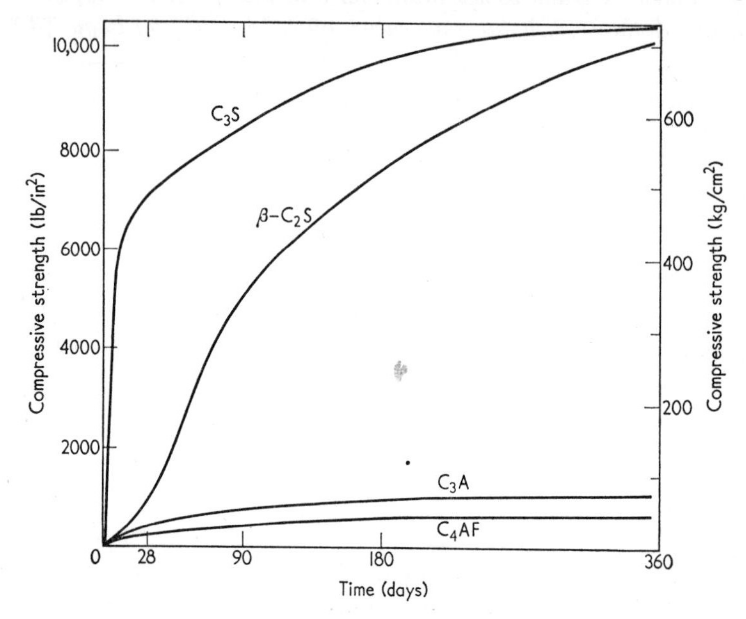

FIG. 1. Curves of compressive strengths against curing time for the pure phases C_3S, β-C_2S, C_3A and C_4AF [8].

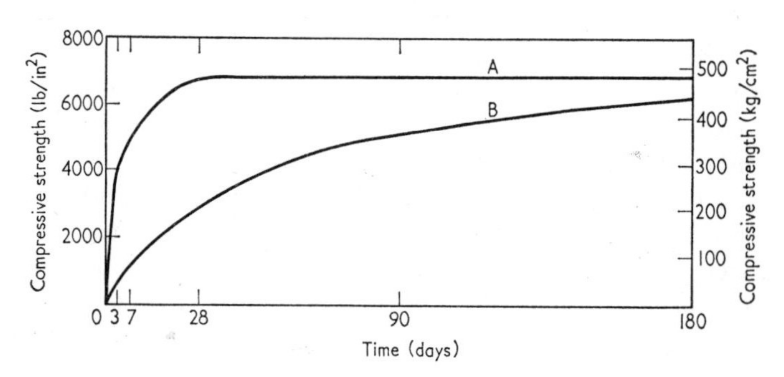

FIG. 2. Curves of compressive strength against curing time for 1:3 mortars made from two different Portland cements [9]. A = 70% C_3S, 10% C_2S. B = 30% C_3S, 50% C_2S.

clinker more finely (Fig. 3). In Britain, ordinary Portland cements are ground to a specific surface not less than 2250 cm^2/g, but rapid hardening Portland cements are ground to not less than 3250 cm^2/g [10].

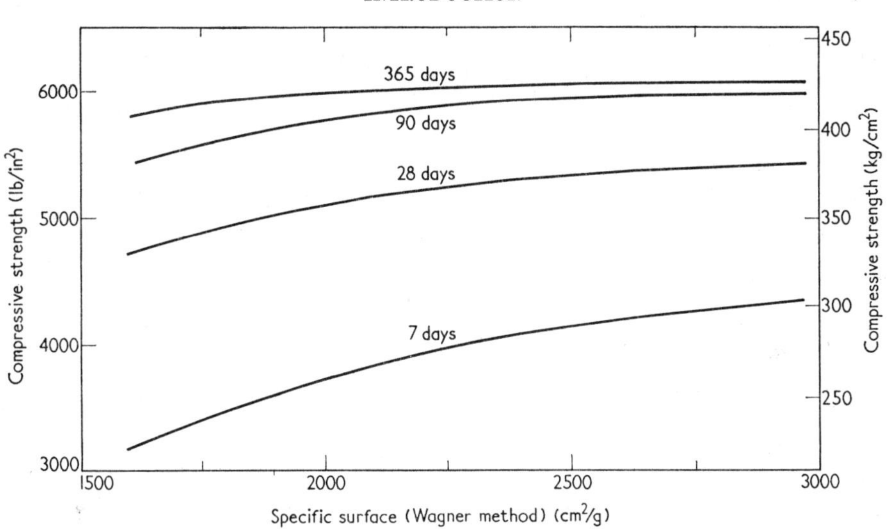

FIG. 3. Effect of fineness of cement on strength of concrete [11].

2. *Evolution of Heat*

The hydration of Portland cement is an exothermic process and the rate and total extent of heat evolution can be important, especially in the construction of dams and other large concrete structures. C_3A and C_3S make the largest contributions to the heat of hydration; in low-heat

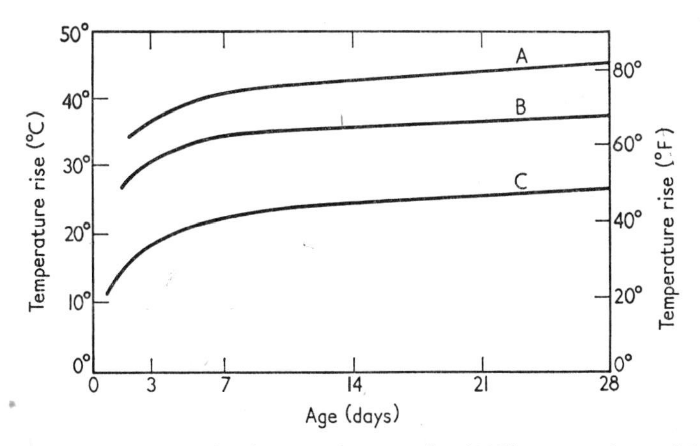

FIG. 4. Temperature rise in concrete samples (10% cement by weight) cured adiabatically. A, Rapid hardening Portland cement. B, Normal Portland cement. C, Low heat Portland cement [1].

Portland cements, the proportions of these compounds are reduced. Some reduction in heat evolution can be effected by lowering the Al_2O_3/Fe_2O_3 ratio and thereby increasing the content of C_4AF and lowering that of C_3A; this does not greatly affect the early strength. Greater reduction in heat evolution can be obtained only by lowering the C_3S content and increasing that of C_2S; this lowers the early strength, but not the final strength. Figure 4 shows typical results for the temperature rise in concretes cured adiabatically and made with cements of different types. In Britain, the specification for low-heat Portland cement [*12*] includes a requirement that the heat of hydration shall not exceed 60 cal/g at 7 days, nor 70 cal/g at 28 days. For a normal Portland cement, the 28-day value is around 100 cal/g.

3. Sulphate Resistance

Concrete made using Portland cement cured at ordinary temperature is susceptible to attack by sulphate solutions, which can cause it to expand and disintegrate. Some ground waters, as well as sea waters, contain enough sulphate for this to be a serious problem. The ease of attack depends primarily on the C_3A content, and sulphate-resisting cements contain reduced proportions of this compound or none at all; rapid cooling of the clinker as it leaves the kiln improves the sulphate resistance, either because it causes less C_3A to be formed, or because it affects the size or perfection of the crystals. The expansion caused by sulphate attack is generally attributed to the formation of ettringite ($C_3A.3CaSO_4.32H_2O$), which has the low density of 1·73 g/cm^3, though this view has recently been disputed [*13*]. With magnesium sulphate solutions, the calcium silicate hydrates are also attacked, $Mg(OH)_2$, $CaSO_4.2H_2O$, hydrous silica, and possibly also magnesium silicate hydrates being formed. For a more detailed discussion of the reactions of Portland cement pastes with sulphate solutions and other aggressive media, other works must be consulted [*1, 3, 14, 15*].

4. Other Portland Cements

In addition to the types of Portland cements already discussed, some others are manufactured for specialized uses. White cements are made from iron-free materials, such as limestone and china clay; they contain C_3S, β-C_2S and C_3A. Special cements have been developed for sealing oil-well pipes to the surrounding rock [*16, 17*]; the slurry must withstand pumping through regions of high temperature and pressure without risk of premature setting. Three approaches have been employed, singly or in combination. Firstly, the cement can be more coarsely ground than usual. Secondly, the cement can have a very low Al_2O_3/Fe_2O_3 ratio, so that no C_3A is present, and the ferrite phase has a composition less aluminous than C_4AF. Thirdly, organic retarding agents such as carbohydrates or lignosulphonates may be used.

Various products are made by grinding Portland cement, or cement clinker, with other materials in addition to gypsum. Many forms of slag cements (Volume 2, Chapter 13), pozzolanic cements (Volume 2, Chapter 14) and expansive cements (Volume 2, Chapter 15) come into this category. Other additives [18] include water-proofing and air entraining agents, pigments, and various substances to accelerate or retard setting. For the production of mortars, it is desirable to increase the plasticity of the paste. Hydrated lime is often added to the cement for this purpose; more sophisticated masonry cements have also been developed for this purpose and are mixtures of Portland cement with various substances, such as ground limestone and air-entraining agents [*19*].

In general, Portland cements are made from raw materials low in involatile components other than lime, silica, alumina and iron oxides. Components generally considered undesirable for various reasons include MgO, P_2O_5, fluorides, and alkali oxides. In recent years, research has been carried out in order to permit the use of raw materials containing higher proportions of these components than have usually been considered allowable. Examples of this are the development of high-magnesia Portland cements [*20*] and the use of raw materials high in P_2O_5 (Chapter 3).

B. Non-Portland Cements

1. Cements Based on Calcium Silicates or Aluminates

(a) *Aluminous cement* (or high alumina cement) is made by heating a mixture of limestone and bauxite, usually to complete fusion. The chief constituent is monocalcium aluminate (CA). Aluminous cement pastes set about as quickly as those of Portland cement, but harden very rapidly; a high strength is reached in 24 hours. The product has a very high resistance to sulphate attack. Aluminous cement, which is also used for making refractory concrete, is discussed in Volume 2, Chapter 12.

(b) *Slag cements* comprise several products made from blastfurnace slag which has been granulated, that is, melted and subsequently quenched to a glass. The most important types are blastfurnace cements and supersulphate cements. Blastfurnace cements are mixtures of granulated slag and Portland cement clinker, in proportions which can vary widely. Supersulphate cements contain 75% or more of granulated slag, mixed with calcium sulphate and up to 5% of Portland cement clinker or Portland cement. Slag cements are considered in Volume 2, Chapter 13.

(c) *Pozzolanas* are substances which are not in themselves cementitious, but which react with $Ca(OH)_2$ in the presence of water at ordinary temperature and thereby act as cements. Examples include certain naturally occurring materials of volcanic origin, and artificial materials made by calcining certain clays or shales. Pozzolanic cements are mixtures of a pozzolana with Portland cement. Pozzolanas and pozzolanic cements are discussed in Volume 2, Chapter 14.

(d) *Expansive cements* are cements which yield a paste that expands slightly during hydration. In this respect they differ from the other cements so far mentioned, as with all of these slight contraction takes place. The development of expansive cements is still at an early stage and various mixtures, mostly consisting essentially of Portland or aluminous cements with suitable additives, have been investigated. Stressing cements are expansive cements of a type designed to stretch a steel reinforcement and thus to cause prestressing. Expansive cements of all kinds are discussed in Volume 2, Chapter 15.

(e) *Calcium silicate products* are made by autoclave treatment of pastes containing lime together with silica or reactive silica-rich materials; examples include calcium silicate bricks, and various types of lightweight building blocks, heat insulating materials, and calcium silicate asbestos sheets. Chapter 16 (Volume 2) deals with these products and also with calcium silicate fillers and lime-stabilized soils.

2. *Cements Not Based on Calcium Silicates or Aluminates*

(a) *Magnesium oxychloride cement* (*sorel cement*) This is obtained by treating a mixture of MgO and aggregate with a concentrated solution of $MgCl_2$. The MgO is obtained by calcining $MgCO_3$ at a low temperature, so as to yield a reactive product. The product is hard, but does not resist water; it can be protected by polishing with wax in turpentine, and is often used as a flooring material.

The initial product of the reaction is believed to be $Mg_6(OH)_{10}Cl_2 \cdot 7H_2O$, which slowly changes into $Mg_2(OH)_3Cl \cdot 4H_2O$. The hardened paste seems to consist of flaky crystals of $Mg(OH)_2$ and needles or fibres of $Mg_2(OH)_3Cl \cdot 4H_2O$ [*1*, *21*].

The crystal structure of the latter compound was determined by de Wolff and Walter-Lévy [*22*]. It contains infinite chains of empirical formula $[Mg_2(OH)_3(H_2O)_3]^+$, held together with Cl^- ions and the remaining H_2O molecules, making the constitution $[Mg_2(OH)_3(H_2O)_3]Cl.H_2O$. The structure is distantly related to that of the calcium silicate hydrate formed on hydration of Portland cement.

A magnesium oxysulphate cement can be made in a similar way.

(b) *Gypsum plasters* These are produced by the partial or complete dehydration of gypsum ($CaSO_4.2H_2O$). The various forms of anhydrous and hydrated calcium sulphate have been described by Deer, Howie and Zussman [*23*]. At least four distinct forms exist; these are (a) gypsum, (b) hemihydrate or bassanite ($CaSO_4.\frac{1}{2}H_2O$), (c) soluble anhydrite or γ-$CaSO_4$, and (d) anhydrite (also called insoluble anhydrite or dead-burned gypsum, β-$CaSO_4$). The existence of a further polymorph, α-$CaSO_4$, stable above 1200° C, has also been suggested.

Hemihydrate and γ-$CaSO_4$ have closely similar crystal structures. In γ-$CaSO_4$ the Ca^{2+} and SO_4^{2-} ions form an open structure with continuous channels in one direction [*24*]. In hemihydrate, these channels contain water molecules. Some investigators have supposed that γ-$CaSO_4$ and hemihydrate represent points on a continuous solid solution series ranging from $CaSO_4$ to $CaSO_4.\frac{2}{3}H_2O$ or even $CaSO_4.H_2O$, but the evidence seems clearly to favour the view that they are distinct compounds [*23*, *25*]. Certain sites in the channels are energetically favourable for occupation by water molecules and, at least to a good approximation, either all of these sites are occupied or none at all.

The arrangements of the Ca^{2+} and SO_4^{2-} ions in gypsum and in anhydrite are quite distinct from each other and from that existing in γ-$CaSO_4$ and hemihydrate. There is nevertheless a sufficiently close similarity between all of these structures that they can under certain conditions be interconverted without complete breakdown occurring. During the setting of gypsum plasters, interconversion occurs, though not by this mechanism. If gypsum is heated in air above about 60° C it changes slowly to hemihydrate, or above about 100° C to γ-$CaSO_4$; above about 200° C, anhydrite is formed.

Figure 4 (Chapter 8) shows the solubilities of gypsum, anhydrite, and hemihydrate; the last of these is metastable under all conditions. The solubilities are much altered if more soluble salts are also present; thus in saturated $MgCl_2$ solution, gypsum transforms to anhydrite at 11° C.

Table II, which is based on information from Lea and Desch [*1*], describes the principal varieties of gypsum plaster, including methods of production and setting behaviour. The temperatures of formation of the various phases are higher than those mentioned above, because relatively short heating periods are used. In all cases, setting takes place on mixing into a paste with water and is due to rehydration to gypsum. This process is discussed later (p. 18).

TABLE II

Varieties of gypsum plaster

Name	Method of preparation	Phases present	Setting behaviour
Plaster of Paris	Gypsum heated to about 150° C in open vessels	Mainly hemihydrate with some gypsum and some γ-$CaSO_4$	Very rapid
Retarded hemihydrate gypsum plaster	(a) Plaster of Paris+ retarder (0·1% keratin)	As above	Moderate
	(b) Autoclaved treatment of gypsum	Hemihydrate	Moderate
Anhydrous gypsum plaster	Gypsum heated to about 190–200° C	γ-$CaSO_4$	Moderate
Keene's Cement or Parian Cement	Gypsum heated to about 600° C; accelerator added (e.g. 0·5–1% of K alum or K_2SO_4)	Anhydrite	Moderate
Estrich Gips	Gypsum heated to 1100–1200° C	Anhydrite and CaO	Very slow

(c) *Other cements* These may be treated briefly. Strontium aluminate and barium aluminate cements have been developed for use as refractory cements [*26*]. They permit higher temperatures to be reached than with calcium aluminate cements. Cements suitable for curing in the autoclave can be made from a variety of magnesium silicate minerals [*27*, *28*] (see Chapter 11). Barium silicate cements have a possible use for making concrete for use in nuclear shielding, but have so far been little investigated.

III. The Mechanism of Cementing Action

At least three questions need to be answered.

(i) Why do some compounds act as hydraulic cements, while others, of broadly similar composition, do not?

(ii) What kinds of substances are formed in the setting and hardening reactions?

(iii) What causes setting, and what gives the hardened material its strength?

A. Reactivities of Anhydrous Compounds

The various anhydrous calcium silicates and aluminates differ markedly from each other in their reactivities towards water at ordinary temperatures. Thus C_3S and CA, the main constituents of ordinary Portland and aluminous cements respectively, react relatively quickly and are strongly hydraulic. There are various polymorphs of dicalcium silicate; one of these, β-C_2S, reacts slowly but is strongly hydraulic, while another, γ-C_2S, is virtually inert. The aluminates C_3A and $C_{12}A_7$ react rapidly but pastes made from them develop little strength. Some other compounds, such as wollastonite (β-CS), gehlenite (C_2AS) and calcium hexaluminate (CA_6) are inert at room temperature.

Brandenberger [*29*] and Büssem [*30*] noted that, in many compounds stable at room temperature, each Ca atom was co-ordinated to six oxygens. They postulated that, in compounds formed at high temperatures, the co-ordination number was lower, and attributed reactivity to the tendency of the Ca to attain six co-ordination. Büssem believed that irregularity in the arrangement of co-ordinated oxygens was also an important factor. Bredig [*31*] put forward the opposite view: he considered that in hydraulically active compounds, the co-ordination number of the calcium was higher than usual.

All of these hypotheses were necessarily speculative at the time when they were proposed, because none of the crystal structures of the principal cementing compounds had been determined. By 1952, the structures of the important cementing compounds C_3S, β-C_2S and CA, and also of the inert γ-C_2S, were known. Jeffery [*32*] pointed out that in all of the reactive compounds mentioned above, the co-ordination of the calcium was irregular; he regarded irregularity of co-ordination, rather than the actual number of nearest neighbours, as the important factor. With C_3S, the presence of oxide ions surrounded entirely by calcium was an additional factor; large parts of the structure could be regarded as consisting of distorted CaO. He attributed the inertness of γ-C_2S to the fact that the co-ordination of the Ca^{2+} ions in this compound is symmetrical.

Most if not all of the main phases in Portland and aluminous cements have compositions which differ in varying degrees from the ideal on account of isomorphous replacement; thus, the C_3S in Portland cement contains small amounts of Mg^{2+} and Al^{3+}. These variations in composition appear to influence reactivity. The reactivity of the ferrite phase in Portland cement is reported to increase with its Al/Fe ratio [*33*].

Even at the time of writing, the crystal structures of most of the anhydrous calcium silicates and aluminates are not known to such a high degree of accuracy that one can be certain that the details of the calcium co-ordination are correct. Jeffery's views can be accepted as providing the most reasonable explanation of the limited data available.

The fact that an anhydrous calcium silicate or aluminate reacts with water does not necessarily mean that it is a good hydraulic cement; as has been seen, C_3A reacts quickly with water, but pastes so formed develop only low strengths. At least two other factors have an important bearing on hydraulic activity: the characteristics of the hydration products, and the circumstances under which these are formed.

B. Products Formed on Hydration

Table III gives a summary of the main products formed in the hydration reactions of some typical hydraulic cements. All have low or very low solubilities, but in most other respects they vary widely.

1. Chemical Nature

The products mentioned in Table III include a range of hydrated salts, basic salts, simple and complex hydroxides, and other compounds. All contain the elements of water, but these can occur as molecules or as ionic or covalently bonded hydroxyl, or in a combination of two or more of these ways. Infinite chain anions occur in 11 Å tobermorite and perhaps also in tobermorite gel, but not in the other cases mentioned. They are, therefore, not an essential feature.

2. Particle Size

In particle size, the products range from colloids to crystals sometimes visible under the microscope. Hydraulic cements may be grouped into three categories in this respect:

(a) *Cementing action due almost wholly to the formation of colloidal products* Portland cement cured at ordinary temperatures (Chapters 8–10) falls into this category; the hardening of the paste seems to be attributable almost entirely to the formation of tobermorite gel. The most direct evidence for this is as follows. Pastes made from pure C_3S set and harden in much the same way as pastes made from Portland cement (Chapter 7). The only products in a C_3S paste are tobermorite gel and $Ca(OH)_2$; the Al^{3+}, Fe^{3+} and SO_4^{2-} containing phases in the Portland cement paste therefore do not play any essential role, except perhaps at very early ages. Pastes made from pure β-C_2S also set and harden; in this case tobermorite gel with a little $Ca(OH)_2$ is formed. The process is slow, but

TABLE III

Main products of hydration of some typical hydraulic cements

Type of cement	Colloidal products (< 0·1 μ)	Submicrocrystalline products (0·1–1 μ)	Microcrystalline products (> 1 μ)
Portland cement (normal curing at 20° C)	Tobermorite gel ($C_{1\cdot7}SH_x$ approx.) and probably $Ca(OH)_2$	$Ca(OH)_2$ and various Al^{3+}, Fe^{3+} and SO_4^{2-} containing phases	$Ca(OH)_2$
Autoclaved Portland cement–silica and lime–silica materials	Ill-crystallized tobermorites(?)	11 Å tobermorite ($Ca_5(Si_3O_9H)_2 . 4H_2O$)	—
Aluminous cement (normal curing at 20° C)	$Al(OH)_3$	CAH_{10} ($CaAl_2(OH)_8 . 6H_2O$) and some C_2AH_8	—
Supersulphate cements	Tobermorite gel	Ettringite ($Ca_6Al_2(OH)_{12}(SO_4)_3 . 26H_2O$)	—
Hemihydrate plaster	—	—	Gypsum ($CaSO_4 . 2H_2O$)
Sorel cement	—	—	$Mg_2(OH)_3Cl . 4H_2O$ and $Mg(OH)_2$

this is attributable to the slowness with which β-C_2S reacts with water. The final strength is similar to that of a C_3S paste (Fig. 1). The $Ca(OH)_2$ therefore probably also plays no essential role. This last conclusion is supported by electron microscope evidence (Chapter 9), and by the fact that hardening also occurs in the presence of pozzolanas, which inhibit the formation of $Ca(OH)_2$ (Volume 2, Chapter 14).

(b) *Cementing action due mainly or wholly to the formation of crystalline products* Sorel cement and hemihydrate plaster come into this category; microscopic examination of the fully hardened pastes show them to consist of tightly packed masses of crystals. In the early stages of setting of these cements, crystals of colloidal or submicroscopic size are formed, but these appear to be replaced by larger crystals as hardening proceeds.

(c) *Cementing action due to the formation of both colloidal and crystalline products* With some types of cement, both colloidal and crystalline products seem to contribute significantly to hardening. Thus in supersulphate cements, the strength at early ages seems to be due mainly to the formation of ettringite, whereas at later ages tobermorite gel plays an increasingly important role (Volume 2, Chapter 13). In normally cured aluminous cement (Volume 2, Chapter 12), crystalline CAH_{10} and C_2AH_8 and colloidal $Al(OH)_3$ probably all contribute to the strength. There is some uncertainty about the relative importance of crystalline and colloidal products in autoclaved lime–silica or Portland cement–silica materials (Chapter 11 and Volume 2, Chapter 16). Some investigators have assumed that the highest strengths are obtained if the yield of crystalline 11 Å tobermorite is at a maximum, but others consider that an optimum ratio of colloidal to crystalline products exists [*28*].

Hardening can thus be associated with the formation of colloidal products, crystalline products, or both. The particle size of the products affects certain properties of the hardened material. One of these is the moisture movement, that is, the extent to which the material undergoes volume changes on account of variations in the humidity of the atmosphere in which it is placed. Other things being equal, cements containing a high proportion of crystalline products are likely to have lower moisture movements than cements containing mainly colloidal products. They are also likely to show better resistance to chemical attack.

3. *Particle Shape*

The shape of the particles of the hydration products is also very variable. Gypsum, magnesium oxychloride ($Mg_2(OH)_3Cl.4H_2O$), and ettringite form acicular crystals. 11 Å Tobermorite forms platey crystals. Tobermorite gel appears to contain both minute fibrous particles and even smaller irregular platelets (Chapter 9). The morphology of CAH_{10} is not known with certainty. C_3AH_6, which is formed on curing of aluminous cements above about 40° C and is less satisfactory as a cementing agent than CAH_{10}, is cubic and forms roughly equidimensional crystals. It is possible for compounds to form crystals of roughly similar size and shape but to differ markedly in their values as cementing agents. Thus in autoclaved Portland cement–silica products, the formation of the platey 11 Å tobermorite is associated with high strength. On the other hand formation of α-C_2S hydrate ($Ca_2(HSiO_4)(OH)$), which also yields platey, submicroscopic crystals, is associated with low strength.

4. Surface Structure

A factor which may be very important in determining the ability of a hydration product to act as a cementing agent is its surface structure. The forces which promote adhesion between the particles in a hardened paste are not well understood, but it is reasonable to suppose that they are highest for those compounds which have surfaces in which the co-ordination of atoms or ions is unsatisfactory or in which there are marked separations of electric charge. A measure of the extent to which such imperfections occur in the surface structure is provided by the surface energy; this is high for tobermorite gel (Chapter 7). Differences in surface structure possibly explain the differing cementing properties of some otherwise similar compounds, such as 11 Å tobermorite and α-C_2S hydrate.

5. Circumstances of Formation

Perhaps more important than any of the factors so far discussed are the circumstances under which a given hydration product is formed. If formation occurs under conditions which do not entail the disruption of a pre-existing structure, particles of many different sizes and shapes can yield materials of high strength. If, on the other hand, formation causes the disruption of an already hardened material, the same or another hydration product can be associated with low strength. The classicial example of this is provided by ettringite ($C_3A \cdot 3CaSO_4 \cdot 32H_2O$). This is formed at early ages in pastes made from supersulphate cements, and from certain kinds of Portland and of expansive cements; in probably all these cases its formation has a positive effect on the development of strength. In contrast, the formation of ettringite in an already hardened paste can cause complete disruption of the material. This happens when Portland cement pastes are attacked by sulphate solutions.

C. Theories of Setting and Hardening

1. The Crystallization Hypothesis

As early as 1765, Lavoisier [*34*] showed that the setting of Plaster of Paris could be attributed to rehydration of the material to give gypsum; he reported that "a rapid and irregular crystallization occurs, and the small crystals which are formed are so entangled with each other that a very hard mass results".

The first detailed attempt to explain the setting and hardening of hydraulic cements was made by Le Chatelier in 1887 [*35*] and may be regarded as representing an extension of Lavoisier's ideas. Le Chatelier supposed that the cement dissolved in water to give a solution which was supersaturated relative to the hydration product. The latter precipitated, forming crystals that interlocked in the manner described by Lavoisier. The strength of the hardened paste depended on the internal cohesion of the crystals of the hydration product and on the adhesion between them. Le Chatelier observed that crystals precipitated from highly supersaturated solutions were often markedly acicular. Crystals of this shape might be expected to felt together well.

An essential feature of the crystallization hypothesis is that the material passes through the solution. The amount of water used in mixing is such that only a small proportion of the solid (about 0·1% for a gypsum plaster) can be in solution at any one time.

The crystallization hypothesis explains the facts well in the case of gypsum plasters. Dissolution of hemihydrate at room temperature gives a solution which is supersaturated relative to gypsum (Fig. 4 of Chapter 8). The crystallization hypothesis is possibly substantially correct for the setting of gypsum plasters though, as is shown later, it is probably oversimplified as regards the initial stages of the process.

Le Chatelier considered that the crystallization hypothesis applied to all hydraulic cements. The anhydrous compounds in Portland cement yield unstable solutions that are supersaturated relative to the hydration products. Le Chatelier considered that the hypothesis accounted quite satisfactorily for the setting and hardening processes if it were assumed that the hydration products other than the $Ca(OH)_2$ were submicrocrystalline.

2. The Gel Hypothesis

In 1893 Michaëlis [*36*] attributed the hardening of hydraulic cements to the formation of a gel. When first formed, this was soft and contained much water; at this stage, unreacted cement grains were still present.

In hydrating, these absorbed water from the gel already formed, which thereby became hard and impermeable. Michaëlis recognized that, in the hydration of Portland cement, crystalline products were formed in addition to gelatinous ones, but he regarded them as playing only a minor part in setting and hardening.

3. The Nature of Gels

Current investigators mostly accept the description of the set and hardened pastes of Portland cement as gels, and it is therefore useful to summarize existing ideas on the structure of gels. Hermans [*37*] has defined gels as bodies satisfying the following criteria.

(i) They are coherent, colloidal disperse systems of at least two components.

(ii) They exhibit mechanical properties characteristic of the solid state.

(iii) Both the dispersed component and the dispersion medium extend themselves continuously through the whole system.

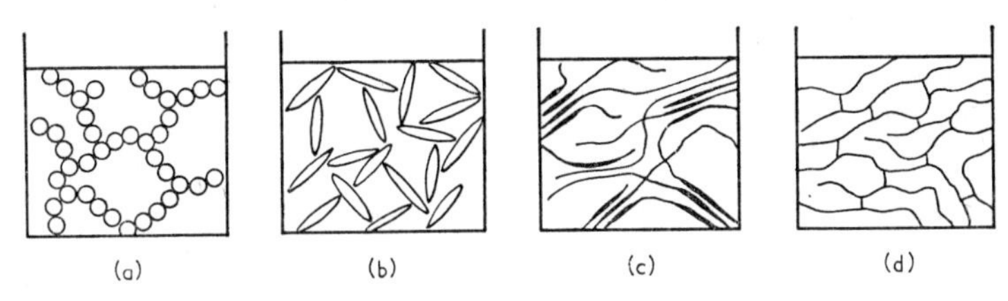

FIG. 5. Four types of gel structure, shown diagrammatically [*37*].

The particles of the dispersed component can be either crystalline or amorphous; they may, for example, be amorphous globules, crystalline particles, short fibres, or long-chain macromolecules (Fig. 5). The essential requirement is that they should be able to adhere together so as to form an open and irregular framework in which the molecules of the dispersion medium are enmeshed. The proportion of the total volume which is occupied by the dispersed component need only be small; in an extreme case, the vitreous material of the eye, it is only 0·1%. The junctions between the particles in the framework can be made in various ways, depending on the nature of the substance; thus ionic or covalent structures may weld into each other, or electrostatic attractions, such as hydrogen bonds, may occur. Van der Waals forces may also be important. With long-chain macromolecules, two or more chains may run parallel for a short distance at a junction, forming crystalline regions in an otherwise amorphous material (Fig. 5(c)). Some gels exhibit thixotropy, that is, the structure can be destroyed mechanically but reforms on standing. This occurs if the cohesive forces at individual junction points are weak, but large numbers of potential junction points exist so that the separated particles can readily join up again. Other characteristic properties, such as sorption and swelling, are readily explained on this model, at least in principle.

4. *Modern Views*

For a long time after Le Chatelier and Michaëlis had put forward their views, controversy existed as to which was correct. Bogue [*3*] has summarized the history of this controversy. More recently, the tendency has been to look rather for a synthesis of the two approaches. Thus Bernal [*38*] pointed out that the particles in tobermorite gel had appreciable crystalline character; Michaëlis' concept of the hardened cement paste as a partly hydrated gel was correct, but the material could also be regarded as a mass of very small, interlocking crystals. Much earlier, in 1926, Baykoff [*39*] had suggested that setting could be attributed to the formation of a gel, but that the subsequent process of hardening was due to the formation of crystalline products. There is some evidence that this is true for gypsum plasters. Some authors have considered that this concept applies also to the hydration of Portland cements, but this is doubtful in the opinion of the present writer. As has been stated earlier (p. 14), the hardening of Portland cement pastes at ordinary temperature is caused almost entirely by the formation of tobermorite gel.

There is still no general agreement about the details of the processes which cause setting and hardening, but the following summary would probably be accepted by the majority of investigators at the present time. It is given in terms of Portland cement hydrated at ordinary temperatures but applies, with differences to be mentioned later, to other cements as well.

The initial result of mixing the cement with water is to produce a dispersion; the water/cement (w/c) ratio needed to produce a paste (0·3–0·7 *w*/*w*) is such that the grains of cement are not close packed (Fig. 6(a)). Reaction with water quickly produces a surface layer of hydration products on each grain. These occupy space partly at the expense of the grains, and partly at that of the liquid (Fig. 6(b)). The particles of the hydration products at this stage are largely of colloidal dimensions (10–1000 Å), but some larger crystals ($Ca(OH)_2$ and Al^{3+}, Fe^{3+} and SO_4^{2-} containing phases) may also be formed. The solution quickly becomes saturated with Ca^{2+}, OH^-, SO_4^{2-}, and alkali cations. With further reaction, the coatings of hydration products extend and begin to meet each other, so that a gel in the classical sense is formed in the spaces between the grains (Fig. 6(c)). This is the stage of setting. With still further reaction, the particles between the clinker grains become increasingly densely packed, until the material can equally well be regarded as a mass of particles in contact with each other. Differentiation of the gell also occurs in that it becomes more densely packed in some regions and less so in others, so that pores are formed (Fig. 6(d)). The

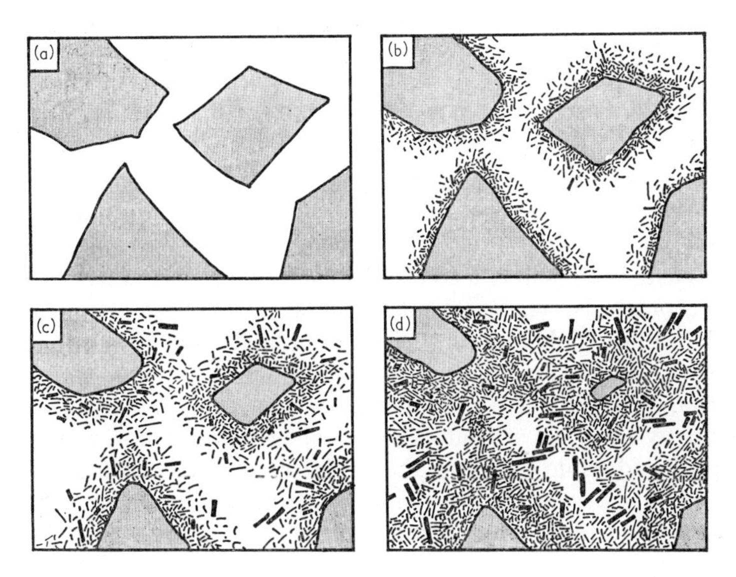

FIG. 6. Four stages in the setting and hardening of Portland cement: simplified diagrammatic representation of the possible sequence of changes. (a) Dispersion of unreacted clinker grains in water. (b) After a few minutes: hydration products eat into and grow out from the surface of each grain. (c) After a few hours: the coatings of different clinker grains have begun to join up, the gel thus becoming continuous (setting). (d) After a few days: further development of the gel has occurred (hardening).

broad structure of the gel is discussed in Chapter 10, and the fine structure in Chapter 9. Crystalline particles ($> 1\ \mu$) are disseminated through the gel, and also form in the pores by recrystallization. Where they are in the gel, they appear to have either no effect at all on the latter's structure, or at the most to produce only a local ordering either of the internal structure of the individual particles or of the way in which these are arranged (Chapter 9).

As stated earlier, the forces which bind the colloidal particles together in the gel are not definitely known. Perhaps the most important are hydrogen bonds, van der Waals forces, and ionic attractions of various kinds, caused by the occurrence of unbalanced electrical charges. Si—O—Si bonds may also be formed.

Broadly speaking, the above picture seems to be true also for those cements in which substantial proportions of crystalline products are formed (e.g. autoclaved Portland cement–silica materials). The scale of the phenomena is different, but the fundamental processes are basically similar.

An unsolved problem relating to hydraulic cements of all kinds concerns the relative importance of mechanisms involving dissolution and reprecipitation, and those taking place by the conversion of the solids *in situ*. As already stated, Le Chatelier believed the first of these mechanisms to operate. This may be partly or wholly true during the initial stages of reaction, when transport of material through the solution is relatively easy, although, as Hansen [*40*] has shown, the quantity of material that would have to be transported for a normal Portland cement paste is large in relation to the concentrations in the solution. As reaction proceeds, transport of material must become increasingly difficult, and it would appear even more unlikely that all of the material could then pass through the solution. McConnell [*41, 42*] has shown that hydration of β-C_2S to give a product resembling tobermorite gel can occur in nature by a process not entailing passage through solution, and Funk, Schreppel and Thilo [*43*] and Funk [*44*] have demonstrated that this process can be reproduced in the laboratory. Similar mechanisms may well operate during the hydration of cement pastes.

REFERENCES

1. Lea, F. M. and Desch, C. H. (1956). "Chemistry of Cement and Concrete", 2nd Ed., revised by F. M. Lea. Edward Arnold, London.
2. Bredsdorff, P., Idorn, G. M., Kjaer, A., Munk Plum, N. and Poulsen, E. (1962). *Chemistry of Cement, Proceedings of the Fourth International Symposium, Washington 1960*, p. 749. National Bureau of Standards Monograph 43. U.S. Department of Commerce.
3. Bogue, R. H. (1955). "Chemistry of Portland Cement", 2nd Ed. Reinhold, New York.
4. *Proceedings of the Symposium on the Chemistry of Cements, Stockholm 1938.* Ingeniörsvetensakademie, Stockholm (1939).
5. *Proceedings of the Third International Symposium on the Chemistry of Cement, London 1952.* Cement and Concrete Association, London (1954).
6. *Proceedings of the Fourth International Symposium on the Chemistry of Cement, Washington 1960.* National Bureau of Standards Monograph 43, U.S. Department of Commerce (1962).
7. "Review of the Portland Cement Standards of the World, 1961". Cement Statistical and Technical Association (Cembureau), Malmö.
8. Bogue, R. H. and Lerch, W. (1934). *Industr. Engng Chem.* **26**, 837.

9. Woods, H., Starke, H. R. and Steinour, H. H. (1932). *Engng News Rec.* **109**, 404, 435; (1933). **110**, 431.
10. British Standard Specification 12:1958.
11. Price, W. H. (1951). *Proc. Amer. Concr. Inst.* **47**, 417.
12. British Standard Specification 1370: 1958.
13. Chatterji, S. and Jeffery, J. W. (1963). *Mag. Concr. Res.* **15**, 83.
14. Thorvaldson, T. (1954). *Proceedings of the Third International Symposium on the Chemistry of Cement, London 1952*, p. 436. Cement and Concrete Association, London.
15. van Aardt, J. H. P. (1962). *Chemistry of Cement, Proceedings of the Fourth International Symposium, Washington 1960*, p. 835. National Bureau of Standards Monograph 43. U.S. Department of Commerce.
16. Hansen, W. C. (1954). *Proceedings of the Third International Symposium on the Chemistry of Cement, London 1952*, p. 598. Cement and Concrete Association, London.
17. Budnikov, P. P., Royak, S. M., Lopatnikova, L. Ya. and Dmitriev, A. M. (1960). *Dokl. Akad. Nauk SSSR* **134**, 591.
18. Vivian, H. E. (1962). *Chemistry of Cement, Proceedings of the Fourth International Symposium, Washington 1960*, p. 909. National Bureau of Standards Monograph 43. U.S. Department of Commerce.
19. Wuerpel, C. E. (1954). *Proceedings of the Third International Symposium on the Chemistry of Cement, London 1952*, p. 633. Cement and Concrete Association, London.
20. Dolezsai, K. (1963). *Proc. 7th Conf. Silicate Industry, Budapest, 1963.* Akadémiai Kiadó, Budapest, in the press.
21. Lukens, S. (1932). *J. Amer. chem. Soc.* **54**, 2372.
22. de Wolff, P. M. and Walter-Lévy, L. (1953). *Acta cryst.* **6**, 40.
23. Deer, W. A., Howie, R. A. and Zussman, J. (1962). "Rock Forming Minerals", Vol. 5, p. 202. Longmans, London.
24. Flörke, O. W. (1952). *Neues. Jb. Min.* (*Abhand.*) **84**, 189.
25. "Gmelins Handbuch der anorganischen Chemie", Vol. 28, "Calcium". Verlag Chemie (1961).
26. Braniski, Al. (1962). *Chemistry of Cement, Proceedings of the Fourth International Symposium, Washington 1960*, p. 1075. National Bureau of Standards Monograph 43. U.S. Department of Commerce.
27. Mchedlov-Petrosyan, O. P. and Worobjow, J. H. (1960). *Silikattechnik* **11**, 466.
28. Bozhenov, P. I., Kavalerova, V. I., Salnikova, V. S. and Suvorova, G. F. (1962). *Chemistry of Cement, Proceedings of the Fourth International Symposium, Washington 1960*, p. 327. National Bureau of Standards Monograph **43**. U.S. Department of Commerce.
29. Brandenberger, E. (1936). *Schweizer Archiv.* **2**, 45.
30. Büssem, W. (1939). *Proceedings of the Symposium on the Chemistry of Cements, Stockholm 1938*, p. 141. Ingeniörsvetenskapsakademien, Stockholm.
31. Bredig, M. A. (1943). *Amer. Min.* **28**, 594; *J. phys. Chem.* **49**, 537.
32. Jeffery, J. W. (1954). *Proceedings of the Third International Symposium on the Chemistry of Cement, London 1952*, p. 30. Cement and Concrete Association, London.
33. Copeland, L. E., Kantro, D. L. and Verbeck, G. (1962). *Chemistry of Cement, Proceedings of the Fourth International Symposium, Washington 1960*, p. 429. National Bureau of Standards Monograph 43. U.S. Department of Commerce.

34. Lavoisier, A. L. (1765). *C. R. Acad Sci., Paris,* February 17.
35. Le Chatelier, H. (1904). Thesis 1887: "Récherches experimentales sur la Constitution des Mortiers hydrauliques" 2nd Ed. Dunod, Paris.
36. Michaëlis, W. (1893). *Chem. Ztg.* 982.
37. Hermans, P. H. (1949). "Colloid Science", ed. by H. R. Kruyt, p. 483. Elsevier, Amsterdam.
38. Bernal, J. D. (1954). *Proceedings of the Third International Symposium on the Chemistry of Cement, London 1952,* p. 216. Cement and Concrete Association, London.
39. Baykoff, H. T. (Baĭkov) (1926). *C. R. Acad. Sci., Paris,* **182**, 128.
40. Hansen, W. C. (1954). *Proceedings of the Third International Symposium on the Chemistry of Cement, London 1952,* p. 318. Cement and Concrete Association, London.
41. McConnell, J. D. C. (1955). *Min. Mag.* **30**, 672.
42. Long, J. V. P. and McConnell, J. D. C. (1959). *Min. Mag.* **32**, 117.
43. Funk, H., Schreppel, B. and Thilo, E. (1960). *Z. anorg. allg. Chem.* **304**, 12.
44. Funk, H. (1962). *Chemistry of Cement, Proceedings of the Fourth International Symposium, Washington 1960,* p. 291. National Bureau of Standards Monograph 43. U.S. Department of Commerce.

Part I

Chemistry of Anhydrous Cement Compounds and the Production of Portland Cement

CHAPTER 1

Raw Materials and Processes for Portland Cement Manufacture

H. W. W. Pollitt

Associated Portland Cement Manufacturers Ltd., Research Department, Greenhithe, Kent, England

CONTENTS

I. Introduction

The history of the development of Portland cement and the materials and processes used in its manufacture were comprehensively described by Bogue and by Lea as recently as 1955 [*1*] and 1956 [*2*]. As in several respects there has since been little advance, there will be occasional repetition of these authorities.

Portland cement consists principally of four compounds: tricalcium silicate, C_3S, dicalcium silicate, C_2S, tricalcium aluminate, C_3A, and a phase approximating to tetracalcium aluminoferrite, C_4AF. These compounds are formed by a series of reactions at temperatures rising to the region of 1300°–1500° C between lime on the one hand and silica, alumina and iron oxide on the other. The lime is obtained by decarbonating calcareous materials such as chalk or limestone; the alumina, silica and iron oxide are obtained by heating argillaceous materials such as clay, shale or schist. Optimum cement quality is obtained when the required proportions of the four oxides are consistent throughout the cement.

The basic cement making process consists therefore in quarrying calcareous and argillaceous materials in the required proportions, reducing these materials to a finely divided state, blending them to a uniform composition and heating them—first, to drive off any water and carbon dioxide, and secondly, to the point of incipient fusion when the principal cement compounds are formed. Finally, the resulting clinker, as it is termed, is cooled and ground with gypsum to a fine powder, which is Portland cement.

II. Raw Materials

The calcareous and argillaceous materials most commonly used are those which occur naturally.

A. Calcareous Materials

The most important calcareous material is limestone, which occurs in a variety of forms, viz. chalk, other sedimentary limestones, metamorphic, coral and secondary limestones, and carbonatite.

Chalk occurs widely in England and Europe and is also used in the United States. It is commonly found in horizontal or slightly sloping beds and is characteristically of uniform composition within a given horizon but may vary in its calcium carbonate content from horizon to horizon. Common impurities are sand and flints. These, together with other overburden material, sometimes fill vertical cylindrical or conical cavities known as "pipes". Chalk is usually comparatively soft and can be easily reduced to a fine state. An exception is indurated chalk, on which a works in Northern Ireland is based. The hardness of this is due to the more recent crystallization of calcite in the pores of the chalk. The ease with which chalk can usually be comminuted and the fact that as dug it can contain of the order of 25% by weight of water explain the popularity of the wet process of manufacture where chalk occurs, as, for example, in the cradle of the world cement industry on the banks of the Thames. Chalk is, strictly speaking, a sedimentary limestone, but in view of its characteristic properties it is usually classified separately.

Sedimentary limestones other than chalk occur in a variety of ways throughout the world. The beds may be comparatively flat and well defined, as chalk usually is, or contorted in complex folded systems. They may, as in the case of oolitic limestone, be comparatively soft, but can be extremely hard. Chemical composition is in certain cases uniform but more usually varies laterally and in depth according to the nature of the original deposition and any more recent geological movements such as faulting and folding. A very wide range of impurities is known; among the more important are magnesia, fluorine compounds, metallic oxides and sulphides, phosphates, and certain siliceous materials such as chert. Sand, clay and shale may also be found but are not necessarily harmful. Indeed, when the correct proportion of these is sufficiently well dispersed in the limestone the material is known as "natural cement rock". On such rock is based the concentration of cement works in the Lehigh Valley of the United States.

Metamorphic limestones occur widely but are most common in the geologically old "continental shields". They are usually ancient sedimentary limestones which have been subjected to geological processes more extensive than comparatively recent faulting and folding. They are often coarsely crystalline, relatively easily crushed and ground, and chemically broadly similar to sedimentary limestones, but in detail can vary very rapidly both laterally and in depth. This is of importance insofar as it affects the distribution of impurities such as magnesia. Intrusions and intercalations of schist, ironstone and granite may occur.

Coral and secondary limestone are of less importance, although both are used in cement manufacture. Coral limestone is of obvious derivation and may be variably magnesian and siliceous. Secondary limestones are of geologically recent origin, occurring usually as thin superficial deposits or even as nodules in soil, e.g. in Africa as *kankar*, as a result of leaching and redeposition of calcareous material. They may be magnesian, phosphatic or siliceous, and are frequently contaminated with clay.

Carbonatite is a form of limestone associated with volcanic structures. It may contain a very wide range of minerals; apatite, fluorspar and magnesia are important impurities. A carbonatite deposit is the basis for the cement works at Tororo, Uganda, in connection with which Nurse [*3*] studied the effects of phosphate on cement manufacture and properties.

Other calcareous materials include shell deposits, which may be dredged from the sea, lake or river beds, and marls, which are calcareous muds whose composition is sometimes suitable for cement making without admixture of other materials.

B. Argillaceous Materials

Natural weathering processes tend to remove soluble components of rocks, leaving the relatively insoluble oxides of silicon, iron and aluminium. As a result, many weathered rocks, either *in situ* or after redeposition, can provide argillaceous materials for cement manufacture.

The most common materials are clays, shales and marls, but mudstones, slate, schist, some volcanic rocks and ash, estuarine mud, and alluvial silt are also used.

Clays, shale, mudstone and slate represent different stages in rock-formation from muds. Chemically similar, they have markedly different physical properties requiring different methods of size reduction. Clays, mud and silt can be converted to slurries by vigorous stirring in water; marl and shale sometimes, volcanic rocks, mudstone, slate and schist always, require crushing and grinding. Usual impurities are free silica, sulphur and alkalis, and magnesia is sometimes encountered. The behaviour of raw materials throughout the manufacturing process is much influenced by the clay minerals present in clays and shales. Indeed, a useful guide to probable behaviour in the process can be obtained by a methylene blue adsorption test [*4*] which is used for assessing clays in the ceramic and pharmaceutical fields.

Kaolin is a white or lightly coloured clay which may be sedimentary but is frequently the result of decomposition of granite *in situ*. It is relatively rich in alumina but normally contains little iron oxide, for which reason it is used in the manufacture of white cement.

Schists are formed by recrystallization of the minerals in sedimentary clays and shales, and contain mica. Where exposed they are soft. While the overall chemical composition is often suitable, marked local variations in the relative proportions of alumina, iron oxide and silica can cause difficulty in process control. The silica is, moreover, sometimes present as coarse-grained quartz, which, if not specially catered for, can cause additional difficulties. Schists are frequently rich in alumina and may also have high contents of alkalis and sulphur.

Volcanic ash is normally soft, fine-grained and friable as deposited. Secondary geological processes can produce from it materials ranging from a plastic clay to an indurated rock. High alkali contents are common. Volcanic rocks include tuff, agglomerate and greenstone, and are usually hard.

Estuarine mud and alluvial clay and silt from rivers and lakes are characteristically siliceous and friable. Lake alluvium is usually of uniform composition but the other two types of material are frequently variably siliceous. Estuarine mud can contain sodium chloride in sufficient quantities to promote corrosion of plant, and organic matter which in variable amounts can cause difficulties in control of composition.

Marl is compact calcareous mud; an indurated form is known as marlstone. Certain types are sufficiently rich in calcium carbonate to approximate to a natural cement mix. While phosphate contamination sometimes occurs, the natural intimate mixture of finely divided calcareous and argillaceous material provides a convenient basis for a homogeneous cement mix.

C. Other Materials

Many cement works are based on one calcareous and one argillaceous material. However, at least as many use three or more component materials in order to obtain the required control of composition. Usual "correcting" materials are iron oxide in various forms, bauxite or kaolin to make good deficiencies in alumina content, and sand or sandstone to give sufficient silica. An additional calcareous material of high calcium carbonate content is frequently used as a "sweetener" to raise the lime content of the raw materials mixture.

Certain industrial waste products containing one or more of the four basic oxides may also be regarded as raw materials. Blast-furnace slag from steelworks and pulverized fuel ash from coal-fired power stations are the more important of argillaceous wastes; precipitated calcium carbonate from industrial chemical processes is the principal calcareous waste. In fact, the use of these wastes for the manufacture of Portland cements is extremely small compared with that of naturally occurring materials. Clausen [*5*] has listed for North America the materials used and the frequency of their use.

Anhydrite is a special case; from it and added argillaceous materials sulphuric acid is made, Portland cement clinker being a by-product.

Gypsum is added during the grinding of clinker to control the setting properties of the cement and can thus be considered a raw material. It may contain natural anhydrite, calcium carbonate or clay as impurities.

D. Physical and Chemical Properties of Raw Materials

Behaviour during processing and the properties of the cement produced are much influenced by the fineness and intimacy of mixture of the raw materials and by the presence in them of impurities.

Differences in particle size distribution and specific gravity of the component raw materials may promote segregation. Clinker rings, which impede production and which are accumulated deposits of sintered or molten material formed in the kiln at or beyond the tail of the flame, may be caused by segregation of a mixture having a melting point lower than that of the raw mix as a whole. The temperature at which the raw materials will satisfactorily combine is a function of the maximum size of particles present in them: the larger the particles the higher the temperature required.

The nature of the largest particles is also important. Large grains of lime or silica produce zones in the clinker which are deficient in silica or lime, respectively, with, of course, corresponding zones elsewhere with an excess of silica or lime. Lime deficiency is accompanied by low potential cement strength. Excess of lime, on the other hand, leads to risk of unsoundness, i.e. expansion which can occur due to hydration of the lime after the cement has set. This risk is particularly great with lime burned at a high temperature, the need for which may well be created by the coarse lime.

Bogue [*6*] described experiments proving that in a relatively low-burned mix with a lime saturation factor† of 85% it was necessary to reduce the size of silica to less than 50 μ in order to combine all but 1% of the lime at 1350° C, but that about 90 μ was an acceptable maximum size for the limestone and aluminous material. In mixes containing more normal proportions of lime, e.g. with lime saturation factors of 92–98%, it is probable that even greater finenesses would be required for the same degree of combination at 1350° C. On the other hand, it is common to burn ordinary Portland clinker at temperatures up to 1450° C, and in Britain and Europe as much as 2 or 3% of free lime is regarded as acceptable in normal applications of cement. Certainly, few if any British or European works produce raw materials for the kiln with no particles greater than 90 μ in size.

Much depends on the relative proportions of silica, alumina and iron oxide and on the form of the silica. The silica ratio, i.e. the fraction

$$\frac{SiO_2}{Al_2O_3+Fe_2O_3},$$

usually lies between 1·5 and 3, and as it increases combination becomes more difficult, particularly when the ratio exceeds 3, as it may do in, for example, white cement. The alumina/iron oxide ratio is normally between 1·5 and 3·5; again, as it increases combination becomes more difficult. On the other hand, when it is low, especially in sulphate-resisting Portland cement, difficulties can arise through over-production of flux or liquid in the kiln, which has the effect of narrowing the range of temperature over which the kiln can be operated. Bogue [*7*] discusses the factors affecting liquid production. Free silica, e.g. sand, is more difficult to combine than silicates or aluminosilicates, e.g. clays. This effect is independent of that of silica ratio, so that a sandy mix is more difficult to combine than a clayey mix of the same silica ratio.

† Lime saturation factor is defined in British Standard 12: 1958 as

$$LSF = \frac{CaO-0{\cdot}7SO_3}{2{\cdot}8SiO_2+1{\cdot}2Al_2O_3+0{\cdot}65Fe_2O_3}$$

Thus, the fineness required for combination of the four principal oxides depends largely on their relative proportions in the mix and in particular on the form of the silica. Adequate combination would also appear to require a reasonable homogeneity of the composition of the mix. Theoretically it would seem ideal to arrange that each particle of a given raw material should be in close touch with the corresponding numbers of particles of the other raw materials. From this ideal arrangement a uniform clinker composition should follow. In fact, clinker contains essentially two distinct crystalline calcium silicates, C_3S and C_2S, and flux. The most uniform clinker that can be hoped for is therefore a well dispersed mixture of these three components. Experience suggests that a more uniform appearance is associated with better formed crystals and a higher temperature of burning. These factors do not necessarily yield the best cement; indeed, there is increasing interest in the suggestion that the activity of cement may be bound up with imperfections in its crystal structure which might be supposed to be produced by the minimum heat treatment compatible with combination. The demand for homogeneity therefore rests on, first, the need to avoid uncombined lime and silica, and, second, the possibility of achieving combination with a minimum of burning.

The principal impurities occurring in raw materials and having significant effects on the properties of cement are magnesia, fluorine compounds, phosphates, lead oxide and zinc oxide, alkalis and sulphides.

Magnesia is considered harmful because in normal processing with relatively slow cooling of the clinker it can emerge in the cement as periclase, which is capable of expansion on hydration long after the cement has hardened and can thus disrupt the mortar or concrete. In British Standard 12:1958, "Ordinary and Rapid Hardening Portland Cement", the maximum permitted magnesia content is 4%. In ASTM Specifications for Portland cements the limit is 5%, but an autoclave test for soundness is also specified as an added precaution. Although rapid cooling of clinker is known to render the magnesia harmless by freezing it into the flux or by producing only small, easily hydrated crystals, the specified limits of magnesia content must be observed.

Fluorine compounds have been found variably successful as fluxing agents according to the type of compound and conditions of use [*8*]. Beneficial effects of fluorine on cement strength have also been reported on the basis of experiments in which clinker was formed during long periods of heating [*9*]. On the other hand, the presence of more than 0·1% or so of fluorine, usually as calcium fluoride, in the raw materials has been proved in Britain to produce disastrous reductions in cement strength below that obtainable if the fluorine is absent. Therefore, unless there is a special need for its fluxing action it is regarded as undesirable and removed from the raw materials.

Phosphates were shown by Nurse to affect adversely both kiln operation and the setting and hardening of the cement produced. It has more recently been found that 0·20–0·25% in the cement has an optimum beneficial effect on strength but that more than 0·5% is deleterious [*10*].

Lead and zinc oxides, like the fluorine compounds, have a mineralizing or fluxing action; thus, 1% of zinc oxide has been shown to reduce from 7·5 to 2·7% the uncombined lime in difficult raw materials burnt at 1500° C. Both oxides have a deleterious effect on cement properties. Fortunately for cement making, they are usually concentrated in mineralized zones of limestone which are likely to be rejected for other reasons.

Alkalis affect plant operation and the properties of the finished cement. In wet processes, corrosion of the kiln shell, ducting and dust precipitator housing are promoted by alkali chlorides in association with sulphur gases. In semi-dry processes alkali sulphates accumulate in inconvenient deposits which must be removed at regular intervals. In dry processes of high thermal efficiency, such as the Humboldt process, inconvenient deposits of alkali sulphates may again form, but in addition most of the alkali may emerge in the cement as alkali sulphate if sufficient sulphur is available in the fuel or raw materials. This need have no adverse effect on strength; indeed, alkali sulphates are recognized accelerators of the cement hardening process. On the other hand, in certain countries, notably in North America, it has been reported [*11*] that certain opaline rocks used as concrete aggregates are capable of long-term reaction with alkalis in cement, producing disruption of the concrete. This alkali–aggregate reaction has led to the introduction of specified limits which the user may invoke when suspect aggregates are to be used. The U.S. Federal Specification limit is 0·60% of total alkali expressed as Na_2O. It is interesting that a variety of reactive aggregates from the U.S. showed no reaction with a range of British cements containing more than the permitted amount of alkali, when tested both according to ASTM Method C227–58T and following a method [*12*] proposed by Jones. This is in accord with the view that the reaction is critically affected by the ratio of alkali to reactive opaline material, more alkali being required to promote expansion with high opaline silica contents. Fortunately, in many countries reactive aggregates are not common, and in Britain they are practically unknown, so that alkalis are not so seriously regarded there.

Sulphides are important for three principal reasons. First, they may oxidize to sulphates, producing calcium sulphate in the clinker. This is likely to be of the nature of hard-burned anhydrite, which, being less soluble than gypsum, itself contributes less to control of setting and reduces the amount of gypsum which must be ground into the cement for that purpose. Secondly, they may oxidize to produce alkali sulphates with the consequences discussed above. Thirdly, they may help to promote reducing conditions in the kiln with adverse effects on cement strength under certain conditions.

III. Processes

The processes involved in Portland cement manufacture can conveniently be divided into the three stages of raw material preparation, heat treatment, and clinker cooling and grinding. At all stages control is essential.

A. Raw Material Preparation

The object of raw material preparation is to produce for heat treatment a properly blended material, intimately mixed and of sufficient fineness to ensure that proper combination can take place in the kiln. The required uniformity of composition and fineness are obtained by methods which vary widely according to the nature of the raw materials. Blending often begins during quarrying. A limestone face of vertically varying composition may be blasted so that the debris constitutes a mixture or may be worked simultaneously at different levels. A laterally varying face may be worked by two or more quarrying units whose outputs can be taken in proportion. A number of different limestone faces may be worked simultaneously, or in rotation with stockpiling, to produce a blend of the correct lime content. The winning of argillaceous materials may similarly be controlled.

The first step in quarrying is removal of overlying material, known as overburden, which may be thrown away or sometimes used as a correcting material, more usually the former. The economics of quarrying are much affected by the overburden ratio, i.e. the ratio of depth of overburden to that of the workable deposit; however, no hard and fast rule can be stated. Most British works operate with a very small overburden ratio, but it may approach and occasionally exceed 1:1. Thus, when the only calcareous deposit within convenient reach of a cement market has a heavy overburden there is no choice but to remove it. For example, one British works removes 150–200 ft (45–60 m) of hard basalt from above the indurated chalk deposit used.

The next step with the hardest rocks, such as limestone, slates and certain shales, is to loosen them by blasting. Careful control of the spacing and depth of placing explosive charges and of the sequence of detonation permits the desired degree of fragmentation to be obtained. Softer rocks, such as chalk, marl, schist, clay and shale, can be dug from the face directly by excavators. Certain clays and estuarine muds when under water may be dredged.

Transport of the materials from the quarry face is by a variety of methods. Blasted rock may be up to 4–5 ft (1·2–1·5 m) in size and is then usually dealt with by large excavators feeding dump trucks capable of carrying up to 30 yd^3 (23 m^3) or more. Directly dug materials are usually handled by excavators into dump trucks or railway wagons or on to conveyor belts. Dredged materials are frequently transported wet by pipeline. The destination of the transported materials is the crushing and grinding plant.

All but the softest materials are crushed, commonly in two stages. The ratio of input to output size at each stage is of the order of from 4:1 to 10:1, depending on the type of crusher and material. Crushers and their performance are described by Taggart [*13*]. A bibliography on crushing has been prepared [*14*]. The crushed rock is then ground, often in ball and tube mills, which are used for both wet and dry processes.

A novel mill of large diameter has shaped liners which lift and drop the coarse material, causing it to grind itself with the aid of few or no balls. Edge-runner mills, which are essentially rotating tables on which run loaded rollers or balls, are often used in dry processes. Soft materials are usually associated with the wet process and are broken down by vigorous stirring with water in washmills which discharge a fine slurry through screens arranged vertically around them. The nature and action of ball and tube mills are described later.

The products of the mills are frequently classified in order (a) to reject unwanted coarse material in the case of open-circuit grinding, or (b) to obtain the desired freedom from coarse particles, without over-production of fines and without discarding material, in closed-circuit (feedback) grinding. There is, of course, no point in feeding back coarse material unless the mill is able to grind it. Thus, coarse sand in the product from a washmill cannot usefully be returned to the washmill, but it can after separation be taken to a more suitable mill, e.g. a tube mill.

Classification in wet process milling has been effected by a variety of means. Recent developments of importance are the hydrocyclone [*15*] and the sieve-bend [*16*], both of which are small in relation to their handling capacity. The hydrocyclone is basically similar to the familiar dust collecting cyclone, but has water instead of gas as the carrier fluid and a slurry pump instead of a fan. A typical hydrocyclone of 6 in (15 cm) maximum diameter and 40 in (1 m) maximum height can pass 50–100 tons of slurry per hour while reducing the quantity coarser than 300 μ from 3·0% to 0·5%. The sieve-bend is essentially a short length of cylinder of relatively large diameter, the narrow curved surface consisting of a ribbed screen, the vertical flat end surfaces being of sheet metal. A nozzle in the base injects slurry tangentially so that it flows circumferentially along the inside of the screen. Fine material passes out through the screen, whereas coarse material travels round inside the screen and flows out through a port in the base. The sieve-bend can also treat large volumes but, especially with finer slurries, is a less efficient separator than is a well designed hydrocyclone.

Classification in dry process grinding is usually obtained by elutriation, in which the air velocity is sufficient to carry ground material of approximately the required fineness, assisted by centrifugal action, which tends to throw coarser particles out of the air stream and is often obtained by fixed or rotating vanes. Certain air-swept mills incorporate rotating vane classifiers.

By careful control of the feeeding of the raw materials to the crushing and grinding processes a first approximation to the required chemical composition is obtained. The need for further blending depends on the success achieved thus far. A notable example of the control that can be achieved is provided by a works in the U.S.A. [*17*] with very variable raw materials. The two limestone quarries in operation are drilled in detail ahead of the working faces. Chemical analyses of the cores, showing the variation in contents of magnesian and siliceous rock, are fed to a digital computer which, by linear programming techniques, optimizes the quarrying programme in such a way as to minimize the need for (a) expensive additional materials such as iron ore, and (b) switching the quarry plant too frequently. The optimized programme also ensures that the maximum of the least satisfactory limestone is used, thus husbanding reserves, and, of course, produces the theoretically correct composition in the mix. The crushed and imported materials are laid down in thin layers on long stockpiles. Samples are taken automatically and analyses of these are fed back to the computer to correct the quarrying programme. It is claimed that the uniformity of the stockpiled materials, after recovery by a Ferris wheel and grinding, is such that no further blending is required, although provision for this has been made.

At the other extreme is the works based on natural cement rock alone. More common is the use of two materials; in this case a favourite practice is to produce a high-lime and a low-lime mix in separate mixers and then to combine these in a further mixer in proportions dictated by periodic chemical analyses. Mixing in the wet process is effected by stirring the slurry in large cylindrical tanks by mechanical and/or compressed air action. In the dry process the dry powdered mix is often blended in a silo by a vigorous circulation induced by fluidization [*18*].

In most cases the procedure described results in a slurry or powder which can be fed to the kiln. In certain works, however, the best or only materials available are not suitable as quarried for cement making, owing, for example, to the presence of too little calcium carbonate in the calcareous material or to an excess of undesirable matter in one of the component materials. In such cases enrichment processes are employed. The procedure adopted depends on the nature of the problem. The simplest is perhaps where soft material rich in impurities is removed from coarse hard rock by washing and screening. The most complex is the flotation process [*19*]. Briefly, calcium carbonate crystals can be attached to a froth formed by blowing air through water containing a fatty acid or soap. The froth is sometimes stiffened by addition of an alcohol. The acidic or siliceous matter remains in the underlying water. Collection of the froth produces a lime-enriched material. Alternatively, cationic surface-active agents can be used for floating off acid or siliceous particles. In both types of flotation process the material must, of course, be sufficiently finely ground to break down aggregations of the desired and undesired matter and to permit flotation. The fineness and the reagents required are determined by preliminary trial. Flotation produces slurries of high water content and is therefore usually followed by sedimentation, as in a thickener, to reduce the amount of water to be boiled off and thus improve heat economy.

The demand for economy in fuel has led to the development of other means of reducing the water content of slurry from its normal level of 35–45% by weight. The use of viscosity-reducing agents, notably sodium carbonate, sodium tripolyphosphate, sodium silicate and sulphite lye, permits a reduction of the order of 3–4% in the water content of certain slurries. In many cases, however, the cost of the additive offsets the saving in fuel and it is necessary to establish by trial whether the addition is worth while. In vacuum filtration, slurry is picked up by continuously rotating drums or disks covered with filter cloth, inside which a partial vacuum is maintained. Water is sucked from the slurry, which cakes and is removed by scrapers. The water content can be reduced to 20–30%, but the resulting cake is somewhat sticky. A firmer cake with a water content of about 20% or less has been obtained by pressure filtration. This is essentially a batch process in which slurry is squeezed between plates lined with filter cloth. After pressing, the plates are parted, allowing the cakes to be removed and cut up for feeding to the kiln. In both vacuum and pressure systems filtration can be accelerated by the use of heated slurry and by the addition of flocculating agents, of which lime and certain flue dusts are economical examples.

Dry powder, as normally fed to a dry process kiln, is sometimes nodulized, i.e. formed into roughly spherical aggregations held together by water. The powder is fed into a rotating dish whose axis is inclined from the horizontal so as to retain the feed. Cylindrical and conical drum nodulizers are also used. A fine water spray causes the powder to form nodules which become compacted and grow by further rolling until they roll over the lip of the dish. The water content of the nodules is commonly 10–15% but depends on the nature of the materials, as do the strength and behaviour during heat treatment.

B. Heat Treatment

Reactions and temperatures in rotary kilns are discussed in Chapter 3; they have also been discussed by Lea [*20*] and by Lyons *et al.* [*21*].

The basic steps in the heat treatment are boiling off any slurry water at up to 100° C, combustion of any organic matter and decarbonation of the calcium carbonate at up to 1000° C, heating the decarbonated feed to 1300–1500° C according to its composition and fineness, maintaining this temperature sufficiently long for the cement compounds to form, and finally cooling the resulting clinker. These operations may proceed in one kiln, as in the case of shaft kilns; in a kiln and cooler, as in the majority of wet and dry processes; and, more recently, in a preheater, kiln and cooler.

Shaft kilns were the first to be used for cement manufacture in England, and in improved form are still used, particularly in Germany. The raw material is mixed with the fuel, usually coke, briquetted or nodulized, and introduced at the top of the vertical cylindrical kiln. An upward flow of combustion air from the bottom is maintained by blowers. Combustion occurs in the lower half, the descending feed thus being heated by the rising products of combustion until it passes through the combustion zone. After this it is cooled by the air for combustion. Although the heat transfer in a deep granular bed of this nature is theoretically good, the possible formation of air and gas channels and losses at the walls may militate against uniformity of the combustion and the product. The heat requirement for this type of kiln is of the order of 900–1000 kcal/kg of clinker.

By far the majority of the world's cement works operate the rotary kiln. This is a refractory-lined steel cylindrical shell which rotates slowly about an axis which is inclined a few degrees from the horizontal. The prepared raw material feed is introduced at the upper, "back", end, from which it is transported by the slope and rotation of the kiln to the lower, hotter, end. Here a flame is maintained by injecting fuel through a pipe, usually with part of the combustion air known as "primary air", the remaining "secondary" air being drawn into the kiln by the suction of a fan beyond the back end. The flame creates in its immediate vicinity the burning zone, where the cement compounds are formed. From the burning zone the combustion gases pass up the kiln, losing heat to the incoming feed, until they pass from the kiln to a dust collector system, after which they are released to the atmosphere through a tall chimney. In order to avoid waste of heat in the gases it is usual to incorporate some form of heat exchanger at the back end of the kiln or just outside it, although in certain dry process kilns none is used. The commonest heat exchanger in wet process kilns is a system of chains, "curtain" chains which hang vertically just inside the back end, and "festoon" chains which hang in catenaries and which occupy from one-quarter to one-third of the length of the kiln. As the kiln rotates, the curtain chains dip into the slurry and then hang vertically in the gases, trapping some dust and partially drying the slurry. The festoons help to dry the slurry further through a thick plastic consistency to the state where it will nodulize; at the same time the swing of the festoons as the kiln rotates helps to drive the stickier material forward. Various types of grid, plate and scroll structures have also been used to increase the heated surface area in both wet and dry process rotary kilns. The Miag calcinator has been used to dry the slurry before this passes through the kiln. It is a rotating drum loaded with small steel bodies, the whole being heated by the gases from the kiln.

Rotary kilns are up to 575 ft (175 m) in length and up to 19 ft (6 m) in diameter [*22*]. With proper recuperation from the clinker, the heat requirement for the wet process is of the order of 1500–2000 kcal/kg. Similar figures obtain for the traditional dry process and they reflect how much heat is wasted in the kiln gases. Starting from the dry and wet processes, preheaters have been developed to do more efficiently the work of the back end of the normal kiln. The Humboldt dry process, the Lepol semi-dry process and the Davis semi-wet process preheaters are all in use.

The Humboldt preheater consists of a set of four cyclones through which pass the kiln gases, initially at about 900–1000° C. The dry powder feed is introduced into the uppermost cyclone and by normal cyclone action is rejected from the gas stream. The feed passes successively through the other cyclones and finally enters the kiln at about 750° C. The gas passes upwards through the four cyclones and then to a dust collector, the dust being returned with the feed. The final gas temperature is about

300–350° C. Extremely efficient heat and mass transfer is obtained in this process, and special measures have had to be taken in certain installations to reduce the incidence of alkali sulphate deposits in the plant [*23*]. Operated normally the process uses about 900 kcal/kg.

The Lepol preheater is analogous in construction to a chain grate stoker. The moving grate carries a bed of nodules which are heated by gases passing vertically through the bed and grate. The grate housing is divided into two sections by a hanging vertical wall which touches the top of the bed. In the modern version, the "double-pass" grate, the kiln gases pass first through the bed and grate of one section, are removed by an interstage fan and then fed to the second section. This yields better heat economy than the "single-pass" grate, in which some hot gas is wasted. The heating of the nodule bed is gradual, reducing the risk of nodule breakdown due to thermal shock. The process requires about 1100 kcal/kg of clinker.

The Davis preheater has been developed to the design of Mr. Geoffrey Davis, of the A.P.C.M. Ltd. in Britain, to achieve better heat economy and output on wet process works. The preheater consists of a rotating horizontal hearth, with a central discharge orifice, above which are suspended a domed roof and cylindrical walls. The net effect is to produce an annular bed of L section, fed from above and discharging through the central orifice of the hearth. Through this orifice pass the upward-moving gases from the kiln; they are checked by the domed roof and forced to pass through the annular nodule bed, where efficient heat exchange occurs. The nodules are obtained by filter-pressing of slurry and a shaping process. The preheater and kiln consume about 1200–1300 kcal/kg.

Finally, a most interesting form of dry process, the Pyzel, is being developed in the U.S.A. A vertical cylindrical reactor vessel is fed from the bottom with dry powder feed, fuel and preheated combustion air. The result is a fluidized bed into which are injected previously processed fine clinker particles which are said to "seed" the bed, promoting the formation at the high prevailing gas temperature of spherical clinker which flows by gravity through a port in the side of the reactor. The successful exploitation of the process requires special heat exchangers to recuperate the high-level heat in the gases passing from the top of the reactor by heating the raw material feed, and the combustion air. The Pyzel process, like the shaft kiln, involves gases at flame temperature passing through the feed and thus contrasts with the rotary kiln, in which the flame passes over a bed of feed.

The flame in rotary kilns is maintained by injecting fuel and air through a burner pipe. The two principal fuels are pulverized coal and oil; natural gas is also used. Coal is ground in mills that are often of the type suitable for raw material grinding; mills specially designed for, e.g., power station boiler firing are also used. Air-swept mills are common, since air is necessary to transport the ground coal and hot air can dry damp coal. Two systems of coal firing are used: direct, where the air carries the coal from the mill into the burner pipe, and indirect, where the coal is delivered into a hopper from which it is drawn at the required rate. Indirect firing offers in principle closer control of coal feed rate but direct firing has the important advantages of simpler operation and maintenance and reduced fire and explosion risk.

Oil for firing kilns often requires heating to reduce its viscosity to a level suitable for pumping and for atomization, i.e. production of droplets by oil pressure with a special nozzle (pressure jet), or by air or steam jets. Gas is supplied by pipeline and after necessary reduction in pressure is fed straight to the kiln.

The characteristics of flames are influenced by the fineness of ground coal or oil droplets, by the proportion, temperature and velocity of the primary air, i.e. the air introduced with the fuel through the burner pipe, and by the temperature of the secondary air, i.e. the remaining air necessary for combustion, which is usually drawn into the kiln by way of the clinker cooler. The nature of the fuel is also important. The volatile, carbon and ash contents of coal help to determine the distribution of temperature and emissivity along the flame. The carbon/hydrogen ratio and the possibility that soot may form in the early stages of the flame are similarly important for oil. Coal produces a high emissivity and therefore greater radiant heat transfer than does oil. It is for this reason that the kiln exit gas temperature rises appreciably when a rotary kiln is changed from coal to oil firing. The change also commonly increases the heat requirement in kcal/kg by about 10%, owing to the above effect and to the fact that the contribution of a part of the coal ash to production is lost. On the other hand, coal ash, particularly when it varies, creates problems in controlling clinker composition. Heilmann [*24*] has discussed these.

Research on the properties of flames is in progress at the International Flame Research Foundation's experimental station near Ijmuiden, Holland. Convenient summaries of the present state of knowledge have been given by Thring *et al.* [*25*] and Loison and Kissel [*26*].

Choice of fuel is normally based on economics, and in the U.S.A. it is known for a works to vary the fuel according to the supply and price position. Oil, while reasonably consistent in composition, may contain light fractions capable of volatilization in the pipes to and from the burner and, thus, of causing difficulty in metering. Gas may vary in composition and supply pressure. On balance, however, better control can be obtained with oil than with coal, unless the latter is of specially consistent quality.

Control is effected by simultaneous adjustment of fuel and air inputs in relation to the raw material feed rate. This is normally done manually, but automatic control is increasingly used. A system popular in Britain has been described by Field and Barker [*27*].

The use of an electric arc to burn clinker has been tried in Switzerland [*28*], and wider application is forecast in the U.S.S.R. [*29*].

Before leaving the kiln, mention should be made of refractories, ring formation and dust. Apart from specific shut-down periods in certain countries, cement production is continuous throughout the year. Stoppages to replace refractories or to remove rings are therefore unwelcome and expensive. It is common to use different grades of refractory to meet the differing requirements in the drying, decarbonating, burning and cooling zones of the kiln. The bricks in the burning zone, by reacting with the clinker, acquire a coating which helps them to resist the severe conditions. They are usually of chrome-magnesite or high-alumina composition (60–80% Al_3O_3) although dolomite has been used in Europe. The remaining bricks are usually of low (40%) alumina content, or may be silica bricks. The life of refractories is much affected also by operating conditions.

Ring formation occurs in three zones. Slurry or mud rings occur at the back end of wet process kilns, usually at a point where the feed is plastic. Concentrations of sulphates have been observed, and it seems possible also that clinker dust may be caught, thus promoting a stiffening and accumulation of the feed. Clinker rings occur where the feed is about to enter the burning zone. Much has been written [*30, 31, 32, 33*] on probable causes; it seems probable that a variety of causes may operate in different kilns. The spraying of molten clinker dust and coal ash from the flame, segregation from the feed of a material with low melting point, and local high temperatures and reducing conditions created by incompletely burned particles of coal or oil, are perhaps among the more important possibilities. Refractory manufacturers suggest also that correct choice of brick can avoid clinker rings. Ash rings occur close to the front end of the kiln and in spite of their name often prove to consist of sintered clinker.

Dust is an important problem, not only as regards atmospheric pollution and public relations but also as it affects the economy of the process. Every pound of dust lost from the chimney has been quarried, ground, blended and dried, and represents wasted money. Naturally, therefore, every effort is made to minimize the loss. Settling chambers and cyclones have been used to deal with relatively coarse dust; modern works are equipped with electrostatic precipitators or bag-filter systems. In electrostatic precipitators the dust-laden gas passes slowly between electrodes maintained at a potential difference of normally 40–60 kV. Ionization occurs, and the dust migrates to the earthed electrode, where it accumulates and is removed by periodic rapping. Factors affecting efficiency of precipitation have been discussed by Rose and Wood [*34*]. Bag-filters consist of tubular "stockings" of suitable material into which the dusty gas is slowly passed from one end. The other end is sealed, so that the gas must emerge through the wall of the bag or "stocking". The dust caught is removed periodically by a variety of means and, as in the electrostatic precipitators, is either returned to the process or may be sold as a fertilizer or filler.

Dust in rotary kilns is produced by the passage of gases at velocities of the order of 20 ft/sec (6 m/sec) over the bed of feed. With a very friable feed, kiln output may be limited by the fact that increased feed, fuel and air rates lead to increased dust loss and decreased clinker production.

C. Clinker Cooling and Grinding

There are three main types of clinker cooler associated with rotary kilns: integral, rotary and grate. In each, secondary combustion air at atmospheric temperature is drawn through the clinker by the fan at the back end of the kiln, thus heating the air and cooling the clinker. Integral coolers are cylinders built over slots in the kiln shell, each having a simple labyrinth to promote counterflow heat exchange. Rotary coolers are essentially like small kilns but may have lifters which raise the clinker and drop it through the air stream to improve heat transfer. Grate coolers are in principle grate preheaters in reverse. They are sometimes equipped with their own fans and may use more air than is required for the kiln. In this case some has to be discarded, with a loss of overall thermal efficiency, although not of cooling efficiency.

Apart from thermal efficiency, another reason for clinker cooling is that cooling rate affects the properties of cement (Chapter 3). C_3S is unstable below 1250° C. It has also been demonstrated that slow cooling produces γ-C_2S, which dusts and gives less strength than β-C_2S, and permits the flux to crystallize out, leading to setting time difficulties due to crystalline C_3A and unsoundness due to crystalline MgO. Rapid cooling, on the other hand, preserves the β-form of C_2S and tends to freeze the flux as glass, so minimizing the effects of C_3A and MgO, any crystals of which are small. Finally, cooled clinker is easier to convey and grind.

Cooled clinker is ground with gypsum in a mill which consists essentially of a horizontal steel cylinder with ends closed except for orifices for feeding and discharging. The cylinder is lined with hardened steel or other suitables plates and is normally loaded with wear-resisting steel balls, usually filling about one-third of the mill volume. For special cements ceramic balls or flints may be used. Rotation of the mill inclines the bed of balls and causes them to pound and abrade the clinker and gypsum, which are normally fed in at one end and discharged at the other.

It has been shown that the ball size required at a given stage of grinding is related to the particle size to be ground [*35*]. For this reason the mill is usually divided into three chambers with balls graded from 3 to 5 in (7·5–12·7 cm) in the first chamber down to 0·5–1 in (1·3–2·5 cm) in the third. The chambers are separated by slotted diaphragms, although in some cases specially shaped liner plates are used to segregate the different sizes of ball in two chambers, thus avoiding one diaphragm. Typical mills range up to 8 ft (2·5 m) or so in diameter by 50 ft (15 m) or so in length, with drives up to 1500 h.p.

The function of the mill described may of course be carried out in two separate mills; usually one with a length/diameter ratio less than 2 replacing the first chamber, and another with a ratio greater than 2 replacing the second and third chambers. While each of the three mills is in fact a ball mill, this term is by convention applied to the short fat mill, the longer mills being known as tube mills.

The rate of rotation of the mill is related to the mill diameter, and in practice it is kept at about 70–80% of the "critical" value at which, assuming sufficient friction, the balls would rotate with the mill as a solid mass. Advantages in grinding rock at supercritical speeds have been claimed [*36, 37*]. The energy consumed in grinding is related to the nature of the material and the fineness produced. Bond [*38*] has proposed a standard test yielding a "work index" from which the energy for grinding may be estimated in terms of the square roots of representative particle sizes. As regards specific surface, the energy is an approximately linear function up to 2000–2500 cm^2/g (Lea and Nurse) but deviates thereafter owing to cushioning and coating, i.e. agglomeration of fines on the balls and lining. Papadakis [*39*] has considered this problem; grinding aids such as triethanolamine have been used to overcome it but are not permitted in certain national standards. Closed-circuit grinding is sometimes used to avoid over-production of fines and also tends to prevent coating, the desired fines being separated from the mill product and the remainder returned with the coarse feed. A new development is a three-compartment mill in which the product from the first two chambers is classified, the rejects being fed to the far end of the third chamber and, after grinding there, to the same centrally placed classifier.

Beke [*40*] describes methods of optimizing the proportion of particles between 3 μ and 30 μ which have been claimed to represent the limits of effective cement particle size. A comprehensive bibliography on grinding has been prepared [*14*]. Ball milling generates considerable heat. Clausen [*41*] estimates that over 90% of the input energy is consumed in this way and that the ground cement is some 75–80° C hotter than the clinker feed. Gypsum can be dehydrated to the hemihydrate or soluble anhydrite forms of calcium sulphate in the mill by heating above about 100° C, with consequent possibilities of setting time irregularities. It is therefore usual to cool cement mills by external wetting, and the use of fine water sprays in the last chamber has also been advocated. Some cooling is also provided by the air drawn by the mill's dust collecting plant. Cement has also been cooled, after leaving the mill, in special heat exchangers; however, this is more relevant to hot-weather concreting problems than to the condition of the gypsum.

D. Control

Production of cement of consistent quality involves close control throughout the process. Selective quarrying based on raw material analyses is followed by feeding of controlled proportions to the crushing, grinding and washing plant. Regular sampling at this stage and in the storage tanks ensures the correct fineness and composition in the mixers or blending silos. An outstanding example of control up to this point has been referred to. The feed is metered into the rotary kiln at a constant rate. The burner who controls the kiln has at his disposal indicators showing the temperature and the oxygen and carbon monoxide contents of the gas leaving the kiln, and the suction in the kiln hood (a measure of secondary air quantity) and elsewhere. In some cases optical pyrometers measure and record the burning zone temperature and thermocouples may be inserted into the kiln at points of lower temperature. With this information and that obtained by close observation of the interior of the kiln, the burner controls the fuel and air supplies and, when required, the rates of raw material feed and rotation of the kiln. Preheaters are usually equipped with additional instruments for recording gas flow and temperature. Automatic combustion control is increasingly common, and in the U.S.A. the best method of kiln control is being sought by applying data logging and processing techniques to the recorded information.

Cement grinding mills are equipped with ammeters or kWh-meters and various forms of feed control, of which the weigh feeder is increasingly used. Acoustic control has been applied; the noise made by the mill is electronically filtered and used to control the feed rate. The miller uses the results of sieving or specific surface determinations to set the controls.

Finally, the cement is regularly sampled and analysed, and tests of setting time, soundness, fineness and strength are made according to the appropriate standards.

REFERENCES

1. Bogue, R. H. (1955). "The Chemistry of Portland Cement". Reinhold, New York.
2. Lea, F. M. and Desch, C. H. (1956). "The Chemistry of Cement and Concrete", 2nd Ed., ed. by F. M. Lea. Arnold, London.
3. Nurse, R. W. (1952). *J. appl. Chem.* **2** (12), 708.
4. Robertson, R. H. S. and Ward, R. M. (1951). *J. appl. Pharmacol.* **3**, 27.
5. Clausen, C. F. (1960). Portland Cement Association, Research and Development Division, Report MP-95. Skokie, Ill.
6. Bogue, R. H. (1955). "The Chemistry of Portland Cement", p. 206. Reinhold, New York.
7. Bogue, R. H. (1955). "The Chemistry of Portland Cement", p. 209. Reinhold, New York.
8. Bogue, R. H. (1955). "The Chemistry of Portland Cement", p. 217. Reinhold, New York.
9. Welch, J. H. and Gutt, W. (1962). *Chemistry of Cement, Proceedings of the Fourth International Symposium, Washington 1960*, p. 59. National Bureau of Standards Monograph 43. U.S. Department of Commerce.
10. Ershov, L. D. (1955). *Tsement* **4**, 19.
11. Jones, F. E. (1952). D.S.I.R., Building Research Station, National Building Studies, Res. Paper No. 14. HMSO, London.
12. Jones, F. E. (1953). D.S.I.R., Building Research Station, National Building Studies, Res. Papers Nos. 15, 17. HMSO, London.
    Jones, F. E. and Tarleton, R. D. (1958). D.S.I.R., Building Research Station, National Building Studies, Res. Papers Nos. 20, 25. HMSO, London.
13. Taggart, A. F. (1950). "Handbook of Mineral Dressing". Wiley, New York.
14. D.S.I.R. (1958). "Crushing and Grinding—A Bibliography". HMSO, London.
15. Symposium on Recent Development in Mineral Dressing, 1953. Institution of Mining and Metallurgy, London.
16. Fontein, F. J. (1957). *Zement-Kalk-Gips* **10** (4), 121.
17. Nalle, P. B. and Weeks, L. W. (1960). *Min. Engng* **12** (9), 1001.
18. Manecke, H. (1954). *Zement-Kalk-Gips* **7** (12), 464.
19. Craddock, Q. L. (1952). "Cement Chemists' and Works Managers' Handbook", p. 78. Concrete Publications, London.
20. Lea, F. M. (1956). "The Chemistry of Cement and Concrete", 2nd Ed., ed. by F. M. Lea, p. 112. Arnold, London.
21. Lyons, J. W., Min, H. S., Tarisot, P. E. and Paul, J. F. (1962). *I. & E.C. Process Design and Development* **1** (1), 29.
22. Comte, J. M. A. (1959). *Rock Products* **62** (5), 128.
23. Mussgnug, G. (1962). *Zement-Kalk-Gips* **15**, 197.
24. Heilmann, T. (1962). *Chemistry of Cement, Proceedings of the Fourth International Symposium, Washington 1960*, p. 87. National Bureau of Standards Monograph 43. U.S. Department of Commerce.
25. Thring, M. W. *et al.* (1953). *J. Inst. Fuel* **26** (153), 189.
26. Loison, R. and Kissel, R. R. (1962). *J. Inst. Fuel* **35** (253) 60.
27. Field, G. and Barker, S. R. (1960). *Control* **3** (21), 91.

28. Gygi, H. (1947). *Rev. Matér. Constr.* (C) **381**, 247.
29. Lysenko, V. D. (1962). *Tsement* **2**, 7.
30. Matouschek, F. (1951). *Zement-Kalk-Gips* **4**, 67.
31. Konopicky, K. (1951). *Zement-Kalk-Gips* **4**, 240.
32. Alègre, R. (1958). *Rev. Matér. Constr.* (C) **509**, 33.
33. Alègre, R. and Terrier, P. (1959). *Rev. Matér. Constr.* (C) **523**, 89.
34. Rose, H. E. and Wood, A. T. (1956). "An Introduction to Electrostatic Precipitation in Theory and Practice". Constable, London.
35. Slegten, J. A. (1946). *Rock Products* **49** (9), 60.
36. Hukki, R. T. (1958). *Trans. Amer. Inst. mech. Engrs*, May, 581.
37. Masson, A. (1961). *Rev. Matér. Constr.* (C) **550–1**, 352.
38. Bond, F. C. (1952). *Trans. Amer. Inst. mech. Engrs* **193**, 484.
39. Papadakis, M. (1960). *Rev. Matér. Constr.* (C) **542**, 295.
40. Beke, B. (1958). *Zement-Kalk-Gips* **12**, 529.
41. Clausen, C. F. (1960). Portland Cement Association, Research and Development Division, Report MP-95, p. 53. Skokie, Ill.

CHAPTER 2

Phase Equilibria and High-temperature Chemistry in the $CaO–Al_2O_3–SiO_2$ and Related Systems

J. H. WELCH

Building Research Station,
Garston, Watford, Hertfordshire, England

CONTENTS

I. Introduction

This chapter deals with phase equilibria in the oxide systems relevant to the manufacture of cements. Only systems at high temperatures and atmospheric pressure will be considered; no attempt will be made to discuss hydrothermal or high-pressure conditions.

The most important of these systems for cement chemistry is the ternary system $CaO-Al_2O_3-SiO_2$. Its geological and technological significance led to the pioneering studies at the Geophysical Laboratory in Washington over half a century ago, which still serve as a model for all other phase equilibrium studies on oxide systems at high temperatures [*1*]. This chapter will deal more specifically with the $CaO-Al_2O_3-SiO_2$ system and with portions of the quaternary systems formed by these three components with either MgO or Fe_2O_3. The opportunity has been taken to include some hitherto unpublished observations. Systems containing other components, such as FeO or alkali oxides, will not be discussed.

An understanding of the $CaO-Al_2O_3-SiO_2$ system requires, first, a consideration of the three bounding systems $Al_2O_3-SiO_2$, $CaO-SiO_2$ and $CaO-Al_2O_3$. The first of these is relevant to refractories rather than cement, but brief mention of it is necessary.

II. The System $Al_2O_3-SiO_2$

Because of its importance for refractories and other ceramic products, this system has been often studied. The results, even in recent studies, have been conflicting in some respects. An excellent review of work up to 1959 is given by Grofcsik and Tamás [*2*].

The various polymorphs of silica and alumina will be only briefly considered. Alumina crystallizes from melts as corundum (α-Al_2O_3), which takes only a very small proportion (under 0·2% by weight) of SiO_2

into solid solution. β-Al_2O_3 is not pure Al_2O_3 but a binary phase containing essential alkali, e.g. NA_{11} or KA_{11}. CA_6, which is discussed later, has a closely related structure and is frequently also called β-Al_2O_3. The other known polymorphs of Al_2O_3, such as γ-Al_2O_3, are normally formed by the dehydration of the various hydrates. They are not formed from α-Al_2O_3 or oxide melts on cooling and will not be discussed here.

Many polymorphs and stacking variants of SiO_2 are known [*3*]; the crystalline modifications stable under the conditions treated in this chapter comprise the various forms of quartz, tridymite and cristobalite. The more important of the transformations that occur on cooling or subsequent reheating are shown below:

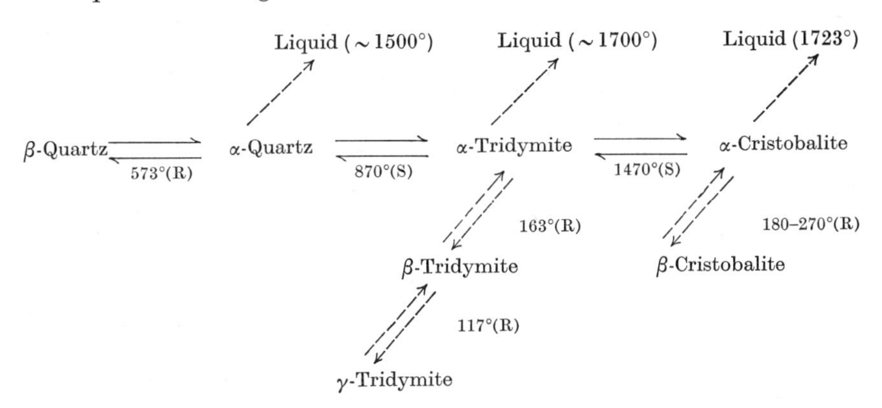

There is confusion in the literature over the nomenclature to be adopted for high–low transitions. Here α- is used for the highest, and γ- for the lowest-temperature form; the letters R and S denote rapid and sluggish, respectively. In general, the sluggish transformations can only be accomplished in the presence of suitable mineralizers. There is dispute as to whether tridymite is a stable polymorph of pure SiO_2, or whether alkali or water is needed to stabilize the structure. Hill and Roy [*4*] concluded that tridymite is a stable phase, but their study was a hydrothermal one. The contrary view, that impurities are needed, has been advanced by Holmquist [*5*], Flörke [*6*] and Wahl, Grim and Graf [*7*]. It should be noted that the amount of impurity need only be very small and also that the tridymite and cristobalite structures are closely related to each other [*6*].

Four anhydrous aluminium silicate phases occur in nature. Three (sillimanite, kyanite and andalusite) are polymorphs of AS, and the fourth, mullite, has a composition close to A_3S_2. Only mullite is stable at high temperatures. Synthetic mullites vary in Al/Si ratio, and it has recently been shown that at high temperatures the composition approximates to A_2S rather than A_3S_2 [*2*, *8*]. This accounts for the apparent anomalies in the melting behaviour of mullite of composition A_3S_2, which have been a source of controversy. There is still some disagreement about the details of the phase equilibrium diagram. Figure 1 gives Welch's results [*8*], other recent versions are given by Grofcsik and Tamás [*2*].

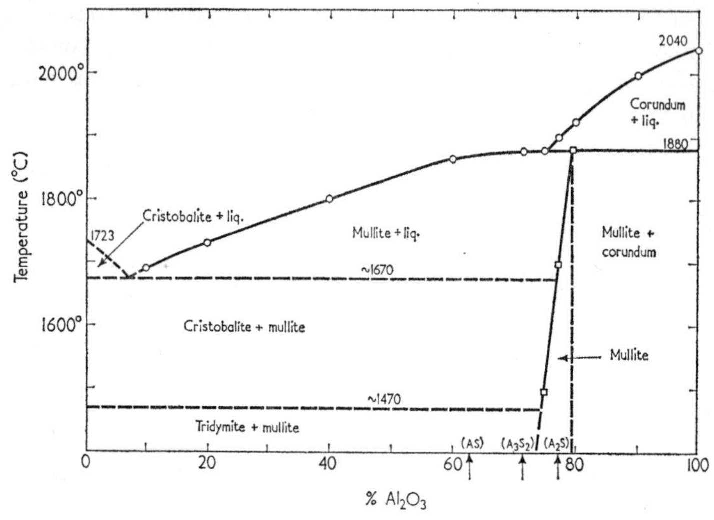

FIG. 1. The system SiO_2–Al_2O_3 (Welch [*8*]).

III. The System CaO–SiO_2

The phase diagram for this system (Fig. 2) derives from the original work of Rankin and Wright [*1*] with later modifications introduced by Greig [*9*], Muan and Osborn [*10*], Nurse [*11*] and Welch and Gutt [*12*].

A. Lime and Tricalcium Silicate

Lime (CaO) has a cubic NaCl-type structure, which remains unchanged up to the melting point of about 2570° C. A reported inversion at 420° C [*13*] between two cubic forms has never been substantiated. Differential thermal analysis (DTA) runs on CaO have sometimes shown an endothermic effect at about 450° C, but this is probably attributable to the expulsion of water from partially hydrated material.

Tricalcium silicate (C_3S) is the most important constituent of Portland cement. Pure C_3S is unstable relative to CaO and C_2S below 1250° C; reversible minor transformations in its structure have been detected in this region by DTA and high-temperature X-ray methods (see Chapter 3).

Tricalcium silicate was once thought to dissociate again into CaO and C_2S above 1900° C, but it has recently been shown that it melts incongruently at 2070° C, giving CaO and liquid [12]. Dissociation on cooling below 1250° C is slow with pure C_3S but may be hastened by inclusion of

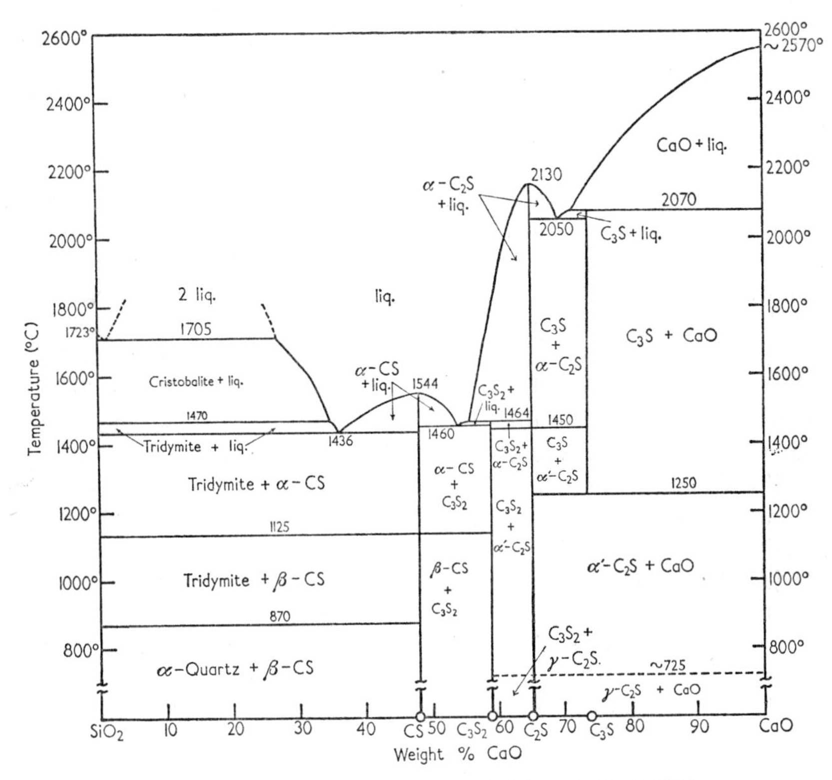

FIG. 2. The system $CaO–SiO_2$ (Welch and Gutt [12]).

foreign ions such as F^- into the structure. The incorporation of Al^{3+} and Mg^{2+} may also prevent or modify the minor structural changes which otherwise occur. The effects of solid solution in C_3S are discussed in Chapters 3 and 4.

B. DICALCIUM SILICATE

Dicalcium silicate (C_2S) is known to exist in at least four polymorphic forms, called α-C_2S, α'-C_2S or bredigite, β-C_2S or larnite, and γ-C_2S. β-C_2S is the second most important constituent of Portland cement. The following inversion sequences have been determined from high-temperature X-ray diffraction [*14*] and DTA [*11*].

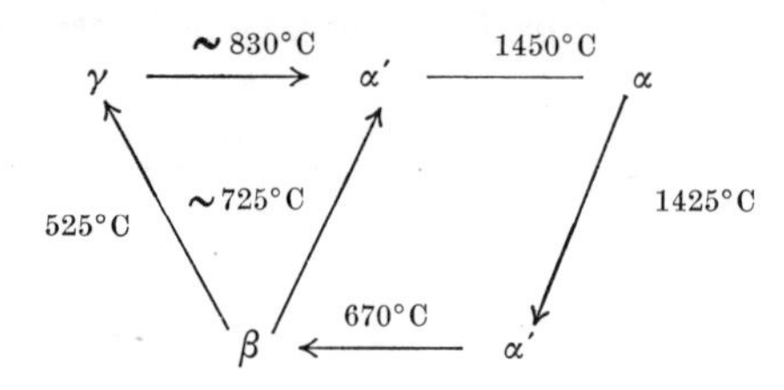

β-C_2S is formed during cooling and is metastable with respect to γ-C_2S. The $\beta \rightarrow \gamma$ change occurs with a large increase in volume which causes lumps of material to shatter; this phenomenon is called "dusting".

The $\beta \rightarrow \gamma$ inversion is reported to occur much more readily in samples of C_2S cooled from above the 1450° C transformation temperature than from below. This has been interpreted in various ways by different authors [*15, 16*], and has led to the suggestion that α'-C_2S obtained by heating γ-C_2S is slightly different structurally from that obtained by cooling α-C_2S (see also Chapter 3). Neither this view nor the existence of a β'-C_2S, as proposed by Vasenin [*17*] and by Toropov, Volkonskii and Sadkov [*18*], can yet be considered more than tentative. The $\alpha' \rightarrow \alpha$ inversion temperature is raised to 1465° C in the presence of excess CaO [*11*], a very small amount of which presumably goes into solid solution.

The stabilizing effect of impurity atoms on the various forms of C_2S has been extensively studied [*19*]; the compound shows a marked variation in its hydraulic activity, depending upon which polymorphic modification has been stabilized [*20*].

C_2S is formed as the initial product in solid state reactions between lime and silica over a wide range of compositions and reacts relatively slowly with excess of one or the other component to form CS, C_3S_2 or C_3S.

C. Rankinite, Wollastonite and Pseudowollastonite

Rankinite (C_3S_2) melts to form C_2S and liquid at 1464° C. This temperature is a revision by Muan and Osborn [*10*] of the melting point originally reported. Rankinite is found in blastfurnace slag but not in Portland cement. It has not been found to undergo any inversions, though another polymorph, known as kilchoanite or Phase Z, has been prepared hydrothermally and is also known as a natural mineral (see Chapter 5).

Monocalcium silicate (CS) melts congruently at 1544·5° C. Two polymorphs are known: wollastonite or β-CS, and pseudowollastonite or α-CS. β-CS is the low-temperature form and inverts to α-CS above about 1125° C. The transformation is reversible but is sluggish in the $\alpha \rightarrow \beta$ direction. β-CS exhibits polytypism; the name parawollastonite has been used for one of the polytypes (see Chapter 4). The high-silica end of the system is notable for liquid immiscibility over a substantial range of compositions at temperatures above 1705° C [*9*]. Neither rankinite nor α-CS nor β-CS shows any hydraulic activity.

IV. The System $CaO–Al_2O_3$

The $CaO–Al_2O_3$ system was first studied by Rankin and Wright [*1*]. Many later workers have since studied the system, and the original phase diagram has been much modified, but there is probably still no definitive

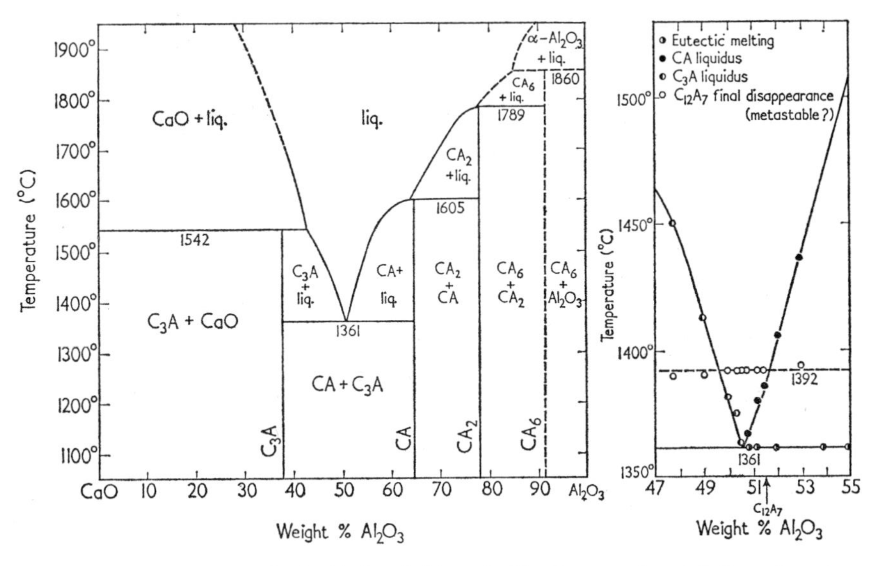

FIG. 3. The system $CaO–Al_2O_3$: (*left*) probable diagram for strictly anhydrous conditions; (*right*) detail for compositions around $C_{12}A_7$ [*109*].

version. Figure 3 (*left*) gives the author's version, which is based partly on new data reported in this chapter. According to this diagram, the compound $C_{12}A_7$ has no place in the system; previous workers have considered that this compound melts congruently approximately as shown in Fig. 4.

The sequence of reactions that occur when mixture of lime and alumina are heated together has recently been studied by Williamson and Glasser [*21*], who include a critical discussion of earlier work.

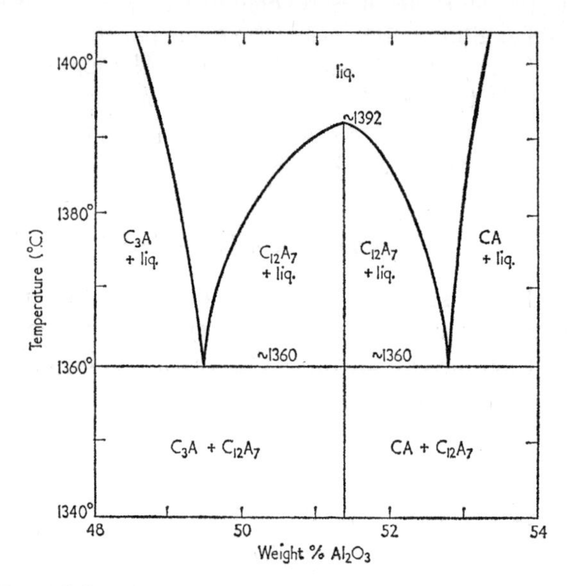

FIG. 4. Part of the system $CaO–Al_2O_3$, considerably modified from the diagram of Rankin and Wright [*1*]. $C_{12}A_7$ is shown as melting congruently; this diagram represents approximately the behaviour of this part of the system in atmosphere of ordinary humidity.

A. THE COMPOUNDS C_3A, CA, CA_2 AND CA_6

Tricalcium aluminate (C_3A) is an important constituent of most Portland cements and reacts rapidly with water at room temperature. It melts incongruently at 1542° C to CaO and liquid. Eitel [*22*] claims that, in the presence of CaF_2, tricalcium aluminate is not formed.

Monocalcium aluminate (CA) is the most important constituent of high-alumina cement and is responsible for the rapid hydration and high early strength of this material. Rankin and Wright reported that CA melts congruently at 1600° C but the author, and also Auriol, Hanser and Wurm [*23*], found that it melts just incongruently to CA_2 and liquid at 1605° C. CA is monoclinic but the crystals are markedly pseudo-hexagonal. Some investigators [*24, 25*] have reported the occurrence of a form giving a uniaxial interference figure, but Welch [*109*] has found by high-temperature microscopy that this is a spherulitic quench-growth produced when liquids of composition close to CA are rapidly chilled.

Rankin and Wright reported the existence of a compound C_3A_5 which they considered to exist in stable and metastable forms. The existence of the metastable form has not been confirmed, and Tavasci [*26*] concluded that the formula of the stable form was CA_2. This view is now generally accepted. Melting has been stated to occur incongruently at 1765° [*26*] or 1770° C [*23*], or congruently at 1770° C [*27*]. From high-temperature microscopy, Welch [*109*] concluded that it melts just incongruently to CA_6 and liquid at 1789° C. It has been stated that CA_2 is non-hydraulic [*28*], but Buttler and Taylor [*29*] showed that it reacts with water if $Ca(OH)_2$ or CA is also present.

Since the original publication by Rankin and Wright, a new alumina-rich compound has been reported. It is generally described as calcium hexaluminate, CA_6 [*30*], although the formula C_3A_{16} has also been proposed [*31*]. CA_6 melts incongruently to corundum and liquid at 1860° C. It resembles corundum very closely in optical properties, and the two are best distinguished by X-ray diffraction.

B. The Compounds $C_{12}A_7$ and C_5A_3

Rankin and Wright [*1*] reported the existence of two polymorphs of a compound to which they assigned the formula C_5A_3. One of these was a metastable, probably orthorhombic, phase obtained by rapid quenching of melts. Aruja [*32*] examined single crystals that J. H. Welch had prepared using the high-temperature microscope. He confirmed the orthorhombic symmetry and from the density and cell parameters concluded that the formula C_5A_3 was probably correct.

The second phase obtained by Rankin and Wright is more important. It is cubic and was reported by them to have a stable field of existence in the system (Fig. 4). Büssem and Eitel [*33*] concluded from determinations of density, cell volume, and probable space group that the true formula was $C_{12}A_7$, and this view is now generally accepted. Büssem and Eitel also determined the crystal structure, which they found to be unusual in that 2 out of the 66 oxygen atoms in the unit cell could not be given definite sites but had to be statistically distributed.

Even today the formula C_5A_3 is sometimes wrongly used to describe cubic $C_{12}A_7$. The X-ray powder patterns of the two phases have also been confused [*21*]. Orthorhombic C_5A_3 has also been confused with a different phase occurring in high-alumina cements. $C_{12}A_7$ reacts readily with water, and pastes made from it set rapidly.

Investigations by Nurse and his colleagues [*34*], Roy and Roy [*35*] and Glasser and Jeevaratnam [*36*] have shown that $C_{12}A_7$ takes up water from the atmosphere when it is heated in air of ordinary humidity.

Nurse and Welch found that if an anhydrous, or nearly anhydrous, specimen was gradually heated, water was sorbed until a content of about 1·4% was reached at about 1000° C. This water content corresponds approximately to the formula $C_{12}A_7H$. If the temperature was raised further, the water was progressively expelled, until the melting point was reached at 1391·5° C. If a sample containing 1·4% of water was slowly cooled from 1000° C, still more water was taken up. Roy and Roy [*35*] found a maximum water content of about 1·4%. They reported a reversible sorption curve, which showed a gradual variation in water content over the range 200–1200° C. Glasser and Jeevaratnam [*36*] found that if an anhydrous sample, made at 1390° C, was maintained at 960° C, a weight increase of 1·12% occurred.

The main sorption of water by $C_{12}A_7$ does not seem to be a mere surface effect, because it causes changes in the cell size, density and refractive index. There is, however, a diagreement about the direction of these variations (Table I). The uptake of water which occurs on heating

TABLE I

Properties of anhydrous and hydrated forms of "$C_{12}A_7$"

Reference	Refractive index		Density (g/cm³)		Unit cell dimension (Å)	
	Anhydrous	Hydrated	Anhydrous	Hydrated	Anhydrous	Hydrated
Nurse, Welch and Majumdar [*34*]	1·61	1·62	2·68	2·73	11·9880	11·9747
Roy and Roy [*35*]	1·616	1·610	—	—	11·977[a]	11·987[a]
Glasser and Jeevaratnam [*36*]	1·61	1·62	2·68	2·72	11·983	11·976

[a] Recalculated from d_{642} values given in original paper.

anhydrous material probably explains the results of Bonnickson [*37*], who found in the heat capacity/temperature curve at 1000° C an inflexion which he attributed to an inversion.

The sorption of water by $C_{12}A_7$ markedly affects its melting behaviour. Early investigators evidently had difficulty in determining the melting point; Shepherd, Rankin and Wright [*38*] originally reported that melting occurred congruently at 1386° C but Rankin and Wright [*1*] later stated the value to be 1455° C. Nurse and Welch [*34*] found that in air of normal humidity the compound melts at 1391·5° C but that in air dried by passage over $Mg(ClO_4)_2$ it melts incongruently to CA and liquid at 1374° C. From high-temperature microscopy they found that C_3A and CA form an apparent eutectic at 1361° C (Fig. 3) and that at 1361–1391·5° C either can coexist with $C_{12}A_7$ and liquid. The apparent violation of the phase rule can be explained by assuming that ternary phase relations exist. There is no doubt that this is so at 1374–1391·5° C, but further work is needed to establish whether some water remains in the structure even when the melting point has fallen to 1374° C. If this is the case, the already diminished stability region at 1361–1374° C for anhydrous $C_{12}A_7$ would probably vanish altogether for the true, binary system. This question is further discussed in Chapter 4.

In Fig. 3 (*left*,) as already stated, the stability field for $C_{12}A_7$ has been omitted; this diagram represents the possible behaviour of the system under strictly anhydrous conditions. Figure 3 (*right*) shows the part of the system around the $C_{12}A_7$ composition in more detail. Behaviour of a pseudobinary type (Fig. 4) is encountered when melting occurs in air of ordinary humidity. Since this is the condition normally encountered, the primary phase field of $C_{12}A_7$ will be included in the various ternary and quaternary systems containing lime and alumina which are discussed later in this chapter. Further study of the role of $C_{12}A_7$ in these systems is obviously required.

V. The System $CaO–Al_2O_3–SiO_2$

A. Ternary Compounds

Now that the three limiting systems have been discussed, the ternary system $CaO–Al_3O_2–SiO_2$ will be considered. Two ternary compounds, gehlenite and anorthite, are stable at liquidus temperatures.

1. Gehlenite (C_2AS)

Gehlenite has the melilite structure and forms extensive solid solutions, though not with other compounds in the $CaO–Al_2O_3–SiO_2$ system. It melts congruently at about 1590° C; Welch [*109*] found 1582° C. It is not hydraulically active but some of its solid solutions may be; this may account for conflicting reports in the literature concerning its latent cementing ability.

2. Anorthite (CAS_2) *and its Polymorphs*

Several polymorphs of CAS_2 are known. The only one which is stable in contact with liquid in the system $CaO–Al_2O_3–SiO_2$ is anorthite, a triclinic feldspar which melts congruently at 1553° C. Like gehlenite, it enters extensively into solid solutions, of which the plagioclase series between anorthite and albite is probably the best known. In recent years hexagonal and orthorhombic polymorphs of CAS_2 have been described and investigated [*39*, *40*]. The hexagonal modification melts metastably at 1405° C. It can be grown under the high-temperature microscope from melts of anorthite composition following suitable heat treatment. The crystal habit is distinctive in that large thin sheets are formed which crack and buckle under the internal stresses set up during growth. The orthorhombic modification melts, again metastably, at 1426° C. Under the high-temperature microscope the crystals are seen to develop initially as small rhombic plates. When subjected to moderate supercooling the rhombs generally develop into six-sided platelets which exhibit remarkable melting behaviour if the temperature is again raised. Instead of starting normally from the outside edge, melting proceeds from within the crystal, so that portions of the central part of the crystal are melting while at the same time the outside edge may actually be growing. The effect is to produce a mosaic pattern, which is a diagnostic feature of this polymorph. The cause of the phenomenon has not been investigated but has been attributed to order–disorder effects within the crystal taking place during cooling and subsequent heating.

3. *Grossular* (C_3AS_3)

One further anhydrous calcium aluminosilicate is definitely known, not counting dehydrated zeolites; this is grossular, which has the garnet structure. It can be synthesized hydrothermally (Chapter 6) but has no stability field in contact with melts in the $CaO–Al_2O_3–SiO_2$ system.

B. The Phase Diagram

Figure 5 shows the ternary system $CaO–Al_2O_3–SiO_2$ and is based on the diagram of Osborn and Muan [*41*], with further revisions in accordance with the work on the bounding systems already described. It differs from Rankin and Wright's original diagram mainly in the phase field of C_3S and the mullite–corundum boundary and in the inclusion of the field of CA_6. The primary phase field of C_3S is a long sliver extending away from the $CaO–SiO_2$ edge of the triangle. Its location and shape are of great importance to cement chemistry and some applications are discussed in the following paragraphs.

C. Crystallization Paths for Cement Compositions

Figure 6 shows in more detail the part of the system which is of most importance for the formation of Portland cement clinker. Modern Portland cement compositions lie in the area shown. The presence of Fe_2O_3

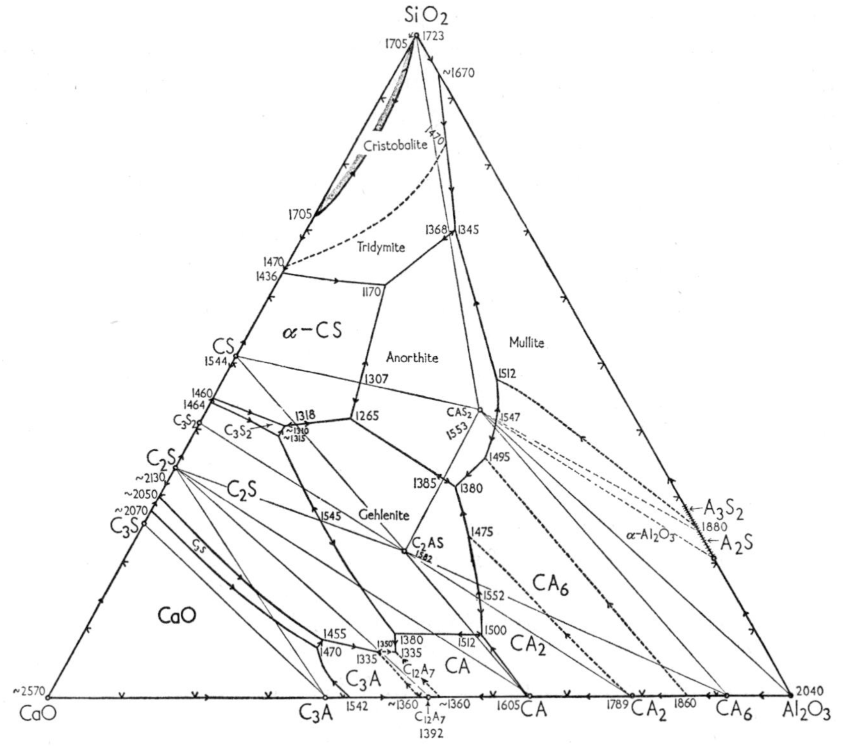

Fig. 5. The system $CaO-Al_2O_3-SiO_2$; based on the diagram of Osborn and Muan [*41*], with later modifications. In this and subsequent figures, shaded boundaries denote limits of liquid immiscibility.

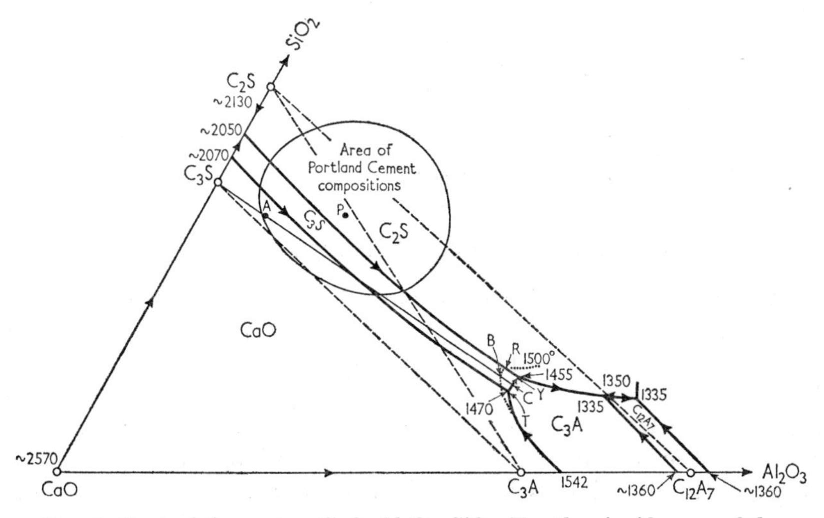

Fig. 6. Part of the system $CaO-Al_2O_3-SiO_2$. For the significance of the line C_3S–A–B–C see text. The dotted line represents part of the 1500° C isotherm.

and other components is, of course, ignored; considerations based on the ternary system nevertheless serve as an introduction for understanding the more complex processes that occur when these other components are present. They apply with relatively little modification to actual cement mixes of high Al_2O_3/Fe_2O_3 ratio.

The mix forms at the clinkering temperature of, say, 1500° C as a liquid and either one or two solids. A mix of composition A will at 1500° C consist of solid C_3S and a liquid of composition given by the point B, at which the prolongation of the C_3S–A line intersects the 1500° C isotherm. On cooling the clinker, the liquid will precipitate more C_3S and the composition of the liquid will move along the line C_3S–A–B–C until the line TY, which forms the boundary between the phase fields of C_3S and C_3A, is reached. At this point (C), C_3A will be coprecipitated and the composition of the liquid will proceed along the junction line TY until point Y is reached. C_2S will then be precipitated and the liquid will solidify.

A mix of a lower lime content, such as that represented by point P, will at 1500° C be composed of a mixture of C_3S, C_2S and a liquid having a composition lying at the intersection R of the 1500° C isotherm and the junction of the C_3S and C_2S phase fields. As the mix cools, the two phases C_3S and C_2S will go on being precipitated and the liquid will change in composition along RY until Y is reached, when C_3A will be precipitated.

These paths will be followed only if equilibrium is maintained during cooling. In practice, however, two other courses are possible; either the clinker is quenched or it is partially quenched and then annealed. In the first case the residual liquid will remain on quenching as glass, and in the second case the liquid will crystallize separately with no relation to the solids already present. If we take the points B, R, Y and C, the liquids of these compositions all lie outside the triangle C_3S, C_2S and C_3A and thus cannot give these phases on crystallization. The liquids will give on independent cooling C_2S, C_3A and $C_{12}A_7$.

VI. The System CaO–MgO–Al_2O_3

MgO (magnesia or periclase: synonyms) melts at about 2825° C [*41a*] and is thus one of the most refractory of oxides; its primary phase field dominates the liquidus surface of many of the ternary systems of which MgO is a component. Of the three binary systems that bound the CaO–MgO–Al_2O_3 system, CaO–Al_2O_3 has already been discussed. No intermediate compounds are formed between CaO and MgO, but CaO can take a little MgO into solid solution. MgO and Al_2O_3 form a single compound, spinel (MA), which melts congruently at 2105° C. Spinel can take an appreciable amount of Al_2O_3 into solid solution [*42*], while spinel and MgO show limited mutual miscibility [*41a*].

The ternary system CaO–MgO–Al_2O_3 was first studied by Rankin and Merwin [*25*], who concluded that there were no ternary compounds stable in contact with the melt. A recent reinvestigation [*43*] has shown, however, that two ternary compounds exist, with the approximate compositions $C_{25}A_{17}M_8$ and C_7A_5M, of which the first has a very small primary phase field.

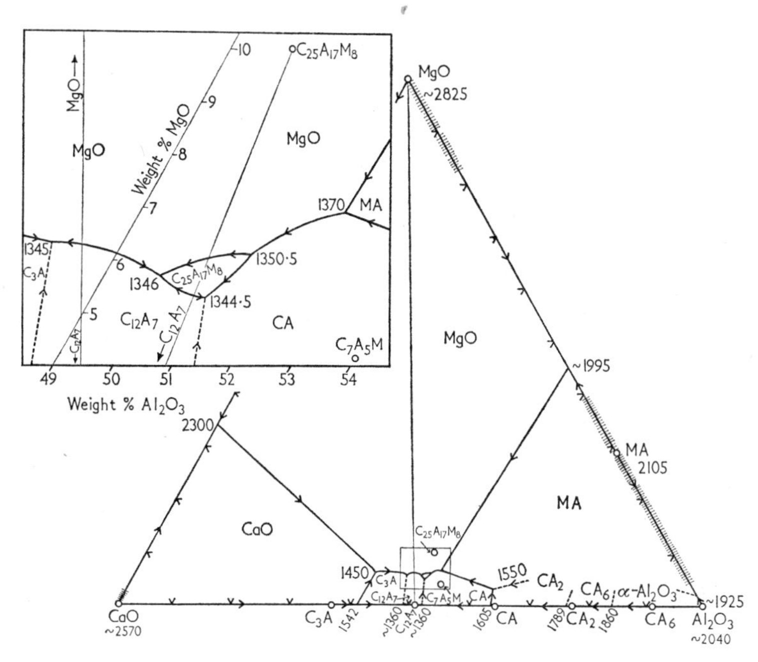

FIG. 7. The system CaO–MgO–Al_2O_3. Inset: region surrounding the primary phase field of $C_{25}A_{17}M_8$ on a greatly enlarged scale.

Figure 7 shows the system CaO–MgO–Al_2O_3, revised in the light of this study and of recent reinvestigations of the CaO–Al_2O_3 system. The compound $C_{25}A_{17}M_8$ melts incongruently at 1350·5° C to a mixture of CA, MgO and liquid, and its primary phase field lies between those of CA, MgO and $C_{12}A_7$. It is still uncertain whether this compound has the formula previously stated or the very similar one C_3A_2M.†

The second compound, of approximate composition C_7A_5M, closely resembles the first in optical properties but is readily distinguished by its

† The most recent work supports the formula C_3A_2M (see Chapter 4).

X-ray powder pattern. It is a metastable product in the system and melts at 1332° C; it was studied with the high-temperature microscope. C_7A_5M is discussed further in connection with the quaternary system $CaO–MgO–Al_2O_3–SiO_2$.

VII. The System $CaO–MgO–SiO_2$

A. The System $MgO–SiO_2$

This system was investigated by Bowen and Andersen [*44*] and later by Greig [*9*]. Two binary compounds, MS and M_2S, are formed and extensive liquid immiscibility occurs at silica-rich compositions.

Forsterite (M_2S) has the olivine structure. It melts congruently; the melting temperature has been given as 1890° C [*44*] or 1860° C [*45*] or 1830° C [*109*].

Magnesium metasilicate (MS) melts incongruently at 1557° C to form forsterite and liquid. It exists in three polymorphic forms, known as clinoenstatite, mesoenstatite [*46*] (called protoenstatite by U.S. workers) and enstatite. All three forms belong to the pyroxene group and there has been much controversy about their relative stabilities. The polymorph stable below about 900° C is enstatite; it occurs as a natural mineral. Preparations heated to a high temperature and quenched contain clinoenstatite, which was formerly supposed to be the high-temperature polymorph. However, high-temperature X-ray evidence obtained by Foster [*47*], and further work by Atlas [*48*], showed that the form which actually exists above 900–1000° C is mesoenstatite and that clinoenstatite is a metastable phase which is formed on cooling. The structural relations between the three polymorphs have been explored [*48, 49*].

B. The Join $CaSiO_3–MgSiO_3$

The ternary system $CaO–MgO–SiO_2$ has been the subject of several investigations; Bowen [*51*] studied the sub-system forsterite–diopside–silica as early as 1914. The system is represented in Fig. 8, which is based on the diagram of Osborn and Muan [*41*]. This diagram incorporates the recent data provided by the work of Ricker and Osborn [*50*] and of Welch and Gutt [*12*]. Four ternary compounds having primary phase fields in the system have been established.

On the metasilicate join (CS–MS) there occurs the pyroxene, diopside, CMS_2. This compound melts congruently at 1391·5° C. At high temperatures there is a nearly continuous range of solid solutions between diopside and mesoenstatite, which gives to primary phase fields mainly located away from the join. Most compositions on the join thus show incongruent melting behaviour. At lower temperatures considerable exsolution occurs [52]. No solid solution of CS in diopside has been detected but both α-CS and β-CS can accommodate Mg^{2+}, especially at high temperatures. The α–β inversion temperature of CS is raised sufficiently by the introduction of Mg^{2+} to bring a stability field for the β-CS solid solution to the liquidus surface [53].

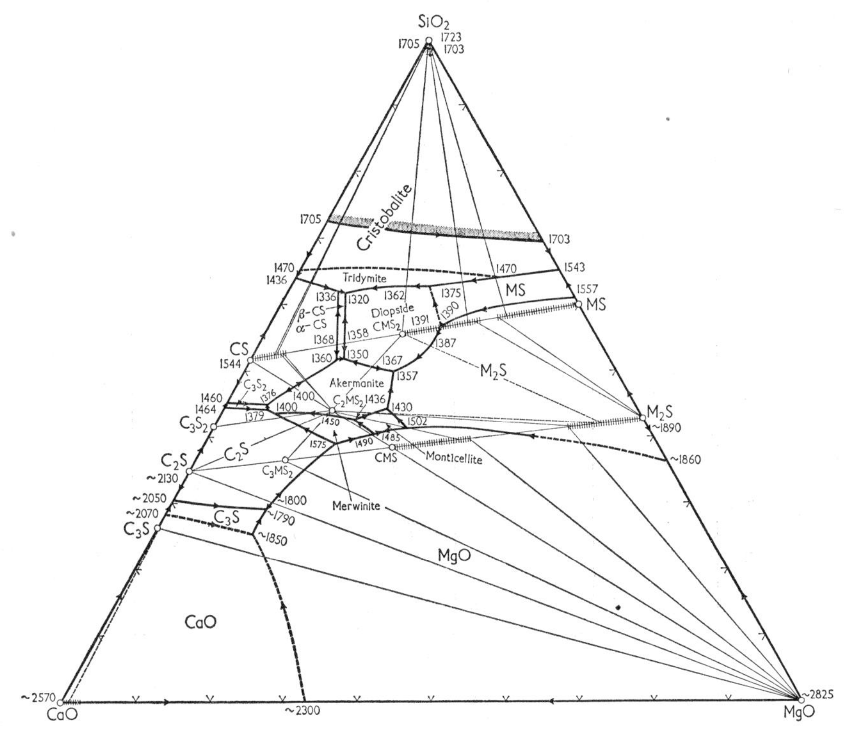

FIG. 8. The system CaO–MgO–SiO_2 (after Osborn and Muan [41]).

C. The Join Ca_2SiO_4–Mg_2SiO_4

Phase relations on the orthosilicate join (C_2S–M_2S) are complex; the system is pseudobinary and is shown in Fig. 9. There are two well established compounds, merwinite (C_3MS_2) and monticellite (CMS). Merwinite resembles α′-C_2S in structure and is also difficult to distinguish from it optically; Bredig [54] postulated that merwinite was merely a form of α′-C_2S stabilized by partial replacement of Ca^{2+} by Mg^{2+}. However, the results obtained by other investigators [50, 55–57] show that merwinite has a definite composition and that it does not form solid solutions, either with C_2S or with CMS. Monticellite has a structure of the olivine type and is thus related to γ-C_2S and forsterite; Bredig considered that extensive solid solution occurs between monticellite and γ-C_2S. This also has not been confirmed, although solid solutions are formed between monticellite and forsterite.

Roy [*57*] studied the join C_2S–CMS at sub-solidus temperatures and a water pressure of 140 kg/cm^2. She found no evidence of either solid solution or the formation of intermediate compounds. In contrast,

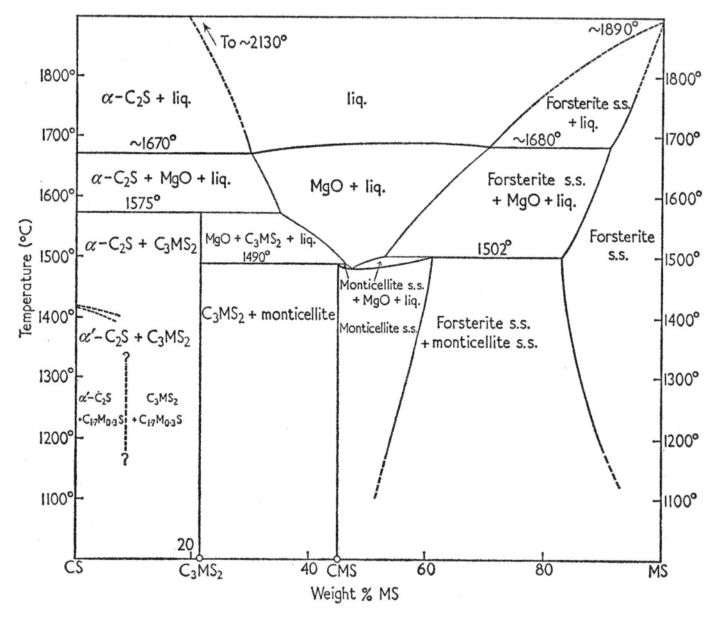

FIG. 9. The pseudobinary system Ca_2SiO_4–Mg_2SiO_4. Heavy lines denote binary equilibria; in other regions ternary equilibrium obtains. For discussion of the compound $C_{1\cdot7}M_{0\cdot3}S$ see text.

Gutt [*58*] obtained a new compound, of approximate composition $5{\cdot}6C_2S.4{\cdot}4C_3MS_2(C_{1\cdot7}M_{0\cdot3}S)$, by prolonged heating at 1250° C of very finely ground reactants. This compound has so far been formed only by solid state reaction, and melts incongruently. Whether it resembles merwinite and monticellite in possessing a primary phase field in the ternary system CaO–MgO–SiO_2 has yet to be determined. X-Ray evidence suggested that it does not form solid solutions with either C_2S or merwinite. Gutt showed that $C_{1\cdot7}M_{0\cdot3}S$ is probably identical with Phase T, which was found by Sharp, Johnson and Andrews [*58a*] to occur in basic arc furnace slags.

D. ÅKERMANITE

The remaining ternary compound with an established phase field in the system is åkermanite (C_2MS_2). Like gehlenite (C_2AS), it has the melilite structure. It is completely miscible with gehlenite but does not form solid solutions with any other phase in the CaO–MgO–SiO_2 system.

Åkermanite melts congruently at 1454° C. Its stability at sub-solidus temperatures has been a subject of dispute. Early work appeared to indicate that it was unstable below about 1300° C; Bowen, Schairer and Posnjak [*59*] found that åkermanite was stable down to 1375° C but that a glass of the same composition crystallized at 1050° C to a mixture of phases which included C_2S. A slowly cooled melt of åkermanite composition was found by Carstens and Kristoffersen [*60*] to yield a little diopside. Osborn and Schairer [*61*] found that åkermanite crystals remained homogeneous on cooling to 1325° C but that below this temperature a little diopside was probably formed.

In contrast, Nurse and Midgley [*62*] reported that åkermanite remained stable on cooling and suggested that the apparent decomposition might be caused by exsolution from slightly impure åkermanite. DeWys and Foster [*63*], in a study of the system anorthite–åkermanite, showed that åkermanite must be stable at least down to the eutectic temperature of 1234° C, and almost certainly down to 1125° C. In a hydrothermal study at water pressures of 2000–4000 kg/cm^2 Harker and Tuttle [*64*] found that åkermanite was stable down to 700–750° C, below which it decomposed to wollastonite and monticellite. The decomposition temperature was essentially independent of pressure over the range studied, and the authors considered that the water acted solely as a flux, not entering into the reaction in any way. Thus, the more recent evidence supports the view of Nurse and Midgley.

Because of the solid solution relations between monticellite and forsterite and in the pyroxene series, crystallization sequences in the CaO–MgO–SiO_2 system are rather complex. They are discussed by Ricker and Osborn [*50*] and by Bowen [*51*].

VIII. The System MgO–Al_2O_3–SiO_2

This system, remote from the interests of the cement chemist but of interest to the refractory technologist, will be briefly considered to complete the survey of systems bounding the quaternary system CaO–MgO–Al_2O_3–SiO_2. Figure 10 is based on the recent diagram due to Osborn and Muan [*41*].

The original study, by Rankin and Merwin [*42*], showed a single compound, cordierite ($M_2A_2S_5$), having a primary phase field. It has a structure related to that of beryl. Subsequently the compound sapphirine ($M_4A_5S_2$ approx.), has been discovered and found to have a small primary phase field [*65–67*]. The mullite–corundum boundary has been revised [*68*]. Recently it has been found that there are several polymorphs of $M_2A_2S_5$; Schreyer and Schairer [*69, 70*] studied this problem

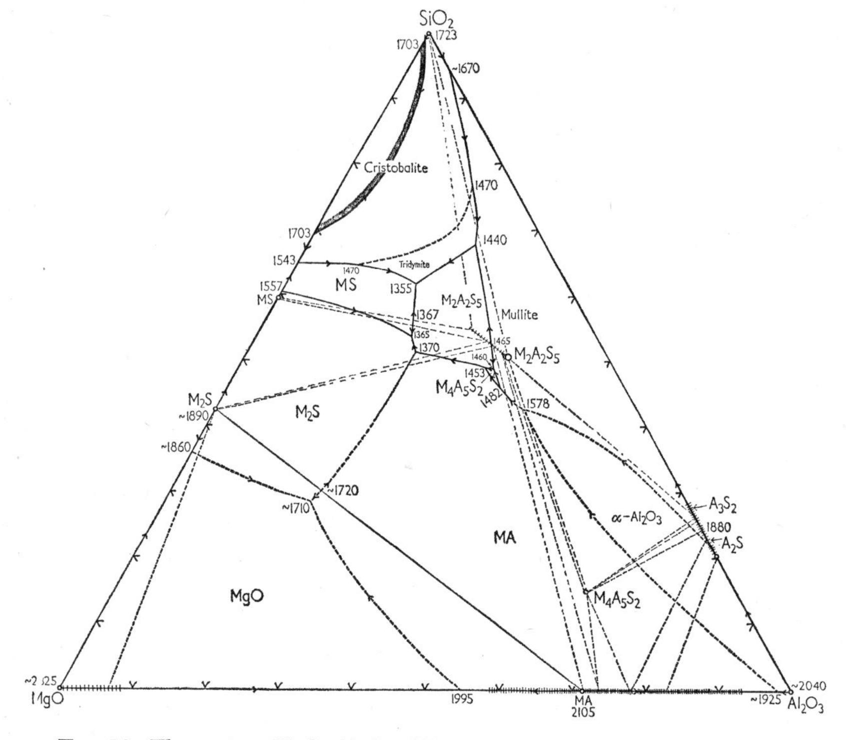

FIG. 10. The system $MgO–Al_2O_3–SiO_2$; based on the diagram of Osborn and Muan [*41*], with later modifications.

and made further revisions of invariant compositions and temperatures in the system.

Several further anhydrous magnesium aluminosilicates are known, but none of these has any stable existence in contact with melts in the system. The best known is pyrope (M_3AS_3), which is a member of the garnet group. It can be synthesized hydrothermally. Recently Schreyer and Schairer [*70*] have discovered two further phases which are formed by devitrification of glasses of appropriate compositions. One had a composition probably close to MAS_4 and a structure related to that of osumilite.

The second was of unknown composition; its structure was related to that of petalite ($LiAlSi_4O_{10}$).

IX. The System $CaO–MgO–Al_2O_3–SiO_2$

A. General Points

This quaternary system has been widely studied because of its importance in geology and silicate technology. Figure 11 shows the more important phases, represented on a tetrahedron. They include the oxides

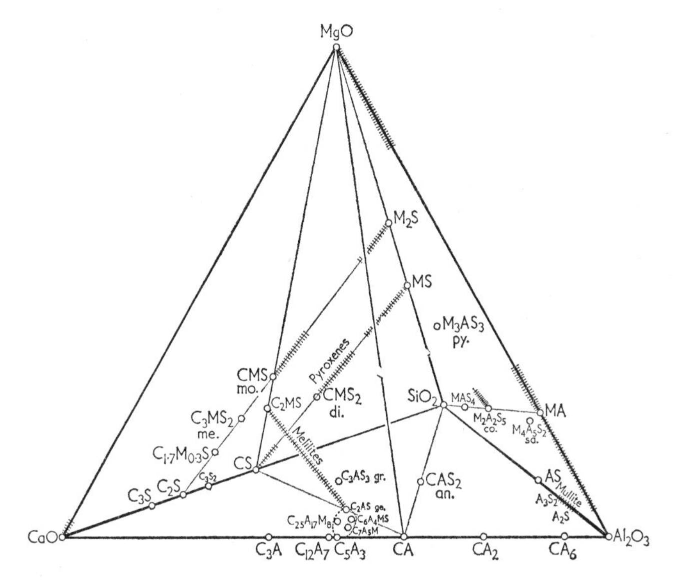

Fig. 11. Phases within the system $CaO–MgO–Al_2O_3–SiO_2$. For significance of the planes MgO–CS–CA and $MgO–SiO_2–CA$ see text. Dotted line encloses a group of phases which may be structurally related to each other; see text and Chapter 4.

and the binary and ternary phases already described, as well as joins exhibiting extensive solid solution. The most important solid solutions are the melilites, which form a complete series ranging from åkermanite (C_2MS_2 or $Ca_2MgSi_2O_7$) to gehlenite (C_2AS or $Ca_2Al_2SiO_7$). Several other phases, such as C_3S and the pyroxenes, can accommodate limited proportions of other cations, thus becoming quaternary in composition.

Parker [*71*] reported that the system contained a quaternary compound C_6A_4MS. He considered that it had a primary phase field and that it was isomorphous with orthorhombic C_5A_3. Subsequent work [*43*] failed to confirm the composition C_6A_5MS and indicated that the phase should perhaps be regarded as a solid solution of C_7A_5M and gehlenite. Still more recent work (Chapter 4) suggests that the substance is a definite compound, though the composition is probably not C_6A_4MS. No other reports of quaternary compounds have been substantiated.

Investigators have studied this system mainly by examining ternary or pseudoternary sub-systems, or by taking planes at arbitrary, constant levels of either Al_2O_3 or MgO. These studies are listed in Table II. Those

TABLE II

Investigations on the quaternary system CaO–MgO–Al_2O_3–SiO_2

1. Sub-systems in the high-silica region (within the tetrahedron MgO–CA–CS–SiO_2)	
Silica–forsterite–anorthite	Andersen [*72*]
Åkermanite–gehlenite	Ferguson and Buddington [*73*]
Åkermanite–gehlenite–pseudowollastonite	Osborn and Schairer [*61*]
Pseudowollastonite–anorthite–diopside	Osborn [*74*]
Forsterite–anorthite–diopside	Osborn and Tait [*75*]
Anorthite–åkermanite	DeWys and Foster [*63*]
Pseudowollastonite–$MgSiO_3$–alumina	Segnit [*76*]
Anorthite–åkermanite–diopside	DeWys and Foster [*77*]
$MgSiO_3$–diopside–anorthite	Hytönen and Schairer [*78*]
2. Sub-systems in the high-alumina region (mainly within the tetrahedron MgO–CA–Al_2O_3–SiO_2)	
Gehlenite–spinel	Nurse and Stutterheim [*79*]
Anorthite–spinel	Welch [*80*]
Anorthite–gehlenite–spinel	DeVries and Osborn [*81*]
Gehlenite–spinel–corundum	DeVries and Osborn [*81*]
Anorthite–periclase–corundum	DeVries and Osborn [*81*]
Anorthite–forsterite—spinel	DeVries and Osborn [*81*]
Anorthite–periclase–forsterite	DeVries and Osborn [*81*]
3. Sub-systems in the high-lime region (mainly within the tetrahedron CaO–MgO–CA–CS)	
MgO–C_2S–C_5A_3	Hansen [*82*]
CaO–MgO–C_2S–C_5A_3	McMurdie and Insley [*83*]
C_2S–MgO–Al_2O_3	Prince [*84*]
C_2S–M_2S–C_5A_3	Segnit and Weymouth [*85*]
4. Cuts through the system at constant levels of one component	
Al_2O_3: 5, 10, 15, 20, 25, 30 and 35%	Osborn *et al.* [*86*]
MgO: 5%	McMurdie and Insley [*83*]
5%	Parker [*71*]
5%	Cavalier and Sandrea-Deudon [*87*]
4–15% at 1% intervals[1]	Cavalier and Sandrea-Deudon [*87*
10%	Prince [*88*]

(1) Reconstructed from data of reference 86.

dealing with sub-systems bounded by specific compounds have been grouped according to the part of the tetrahedron within which they fall. It is convenient for this purpose to define three regions separated by the planes MgO–CS–CA and MgO–CA–SiO_2, as shown in Fig. 11.

The high-silica part of this system is of chiefly geological importance. The high-alumina part is important for refractories, including refractory concretes made from aluminous cement. The high-lime part is important for Portland cement manufacture and also for blastfurnace slags.

The studies of Osborn and his collaborators especially show that the tetrahedron as a whole is dominated by the primary phase volumes of periclase and spinel. The primary volumes of the pyroxenes and the melilites are also of moderate extent. All the phase fields occurring on the CaO–MgO–SiO_2 face persist up to the 10% Al_2O_3 level; by 15% Al_2O_3, however, the fields of rankinite, monticellite and wollastonite have vanished and those of spinel and anorthite have appeared. By 35% Al_2O_3 the fields of periclase and spinel are predominant.

B. THE SECTION AT 5% MgO

Figure 12 shows the intersections of the primary phase volumes with the 5% level of MgO, as represented by Parker [*71*] on the basis of his own work and that of McMurdie and Insley [*83*]. Comparison with Fig. 5

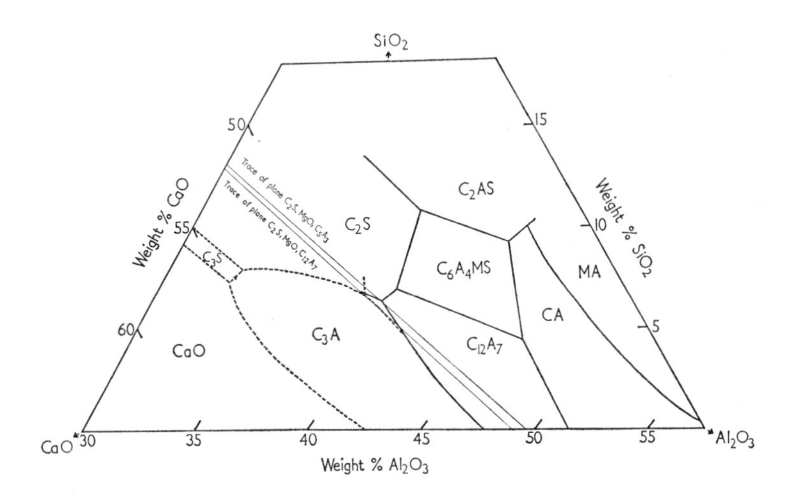

FIG. 12. Part of the system CaO–MgO–Al_2O_3–SiO_2; showing intersections of primary phase volumes with the 5% level of MgO (after Parker [*71*]).

shows that the fields of C_3S, C_2S and C_3A, which are the three most important constituents of most Portland cements, occupy areas not very different from those found in the absence of MgO. Figure 12 includes a primary phase field for C_6A_4MS. As has already been stated, the precise composition of this phase is now uncertain; further work on this part of the quaternary system is required.

McMurdie and Insley [*83*] located an invariant point between liquid and C_3S, C_2S, C_3A and MgO; the liquid contained 5·5% MgO. This result is significant for cement manufacture, because it sets an approximate upper limit to the proportion of MgO that can be tolerated in the raw materials. If there is more than enough MgO to saturate the liquid, periclase would be expected to separate from the melt as comparatively large crystals which are likely to hydrate slowly after the concrete has hardened. The consequent formation of $Mg(OH)_2$, which is less dense than periclase, is likely to cause disruption. The mix contains about 30% of liquid at the clinkering temperature; to avoid saturation of the liquid with MgO the content of the latter in the clinker as a whole must not exceed 1·5–2·0%. If a clinker containing less than 1·5–2·0% MgO is cooled rapidly, the MgO will either remain as a glass, or more probably crystallize out as small crystals which may be expected to hydrate relatively quickly.

C. Relation between Compositions of Cement and Blastfurnace Slags

The studies of Osborn *et al.* [*86*] on sections through the tetrahedron at 5–35% levels of Al_2O_3 had as their main object the delineation of optimum compositions for blastfurnace slags. It is important that these should have low liquidus temperatures and viscosities and a high desulphurization potential, that is, ability to extract sulphur from the metallic phase. This last property is favoured by high proportions of the basic components. The proportions of basic components cannot, however, be as high as in Portland cements because of the requirement that the slag must be completely liquid at some reasonable temperature (e.g. 1550° C). It is also important that moderate variations in composition should be possible without causing any marked worsening in the properties of the slag.

Osborn *et al.* showed that good properties were possible with widely differing contents of Al_2O_3 but that for any given content of Al_2O_3 there were optimum values for the proportions of the other constituents; thus, for 10% Al_2O_3 the optimum values are SiO_2 32%, MgO 14%, and CaO 44%.

X. The System $CaO–Fe_2O_3$

Iron(III) oxide (Fe_2O_3) resembles Al_2O_3 in showing many polymorphs. Only one of them, hematite or α-Fe_2O_3, is formed from melts; it is structurally similar to corundum (α-Al_2O_3). The study of systems containing Fe_2O_3 as a component is complicated by the fact that loss of oxygen tends to occur on heating in air, so that compounds such as magnetite (Fe_3O_4), in which the iron is partly or wholly reduced to iron(II), may be formed. This effect takes place mainly at iron-rich compositions.

Several studies of the $CaO–Fe_2O_3$ system have been made [*89–92*]. Figure 13 shows the results of the recent reinvestigation by Phillips and

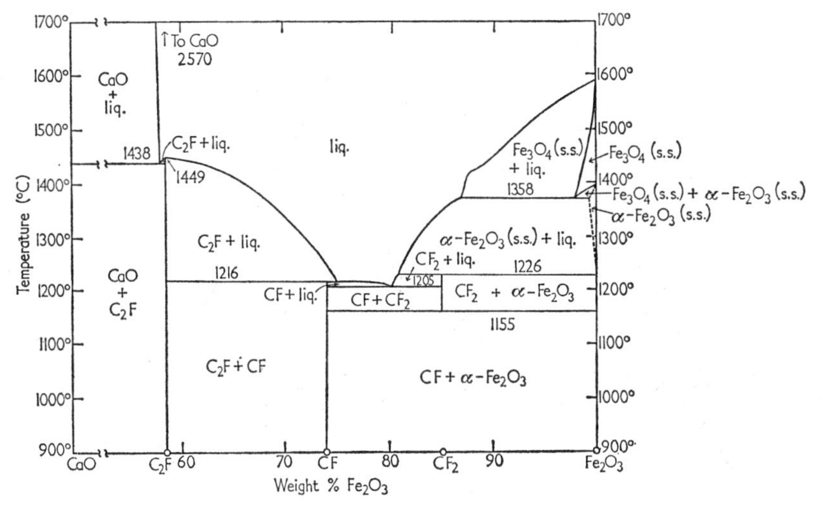

FIG. 13. The pseudobinary system $CaO–Fe_2O_3$ (after Phillips and Muan [*92*]).

Muan [*92*]. The iron-rich portion of the system is only pseudobinary, owing to loss of oxygen; but the lime-rich portion, which is the part of interest to the cement chemist, is essentially binary. There are three compounds, C_2F, CF and CF_2. The first two of there were found by Sosman and Merwin [*89*] in the original study of the system and the third was discovered by Tavasci [*90*].

XI. The System $CaO–Fe_2O_3–SiO_2$

The main studies in this system are due to Burdick [*93*] and to Phillips and Muan [*94*]; Fig. 14 gives the most recent data and is based on the diagram by Osborn and Muan [*41*]. Because of the changes in oxidation state which occur when iron-rich compositions are heated in air, the diagram represents an isobaric surface within the system CaO–FeO–Fe_2O_3–SiO_2. The system is, however, essentially ternary in the lime-rich region.

The primary phase field of C_3S extends from the CaO–SiO_2 edge to two invariant points for CaO, C_2F and liquid at 1412° C, and for C_2S, C_2F and liquid at 1405° C. (There is an error in Phillips and Muan's Table 2,

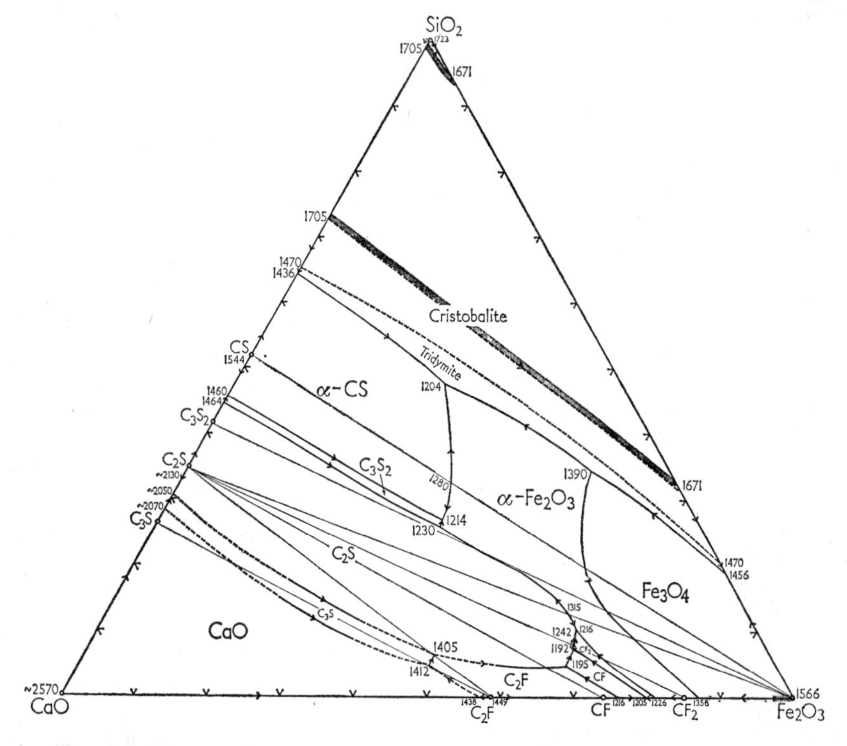

FIG. 14. The pseudoternary system CaO–Fe_2O_3–SiO_2 (after Osborn and Muan [*41*]).

involving a transposition of these points.) Burdick [*93*] gave corresponding temperatures of 1411° and 1414° C but considered that both points were eutectics and that there was a temperature maximum of 1428° C on the boundary joining them.

Burdick concluded that C_2S can accommodate up to 0·5% Fe_2O_3 by forming a solid solution with C_2F and that this lowers the α–α' transition temperature to 1360° C. Newman and Wells [*19*] estimated the maximum solubility of Fe_2O_3 in C_2S at 1·0% but found that at this content of

Fe_2O_3 the α–α' transition temperature was lowered only from 1457° C to 1437° C. Further studies of this system are needed to resolve some of the conflicting evidence.

XII. The System $CaO–Al_2O_3–Fe_2O_3$

A. The System $Al_2O_3–Fe_2O_3$

Two of the binary systems which bound the ternary system $CaO–Al_2O_3–Fe_2O_3$ have already been considered. The third, $Al_2O_3–Fe_2O_3$, is interesting because of the formation of an ordered solid solution, whose composition lies at or near $Al_2O_3.Fe_2O_3$ and which is stable over a narrow range of temperature [*95*]. The two oxides also show limited mutual miscibility.

B. The Ternary System: General Points

A knowledge of the ternary system is important for understanding the iron-containing phase in Portland cement. The first study was made by Hansen, Brownmiller and Bogue [*96*] in 1928. They showed its most important feature to be a series of solid solutions extending from the compound C_2F towards a hypothetical end-member, C_2A. They found that complete miscibility existed along this join from C_2F as far as C_4AF. They considered the latter formula to represent a definite compound which terminated the series; this compound was later called brownmillerite. They also considered that all commerical Portland cement clinkers for which the Al_2O_3/Fe_2O_3 weight ratio exceeded 1·0 would contain a ferrite phase having the terminal composition C_4AF.

Subsequent work has shown that, while normal Portland cement clinkers indeed contain a ferrite phase belonging to the $C_2F–C_2A$ series, this phase does not necessarily have the composition C_4AF. It has also been shown that the series extends to compositions more aluminous than C_4AF. Yamauchi [*97*] placed the limiting molar Al_2O_3/Fe_2O_3 ratio at 2·2 and concluded that slight variation in the CaO/R_2O_3 ratio could also occur; Toropov, Shishakov and Merkov [*98*] found a similar limiting ratio, though they regarded the solid solution as extending towards C_5A_3 and not C_2A. Swayze [*91*] placed the limiting ratio at 2·0 and believed that the new terminal composition C_6A_2F represented a definite compound. More recent investigations, by Malquori and Cirilli [*27*] and by Newkirk and Thwaite [*99*], have, however, supported the view that the limiting ratio is around 2·2–2·3. This corresponds approximately to the formula $C_2A_{0{\cdot}69}F_{0{\cdot}31}$.

Further studies on the system were made by McMurdie [*100*] and by Tavasci [*101*]. Tavasci claimed to have found several additional ternary compounds, but this has not been confirmed by other workers. The most recent study is that of Newkirk and Thwaite [*99*]. Figure 15 shows their results for the sub-system CaO–CA–C_2F.

Newkirk and Thwaite found that ferrite solid solution coexisting with C_3A, CaO and liquid had a composition slightly more aluminous than C_4AF (point B in Fig. 15). In this respect they differ from Swayze, who

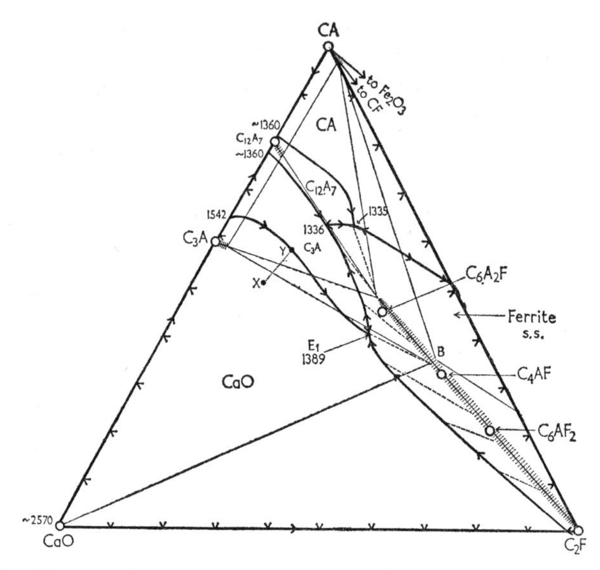

FIG. 15. The pseudoternary system CaO–CA–C_2F, after Newkirk and Thwaite [*99*]. Thin, broken lines denote tie-lines. For the significance of points B, X and Y see text.

considered that ferrite coexisting with C_3A, CaO and liquid had a composition lying between C_4AF and C_6AF_2.

Swayze found that equilibrium was often difficult to attain in this system and that this could markedly affect the composition of the ferrite phase. Thus, if a liquid of composition X (Fig. 15) was cooled, CaO would first be deposited and the liquid composition would move towards the C_3A–CaO boundary at Y. Assuming equilibrium crystallization, it would then move along this boundary towards the invariant point at E_1. This process implies the dissolution of CaO and crystallization of C_3A. In practice it is likely that only a little of the CaO will redissolve, because the C_3A quickly forms a protective coating around it. The liquid composition therefore moves across the C_3A field to the C_3A-ferrite boundary, and the composition of the ferrite phase ultimately precipitated differs considerably from that which would be obtained under equilibrium conditions. Phases thus isolated from the liquid were called "protected phases" by Swayze. A further complication arises from the possibility of zoning in the ferrite crystals.

C. Reported Discontinuities in the Ferrite Series

None of the recent investigations has shown any discontinuities near to the composition C_4AF which might suggest the existence of a definite compound; quench results show a continuous solid solution band between solidus and liquidus, while the X-ray results show a gradual variation in cell parameters with composition which contains no discontinuities near C_4AF. Newkirk and Thwaite's X-ray results, however, showed a slight discontinuity near the composition C_6AF_2 [*99*]; Smith [*102*] has recently confirmed this. The X-ray work is further discussed in Chapters 3 and 4.

Certain minor features of the published phase diagrams appear to be inconsistent with these facts. Swayze [*91*] published a profile of temperature versus composition along the ferrite–C_3A boundary which showed maxima at the intersections of the latter with the joins C_3A–C_4AF and C_3A–C_6A_2F. These maxima are also inconsistent with the probable crystallization paths along the boundary. Newkirk and Thwaite, on the other hand, drew a maximum (shown on Fig. 15) at the intersection of the boundary with the join C_3A–B (as already explained, B represents ferrite having a composition near to C_4AF and which can coexist with C_3A, CaO and liquid). This seems equally illogical and again causes difficulty in visualizing crystallization paths. The need to postulate a temperature maximum seems to have arisen from the fact that early work required the invariant point ferrite–C_3A–CaO–liquid to be a eutectic. It seems more likely from what is now known about the ferrite series that this point is really a reaction point at which CaO should disappear, and that there are no temperature maxima along the ferrite–C_3A boundary.

XIII. The System CaO–Al_2O_3–Fe_2O_3–SiO_2

A. The Bogue Calculation

The high-lime region of this system has been investigated intensively because of its importance in understanding the constitution of Portland cement. From a knowledge of the most important of the bounding systems, Bogue [*103*] put forward in 1929 a method of estimating the compound composition of a Portland cement or cement clinker from its chemical analysis. The calculation initially involves the deduction from the total CaO content of any free CaO which a separate determination has shown to be uncombined. With cements, as opposed to clinkers, a further amount of CaO is also deducted equivalent to the content of SO_3, which is assumed to be present as gypsum, added after firing. The Fe_2O_3 is then assumed to occur as C_4AF and the remaining Al_2O_3 is calculated as C_3A. The CaO which remains after allowing for the amounts in these two phases is apportioned to the SiO_2 to give C_3S and β-C_2S. A specimen calculation is given in Appendix 2 (Volume 2) and typical results are given also in Chapter 3.

B. The Sub-system CaO–C_2S–C_5A_3–C_4AF

1. Introduction

Lea and Parker [*104*] made what is now considered to be a classic study of this sub-system and confirmed the correctness of Bogue's assumption that no quaternary compounds occur within it; the existence of such compounds could vitiate his method of compound computation. The most important contribution of their work was in showing the effect of deviations from complete equilibrium on the compound composition of the cement clinker; these deviations are a cause of discrepancies between the actual phase compositions of cement clinkers and those given by the Bogue calculation.

Lea and Parker chose C_5A_3 as a component because, when the work was carried out, this formula was believed to represent the compound now considered to be $C_{12}A_7$. In the following discussion this phase will be called $C_{12}A_7$; the small difference in composition between C_5A_3 and $C_{12}A_7$ will be ignored. The ferrite phase was assumed to have the fixed composition C_4AF.

2. The Sub-systems C_5A_3–C_2S–C_4AF and CaO–C_2S–C_4AF

Lea and Parker began by studying the two sub-systems C_5A_3–C_2S–C_4AF and CaO–C_2S–C_4AF. The second of these is particularly important in connection with Portland cement clinkers of low Al_2O_3/Fe_2O_3 ratio. Figure 16 shows Lea and Parker's results modified to take account of the extension of the primary phase field of C_3S which has since been discovered [*12*]. The modification in the C_3S phase field occurs at temperatures higher than those encountered in the production of cement clinker. The C_3S phase field is a long sliver, which ends in a eutectic X for CaO, C_3S, C_4AF and liquid and a reaction point W for C_2S, C_3S, C_4AF and liquid.

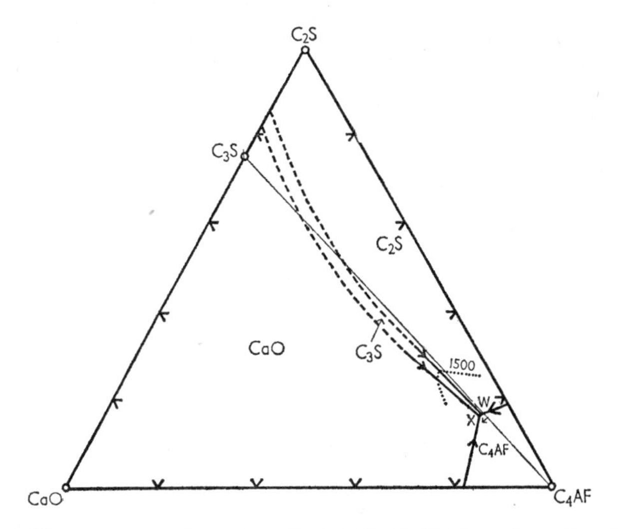

Fig. 16. The ternary sub-system CaO–C_2S–C_4AF; based on the diagram of Lea and Parker [*104*] and modified as regards the field of C_3S. The dotted line represents part of the 1500° C isotherm.

3. The Quaternary System

Figure 17 gives a perspective view of Lea and Parker's results for the quaternary sub-system CaO–C_2S–C_5A_3–C_4AF. The most prominent feature is the thin and roughly triangular primary phase volume of C_3S, which is bounded on one of its broad surfaces by that of CaO and on the other by that of C_2S. As in Fig. 16, Lea and Parker's orginal diagram has been modified to show the extension of the C_3S volume to meet the CaO–C_2S edge.

The part of the system which is of most importance to the cement chemist is the narrow face of the C_3S volume remote from the CaO–C_2S edge. The upper part of this face consists of the C_3S–C_4AF surface and the lower part consists of the C_3S–C_3A surface. These two surfaces are separated by the boundary curve common to C_3S, C_3A and C_4AF, which terminates in two invariant points at 1341° and 1338° C. The first of these is a reaction point; Lea and Parker considered the second to be a eutectic, but it, too, is possibly a reaction point. Which it is depends on its position relative to the C_2S–C_3A–C_4AF plane: if it is on the high-lime side, it is a eutectic, and if it is on the low-lime side, it is a reaction point. Lea and Parker considered that it was on the high-lime side but stated that it was so near the plane that a very slight displacement would transfer it to the other side.

From the positions of these points and of the C_2S–C_3A and C_2S–C_4AF edges of the C_3S primary phase volume, the liquid content of a cement mix at any given temperature can be calculated, as can the corrections

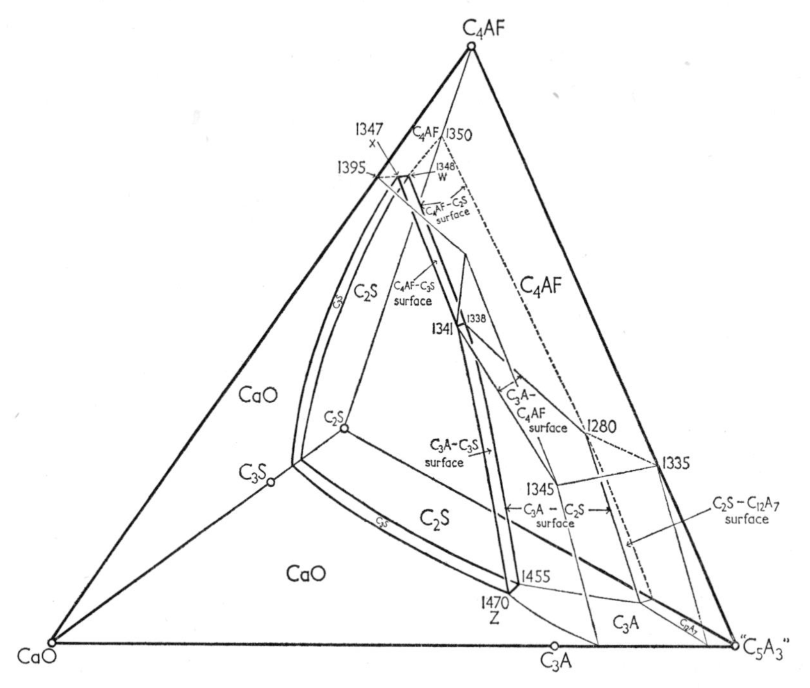

FIG. 17. The quaternary sub-system CaO–C_2S–C_5A_3–C_4AF, after Lea and Parker [*104*], modified as regards the phase volume of C_3S.

that must be applied to the Bogue calculation to allow for various types of non-equilibrium cooling [*105, 106*]. This is discussed further in Chapter 3.

4. Maximum Lime Content in a Cement Mix

A second important result of Lea and Parker's work was the calculation of the maximum CaO content that can be tolerated in the raw materials for the manufacture of Portland cement. The occurrence of free CaO in the product is undesirable for the same reason as is the occurrence of MgO (p. 72). If it could be assumed that equilibrium was reached in the kiln and maintained during cooling, any composition falling on the low-lime side of the plane C_3S–C_3A–C_4AF would be satisfactory in this respect; compositions within the tetrahedron C_3S–C_2S–C_3A–C_4AF would give a mixture of these four compounds in accordance with the results of the Bogue calculation. In practice the occurrence of non-equilibrium conditions during cooling may prevent combination of all the available CaO, and this makes it necessary to restrict the allowable CaO content of the raw mix. The essential condition is that the composition of the liquid at the clinkering temperature must lie on the C_3S side of the CaO–C_3S boundary surface; if it lies in the primary phase volume of CaO, the latter will be present in the solid at the clinkering temperature and its resorption on cooling cannot be guaranteed. The CaO–C_3S boundary surface at the temperatures to be considered is approximated to by the plane passing through C_4AF, C_3S and the invariant point Z for CaO–C_3A–C_3S–liquid (Fig. 17). On this assumption Lea and Parker showed that the maximum lime content is given by

$$CaO = 2{\cdot}80SiO_2 + 1{\cdot}18Al_2O_3 + 0{\cdot}65Fe_2O_3$$

where the formulae represent concentrations in weights percent.

Lea and Parker considered that this result was only valid for Al_2O_3/Fe_2O_3 weight ratios greater than 0·64, but it would appear that it is equally valid for lower Al_2O_3/Fe_2O_3 ratios. This follows from the choice of the restrictive plane C_4AF–C_3S–Z already mentioned. The line C_3S–Z is a close approximation to the join C_3S–"C_2A". The restrictive plane is therefore a close approximation to the plane C_3S–"C_2A"–C_4AF. The join "C_2A"–C_4AF, when extended, includes the remainder of the ferrite solid solution series. The argument is therefore equally valid for Al_2O_3/Fe_2O_3 ratios below that in C_4AF, provided that the extension of the CaO–C_3S boundary surface remains approximately coincident with that of the C_3S–"C_2A"–C_4AF plane. The work of Swayze [*107*], discussed in the next section, confirms that Lea and Parker's results are approximately valid for mixes with Al_2O_3/Fe_2O_3 ratios below 0·64.

C. Variable Ferrite Composition

Swayze [*91, 107*] extended the findings of Lea and Parker on the sub-system CaO–C_2S–C_5A_3–C_4AF in two respects: first, by considering the effects of variable ferrite composition and secondly, by studying the influence of magnesia. His diagram for the CaO–C_2S–C_5A_3–C_2F sub-system is generally similar to that of Lea and Parker. Figure 18 is based on this diagram, modified to show the extension of the C_3S volume to the CaO–C_2S edge of the tetrahedron. Invariant point 1, at 1342° C, for CaO–C_3S–C_3A–ferrite–liquid, corresponds to Lea and Parker's point at 1341° C. Invariant point 2, at 1338° C, for C_2S–C_3S–C_3A–ferrite–liquid, corresponds to their point at the same temperature. The two investigations thus agree almost exactly as regards the temperatures of corresponding points, but the liquid compositions were slightly different. Swayze considered both points to be reaction points. The composition of the ferrite phase coexisting with liquid at point 1 was $C_2A_{0\cdot44}F_{0\cdot56}$ (Al_2O_3/Fe_2O_3 weight ratio 0·5) and that of the ferrite phase coexisting with liquid at point 2 was about $C_2A_{0\cdot57}F_{0\cdot43}$ (Al_2O_3/Fe_2O_3 weight ratio 0·85). If equilibrium cooling could be assumed, a mix approximating to

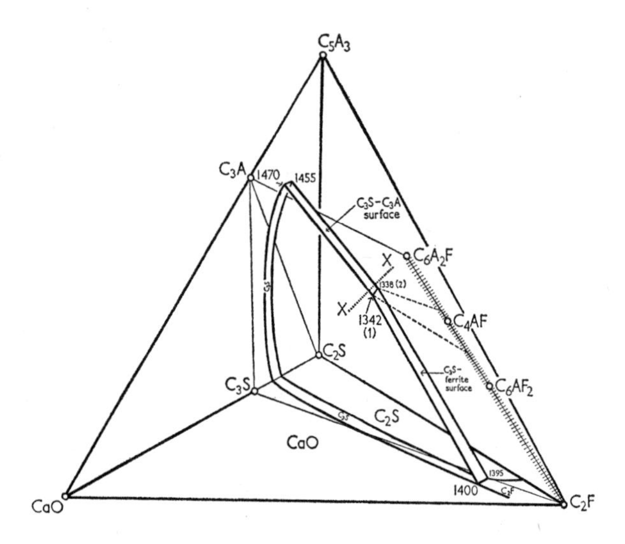

Fig. 18. The quaternary sub-system $CaO–C_2S–C_5A_3–C_2F$, after Swayze [*91*], modified as regards the phase volume of C_3S. X...X denotes the intersection of the $C_3S–C_2S–C_6A_2F$ plane with the edges of the C_3S phase volume.

Portland cement composition would undergo final crystallization at point 2.

Points 1 and 2 both lie outside the tetrahedron $C_3S–C_2S–C_3A–C_6A_2F$. It follows that a ferrite phase of composition C_6A_2F, which is formed in the ternary system $CaO–Al_2O_3–Fe_2O_3$, cannot be a stable crystallization product in the quaternary system. This feature can be seen more readily if one examines the intersection of the plane $C_3S–C_2S–C_6A_2F$ with the edges of the C_3S primary phase volume. It intersects the $C_3S–C_3A$ boundary surface *above* points 1 and 2. It would be necessary for the intersection to occur *below* points 1 and 2 for crystallization of C_6A_2F to be possible.

D. Influence of Magnesia

1. Phase Equilibria

Cement clinkers normally contain significant quantities of magnesia, and Swayze [*107*] therefore also studied the effect of adding MgO to the quaternary system. He studied the quinary system CaO–C_5A_3–C_2F–C_2S–MgO at a constant MgO content of 5%. This amount of MgO was slightly more than enough to saturate the liquid phase over the range of

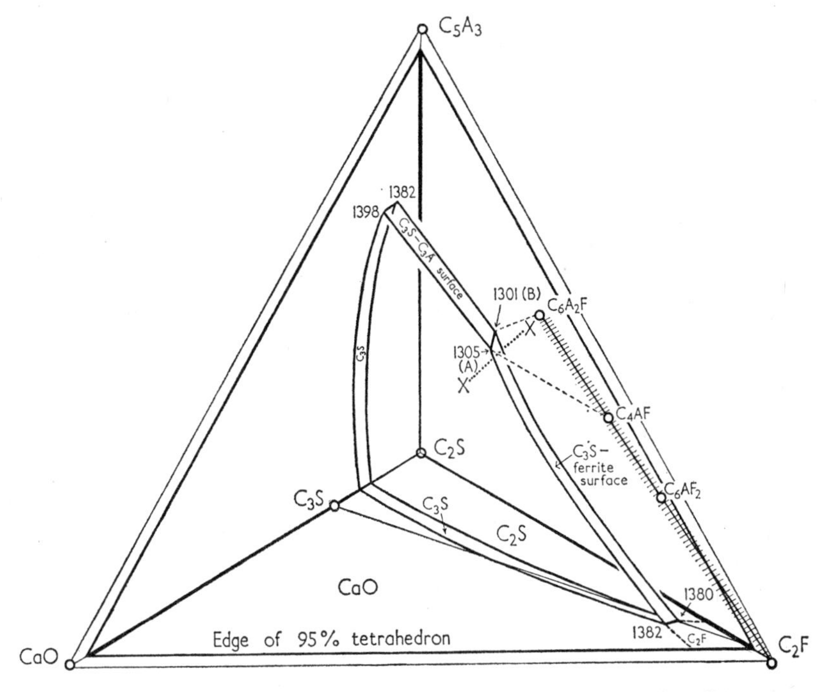

FIG. 19. The pseudosystem CaO–C_2S–C_5A_3–C_2F, modified by the presence of a constant level (5%) of MgO, after Swayze [*107*]. Diagram modified as regards the phase volume of C_3S. The line X...X has the same significance as in Fig. 18.

compositions studied, the small excess of MgO crystallizing as periclase. The results were represented on a tetrahedral model, the system thus being treated as pseudoquaternary. Figure 19 is based on Swayze's diagram, modified to show the extension of the C_3S primary volume to the CaO–C_2S edge.

The primary phase volumes are generally similar to those found for the MgO-free quaternary system (Fig. 18). An important difference is that the two invariant points 1 and 2 of Fig. 18 have shifted in position to A and B, which brings them to the high-alumina side of the plane C_3S–C_2S–C_6A_2F and thus within the tetrahedron C_3S–C_2S–C_6A_2F–C_3A. It follows that, in the presence of 5% MgO, C_6A_2F can be a stable product of crystallization at either point A or point B. As for the quaternary system, Swayze found both A and B to be reaction points. The final crystallization of a mix approximating to Portland cement clinker composition will occur at point B (C_3S–C_2S–C_3A–ferrite–liquid) if cooling is sufficiently slow for equilibrium conditions to be maintained.

The tie-lines between the C_3S–C_2S–ferrite boundary and the solid solution were not established. It was, however, shown that for liquid at point B the ferrite composition was close to C_6A_2F. For liquid at point A the ferrite composition was about $C_2A_{0 \cdot 47}F_{0 \cdot 53}$ (Al_2O_3/Fe_2O_3 weight ratio 0·57).

2. *Composition of the Ferrite Phase in Cement Clinker*

The composition of the ferrite phase, which is formed from a mix of a given Al_2O_3/Fe_2O_3 ratio, depends on whether the liquid crystallizes along the CaO–C_3S–ferrite boundary to point A or along the C_2S–C_3S–ferrite boundary to point B. This, in turn, depends on the overall composition of the mix. For the former case, when C_2S is absent and free CaO present as the ferrite phase begins to crystallize, Swayze showed that for mixes with Al_2O_3/Fe_2O_3 weight ratios of 0·5–1·43 the ferrite phase would be less aluminous than C_4AF and that C_3A would be a final product of crystallization. The Bogue calculation will thus underestimate C_3A and overestimate ferrite.

In the latter case, when some C_2S is present as the ferrite phase begins to crystallize and provided that free lime has been eliminated by thorough grinding and burning of the raw materials, C_3A will not be formed from mixes having Al_2O_3/Fe_2O_3 weight ratios below 0·84, or possibly even higher. The limiting Al_2O_3/Fe_2O_3 ratio cannot be estimated more precisely because the relevant tie-lines were not determined. Under similar conditions, but with mixes having Al_2O_3/Fe_2O_3 weight ratios of 0·84–1·43, some C_3A will be formed, along with a ferrite phase more aluminous than C_4AF. The Bogue calculation will in this case overestimate C_3A and underestimate ferrite.

Uncertainties about the tie-lines relating the compositions of the liquids along the C_2S–C_3S–ferrite boundary to those of the ferrite phase make it impossible as yet to devise quantitative theoretical corrections to the Bogue calculation. Even if such corrections were to be devised, further complications would result from failure to reach equilibrium, due to the formation of protected phases, as described in connection with the $CaO-Al_2O_3-Fe_2O_3$ system. For Al_2O_3/Fe_2O_3 weight ratios below 0·84, assuming that the overall composition is such that the liquid crystallizes along the C_2S-C_3S-ferrite boundary and that equilibrium conditions are maintained, a modified Bogue calculation can, however, be suggested; C_3A and free CaO should be absent and all of the Al_2O_3 and Fe_2O_3 assigned to a ferrite phase having an Al_2O_3/Fe_2O_3 ratio equal to that of the mix. The residual lime and the silica are then apportioned to C_3S and C_2S in the normal way.

XIV. Conclusion

In this chapter it has been possible to give only a brief résumé of phase equilibrium and other high-temperature studies relevant to the constitution of cements. For a detailed appreciation of the great volume of work that has been undertaken in this field, this chapter can in no way substitute for reading the references cited. In addition, the compilation of phase diagrams [*108*] and the set of large-scale diagrams [*41*], both published by the American Ceramic Society, will be found of value. The intention of the author has been to survey some of the progress that has been made and to indicate some of the outstanding problems awaiting solution, for which existing knowledge is still either inadequate or controversial.

Acknowledgment

Acknowledgment is made to the Director of Building Research for permission to make use in this chapter of some unpublished laboratory data obtained at the Building Research Station, Watford, Hertfordshire, England.

REFERENCES

1. Rankin, G. A. and Wright, F. E. (1915). *Amer. J. Sci.* **39**, 1.
2. Grofcsik, J. and Tamás, F. (1961). "Mullite, Its Structure, Formation and Significance". Akadémiai Kiado, Budapest.
3. Sosman, R. B. (1955). *Trans. Brit. Ceram. Soc.* **54**, 655.
4. Hill, V. J. and Roy, R. (1957). *Acta cryst.* **10**, 835.
5. Holmquist, S. B. (1958). *Z. Kristallogr.* **111**, 71.
6. Flörke, O. W. (1955). *Ber. dtsch. keram. Ges.* **32**, 369.
7. Wahl, F. M., Grim, R. E. and Graf, R. B. (1961). *Amer. Min.* **46**, 196.
8. Welch, J. H. (1960). *Nature, Lond.* **186**, 545; (1961). Trans 7th Int. Ceram. Cong., London, 1960, p. 197.
9. Greig, J. W. (1927). *Amer. J. Sci.* **13**, 1, 133.
10. Muan, A. and Osborn, E. F. Preprint of paper presented before Amer. Iron Stl. Inst., New York, May 1951.

11. Nurse, R. W. (1954). *Proceedings of the Third International Symposium on the Chemistry of Cement, London 1952*, p. 56. Cement and Concrete Association, London.
12. Welch, J. H. and Gutt, W. (1959). *J. Amer. ceram. Soc.* **42**, 11.
13. Winchell, A. N. (1931). "The Microscopic Characters of Artificial Inorganic Substances or Artificial Minerals", 2nd Ed., p. 186. Wiley, New York.
14. Trömel, G. (1949). *Naturwissenschaften* **36**, 88.
15. Yannaquis, N. and Guinier, A. (1959). *Bull. Soc. franç. Minér.* **82**, 126.
16. Smith, D. K., Majumdar, A. J. and Ordway, F. (1961). *J. Amer. ceram. Soc.* **44**, 405.
17. Vasenin, F. I. (1948). *Zh. prikl. Khim., Leningr.* **21**, 429.
18. Toropov, N. A., Volkonskii, B. V. and Sadkov, V. I. (1957). *Dokl. Akad. Nauk S.S.S.R.* **112**, 467.
19. Newman, E. S. and Wells, L. S. (1946). *J. Res. nat. Bur. Stand.* **36, 137.**
20. Welch, J. H. and Gutt, W. (1962). *Chemistry of Cement, Proceedings of the Fourth International Symposium, Washington 1960*, p. 59. National Bureau of Standards Monograph 43. U.S. Department of Commerce.
21. Williamson, J. and Glasser, F. P. (1962). *J. appl. Chem.* **12**, 535.
22. Eitel, W. (1941). *Zement* **30**, 17, 29.
23. Auriol, A., Hanser, G. and Wurm, J. G. Private communication.
24. Carstens, C. W. (1926). *Z. Kristallogr.* **63**, 473.
25. Rankin, G. A. and Merwin, H. E. (1916). *J. Amer. chem. Soc.* **38**, 568.
26. Tavasci, B. (1937). *TonindustrZtg.* **61**, 717, 729.
27. Malquori, G. and Cirilli, V. (1954). *Proceedings of the Third International Symposium on the Chemistry of Cement, London 1952*, p. 120. Cement and Concrete Association, London.
28. Lea, F. M. and Desch, C. H. (1956). "The Chemistry of Cement and Concrete", 2nd Ed., revised by F. M. Lea. Arnold, London.
29. Buttler, F. G. and Taylor, H. F. W. (1959). *J. appl. Chem.* **9**, 616.
30. Filonenko, N. E. (1949). *Dokl. Akad. Nauk S.S.S.R.* **64**, 529.
31. Lagerquist, V. K., Wallmark, S. and Westgren, A. (1937). *Z. anorg. Chem.* **234**, 1.
32. Aruja, E. (1957). *Acta cryst.* **10**, 337.
33. Büssem, W. and Eitel, W. (1936). *Z. Kristallogr.* **95**, 175.
34. Nurse, R. W. and Welch, J. H., quoted by Nurse, R. W. (1962). *Chemistry of Cement, Proceedings of the Fourth International Symposium, Washington 1960*, pp. 9, 35. National Bureau of Standards Monograph 43. U.S. Department of Commerce, and new data by Nurse, R. W., Welch, J. H. and Majumdar, A. J.
35. Roy, D. M. and Roy, R. (1962). *Chemistry of Cement, Proceedings of the Fourth International Symposium, Washington 1960*, p. 307. National Bureau of Standards Monograph 43. U.S. Department of Commerce.
36. Glasser, F. P. and Jeevaratnam, J. Private communication.
37. Bonnickson, K. R. (1955). *J. phys. Chem.* **59**, 220.
38. Shepherd, E. S., Rankin, G. A. and Wright, F. E. (1909). *Amer. J. Sci.* **28**, 293.
39. Davis, G. L. and Tuttle, O. F. (1952). *Amer. J. Sci.* Bowen Volume, 107.
40. Welch, J. H. (1958). *Proc. 3rd International Symposium on Reactivity of Solids, Madrid, 1956*, Vol. 2, p. 677.
41. Osborn, E. F. and Muan A. (1960). "Phase Equilibrium Diagrams of Oxide Systems". American Ceramic Society, Columbus, Ohio, U.S.A.

41a. Alper, A. M., McNally, R. N., Ribbe, P. H. and Doman R. C. (1962). *J. Amer. ceram. Soc.* **45**, 263.
42. Rankin, G. A. and Merwin, H. E. (1918). *Amer. J. Sci.* **45**, 301.
43. Welch, J. H. (1961). *Nature, Lond.* **191**, 559.
44. Bowen, N. L. and Andersen, O. (1915). *Amer. J. Sci.* **37**, 487.
45. Nikitin, V. D. (1948). *Izv. Sekt. Fiz.-khim. Anal.* **16**, 3, 29.
46. Thilo, E. and Rogge, G. (1939). *Ber. dtsch. chem. Ges.* **72**, 341.
47. Foster, W. R. (1951). *J. Amer. ceram. Soc.* **34**, 255.
48. Atlas, L. (1952). *J. Geol.* **60**, 125.
49. Brown, W. L., Morimoto, N. and Smith, J. V. (1961). *J. Geol.* **69**, 609.
50. Ricker, R. W. and Osborn, E. F. (1954). *J. Amer. ceram. Soc.* **37**, 133.
51. Bowen, N. L. (1914). *Amer. J. Sci.* **38**, 207.
52. Boyd, F. R. and Schairer, J. F. (1957). *Ann. Rep. Geophys. Lab., Washington, for 1956–7*, **56**, 223.
53. Schairer, J. F. and Bowen, N. L. (1942). *Amer. J. Sci.* **240**, 725.
54. Bredig, M. A. (1950). *J. Amer. ceram. Soc.* **33**, 188.
55. Phemister, J., Nurse, R. W. and Bannister, F. A. (1942). *Min. Mag.* **26**, 225.
56. Osborn, E. F. (1943). *J. Amer. ceram. Soc.* **26**, 321.
57. Roy, D. M. (1956). *Min. Mag.* **31**, 187.
58. Gutt, W. (1961). *Nature, Lond.* **190**, 339.
58a. Sharp, J. D., Johnson, W. and Andrews, K. W. (1960). *J. Iron. St. Inst.* **195**, 83.
59. Bowen, N. L., Schairer, J. F. and Posnjak, E. (1933). *Amer. J. Sci.* **26**, 193.
60. Carstens, C. W. and Kristoffersen, K. (1931). *Neues. Jb. Min. Geol.* **A62**, 163.
61. Osborn, E. F. and Schairer, J. F. (1941). *Amer. J. Sci.* **239**, 715.
62. Nurse, R. W. and Midgley, H. G. (1953). *J. Iron St. Inst.* **174**, 121.
63. DeWys, E. C. and Foster, W. R. (1956). *J. Amer. ceram. Soc.* **39**, 372.
64. Harker, R. I. and Tuttle, O. F. (1956). *Amer. J. Sci.* **254**, 468.
65. Foster, W. R. (1950). *J. Geol.* **58**, 135.
66. Foster, W. R. (1950). *J. Amer. ceram. Soc.* **33**, 73.
67. Keith, M. L. and Schairer, J. F. (1952). *J. Geol.* **60**, 181.
68. Aramaki, S. and Roy, R. (1959). *J. Amer. ceram. Soc.* **42**, 644.
69. Schreyer, W. and Schairer, J. F. (1961). *J. Petrol.* **2**, 324.
70. Schreyer, W. and Schairer, J. F. (1962). *Amer. Min.* **47**, 90.
71. Parker, T. W. (1954). *Proceedings of the Third International Symposium on the Chemistry of Cement, London 1952*, p. 485. Cement and Concrete Association, London.
72. Andersen, O. (1915). *Amer. J. Sci.* **39**, 407.
73. Ferguson, J. B. and Buddington, A. F. (1920). *Amer. J. Sci.* **50**, 131.
74. Osborn, E. F. (1952). *Amer. J. Sci.* **240**, 751.
75. Osborn, E. F. and Tait, D. B. (1952). *Amer. J. Sci.* Bowen Volume, 413.
76. Segnit, E. R. (1956). *Min. Mag.* **31**, 255.
77. DeWys, E. C. and Foster, W. R. (1958). *Min. Mag.* **31**, 736.
78. Hytönen, K. and Schairer, J. F. (1960). *Ann. Rep. Geophys. Lab., Washington, for 1959–60*, **59**, 71.
79. Nurse, R. W. and Stutterheim, N. (1950). *J. Iron St. Inst.* **165**, 137.
80. Welch, J. H. (1956). *J. Iron St. Inst.* **183**, 275.
81. DeVries, R. C. and Osborn, E. F. (1957). *J. Amer. ceram. Soc.* **40**, 6.
82. Hansen, W. C. (1928). *J. Amer. chem. Soc.* **50**, 3081.
83. McMurdie, H. F. and Insley, H. (1936). *J. Res. nat. Bur. Stand.* **16**, 467.
84. Prince, A. T. (1951). *J. Amer. ceram. Soc.* **34**, 44.

85. Segnit, E. R. and Weymouth, J. H. (1957). *Trans. Brit. Ceram. Soc.* **56**, 253.
86. Osborn, E. F., DeVries, R. C., Gee, K. H. and Kraner, H. M. (1954). *J. Metals*, **6**, 3.
87. Cavalier, G. and Sandrea-Deudon, M. (1960). *Rev. Met.* **57**, 1143.
88. Prince, A. T. (1954). *J. Amer. ceram. Soc.* **37**, 402.
89. Sosman, R. B. and Merwin, H. E. (1916). *J. Wash. Acad. Sci.* **6**, 532.
90. Tavasci, B. (1936). *Ann. chim. appl.* **26**, 291.
91. Swayze, M. A. (1946). *Amer. J. Sci.* **244**, 1.
92. Phillips, B. and Muan, A. (1958). *J. Amer. ceram. Soc.* **41**, 445.
93. Burdick, M. D. (1940). *J. Res. nat. Bur. Stand.* **25**, 475.
94. Phillips, B. and Muan, A. (1959). *J. Amer. ceram. Soc.* **42**, 413.
95. Muan, A. (1958). *Amer. J. Sci.* **256**, 420.
96. Hansen, W. C., Brownmiller, L. T. and Bogue R. H. (1928). *J. Amer. chem. Soc.* **50**, 396.
97. Yamauchi, T. (1937–38). *J. Jap. ceram. Assoc.* **45**, 279, 361, 433, 614, 880; **46**, 66.
98. Toropov, N. A., Shishakov, N. A. and Merkov, L. D. (1937). *Zh. Tsement* (1) 28; see also Toropov, N. A. (1962). *Chemistry of Cement, Proceedings of the Fourth International Symposium, Washington 1960*, p. 113. National Bureau of Standards Monograph 43. U.S. Department of Commerce.
99. Newkirk, T. F. and Thwaite, R. D. (1958). *J. Res. nat. Bur. Stand.* **61**, 233.
100. McMurdie, H. F. (1937). *J. Res. nat. Bur. Stand.* **18**, 475.
101. Tavasci, B. (1937). *Ann. chim. appl.* **27**, 505.
102. Smith, D. K. (1962). *Acta cryst.* **15**, 1146.
103. Bogue, R. H. (1929). *Industr. Engng Chem.* (*Anal. Ed.*) **1**, 192.
104. Lea, F. M. and Parker, T. W. (1934). *Phil. Trans.* **A234**, 1.
105. Lea, F. M. and Parker, T. W. (1935). Building Research Technical Paper No. 16, HMSO, London.
106. Dahl, L. A. (1938–39). *Rock Prod.* **41**, No. 9, 48; No. 10, 46; No. 11, 42; No. 12, 44; **42**, No. **1**, 68; No. 2, 46; No. 4, 50.
107. Swayze, M. A. (1946). *Amer. J. Sci.* **244**, 65.
108. "Phase Diagrams for Ceramists". American Ceramic Society, Columbus, Ohio, U.S.A. Part 1, by Levin, E. M., McMurdie, H. F. and Hall, E. P. 1956; Part 2, by Levin, E. M. and McMurdie, H. F., 1959.
109. Welch, J. H. Unpublished data.

Mr. Welch died before finishing the manuscript. Dr. R. W. Nurse and Dr. A. J. Majumdar (Building Research Station) and Dr. F. P. Glasser and Mrs. S. P. Kelly (University of Aberdeen) helped in various ways to prepare it for publication.

CHAPTER 3

The Formation and Phase Composition of Portland Cement Clinker

H. G. Midgley

Building Research Station, Garston, Watford, Hertfordshire, England

CONTENTS

I. Introduction

Portland cement is produced by the high-temperature reaction of a lime-bearing material with one containing silica, alumina, and some Fe_2O_3. Reaction is effected in kilns of various types, normally under oxidizing conditions. The product, which is known as clinker, is afterwards ground with gypsum to give cement. The main phases in the clinker are tricalcium silicate (C_3S), β-dicalcium silicate (β-C_2S), tricalcium aluminate (C_3A) and a ferrite phase belonging to the C_2F–C_2A solid solution series. Also present in many clinkers are smaller amounts ($<3\%$) of free lime (CaO), periclase (MgO) and alkali sulphates. The presence of glass, in amounts ranging up to 20% or even higher, has also frequently been postulated, but recent work casts doubt on this conclusion.

It is now well established that none of the major phases has an exact composition; all are modified by solid solution, both by the major oxides and by minor components.

II. The Phases in Clinker

A. Tricalcium Silicate

The main phase in most modern Portland cement clinkers is a form of tricalcium silicate, originally described as "alite" by Törnebohm [*1*], who studied it by thin-section microscopy. General agreement that alite was essentially C_3S was reached by about 1930. In 1952 Jeffery [*2*] showed that pure C_3S is triclinic but that small amounts of solid solution cause it to become monoclinic or trigonal; he found the monoclinic form in a Portland cement clinker. The three polymorphs differ only slightly in structure. He suggested that, with pure C_3S, the triclinic form is stable relative to the others at room temperature, the monoclinic form at moderate temperatures and the trigonal form at high temperatures. This was later confirmed by Yamaguchi and Miyabe [*3*], who used a high-temperature X-ray diffractometer. Midgley [*4*] investigated 20 Portland cement clinkers by X-ray diffraction using film, and showed that in all of these the C_3S was monoclinic. More recent work by Midgley and Fletcher [*5*], using X-ray diffractometry, showed that the C_3S in Portland cement is usually monoclinic but that, even in ordinary clinkers, the triclinic and trigonal forms may occur.

Jeffery [*2*] proposed that the alite in Portland cement clinker had the composition $C_{54}S_{16}AM$, but von Euw [*6*] showed that material of this composition contained C_3A as an impurity. He suggested that alite formed in the presence of excess CaO had the composition of C_3S with about 2% C_3A and 1% MgO. Naito, Ono and Iiyama [*7*] reported the maximum substitutions of Mg and Al in the series

$$(Ca_{1-x-y}Mg_xAl_y)_3.(Si_{1-3y}Al_{3y})O_5$$

to be $x = 0{\cdot}025$ and $y = 0{\cdot}0075$. Brunauer *et al.* [*8*] investigated the incorporation of Al and Mg into C_3S in an effort to produce an "alite" suitable for use as a standard in the X-ray diffractometry of clinkers, and concluded that the correct formulation was $52C_3S.C_6AM$; that is, the type of substitution proposed by Jeffery but with one-third the amounts of Al_2O_3 and MgO. Locher [*9*] studied the solid solution of MgO and C_3A in C_3S at various temperatures and found that the solubility of MgO varies from 2·5% at 1500° C to 1·5% at 1420° C; that of C_3A was 2% at 1500° C. He relied mainly on the determination of free lime by chemical means to estimate the limits of solubility and gave no details of the crystal structures.

Midgley and Fletcher [*5*] have investigated the effect of substitution of Mg and Al in C_3S prepared at 1500° C and subsequently studied at room temperature. They showed by X-ray diffractometer studies that Mg can replace Ca directly in C_3S and that as the amount of Mg increases there is a change in the lattice parameters of the triclinic form, until at about 1·7% MgO the lattice becomes monoclinic. Further incorporation of Mg causes changes in the monoclinic lattice parameters until about 2% MgO is reached, beyond which no further substitution takes place. Midgley and Fletcher also showed that the substitution of Al in the attice is not caused by solid solution of C_3A in C_3S, but by solid solution within the series C_3S–$C_{4{\cdot}5}A$; that is, three-quarters of the Al replaces Si directly, the number of oxygens remains constant and the remaining one-quarter of the Al goes into interstitial positions. Substitution of Al causes a change in lattice parameters but not in symmetry. Lafuma [*10*] and Midgley and Fletcher [*5*] have shown that Fe_2O_3 may also enter into solid solution with C_3S to a limited extent.

Midgley and Fletcher [*5*] studied the effects of simultaneous substitution of Mg for Ca and of $C_{4{\cdot}5}A$ for C_3S. There were changes in cell parameters and symmetry similar to those observed when only the first of these substitutions was made. The trigonal form has not so far been observed in this work. These results, together with those of studies on cement clinkers, suggest that the alite in Portland cement can vary considerably in composition and symmetry. The changes in symmetry can affect the strength produced by the phase; experiments have shown that the differences are of the order of 10%.

Substitution in C_3S is further discussed in Chapter 4.

Alite decomposes into C_2S and CaO below about 1275° C, but this reaction is usually sluggish and C_3S can therefore occur metastably at room temperature. Woermann [*11*] has investigated the effect of substitution on the rate of decomposition and has shown that divalent ions of radius greater than 0·7 Å increase the rate of decomposition while smaller ones do not.

Jander [*12*] suggested that the strength produced by the alite in a cement might depend not only on the presence of ions in solid solution but also on the occurrence of structural defects and of cracks and irregularities of colloidal dimensions. No experimental evidence to support or refute this has so far been presented. Grzymek [*13*] believes that small and elongated crystals of alite can hydrate more rapidly, even if ground to the same specific surface as large and more equant crystals. So far this has not been confirmed by other workers in the field.

B. Dicalcium Silicate

Most of the work on dicalcium silicate in recent years has been on laboratory preparations; little work has been done on C_2S from clinker. Dicalcium silicate can occur in four polymorphic modifications: α, α', β and γ. Nurse [*14*] has shown that the γ-form is almost inert, that β hydrates at a rate depending on the kind of stabilizer, that α' gives very poor strength and that α is non-hydraulic. Budnikov and Azelitskaya [*15*] declare that γ-C_2S has hydraulic properties, but the general view is that it has not.

The usual form of dicalcium silicate in Portland cement clinker is the β-modification. Metzger [*16*] has observed small quantities of α'-C_2S by microscopic investigation, and Midgley and Fletcher [*17*] have found this phase by X-ray methods in commercial clinkers. A phase of approximate composition $KC_{23}S_{12}$ has also been formed in commercial clinkers; as shown later, this is probably a form of α'-C_2S.

Dicalcium silicate can take into solid solution many substances, and the presence of these must modify the structure. Nurse [*14*], Kukolev and Mel'nik [*18, 55*] and Welch and Gutt [*20*] have shown that the strength obtained from β-C_2S depends on the nature of the stabilizer. The substances that might most obviously occur in solid solution with the β-C_2S of Portland cement clinker are magnesia and alumina [*8, 21, 22*] P_2O_5 [*23*], Na_2O, CaO and K_2O [*19, 24–26*]. Yannaquis and Guinier [*26*] suggested that β-C_2S may also be stabilized by crystal size alone. Most work on the pure compound has been on material stabilized with B_2O_3, but it is not likely that this form occurs in Portland cement. Midgley,

Fletcher and Cooper [*27*] showed that the X-ray pattern of the β-C_2S present in Portland cement clinkers differs significantly from that given by C_2S stabilized by B_2O_3.

C. The Ferrite Phase

The ferrite phase in Portland cement clinker, also called brownmillerite, is a solid solution which is usually taken to belong to the series C_2F–C_2A. The limiting composition at the iron-rich end is C_2F. There is some disagreement as to the limit at the alumina-rich end but, from the work of Toropov, Shishakov and Merkov [*28*], Yamauchi [*29*], Swayze [*30*] and Malquori and Cirilli [*31*], summarized by Nurse [*32*], it must occur at a composition slightly more aluminous than C_6A_2F. This problem is also discussed in Chapter 2.

The clinker mineral has been studied by X-ray diffraction methods by Midgley [*33–35*] and by Kato [*36*], who showed that it can vary in composition and may show zoning—a variation in composition within each crystal. This work has been extended by Brunauer *et al.* [*8*], Copeland *et al.* [*32*] and Midgley, Rosaman and Fletcher [*37*], who have made quantitative estimations of the minerals in Portland cement.

Midgley [*35*] has shown that the composition of the ferrite phase from clinker may extend to very near the alumina-rich limit. Cirilli and Brisi [*38*] and Santarelli, Padilla and Bucchi [*39*] suggest that the limit may be C_4AF, although the first-named authors showed that in the presence of 3% MgO the limit may extend approximately to C_6A_2F. Kato [*36*] has shown that about 1·5% MgO may substitute for CaO in the ferrite phase, and Royak [*40*] showed that Na_2O may enter into solid solution.

D. Tricalcium Aluminate

Tricalcium aluminate has no polymorphic modifications, and the only effects observed by Volkonskii [*41*] on heating C_3A to 1500° C in the high-temperature X-ray diffraction camera were caused by thermal expansion. Tröjer [*42*] observed zoning in C_3A in a microscope examination.

Two compounds that are closely allied to C_3A are NC_8A_3 [*43*] and KC_8A_3 [*44–46*]; both give X-ray diffraction patterns very similar to but slightly modified from that of pure C_3A. Suzukawa [*44–46*] showed that SiO_2 and MgO also enter into solid solution, resulting in a change in lattice parameter. Müller-Hesse and Schwiete [*47*] give the solubility of MgO in C_3A as 2·5%.

E. Minor Phases

Calcium oxide (CaO) and magnesium oxide (periclase; MgO) are both undesirable phases in a cement clinker, because they are liable to hydrate slowly after the cement has hardened, causing expansion. The composition of the raw mix must therefore be such that these phases are absent or nearly absent in the product; this question is discussed in Chapter 2. CaO is nevertheless probably present in small amounts in all Portland cement clinkers, generally because of incomplete reaction. MgO, where present, is usually derived from the $MgCO_3$ of the original limestone. As shown later, the harmful effect of MgO can be somewhat reduced by rapid cooling of the clinker.

The clays and shales used in the manufacture of Portland cement usually contain small amounts of sulphates and sulphides, and the fuel used frequently contains sulphur compounds, so that in the kiln at the clinkering temperatures SO_3 is produced. This reacts with any alkalis present, either from the raw materials or the fuel, to produce alkali sulphates. Taylor [*48*] detected K_2SO_4 in commercial clinkers; other alkali sulphates which might reasonably be present include Na_2SO_4, $3K_2SO_4.Na_2SO_4$, and K_2SO_4–Na_2SO_4 solid solutions, which are stable at high temperatures. The subject has been reviewed by Newkirk [*49*], who concluded that the alkali sulphate produced probably depends on the Na/K ratio of the raw materials and the amount of SO_3 available. He also considered that there was a tendency for the alkalis to combine with the entire content of SO_3 in the approximate molar ratio $K_2O/Na_2O_3 = 3$, subject to the availability of the three components involved. In contrast, Suzukawa [*44–46*] concluded that the SO_3 reacts with K_2O in preference to Na_2O. In most clinkers the molar ratio $(Na_2O + K_2O)/SO_3$ exceeds unity. The excess of Na_2O and K_2O is likely to enter the silicate or aluminate phases; this is discussed on pp. 97–99.

III. Effect of Minor Components

The minor components in the raw materials of Portland cement clinker affect the product mainly by ionic substitution in the major phases. These modifications may or may not affect the hydraulic properties. Recently the study of the effects of minor components has become prominent because of the increasing use of impure raw materials.

A. Phosphates

The chemical analyses of most Portland cements show small percentages of phosphates, expressed as P_2O_5. In recent years the role of phosphates has assumed greater interest because of the exploitation of difficult limestone deposits, or the desire to use various trade wastes as raw material in cement manufacture. The effect of phosphate in Portland cement has been reviewed by Steinour [*50*], who concluded that the rules put forward by Nurse [*51*] are a sufficiently accurate guide to practice. Nurse showed that most of the P_2O_5 is present in solid solution with the

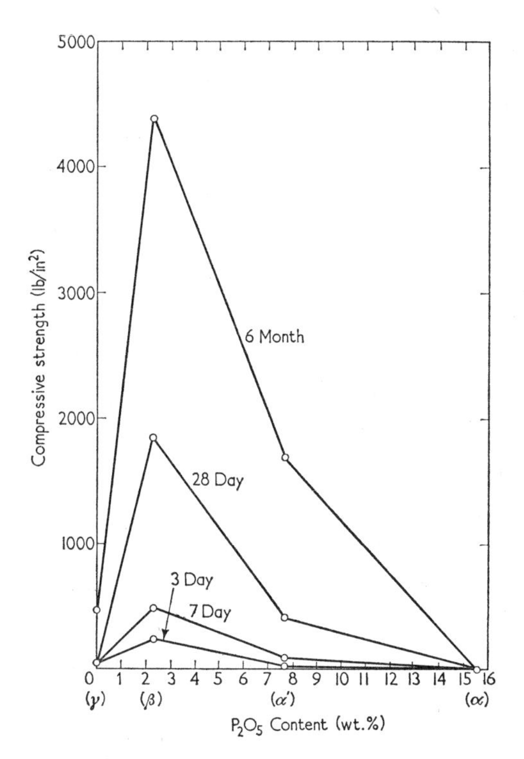

FIG. 1. Compressive strengths on $\frac{1}{2}$-in (12-mm) 1:3 mortar cubes of preparations on the join C_2S–C_3P, showing effect of C_2S polymorphism.

C_2S and that as a consequence the amount of C_3S is markedly reduced. Nurse, Welch and Gutt [*23*] published a diagram showing a series of solid solutions between α-C_2S and a previously unknown high temperature form of C_3P isomorphous with α-C_2S.

Toropov and Borisenko [*52*] also investigated the effect of P_2O_5 on C_3S at high temperatures and found the latter to be decomposed to C_2S and CaO. These authors also found that there was no reaction with the ferrite phase. Simanovskaya and Shpunt [*53*] investigated reactions with P_2O_5 and found the same decomposition of C_3S; they also followed the reaction in a kiln.

Erschov [*54*] found that a rapid-hardening Portland cement could be produced by adding 0·2–0·3% P_2O_5 to the raw material. Kukolev and Mel'nik [*55*] found that addition of 0·3–1·5% P_2O_5 to the raw meal used for making clinker by the wet process increased the hydraulic activity of the resultant cement. They attributed the increased activity to the production of lattice defects in the C_2S due to the substituting ions; as already stated, Jander [*12*] considered that this would increase the activity. Welch and Gutt [*20*] investigated the effect of P_2O_5 on C_3S

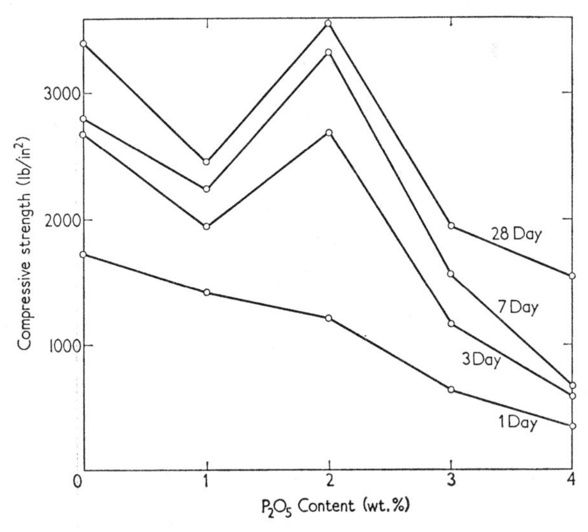

FIG. 2. Compressive strengths on ½-in (12-mm) 1:3 mortar cubes of preparations on the join from C_3S to 24·1 weight % C_3P, 75·9 weight % C_2S, in the system $CaO–SiO_2–P_2O_5$. Mean results of determinations on duplicate preparations containing 1% and 2% P_2O_5 are plotted.

and C_2S; they found that small amounts of P_2O_5 enhanced the strength obtained from C_2S, while large amounts decreased it. As an explanation they showed that with no P_2O_5 the C_2S was in the γ-form; with 2% it was in the β-form; and with 7 and 15% it was in the α'- or α-form (Fig. 1). With C_3S they report that addition of small amounts of phosphate causes a fall in the compressive strength, but that as addition continues the strength rises again, the composition 71·7 CaO, 26·3 SiO_2, 2·0 P_2O_5 (wt %) giving by a small margin the highest strength of the series at 3, 7 and 28 days. Further addition of up to 4% P_2O_5 is detrimental to strength (Fig. 2).

In interpreting the results with C_3S, allowance must be made for the increasing proportions of α'-C_2S occurring in mixes of increasing P_2O_5 content. In the 2% mix, for example, the α'-C_2S content can be estimated as about 30% by weight of the total. So far it has not been possible to isolate the C_3S, which may be modified by the incorporation of P_2O_5, and it has therefore been necessary to use mixes containing the additional phase and to make appropriate corrections. Since the α'-C_2S is comparatively inert hydraulically, the fall in strength of the mixes progressing from 2 to 4% P_2O_5 can be accounted for by the increasing dilution of the C_3S with almost non-hydraulic material. The 2% P_2O_5 mix, although diluted with some 30% of α'-C_2S, gives strengths comparable with those obtained with pure C_3S, so that the form of C_3S present in this mix can be considered to possess enhanced hydraulic value. The authors therefore deduce that limited amounts of P_2O_5 can enhance the hydraulicity of C_3S.

B. Fluorides

Fluorides are occasionally found as impurities in limestone. They may also be added deliberately as a flux or whitening agent or to offset the deleterious effect of P_2O_5.

Yamaguchi *et al.* [*56*] showed that both C_3S and C_3A can exist in the presence of CaF_2 and that CaF_2 can substitute in C_3S, causing a contraction of the lattice parameters. Toropov and his co-workers [*57*, *58*] showed that C_3A is decomposed at high temperatures by 5% addition of CaF_2 and that members of the ferrite solid solution series are also acted upon by CaF_2, being decomposed to $C_{12}A_7$ and a ferrite of composition near to C_6AF_2. Moore [*59*] showed that the addition of 1–3% fluorspar to raw meal ensured assimilation of the free CaO at a much lower temperature; the fluoride is retained by the clinker.

Welch and Gutt [*20*] found that in the presence of phosphatic calcium silicate minerals some of the fluorine is lost on heating. They also found that fluoride ion greatly accelerates the formation of C_3S; increasing fluoride content alters the lattice parameters of the triclinic C_3S, causing the X-ray pattern to become more like that of monoclinic C_3S. They also found that excess CaF_2 decomposed the C_3S to α- and α'-C_2S.

C. Alkali Oxides

The effects of alkali have been widely studied, partly because of reported reactions between the alkali from Portland cement clinker and certain types of aggregate. Newkirk [*49*] summarized work up to 1951.

As already stated (p. 94), the alkali oxides combine preferentially with SO_3 from the fuel or raw materials, forming sulphates. Taylor [*48, 60*] reported the preparation of a compound $KC_{23}S_{12}$, optically similar to β-C_2S but distinguishable from it by X-rays; he considered that any K_2O not present as sulphate was likely to occur in this form. Nurse [*14, 32, 61*] and Suzukawa [*44–46*] concluded, however, that this product was not a distinct compound but a potassium-stabilized form of α'-C_2S.

Brownmiller and Bogue [*62*] described a compound NC_8A_3, the X-ray powder pattern of which was closely similar to that of C_3A. Phase equilibrium studies [*62–65*] indicated that any excess of Na_2O was likely to occur in this form. Newkirk [*49*] concluded that alkalis not present as sulphates occurred mainly as $KC_{23}S_{12}$ and NC_8A_3; he discussed the effects which the formation of these phases had on the potential phase composition of the clinker. Because of the low proportions of alkali in $KC_{23}S_{12}$ and NC_8A_3, small amounts of alkalis in the clinker could markedly affect the potential phase composition; thus, in the absence of SO_3, addition of 1% of K_2O to a CaO–Al_2O_3–Fe_2O_3–SiO_2 mixture could cause the formation of over 20% of $KC_{23}S_{12}$ and decrease the potential C_2S content by a comparable amount. Addition of 1% Na_2O could similarly cause the formation of over 10% of NC_8A_3, with a comparable decrease in potential C_3A. The presence of alkalis could also cause small increases in the potential C_3S content, and in certain cases might lead to the formation of free CaO where this would not otherwise appear. Newkirk considered that the presence of alkalis affected the burning process in three ways: by the formation of new compounds, as mentioned above, by the lowering of the temperature of liquid formation, and by the shifting of the primary phase boundaries. This last effect might cause the CaO primary phase region to be enlarged, and thereby make the complete combination of CaO more difficult to achieve.

The interpretation of the large changes in potential compound composition which are brought about by addition of alkali clearly depends on whether the phases $KC_{23}S_{12}$ and NC_8A_3 differ appreciably in chemical behaviour from β-C_2S and C_3A, respectively. Newkirk stated that the alkali phases hydrate rapidly, and considered that one or more of them might contribute to high early strength. Nurse [*14, 61*], in contrast, found that $KC_{23}S_{12}$ inverted so rapidly to β-C_2S that it was doubtful whether it could exist in clinker; he also considered the phase to be a K_2O-stabilized form of α'-C_2S, and showed that α'-C_2S prepared in other ways had poor hydraulic properties.

The alkali phases in cement clinker have recently been reinvestigated by Suzukawa [*44–46*]. The latter confirmed the existence of the compound NC_8A_3 and also prepared an analogous compound, KC_8A_3. As already stated, he concluded in agreement with Nurse that the phase $KC_{23}S_{12}$ was not a distinct phase but a K_2O-stabilized form of α'-C_2S.

Suzukawa confirmed previous reports that the alkali oxides combine preferentially with SO_3. As already mentioned, he found that the SO_3 combined with K_2O in preference to Na_2O. For rapidly cooled clinkers any excess of K_2O occurred as KC_8A_3 or in solid solution in α- or α'-C_2S; any excess of Na_2O tended to occur as NC_8A_3 which, in his view, contained some SiO_2 due to partial replacement of CaAl by NaSi. Slow cooling led to the formation of C_3A and Na_2SO_4 instead of NC_8A_3. Suzukawa agreed with previous investigators that the presence of Na_2O could cause an increase in the C_3S content of the clinker.

D. Heavy Metal Oxides

Kukolev and Mel'nik [*55*] investigated the effects of Cr_2O_3, P_2O_5, V_2O_5 and BaO on the C_2S in wet process clinker. They found that 0·3–1·5% additions of P_2O_5, V_2O_5 and BaO all increased the hydraulic activity; they attributed this to the production of structural defects in the C_2S.

Mn_2O_3 probably substitutes for Fe_2O_3 in Portland cement. A compound $4CaO.Al_2O_3.Mn_2O_3$ is known to exist; it is analogous to C_4AF, with which it forms a continuous series of solid solutions [*66*]. Newman and Wells [*67*] showed that up to 8% of Mn_2O_3 could also be incorporated in solid solution in C_2S. The α–α' inversion temperature is considerably lowered and the β–γ inversion totally inhibited as a result of the substitution. MnO can be incorporated in C_3S. The limiting solubility is not known; the X-ray pattern is changed in the direction of that of alite [*68*].

IV. The Burning of Portland Cement

The burning of Portland cement begins as a series of reactions between intimately mixed solids, and it is only in the later stages of burning that any liquid is formed, causing the reactions which produce the cement compounds to take place more rapidly. Lea [*69*] has indicated that "the production of Portland cement by a clinkering process in which only a minor proportion of the mix becomes liquid is dependent on three factors: (1) the chemical composition of the mix; (2) the physicochemical state of the raw constituents; (3) the temperature and period of burning; a fourth condition which influences the resulting clinker is the rate of cooling." Most research on the burning of Portland cement has been carried out in the laboratory with small-scale mixes of oxides, but a few authors have reported work with commercial kilns.

A. Reactions in the Kiln

The following discussion applies primarily to rotary kilns, the main features of which were described in Chapter 1. The reactions occur in stages, which were summarized by Lea [*69*] as follows:

Temperature (° C)	Process	Thermal change
100°	Evaporation of free water	Endothermic
500° and above	Dehydroxylation of clay minerals	Endothermic
900° and above	Crystallization of products of clay mineral dehydroxylation	Exothermic
	Decomposition of $CaCO_3$	Endothermic
900–1200°	Reaction between $CaCO_3$ or CaO and aluminosilicates	Exothermic
1250–1280°	Beginning of liquid formation	Endothermic
Above 1280°	Further liquid formation and completion of formation of cement compounds	Probably endothermic on balance

Other authors are in general agreement, though there are some significant differences; Bogue [*70*] considers that reactions giving cement compounds begin at as low a temperature as 600° C. The first stage (evaporation of free water) is quickly completed in a dry process kiln, but with the wet process it occupies roughly the first half of the length of the kiln.

The maximum temperature is reached in the "burning zone" some 5–12 m from the exit end of the kiln, and is normally between 1300° and 1500° C. It is chosen so as to produce a degree of melting sufficient to cause the material to cohere into small balls or lumps of clinker, and is therefore known also as the clinkering temperature. Between 20% and 30% of the material is normally molten at this temperature. Overburning, i.e. operating at a higher temperature, is generally considered undesirable, as it can lead to difficulties in running the kiln and possibly yields a less reactive product, owing to the formation of larger crystals or ones containing fewer defects [*71*]. Beyond the burning zone the temperature falls, and the clinker normally enters the cooler at 1000–1300° C. The mixture takes about 2·5 h to pass through a kiln 60 m long, and the time during which it is at the clinkering temperature is, at the most, 20 min [*69*].

Information on the course of reaction in commercial kilns has been obtained by sampling at various points, either after stopping the kiln and allowing it to cool or, better, during normal operation [*72, 73*]. Temperatures at various points can be determined by inserting thermocouples through the lining. The results of such studies show that the content of free CaO rises to a maximum at a point somewhat before the burning zone. This suggests that at 900–1000° C decomposition of the $CaCO_3$ occurs more rapidly than reaction of the resulting CaO with aluminosilicates. Immediately before the burning zone the free CaO content drops again and the temperature rises sharply to about 1250° C. Both effects are probably caused by the exothermic reaction of the CaO with the aluminosilicates. The mixture then passes through the burning zone at 1300–1500° C, where this reaction is completed and the free CaO content drops almost to zero.

The product at the clinkering temperature consists essentially of crystals of C_3S and C_2S, formation of which is largely completed at this stage, together with a liquid containing CaO together with all or most of the Al_2O_3, Fe_2O_3 and MgO, but relatively little SiO_2. The aluminate and ferrite phases therefore form only during cooling. Other processes that occur, or can occur, during cooling include polymorphic transitions, especially in the C_2S, crystallization of periclase and reaction between the liquid and the C_3S or C_2S crystals already formed. This last process will alter the ratio of C_3S to C_2S.

Alumina and iron oxide are the main fluxes in cement burning; without them the silicates could only be formed at much higher temperatures or in much longer times. The relations between the Al_2O_3/Fe_2O_3 weight ratio of the mix and the amounts of liquid formed at different temperatures are discussed by Lea [*69*]. At a clinkering temperature of 1400° C, rather more liquid is formed for each per cent by weight of Al_2O_3 than of Fe_2O_3, but this situation can be reversed in the earlier stages of liquid formation below about 1300° C. There is also some evidence that Fe_2O_3 can be more effective in promoting solid–solid reactions.

The exact sequence of reactions by which the cement compounds are formed is not well understood. It was at one time generally supposed that they were formed almost wholly by crystallization from the liquid and that little or no interaction occurred before melting began, but Bogue [*70*] has suggested that reactions of solid phases, either with each other or with the melt, may also be important. There is general agreement that, whatever the mechanism, equilibrium conditions are closely approximated to at the clinkering temperature, assuming that grinding, mixing and burning are properly carried out.

B. Effect of Raw Material

As has already been mentioned in Chapter 1, the reactivity of a raw mix depends not only on its chemical composition, but also on the mineral composition and the size of the particles. It may also be affected by the state of crystallinity of individual minerals; defects in a crystal are likely to make it more reactive.

Most work has been concerned with the effects of varying particle size of the limestone and of quartz, which is likely to be one of the least reactive constituents of the clay or other acidic constituents. If the material which is introduced into the kiln is not sufficiently well mixed, reaction will be incomplete. This usually results in a lowering of the C_3S content, C_2S and CaO being formed instead in different parts of the clinker nodules. The commonest cause of incomplete reaction is probably the presence of large particles of quartz, which can produce bands rich in C_3S and C_2S (Plate 1). The remedy is better grinding of the raw materials.

Toropov and Luginina [*74*] investigated the influence of particle size of the raw mix on the processes of combination of calcium oxide in Portland cement burning. They showed that care had to be taken to prevent the fine material from remaining in the clinkering zone too long, as this caused kiln rings to be produced. They later investigated the effect of rapid burning, which they showed to accelerate the combination of the raw materials [*75*]. Heilmann [*76*] showed that not more than 0·5% of silica particles above 0·2 mm nor more than 1% between 0·09 and 0·2 mm should be present in raw mixes with a lime saturation factor (Chapter 1, p. 32) as high as 0·95, but for lower lime saturation factors twice these amounts might be allowed. Up to 5% of pure calcite particles greater than 0·15 mm in size can be tolerated, while impure siliceous limestones of greater size can be used without detrimental effects.

C. Effect of Reducing Atmospheres

In general, oxidizing conditions are maintained in cement kilns, and the iron in the resulting clinker is predominantly in the ferric state. Reducing conditions can occur, however, especially in shaft kilns of certain types. They have several effects on the properties of the clinker, all of which are undesirable.

Woermann [*11*] found the main effect to be a marked acceleration of the decomposition of alite during cooling. C_3S is unstable relative to

C_2S and CaO below 1250° C, but with clinkers prepared under oxidizing conditions the process is much too sluggish to be significant under technical conditions. With kilns run under moderately reducing conditions, microscope examination of the clinker often shows that many of the alite crystals have been decomposed to pseudomorphs containing β-C_2S, CaO and ferrite. Woermann considered that the Ca^{2+} in C_3S was partially replaced by Fe^{2+}. During cooling, the conditions became more oxidizing. The Fe^{2+} was oxidized to Fe^{3+}, which could not be accommodated in the alite structure and was precipitated as ferrite; the C_3S was decomposed by C_2S and CaO.

A second effect, observed by Woermann [*11*] and also by Suzukawa and Sasaki [*77*], was a tendency for the β-C_2S to invert to γ-C_2S on cooling. This can probably be attributed to the fact that Fe^{2+} substitutes for Ca^{2+} more readily in γ-C_2S than in β-C_2S. Woermann noted several other effects. Considerably more periclase was produced; it contained appreciable FeO in solid solution. There were also changes in the properties of the ferrite phase and in the relative amounts of ferrite and aluminate phases. Lastly, sulphides (CaS and iron sulphides) are sometimes formed.

If the conditions are sufficiently strongly reducing, the iron is converted to the metal [*11, 77*]. The clinker formed under these conditions is white and decomposition of the alite does not occur; $C_{12}A_7$ may be produced [*77*].

D. Effect of Cooling Rate

In Chapter 2 it was shown that the rate of cooling of a cement clinker might be expected to influence the compound composition. In the following discussion it will be assumed that the four major constituents have the exact compositions C_3S, C_2S, C_3A and C_4AF. If it could further be assumed that equilibrium is reached at the clinkering temperature and continuously maintained during cooling (except that no decomposition of C_3S into C_2S and CaO occurs), the four products would be formed in the amounts given by the Bogue calculation (p. 77). With sufficiently rapid cooling, the liquid present at the clinkering temperature might be expected either to crystallize independently of the solids already formed or to solidify to a glass. Lea and Parker [*78*] derived methods for calculating the liquid content of a clinker of given composition at any desired temperature, and the corrections that should be applied to the results of the Bogue calculation for each of the types of non-equilibrium cooling mentioned above. Further possibilities of non-equilibrium cooling were considered by Dahl [*79*].

Table I gives the chemical analyses of two cement clinkers, together with the potential compound compositions given by the Bogue calculation and the results of applying Lea and Parker's corrections. It will be seen that independent cooling of the liquid would be expected to produce small amounts of $C_{12}A_7$, while complete vitrification of the liquid would produce substantial amounts of glass.

TABLE I

Calculation of cement composition with varying conditions of cooling

Cement composition	$CaO = 68\%, SiO_2 = 23\%, Al_2O_3 = 6\%, Fe_2O_3 = 3\%$		
	Complete crystallization	1450° C; liquid quenched to glass	1450° C; liquid crystallized independently
C_3S	57·5	59·6	59·6
C_2S	22·6	15·6	21·0
C_3A	10·8	0	9·8
C_4AF	9·1	0	9·1
CaO	0	0	0
$C_{12}A_7$	0	0	0·5
Glass	0	24·8	0
Cement composition	$CaO = 66\%, SiO_2 = 24\%, Al_2O_3 = 7{\cdot}5\%, Fe_2O_3 = 2{\cdot}5\%$		
C_3S	32·4	38·5	38·5
C_2S	44·5	33·3	39·9
C_3A	15·7	0	10·6
C_4AF	7·6	0	7·6
CaO	0	0	0
$C_{12}A_7$	0	0	3·5
Glass	0	28·1	0

The extent to which either of these forms of non-equilibrium cooling occurs in practice is uncertain. There is little doubt that the technical properties of cement clinkers are affected by the rate of cooling; studies by Lerch and Taylor [*80*], Lerch [*81*], Parker [*82*] and others have shown three main effects. First, cements made from certain slowly cooled clinkers and ground with gypsum in the ordinary way show flash set, i.e. rapid setting with marked evolution of heat to give a mixture that is difficult to work and that gives poor early strength; cements made from rapidly cooled but otherwise similar clinkers set normally. Secondly, if the clinker is relatively high (2·5–5·0%) in MgO, slow cooling may give an unsound cement, i.e. one that gradually expands after hardening and thereby becomes weakened or disintegrates. Thirdly, cements made from rapidly cooled clinkers may be more resistant to attack by sulphate solutions. Slow cooling may also cause "dusting", i.e. inversion of the C_2S to the γ-form. In all these respects it would appear that rapid cooling is likely to yield a better product. On the other hand, it has been reported that slowly cooled clinkers may be easier to grind.

These effects on the technical properties of the clinker could reasonably be attributed to differences in glass content, in accordance with the predictions of Lea and Parker. The flash setting and poor sulphate resistance of cements made from the slowly cooled clinkers could be attributed to the formation of too much C_3A, and the unsoundness to the presence of MgO. The contents of C_3A and MgO would both be lowered if glass was present. It would appear reasonable to suppose that, in a rapidly cooled clinker, the position is somewhere between the three extremes of equilibrium crystallization, independent crystallization of the liquid and solidification of the liquid to a glass.

Microscopic studies, discussed later in this chapter, give results which appear to be in general accord with this conclusion. The rapidly cooled clinkers consist largely of crystals of C_3S and β-C_2S embedded in a matrix consisting of the ferrite phase together with glassy material; with slowly cooled clinkers the glassy material is partly or wholly replaced by detectably crystalline phases, including as a rule C_3A and MgO. One recent investigation [*84*] has, however, given a contrary result, in that slow cooling from 1150° C to 750° C was reported to increase the proportion of glass.

Calorimetric evidence has also been considered to support the view that rapidly cooled clinkers contain glass; Lerch and Brownmiller [*83*] used a calorimetric method for the approximate determination of the glass content. They showed that rapidly cooled clinkers had higher heats of solution than the same clinkers which had been reheated and annealed. The difference was attributed to the heat of crystallization of the glass; this quantity was determined separately for a glass of appropriate composition and the percentage of glass in the clinker was thereby obtained. While there are some discrepancies between the results given by the various methods, it would appear both from the effects on the technical properties and from the results of microscopic and calorimetric studies that rapid cooling can lead to the formation of glass at the expense especially of C_3A and MgO.

Recent work nevertheless casts some doubt on the correctness of this view. Quantitative X-ray studies, discussed later in this chapter, do not support the view that commercial clinkers, whether slowly or rapidly cooled, contain any substantial proportions of glass. Moreover, commercial clinkers, including those which have been rapidly cooled, do not normally contain appreciable amounts of $C_{12}A_7$ unless they have been made under reducing conditions. The microscopic detection of glass in clinker is fraught with uncertainty, as material that appears glassy under the microscope may really be microcrystalline. The higher heats of solution and different technical behaviour of rapidly cooled clinkers may be caused, not by the presence of glass, but by the occurrence of disordered or defective crystalline states in the C_3A and other phases separating from the liquid. It is also possible that unstable, though persistent, new phases may be formed as a result of rapid cooling of a liquid; it has been suggested that a crystalline C_3F, structurally similar to C_3A, can be formed in this way [*85*]. The whole question of glass in clinker needs re-assessment after many more systematic, quantitative X-ray studies of the phase compositions of clinkers cooled in different ways have been completed.

V. Qualitative Examination of Clinker

A. Visual Examination

Portland cement clinker as it comes from the kiln forms rounded pellets if a dry process is used and irregular lumps if a wet process is used. Considerable information may be obtained by direct observation of the clinker without resort to apparatus. If the kiln is running satisfactorily, the clinker should be black and dark for a normal Portland cement; if the kiln is running under reducing conditions, however, the clinker will be a reddish brown. If the clinker is being under-burned, the clinker will show white or light coloured patches. Information about the state of the kiln can also be obtained on breaking the clinker nodules. Bad mixing of the raw material and the results of failure to reach equilibrium or of segregation can sometimes be seen with the naked eye.

Some indication of the running of the kiln can be obtained from the bulk density, known in the cement industry by the name of litre weight. The clinker sample is roughly packed into a drum of known capacity and weighed. This shows whether too much dust is coming out from the kiln.

B. Microscopy

The binocular stereoscopic microscope is a most useful instrument for more detailed visual examination of the clinker. The petrological microscope provides further information; it was by the examination of thin sections that the phases in Portland cement were first identified.

Törnebohm [1] named four phases: alite, belite, celite and felite. It was shown later by Insley [86] that alite was C_3S, that belite and felite were two different habits of C_2S and that celite was a ferrite phase then thought to be C_4AF. The greatest use of the thin section today is in the identification of the minor modifications of the phases. Thus, Nurse, Midgley and Welch [87] examined the types of C_3S in Portland cement clinkers and found that optically they were different from those in slags and laboratory preparations.

Most of the work on the microscopic examination of Portland cement clinker has dealt with the examination of polished and etched surfaces under the reflecting microscope. After the pioneering work of Insley [86], Parker and Nurse [88] and Tavasci [89] the method has become almost routine. The methods used have been summarized by Insley and Frechette [90] and in Volume 2, Chapter 20. Plate 2 shows a typical polished surface of a quickly cooled clinker, etched with water followed by 0·25% HNO_3 in alcohol. The material consists essentially of crystals of C_3S and β-C_2S up to 100 μ in size, embedded in a matrix of interstitial material. The various phases detectable optically in this and other clinkers will now be described.

1. Tricalcium Silicate

This occurs mainly as relatively large, euhedral crystals (Plate 2). They frequently appear pseudo-hexagonal, and when etched with water followed by alcoholic HNO_3 are lighter than the crystals of β-C_2S. They are sometimes zoned. In thin section they are colourless.

2. β-Dicalcium Silicate

This also occurs mainly as relatively large crystals (Plate 2). Unlike those of the C_3S, they are anhedral or subhedral; the outlines can be smooth and rounded, as in Plate 2, or fingered. They appear darker than C_3S crystals when seen in polished sections etched with water followed by alcoholic HNO_3. In thin sections they tend to be slightly yellow, brown or green. They usually show striations due to polysynthetic twinning. Several varieties have been distinguished; at one time these were thought to be different polymorphs of C_2S, but Insley [86] showed that all were forms of β-C_2S. The commonest form, called Type I by Insley, shows two or more sets of interpenetrating striations, and probably represents crystals which have originally formed as α-C_2S. The crystals seen in Plate 2 are of this type. Type II crystals show only one set of striations and are rare in commercial clinkers. Type III crystals are untwinned and are sometimes found as overgrowths on crystals of Type I; they have possibly crystallized from the melt during cooling. In a further type, called Ia [*91*], there are inclusions of apparently exsolved material along traces of twinning planes.

β-C_2S also occurs as rims of small crystals on the surfaces of C_3S grains and as very small, rounded particles dispersed in the interstitial material; both forms can be seen in a clinker which has been slowly cooled (Plate 3). The rims have possibly been formed by reaction between C_3S crystals and liquid during cooling, while the dispersed particles have crystallized from the liquid or glass. β-C_2S also occurs as inclusions in C_3S (Plate 4).

In some clinkers the C_3S and β-C_2S crystals tend to occur in separate aggregates which are probably relicts of coarser particles in the raw material. That shown in Plate 2 is of this type. An extreme case of inhomogeneity, attributable to inadequate grinding and mixing, is shown in Plate 1.

3. Interstitial Material

This is the material that was liquid at the clinkering temperature. It is undifferentiated, or only slightly differentiated, in a quickly cooled clinker (Plate 2), but in a slowly cooled clinker (Plate 3) distinct regions are more easily observed. The interstitial phases were reviewed by Insley [*92*]; the following types of material have been recognized.

(a) *Light interstitial material.* This is the ferrite phase. It is unaffected by most etching reagents and has a high reflectance; it therefore appears very bright in reflected light. It is seen as the lightest coloured regions in Plates 2 and 3. The ferrite phase in clinkers is sometimes prismatic but more often forms irregular aggregates. In thin sections it is reddish in colour, birefringent and pleochroic.

(b) *Dark interstitial material.* This has been further subdivided into three types known as rectangular, prismatic and amorphous. The rectangular type is C_3A and is seen most readily in slowly cooled clinkers of high Al_2O_3/Fe_2O_3 ratio. The prismatic form (Plate 3) is probably a form of C_3A modified by incorporation of alkali, perhaps NC_8A_3. The "amorphous" form (Plate 2) is the material generally described as glass; this has been discussed earlier (pp. 103–106).

(c) *MgO.* This usually occurs in the interstitial material and is visible as small, angular, highly reflecting grains in an unetched, polished section (Plate 5). MgO also occurs as inclusions in other phases; the triangular surface of a crystal included in C_3S is visible in Plate 1.

4. CaO

This is only present in quantity in cements that are incompletely burned. It forms rounded crystals which occur singly or in groups and are often as large as the C_2S grains. Plate 6 shows its appearance on an unetched polished section.

The use of the microscope for quantitative estimation of the phases will be discussed later.

C. Chemical Separation

In recent years some attempt has been made to separate the phases chemically. Jander and Hoffman [*93*] evolved a method for obtaining a separation of the non-hydraulic silicates CS and C_3S_2 from a mixture with C_3S and C_2S, but they did not succeed in separating C_3S from C_2S. Midgley [*94*] managed to separate C_2S from a clinker by differential hydration, but although no other crystalline phases were present it seemed likely that a film of alumina gel was left on the surfaces of the grains.

Chemical methods have been sought to separate, or at least concentrate, the ferrite phase sufficiently to make it possible to determine the composition. Fratini and Turriziani [*95*] described a reagent consisting of 25 ml water, 65 ml ammonia, and 10 g ammonium citrate which preferentially dissolves the silicates from Portland cement, leaving a residue consisting mainly of C_3A and the ferrite phase. The treatment as described by them consists of shaking 1 g of cement sample with 100 ml of reagent for 12 h. The solution is decanted from the residue, which is then shaken with a further 100 ml of reagent for a further 12 h. The residue is then filtered off in a sintered glass crucible, washed rapidly with ammonia and dried at 110° C. Midgley *et al.* [*37*] investigated the method and found that the rates of solution of the ferrites C_6AF_2, C_4AF and C_6A_2F differed (Fig. 3). They also found that C_3A dissolves more rapidly even than C_6A_2F. These authors also studied the effect of lowering the sample weight; they found that a single three-hour extraction, with the initial weight of sample reduced to 0·2 g in 100 cm³ of reagent, left a residue of about 22%. X-Ray diffraction showed the complete removal of the crystalline silicates.

More recently two further methods have been put forward for the separation of the aluminate phases. Royak, Nagerova and Kornienko [*96*] suggested the use of 5% boric acid solution to remove the silicate phases and of acetic acid to remove the ferrite phase and the glass, presumably leaving the C_3A. This method does not appear to have been followed up. Takashima [*97*] investigated the dissolution of the silicate phases in solutions of salicylic and picric acids in mixtures of toluene and acetone. The cement sample was stirred with the solvent for 1 h and allowed to stand for 1 day; the residue was washed with toluene, methyl

ethyl ketone or methanol, which dissolved the reaction products. It was suggested that by varying the concentrations of the acid and the final solvent it is possible to remove successively free lime, alite, β-C_2S and amorphous silicates.

From these experiments it seems that the calcium silicate phases can be removed by either acid or alkaline treatment, and it is not easy to see why these methods should work. It has been suggested that the much

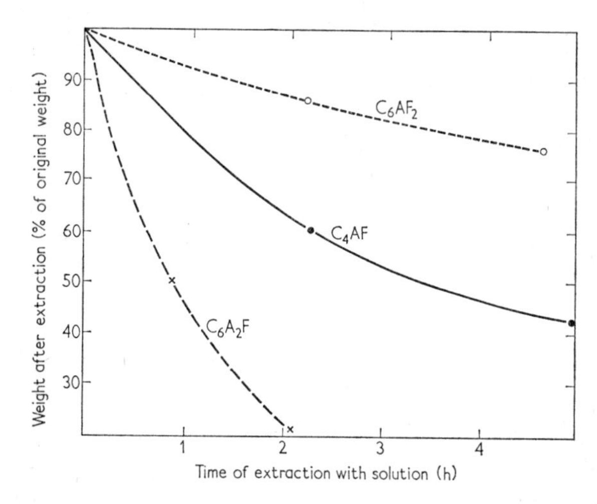

FIG. 3. Rates of solution of ferrites in "Fratini and Turriziani" solution.

more rapid dissolution of the silicate phases can effectively deactivate the reagent before the solution of the other phases has proceeded to any extent.

One very sensitive chemical method exists for detecting free lime; White's reagent (a solution of 5 g phenol in 5 ml nitrobenzene + 2 drops of water) will form calcium phenate in the presence of CaO or $Ca(OH)_2$. Calcium phenoxide occurs in tetragonal, acicular prisms which have a high birefringence and show up on a polarizing microscope under crossed nicols. The crystals form rapidly if much free lime is present, but may require an hour to form if only traces occur.

D. PHYSICAL SEPARATION

The physical separation of the constituent clinker minerals is extremely difficult, since the individual crystals are frequently smaller than 1 μ.

The most obvious method is based on the use of heavy liquids, since the four main minerals have different densities (C_3S 3·13 g/cm^3, C_2S 3·28 g/cm^3, C_4AF 3·77 g/cm^3, and C_3A 3·00 g/cm^3). This means that with liquids of suitable densities, e.g. methylene iodide–benzene mixtures, it should be possible to effect some separation. Guttmann and Gille [*98*] made an effective separation of the alite phase from a Portland cement clinker and showed it to be essentially C_3S.

Attempts at the Building Research Station by Midgley and Smith [*99*] have not been very successful; a complete separation of the silicate phases has never been effected. A nearly complete separation of the ferrite phase was made, but the method was so long and tedious that it was not repeated. These authors have found that one of the major difficulties in separation is the small crystal size, which necessitates the use of powder of less than 300 B.S. sieve size. This is much finer than is generally considered suitable for gravity separation. They found that a contributing cause of non-separation was the non-wettability of the cement grains; this could be overcome by the addition of a wetting agent.

With some Portland cement clinkers it can be seen on microscopic examination that the silicate phases differ greatly in particle size; this suggests that a mixture of particle size separation and gravity separation should be possible. Midgley and Smith [*99*] carried out such a separation on a normal Portland cement clinker in which the C_2S had grown to a much larger crystal size than the C_3S. The sample was ground to less than 10 μ and then separated into two fractions of 5–10 μ and less than 5 μ. Each fraction was then separated by means of the centrifuge and heavy liquids; the parts having a density less than 3·2 g/cm^3 were retained. The amounts of the phases present were estimated microscopically as follows:

Less than 5μ	C_3S 85%	C_2S 7·5%	Interstitial material 7·5%
5–10μ	C_3S 4%	C_2S 91%	Interstitial material 5%

It was thus possible to concentrate but not to separate the phases.

Another property which has been used to separate the phases is magnetic susceptibility; Midgley [*33*] concentrated the ferrite phase in clinkers to enable the weak X-ray lines of the ferrite to be recorded. Malquori and Cirilli [*31*] suggested that the concentration found by Midgley could have been due, not to the differing magnetic susceptibilities of the ferrite and silicate phases, but to included free iron in the ferrite phase. This method has now been superseded by ones based on chemical separation. Other possible methods of separation which have not been investigated are electrostatic separation and froth flotation.

E. X-Ray Examination

1. General Points

The use of X-ray diffraction methods for determining the phase composition dates back to the pioneer work of Bogue and Brownmiller [*100*], who used a powder camera to identify C_3S, C_2S, C_3A and C_4AF in clinker. Various experimental techniques are now available for powder work which give differing degrees of information about the crystalline modifications of the phases. It must be emphasized that with Debye–Scherrer cameras the best results will be obtained only if special care is taken with the preparation of specimens and accurate centring in the cameras. At the Building Research Station it has been found possible to use cellulose acetate capillaries 0·2 mm in diameter for the mounting of specimens, and with these the resulting lines are quite sharp. In much of the more recent work either focusing cameras or diffractometers have been used.

In Portland cement the predominant mineral is tricalcium silicate, which gives a very strong pattern and dominates the diffraction pattern. Midgley [*34*] concluded that, with simple Debye–Scherrer techniques, the limits of detection are C_3S 5%, C_2S 15%, C_4AF 3%, C_3A 5%. By using focusing cameras or diffractometers these limits can be lowered. Midgley *et al.* [*27*], using a proportional counter with pulse height analysis, found that the limits of detection were about C_3S 3%, β-C_2S 5–6%, ferrite 2%, C_3A 1%; phases other than C_3S, β-C_2S, C_3A and ferrite were also detected. Yannaquis [*101*] reported detection of γ-C_2S. Midgley and Fletcher [*17*], using X-ray diffraction methods, have reported the presence of α'-C_2S (bredigite) in cements made from a phosphatic limestone. They identified the bredigite by the occurrence of a reflection at 3·16 Å; the limit of detection in this case was estimated to be about 5%.

Yannaquis [*101*] heated a clinker to 1150–1200° C to decompose the C_3S to β-C_2S and CaO; the X-ray diagram of the residual β-C_2S was identical with that of a synthetic product. He also eliminated the C_3S by hydration; the product so obtained was found to contain β- and γ-C_2S. In an effort to obtain better X-ray diffraction diagrams, Yannaquis also made a mechanical separation under the microscope of both C_3S and C_2S. He found that the X-ray patterns from homogeneous specimens so obtained were simpler than those of the pure synthetic silicates. He also deduced that, because of superposition of the stronger lines, the identification of C_2S in the presence of C_3S could only be made on lines of medium intensity.

2. Tricalcium Silicate

As already stated, C_3S occurs in more than one polymorphic modification. Simons, quoted by Jeffery [*2*], concluded that alite has single lines at 1·761 and 1·485 Å, while pure C_3S has doublets; by examinations of these lines the polymorphic state can be determined. In a more recent study Yamaguchi and Miyabe [*3*] showed that the line at about 1·761 Å ($52°2\theta$ for CuK_α radiation) is a singlet for the trigonal modification, a doublet for the monoclinic one and a triplet for the triclinic one. Midgley and Fletcher [*5*] used a diffractometer to study the profile of this line for a number of cement and clinker samples and also for two samples of alite having the appropriate limiting compositions (Table II). The MgO contents are significant and are included where available. In all these samples alumina was present in excess of the amount required to stabilize the monoclinic modification. The work showed that the C_3S in samples 1–6 and also in sample 10 was monoclinic, in samples 7–9 triclinic, and in samples 11 and 12 trigonal. The 2θ values for samples 1–4 fell within the range defined by the synthetic preparations. Sample 6, which contained about 73% of alite, did not contain sufficient magnesia to stabilize the monoclinic form, but as it was a white cement it may have contained other ions which could have this effect. Samples 7, 8 and 9 did not contain sufficient magnesia to stabilize the monoclinic form and the triclinic modification was present. No explanation for the occurrence of the trigonal modification in a clinker can be put forward. X-Ray diffractometry can therefore be used to determine the type or modification of the C_3S present in the clinker.

3. Dicalcium Silicate

The identification of β-C_2S in mixtures is difficult, since its strong lines almost coincide with those of alite, and the identification therefore depends on weak lines. The most useful lines are the pair at 2·448 and 2·403 Å; one of these (2·448) almost coincides with a weak C_3S line at 2·449 Å, but the line at 2·403 Å is clear except for a weak line of C_3A at 2·39 Å. Since C_3A is usually present in small quantities, this weak line does not show in patterns from a cement clinker.

4. Ferrite

The ferrite phase is a solid solution series ranging in composition from C_2F to just beyond C_6A_2F. X-Ray powder data for the solid solution series show that there is a change in the lattice spacing with composition; a determination of the necessary lattice constants therefore gives the

TABLE II

2θ values of X-ray reflections for alite in some Portland cements

Sample	Modifications of monoclinic C_3S		Portland cements						High C_3S clinkers			Clinkers		
			1	2	3	4	5	6	7	8	9	10	11	12
	$C_{153}M_6S_{52}A$	$C_{156}M_3S_{52}A$	Sulphate resisting	High C_3A	High C_3S	Normal	Normal	White						
Reflections at about 52° 2θ (CuK_α radiation)	51·92	51·87	51·95	51·88	51·86	51·90	51·85	51·78	51·94 51·79	51·91 51·78	51·91 51·82	51·73		
	51·78	51·70	51·73	51·73	51·70	51·71	51·70	51·63	51·69 51·55	51·67 51·52	51·68 51·55	51·64	51·63	51·66
MgO%	1·7	0·9	2·8	2·3	1·3	1·2	1·1	0·5	1·5	0·8	0·6	—	—	—

composition. The relationship between spacing and composition has been studied by several investigators [*22*, *31*, *33–35*, *102–105*]. The strongest line, 141,† occurs at 2·63–2·68 Å but is very close to the 2·70 Å line of C_3A and the 2·60 Å line of the C_3S; the position is complicated further by the fact that a moderately strong line of the ferrite, 200, occurs at 2·66–2·71 Å, thus overlapping the C_3A line. If enough resolution can be obtained, it is nevertheless possible to use the position of the 141 line to determine the composition. Figure 4 shows the relationships

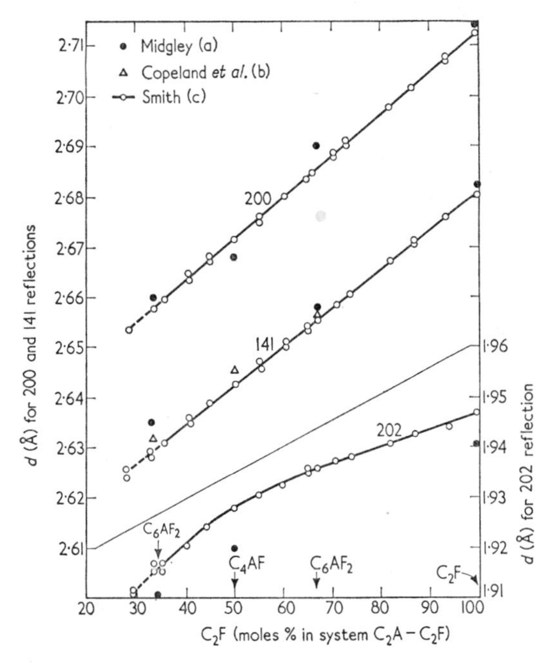

FIG. 4. Relationship between composition and X-ray powder spacings for the ferrite phase. (a) New data; (b) reference [*22*]; (c) reference [*103*].

between composition and spacing for the 141, 200 and 202 reflections. If the resolution is insufficient, the ferrite phase can be concentrated by drawing a bar magnet through the ground clinker. Lines of lower *d* spacing, such as 202, may then be used.

Copeland *et al.* [*22*] determined the composition of the ferrite phase in a number of cements from the position of the 141 line, and also by a calculation based on the intensity of this line and the total iron content

† The 141 line was wrongly indexed as 200 in some of the earlier investigations. There is a drafting error in the relevant figure of reference [*35*].

of the cement. Midgley [*35*] made a similar study, using the 202 line of a magnetic fraction; a Debye–Scherrer camera was employed. Later Midgley *et al.* [*37*] studied the same series of clinkers, using a diffractometer; in this work they used the 141 and 202 reflections from concentrates obtained by Fratini and Turriziani's method, and also the 141 reflections from the untreated clinkers. Table III compares the results of the various methods.

TABLE III

Weights % of C_2F in ferrite phase determined by X-ray diffraction

Clinker	Camera; magnetic fraction; 202 reflection	Diffractometer; untreated clinker; 141 reflection	Diffractometer; product of Fratini and Turriziani extraction	
			141 reflection	202* reflection
A. 9	56	54	52	48
A. 10	65 39 } (52)	56 47 } (51)	60 54 } (57)	50 42 } (46)
A. 20	66	60 48 } (54)	61 51 } (56)	No distinct peak
A. 35	71 39 } (55)	57	58 53 } (55)	50
A. 36	74 39 } (57)	56 48 } (52)	50	48
A. 37	66	63 47 } (55)	56	46 35 } (40)
A. 39	66	50	54 50 } (52)	49 41 } (45)
A. 40	66	63 52 } (57)	56	46
A. 143	65 39 } (52)	59 50 } (55)	56 48 } (52)	48 42 } (45)
A.F.	56	73 59 } (66)	64	62
A. 141	65 39 } (52)	62 50 } (56)	63 55 } (59)	45
A. 7	54 39 } (47)	64 46 } (55)	62 52 } (57)	No distinct peak
A. 35	54 39 } (47)	62 54 } (58)	54	48
A. 33	71 39 } (55)	68 51 } (59)	63 57 } (60)	37 33 } (35)
A. 113	—	70 54 } (62)	70 46 } (58)	66 46 } (56)

* The use of the 202 reflection is not very satisfactory for diffractometric techniques with KBr or Si as an internal standard, as both KBr and Si give peaks close to the ferrite 202 reflection. 2θ values for CuK_α radiation: Ferrites (46·6–47·5)°; KBr 45·60° and 47·74°; Si 47·30°.

5. *Other Phases*

Tricalcium aluminate can usually be detected by its strongest line at 2·70 Å; this is frequently the only detectable line due to this compound. The overlap of this line with the 200 line of the ferrite causes certain difficulties, which were discussed by Copeland *et al.* [*22*].

Midgley and Fletcher [*106*] found a significant variation in the spacing of the strongest X-ray powder reflection given by the C_3A in a number of commercial cement clinkers; the spacing was shifted from 2·699 Å in pure C_3A to 2·692, 2·695 and 2·695 Å in three clinkers studied. The 95% probability range for these results was 0·001 Å. They suggested that the shift might be due to MgO or Al_2O_3 or both. K_2O and Na_2O substitutions shift the reflection to higher *d* values in, for example, NC_8A_3. The shift to lower values in the C_3A of Portland cement may therefore be the algebraic sum of shifts up and down. This would mean that the measurement of the *d* spacing for C_3A in a Portland cement clinker cannot be used alone to determine the composition of the phase. Mather [*107*] found that with some clinkers the strong C_3A reflection occurred at spacings significantly higher than 2·699 Å. She obtained evidence that this was due, not to the occurrence of substituted forms of C_3A, but to overlap of the C_3A peak with a peak of spacing 2·720–2·725 Å attributable to some other phase. This phase was probably a calcium silicate.

MgO can be detected by its line at 2·106 Å and CaO by that at 1·390 Å. The 1·390 Å line is one of the weaker lines of the compound; the strongest line, at 2·405 Å, coincides with a line of β-C_2S. If more than about 5% of CaO is present, the 1·390 Å line becomes very strong.

F. Other Methods

1. *Infra-red Absorption*

The use of infra-red absorption spectrometry in the study of cements is comparatively recent. Hunt [*108*], Lehmann and Dutz [*109*], Lazarev [*110*], Midgley [*111*] and Roy [*112*] have reported data for cement minerals; Midgley's results are given in Fig. 5. Lehmann and Dutz [*109*] and Midgley [*111*] also investigated Portland cements. Lehmann and Dutz report that the cement they examined "consists essentially of β-C_2S which can be identified by the bands at 10·1 and 11·8 μ. Bands at 10·9 and 11·25 μ have the same intensity in β-C_2S as that at 10·05 μ. In Portland cement, however, they are better revealed, an indication that alite is present, whose bands are superimposed on those of β-C_2S at 10·9 and 11·2 μ. All maxima are displaced by 0·05–0·1 μ in the direction of longer waves as compared with pure minerals. Flat absorptions at

13·35 and 14·0 μ are attributable to vibration of AlO_6 octahedra." Midgley [*113*] showed that the four main constituents of Portland cement can be detected by measuring the absorptions at 10·2, 10·8, 13·5

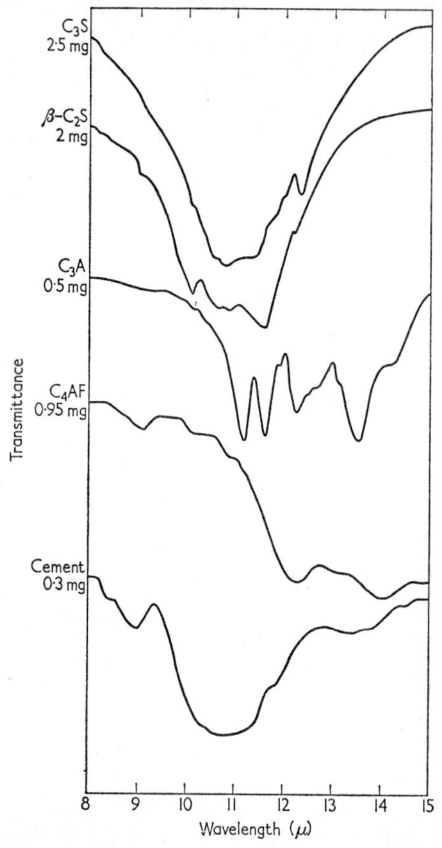

FIG. 5. Infra-red absorption spectra for clinker minerals and for a typical cement.

and 14·0 μ and then solving simultaneous equations. This method can possibly be used quantitatively and is discussed later.

2. Electron Microscopy

Very little work has been done on Portland cement clinker with the electron microscope. The only recent work seems to be that of Fahn [*114*], who examined powder mounts of clinker minerals. The present writer has attempted to make replicas of polished surfaces. The usual tech-

PLATE 1. Thin section of a Portland cement clinker showing bands rich in C_3S and C_2S produced by poor mixing of the raw material. Ordinary light.

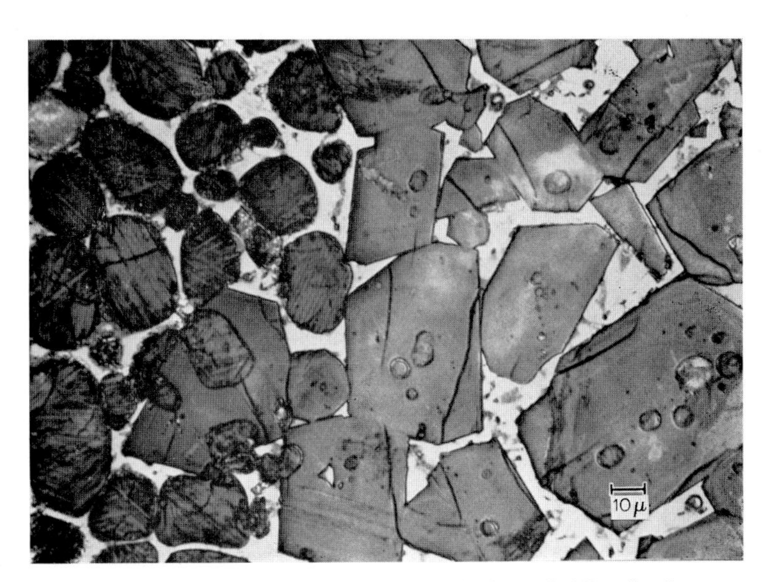

PLATE 2. Polished and etched surface of a quickly cooled Portland cement clinker, showing crystals of C_3S (right) and β-C_2S (left) embedded in a matrix of interstitial material which is only slightly differentiated.

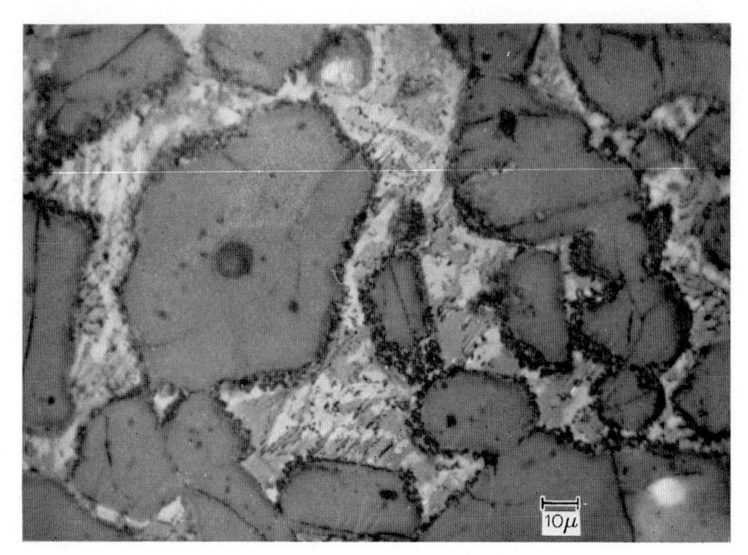

PLATE 3. Polished and etched surface of a slowly cooled Portland cement clinker, showing C_3S crystals with reaction rims of β-C_2S, embedded in a matrix of strongly differentiated interstitial material.

PLATE 4. Polished and etched surface of a Portland cement clinker, showing inclusions of β-C_2S in C_3S (centre and bottom right).

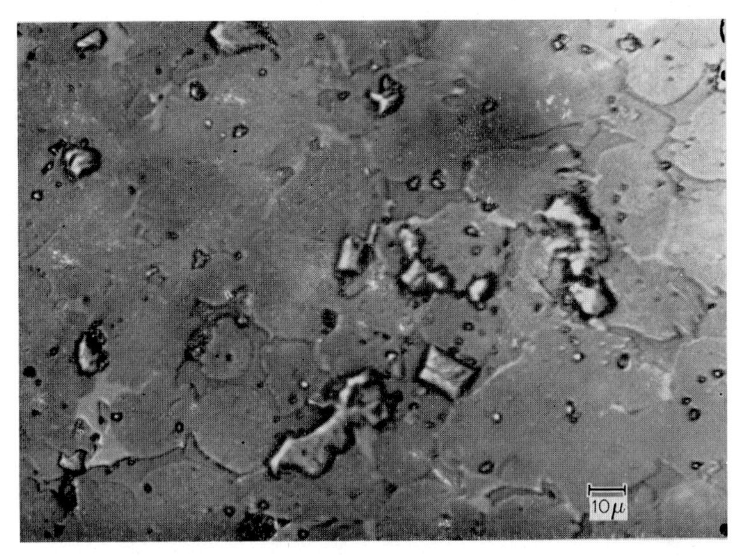

PLATE 5. Polished, unetched surface of a Portland cement clinker, showing MgO.

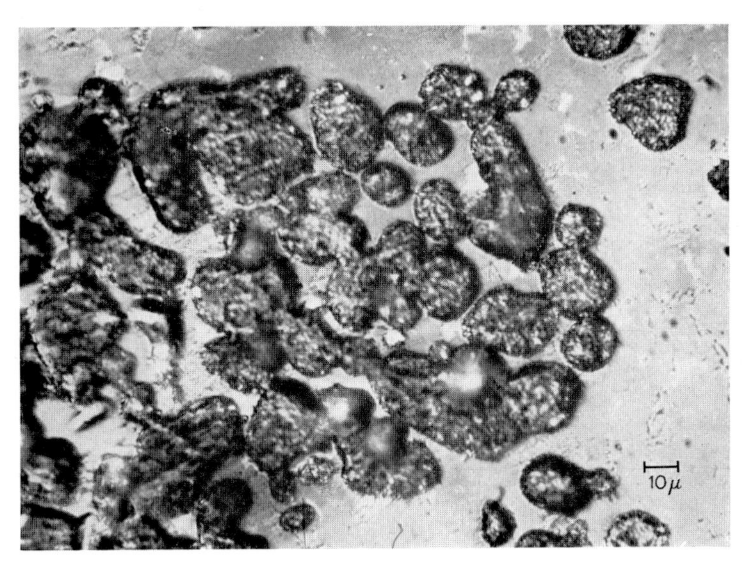

PLATE 6. Polished, unetched surface of a Portland cement clinker, showing CaO.

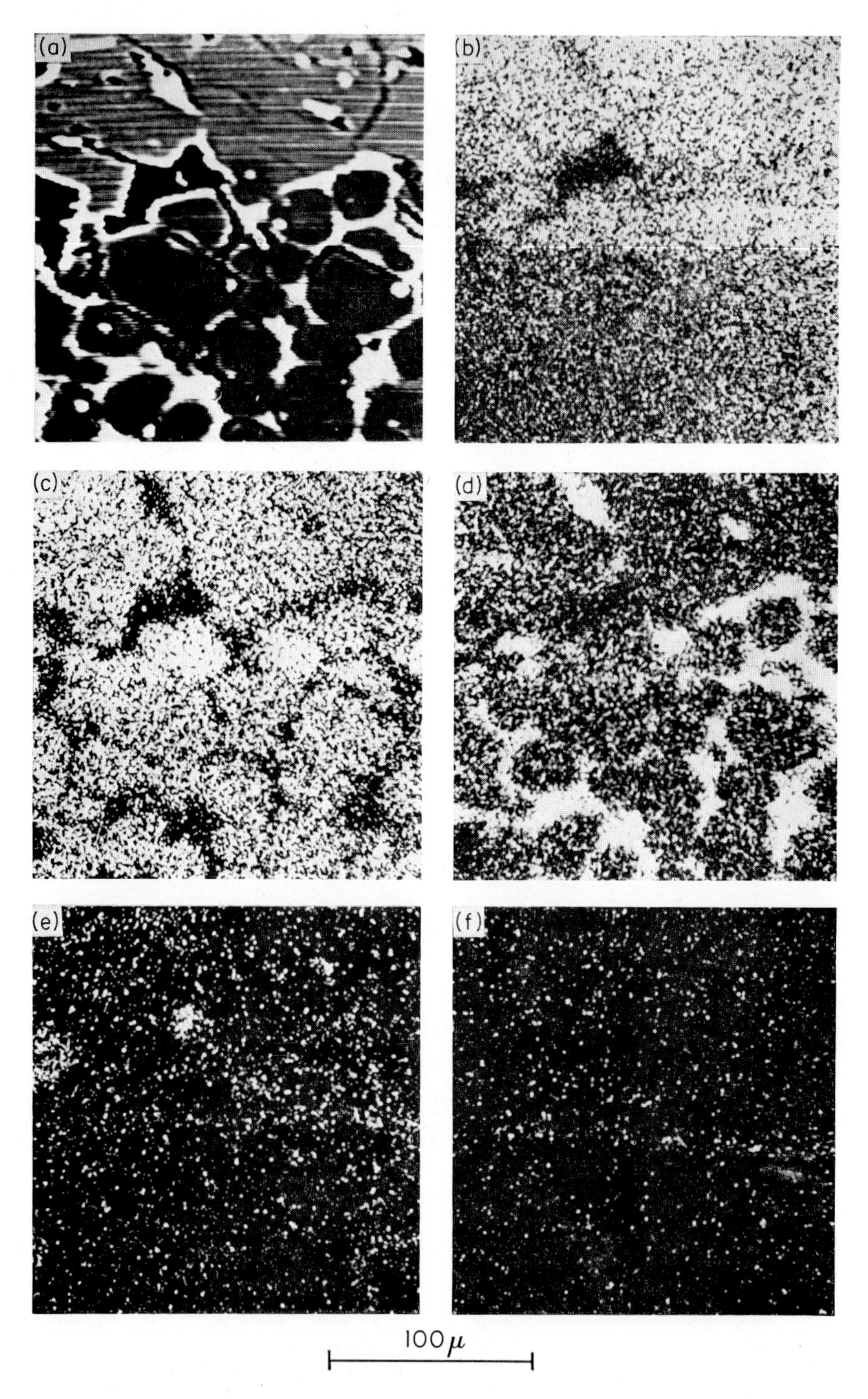

PLATE 7. Electron probe results for a polished surface of cement clinker (Japan Electron Optics Laboratory Co. Ltd.). (a) Absorbed electron image; (b)–(f) characteristic X-ray images for (b) CaK_α, (c) SiK_α, (d) FeK_α, (e) AlK_α, (f) TiK_α.

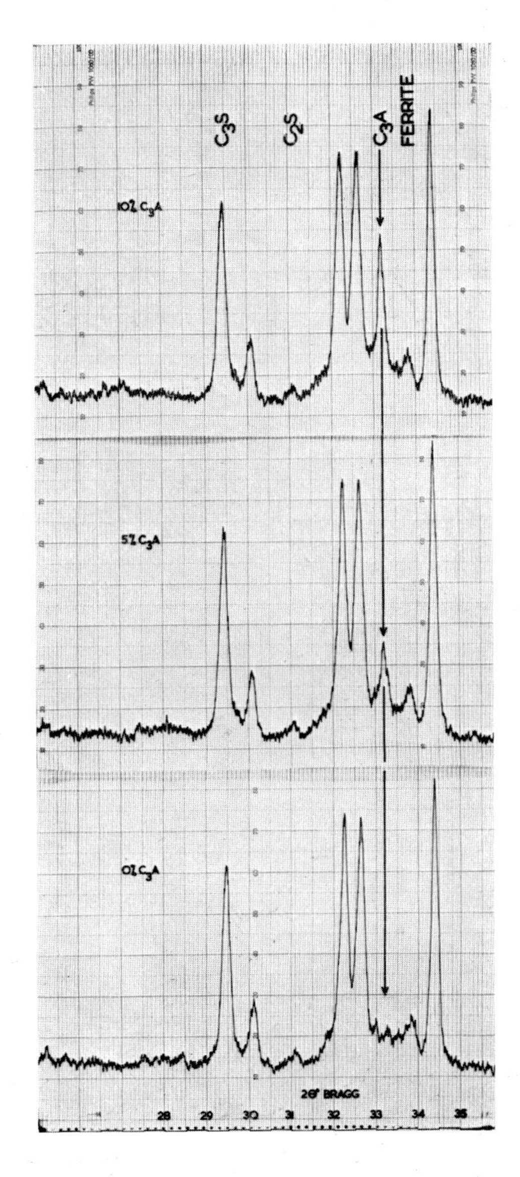

PLATE 8. Diffractometer traces of Portland cements of differing C_3A content.

niques proved unsuccessful but one technique, a modification of that of Hall [*115*], who had used it for teeth, proved possible. A replica is made from collodion in amyl acetate, backed with gelatine in water, and a carbon layer deposited *in vacuo*. The gelatine is removed with warm water and the collodion with amyl acetate. This work is still in its early stages and much additional research is required to perfect the technique.

3. *Electron Probe Analysis*

Electron probe analysis has not been in use long enough to have been applied systematically to research on clinkers. Some preliminary experiments have been carried out for the author by Wright [*116*], using an AEI (Associated Electrical Industries, Britain) X-ray microanalyser, and by the Japan Electron Optics Laboratory, using a JXA-3 instrument. With an electron probe the polished surface of the specimen is scanned by an electron beam. It is possible to measure continuously the intensity of the electrons scattered or absorbed and to display the results on a cathode-ray tube, the beam of which is synchronized with that scanning the specimen. The absorption of electrons by the specimen increases with the mean atomic number of the material, and the image produced in the cathode ray tube is so arranged that material of high mean atomic number appears bright. Mean atomic numbers for clinker compounds are: CaO 14·0; C_4AF 13·2; C_3S 12·7; C_2S 12·3; C_3A 12·2. In Plate 7(a), the dark, rounded areas at the bottom represent C_2S and the lighter areas at the top represent C_3S. The interstitial material, which is rich in ferrite, appears white. There are also indications of free CaO (white inclusions in C_2S) and of C_3A (very dark areas corresponding to the dark areas in Plate 7(b)).

If the beam scanning the specimen is of sufficiently high energy, characteristic X-rays are also emitted and may be analysed with a crystal spectrometer, the output from which can be fed to a cathode-ray tube as in the previous case. Bright areas in the resulting image represent parts of the specimen where the concentration of the element producing the X-rays of the wavelength isolated is high. Plate 7(b–f) shows the results thus obtained for Ca, Si, Fe, Al and Ti, using the same specimen as in Plate 7(a) The Ca concentration is high in the C_3S, lower in the C_2S and lower still in the C_3A. The Si concentration is low and the Fe concentration high in the interstitial material. The Al concentration is high in the C_3A.

If the electron beam is kept stationary, the intensities of the characteristic X-rays can be measured and the results used to give a quantitative chemical analysis of the material in the small region of the specimen irradiated. In this way Wright [116] found 55·2% (theoretical, 52·7%) for Ca in C_3S, and 45·4% (theoretical, 46·5%) for Ca in C_2S.

VI. Quantitative Estimation of Phases

A. Chemical Analysis

As has been shown earlier, it has not been found possible to effect a complete mechanical separation of the phases, and other methods of determining the quantitative phase composition of clinker have therefore been evolved. The most important and the most widely used is the recasting of a chemical analysis into a phase composition by means of the Bogue calculation (p. 77). As has already been indicated, this method, at least in its original form, is subject to errors of two main kinds. First, the phase composition obtained is only a potential one because equilibrium crystallization is assumed. This question was discussed on pp. 103–106. Secondly, it is assumed that the principal phases have the exact compositions C_3S, C_2S, C_3A and C_4AF. This assumption is only approximately true, especially in regard to the ferrite phase.

A refinement is clearly possible if the true composition of the ferrite

Table IV

Typical results obtained by Bogue and modified Bogue calculations

	%		A %	B %
SiO_2	20·15	C_3S	49·9	46·0
Al_2O_3	4·47	C_2S	20·1	23·1
Fe_2O_3	6·44	C_3A	0·9	5·7
TiO_2	0·23	Ferrite	19·6	15·7
MgO	2·81	Gypsum	4·1	4·1
CaO	61·51	CaO	0·8	0·8
Na_2O	0·19	MgO	2·8	2·8
K_2O	0·64	Unassigned (TiO_2, alkalis, part of ignition loss)	2·0	2·0
SO_3	1·95			
Ignition loss	1·80			
Total	100·19		100·2	100·2
Free CaO	0·8			

A. Results of the original type of Bogue calculation, assuming the phases to be C_3S, C_2S, C_3A, C_4AF, $CaSO_4.2H_2O$, CaO and MgO.

B. Results of modified Bogue calculation [37], assuming the ferrite phase to be $C_4A_{0\cdot72}F_{1\cdot28}$.

phase is known [*37*]. Table IV gives the chemical analysis of a typical Portland cement (containing gypsum), with the results obtained from Bogue calculations of the original type and modified to take into account the true composition of the ferrite phase. If the ferrite composition is wrongly assumed to be C_4AF, the main error introduced is usually in the content of C_3A, the calculated value of which will be high if the molar Al_2O_3/Fe_2O_3 ratio of the ferrite phase exceeds unity and low if it is less than unity. Corrections have also been suggested on the basis of assumptions regarding the alkali phases, and also the SO_3 in the case of clinker [*49, 117*].

The uncertainties inherent in the calculation of the phase composition from the chemical analysis have led to the development of independent methods of determination. The most important of these are based on optical microscopy and X-ray diffraction; infra-red spectroscopy provides a possible additional method.

B. Microscopic Examination

As described earlier, it is possible to identify the C_3S, C_2S, ferrite and possibly the C_3A phases on polished surfaces under the microscope. Once this identification has been made, it is a short step to try to measure the amounts. The first method was that of Rosiwal [*118*]: the area of each constituent on a surface is estimated by measuring the intercepts made by each mineral on a series of equally spaced lines. Various types of aid were invented to help with this method, the most popular being the integrating stage which moved the sample along by micrometers. Parker and Nurse [*88*] used this method to determine the constituents in 12 cements. This method became standard practice, but it was subject to a serious error, as the limits of each individual grain had to be accurately located.

Chayes [*119*] in 1949 made popular the Glagolev statistical method. This method substitutes points for lines in the estimation of areas. A point grid is superimposed on the area to be measured, and the mineral under each point is identified and recorded. In practice, the points of the grid are successively presented to the microscope. Various types of apparatus have been invented recently to carry out this task mechanically (see Chayes [*120*]) in which the specimen is moved across the stage and each mineral when identified is recorded on some form of counter. All these methods give volume percentages of the constituents. Conversion to weight percentages is necessary for comparison with other methods.

The reproducibility of the method may be judged from a series of measurements made on one clinker sample at the Building Research Station. This clinker was examined by nine operators and the standard deviations of the mean for one result were for C_3S 1%, for C_2S 1·5% and for interstitial material 2·25%. This clinker gave the following volume percentages: C_3S 60·3%, C_2S 24·5%, interstitial material 15·0%. In weight per cent this is equivalent to C_3S 60·4%, C_2S 25·2%, interstitial material 14·3%.

C. Infra-red Absorption

Infra-red absorption spectrometry has been used extensively for quantitative analysis in organic chemistry, but, although the method has been used more and more for qualitative and structural studies in mineralogy, it has had little use in quantitative mineralogy. Lyon, Tuddenham and Thompson [*121*] have reported a method for the analysis of a granite, using infra-red absorption, and this method could be used for the analysis of Portland cements.

Midgley [*113*] concluded that the infra-red absorption spectra for the four main constituents were well established and that the spectra differed sufficiently to allow analysis to be attempted. To carry out the investigation samples of the pure minerals C_3S, C_2S, C_3A and C_4AF were ground to less than 2 μ; since all the minerals under investigation were anhydrous, the KBr disk method could be used. Various weights of the preground minerals were added to either 0·5 or 1·0 g of KBr and blended in a vibratory shaker. The disks were produced by vacuum die pressing. The cement clinker sample was dealt with in the same way. The absorption curves were run on a Perkin-Elmer Type 137 Infra-cord spectrometer.

The absorption curves for the four constituents are given in Fig. 5. Experiment established that Beer's Law held for these minerals. As four components are being investigated, the absorption at four points, 10·2, 10·8, 13·5 and 14 μ, is measured and then from four simultaneous equations the amounts of the four components may be established. These equations may be solved for the minerals present in cement, a typical curve for which is included in Fig. 5. The results so far obtained show only moderate agreement with those obtained by other methods, but the method is still in an early stage of development.

D. X-Ray Diffraction

1. Methods

The quantitative estimation of the phases, using X-ray diffraction, is based on the fact that the integrated intensity of a reflection is directly proportional to the amount of the substance producing it [*122, 123*]. This may be expressed as $I_1/I_0 = aw_1/w_0$ where I_1 is the intensity of a given peak of one constituent of which w_1 is the weight percentage, I_0 and w_0 similarly relate to a second constituent, and a is a constant. In practice, calibration mixtures are prepared by taking the four principal clinker compounds in varying known proportions and adding to each a fixed percentage of an internal standard. A given peak is then chosen for the determination of each compound and its intensity (I_1) is measured relative to that (I_0) of a selected peak of the internal standard. From a plot of I_1/I_0 against w_1/w_0 the constant a is obtained. Difficulties occur because of the overlapping of peaks from different compounds, which makes it difficult to find suitable peaks, and also because of the possibility that the patterns of the clinker compounds may differ significantly on account of isomorphous replacement from those of the preparations used in the calibration mixtures.

Table V summarizes the experimental methods that have been used by the various investigators and shows the peaks selected for the various compounds. Most of these peaks have d spacings between 3·1 and 2·6 Å (29–34°2θ for CuK$_\alpha$ radiation). Plate 8 shows traces covering this region for three different clinkers of varying C_3A content.

Von Euw [*6*] used a photographic technique and determined intensities with a microdensitometer. He did not correct for interfering peaks and gave the following coefficients of variation: alite ±3%, β-C_2S ±6–9%, C_3A ±0·7%, ferrite ±1·5%. Brunauer, Copeland and co-workers [*8, 22*] used a diffractometer with a Geiger counter and rate-meter. They corrected for overlapping peaks by mathematical equations based on the diffraction intensities of individual phases. The method which they used at first [*22*] was based partly on X-ray intensities and partly on chemical analyses for certain oxides. Later they employed a modified method [*8*] which permitted the direct determination of the four major phases and also that of the composition of the ferrite phase. Possible errors arising from differences between the alite and β-C_2S used for standardization and the phases present in the cement were corrected for by adjustments to the constants in the equations. These were made by a process of successive approximation in which the results of the earlier method based partly on chemical analysis were also employed.

Midgley *et al.* [*37*] used initially a diffractometer with Geiger counter and ratemeter. They corrected graphically for interfering peaks. For the ferrite phase separate calibration curves were obtained for different compositions. Difficulties were encountered in grinding silicon satisfactorily for use as an internal standard, and KBr was used. Subsequently Midgley *et al.* [*27*] improved the technique. A better method of grinding

TABLE V

X-Ray diffraction methods used by various authors

Author	Apparatus	Radiation	Reflections used, spacings (Å) and 2θ values				Standard	
			C_3S	C_2S	C_3A	Ferrite		
Von Euw [6]	Guinier camera + photometer	CoK_α	3·07 (34·2°)	2·88 (36·2°)	2·70 (38·6°)	2·66* (39·4°)*	NaCl	2·82 (37·1°)
		CuK_α	3·07 (29·1°)	2·88 (31·1°)	2·70 (33·2°)	2·66* (33·7°)*	NaCl	2·82 (31·8°)
Midgley *et al.* [37]	Geiger counter + ratemeter	CuK_α	3·04 (29·4°) 2·98 (30·0°)	2·88 (31·1°)	2·70 (33·2°)	2·66* (33·7°)*	KBr	3·29 (27·1°)
Copeland *et al.* [22]	Geiger counter + ratemeter	CuK_α	2·19 (41·2°) 1·78 (51·2°) Both reflections common to both phases		2·70 (33·2°)	2·66* (33·7°)*	Si	3·14 (28·4°)
Smolczyk [*124*]	Geiger counter + counting rate computer	CuK_α	1·76 (51·8°)	2·88 (31·1°)	2·70 (33·2°)	2·66* (33·7°)*	Al_2O_3	1·74 (52·6°)
Midgley *et al.* [27]	Proportional counter + discriminator	CuK_α			2·70 (33·2°)	2·66* (33·7°)*	Si	3·14 (28·4°)

* Approximate values.

permitted the use of silicon, and the Geiger counter was replaced by a proportional counter with pulse height analyser. A 30% increase in peak-to-background ratio was obtained, and the background stability improved. They concluded that the graphical method of correction for interfering peaks was still the best available. The coefficients of variation claimed were as follows (values claimed by Midgley *et al.* [*37*] in parentheses): alite ±5(5)%, β-C_2S ±5(11)%, C_3A ±1(3)%, ferrite ±2(6)%.

Smolczyk [*124*] used a diffractometer with Geiger counter and counting rate computer. He used Al_2O_3 as internal standard and found an improvement in the determination of C_3A. He gave the following coefficients of variation: alite ±2%, β-C_2S ±3%, C_3A ±1%, ferrite ±1%. Kheĭker, Konstantinov and Alekseev [*125*] have suggested the use of a scintillation counter.

2. Results

Brunauer, Copeland and co-workers [*8*, *22*] found a reasonably close agreement between the two methods employed. They reported that the C_3A contents were always lower than those predicted by the Bogue calculation. The totals for the four main phases in 20 cements, as determined by their direct method, were on the average 0·8% greater than the sums of the contents of the major oxides (usually 90–95%). They concluded that some of the minor oxides were dissolved in the four main phases and also that no significant amounts of glass were present.

Midgley *et al.* [*37*] found that, in general, the phase compositions determined by X-rays agreed quite well with those given by a modified Bogue calculation in which the observed composition of the ferrite phase was taken into account. They agreed less well with those given by the normal Bogue calculation. Like Copeland *et al.* [*22*], they found no evidence of the presence of appreciable amounts of glass; the sums of the four main phases in 17 clinkers of differing kinds ranged from 90 to 104%. In contrast to Copeland *et al.*, they found no fixed bias between the X-ray results and those of the normal Bogue calculation, except that the latter tended to underestimate C_3A in cements low in this phase. This could be attributed to the fact that in such cements the ferrite phase tended to be rich in C_2F. Smolczyk [*124*] also compared his results with those given by the normal and modified forms of the Bogue calculation; his conclusions were similar to those of Midgley *et al.*

To sum up, the established method of compound determination is the Bogue calculation. This method has been shown to be in error, owing to the very variable ferrite composition; if the composition of the ferrite can be determined separately, however, the modified calculation can be of use. If independent estimates of the compounds are required, the silicate phases can be determined by point counting with a microscope, but the aluminate phases cannot be determined accurately by this method. X-Ray diffractometry can be used for all four phases; it is most accurate for the aluminate phases and less accurate for the silicate phases. Infra-red absorption can be used, but does not yet appear to give accurate results.

REFERENCES

1. Törnebohm, A. E. (1897). *TonindustrZtg* 1148.
2. Jeffery, J. W. (1952). *Acta cryst.* **5**, 26; (1954). *Proceedings of the Third International Symposium on the Chemistry of Cement, London 1952*, p. 30. Cement and Concrete Association, London.
3. Yamaguchi, G. and Miyabe, H. (1960). *J. Amer. ceram. Soc.* **43**, 219.
4. Midgley, H. G. (1952). *J. appl. Phys.* **3**, 277.
5. Midgley, H. G. and Fletcher, K. E. (1963). *Trans. Brit. Ceram. Soc.*, in press.
6. von Euw, M. (1958). *Silicates industr.* **23**, 647.
7. Naito, R., Ono, Y. and Iiyama, T. (1957). *Semento Gijutsu Nenpo* **11**, 20.
8. Brunauer, S., Copeland, L. E., Kantro, D. L., Weise, C. H. and Schulz, E. G. (1959). *A.S.T.M. Bull.* **59**, 1091.
9. Locher, F. W. (1962). *Chemistry of Cement, Proceedings of the Fourth International Symposium, Washington 1960*, p. 99. National Bureau of Standards Monograph 43. U.S. Department of Commerce.
10. Lafuma, H. (1961). Centre d'Études et de Récherches de l'Industrie des Liants hydrauliques, Note d'information 17.
11. Woermann, E. (1962). *Chemistry of Cement, Proceedings of the Fourth International Symposium, Washington 1960*, pp. 104, 119. National Bureau of Standards Monograph 43. U.S. Department of Commerce.
12. Jander, W. (1938). *Angew. Chem.* **51**, 696.
13. Grzymek, J. (1953). *Cement-Wapno-Gips.* **9**, 162.
14. Nurse, R. W. (1954). *Proceedings of the Third International Symposium on the Chemistry of Cement, London 1952*, p. 56. Cement and Concrete Association, London.
15. Budnikov, P. P. and Azelitskaya, R. D. (1956). *Dokl. Akad. Nauk S.S.S.R.* **108**, 515.
16. Metzger, A. (1953). *Zement-Kalk-Gips* **6**, 269.
17. Midgley, H. G. and Fletcher, K. E. (1960). D.S.I.R. Building Research Station Note D707. HMSO, London.
18. Kukolev, G. V. and Mel'nik, M. T. (1956). *Dokl. Akad. Nauk S.S.S.R.* **109**, 1012.
19. Funk, H. (1958). *Angew. Chem.* **70**, 655.
20. Welch, J. H. and Gutt, W. (1962). *Chemistry of Cement, Proceedings of the Fourth International Symposium, Washington 1960*, p. 59. National Bureau of Standards Monograph 43. U.S. Department of Commerce.
21. zur Strassen, H. (1961). *TagungsBer. Zementindustr.* **21**, 1.
22. Copeland, L. E., Brunauer, S., Kantro, D. L., Schulz, E. G. and Weise, C. H. (1959). *Analyt. Chem.* **31**, 1521; Kantro, D. L., Copeland, L. E. and Brunauer, S. (1962). *Chemistry of Cement, Proceedings of the Fourth International Symposium, Washington 1960*, p. 75. National Bureau of Standards Monograph 43. U.S. Department of Commerce.

23. Nurse, R. W., Welch, J. H. and Gutt, W. (1959). *J. chem. Soc.* **220**, 1077.
24. Thilo, E. and Funk, H. (1953). *Z. anorg. Chem.* **273**, 28; (1953). *Naturwissenschaften* **40**, 241.
25. Funk, H. and Thilo, E. (1955). *Z. anorg. Chem.* **281**, 37.
26. Yannaquis, N. and Guinier, A. (1959). *Bull. Soc. franç. Minér. Crist.* **82**, 126.
27. Midgley, H. G., Fletcher, K. E. and Cooper, A. C. (1963). Symp. Anal. Calcareous Materials, London.
28. Toropov, N. A., Shishakov, N. A. and Merkov, L. D. (1937). *Tsement, Moscow* **5**, No. 1, 28.
29. Yamauchi, T. (1937). *J. Jap. ceram. Soc. (Ass.)* **45**, 279, 361, 433, 614.
30. Swayze, M. A. (1946). *Amer. J. Sci.* **244**, 1, 65.
31. Malquori, G. and Cirilli, V. (1954). *Proceedings of the Third International Symposium on the Chemistry of Cement, London 1952*, p. 120. Cement and Concrete Association, London.
32. Nurse, R. W. (1962). *Chemistry of Cement, Proceedings of the Fourth International Symposium, Washington 1960*, p. 9. National Bureau of Standards Monograph 43. U.S. Department of Commerce.
33. Midgley, H. G. (1954). *Proceedings of the Third International Symposium on the Chemistry of Cement, London 1952*, p. 140. Cement and Concrete Association, London.
34. Midgley, H. G. (1957). *Mag. Concr. Res.* **9**, 17.
35. Midgley, H. G. (1958). *Mag. Concr. Res.* **10**, 13.
36. Kato, A. (1958). *Semento Gijutsu Nenpo* **12**, 17.
37. Midgley, H. G., Rosaman, D. and Fletcher, K. E. (1962). *Chemistry of Cement, Proceedings of the Fourth International Symposium, Washington 1960*, p. 69. National Bureau of Standards Monograph 43. U.S. Department of Commerce.
38. Cirilli, V. and Brisi, C. (1953). *Industr. ital. Cemento* **23**, 289.
39. Santarelli, L., Padilla, E. and Bucchi, R. (1954). *Industr. ital. Cemento* **24**, 55.
40. Royak, G. S. (1958). *Tsement, Moscow* **24**, No. 5, 21.
41. Volkonskii, B. V. (1956). *Rep. Symposium on the Chemistry of Cement, Moscow*, p. 83.
42. Tröjer, F. (1953). *Zement-Kalk-Gips* **6**, 312.
43. Brownmiller, L. T. and Bogue, R. H. (1935). *Amer. J. Sci.* **26**, 260.
44. Suzukawa, Y. (1956). *Zement-Kalk-Gips* **9**, 345.
45. Suzukawa, Y. (1956). *Zement-Kalk-Gips* **9**, 390.
46. Suzukawa, Y. (1956). *Zement-Kalk-Gips* **9**, 433.
47. Müller-Hesse, H. and Schwiete, H. E. (1956). *Zement-Kalk-Gips* **9**, 386.
48. Taylor, W. C. (1942). *J. Res. nat. Bur. Stand.* **29**, 437.
49. Newkirk, T. F. (1954). *Proceedings of the Third International Symposium on the Chemistry of Cement, London 1952*, p. 151. Cement and Concrete Association, London.
50. Steinour, H. H. (1957). Portland Cement Assoc. Res. Dev. Labs (Chicago), Bull. 85.
51. Nurse, R. W. (1952). *J. appl. Chem.* **2**, 708.
52. Toropov, N. A. and Borisenko, A. I. (1954). *Tsement, Moscow* **20**, No. 6, 10.
53. Simanovskaya, R. E. and Shpunt, S. Ya. (1955). *C.R. Acad. Sci. U.R.S.S.* **101**, 917.
54. Erschov, L. D. (1955). *Tsement, Moscow* **21**, No. 4, 19.

55. Kukolev, G. V. and Mel'nik, M. T. (1956). *Tsement, Moscow* **22**, 16.
56. Yamaguchi, G., Ikegami, H., Shirasuka, K. and Amano, K. (1957). *Semento Gijutsu Nenpo* **11**, 24.
57. Toropov, N. A., Volkonskii, B. V. and Sadkov, V. I. (1955). *Tsement, Moscow* **21**, No. 4, 12.
58. Toropov, N. A. and Skue, E. R. (1954). *Dokl. Akad. Nauk S.S.S.R.* **98**, 415.
59. Moore, R. E. (1960). *Rock Prod.* **63**, 108.
60. Taylor, W. C. (1941). *J. Res. nat. Bur. Stand.* **27**, 311.
61. Nurse, R. W. (1954). *Proceedings of the Third International Symposium on the Chemistry of Cement, London 1952*, p. 169. Cement and Concrete Association, London.
62. Brownmiller, L. T. and Bogue, R. H. (1932). *Amer. J. Sci.* **23**, 501.
63. Greene, K. T. and Bogue, R. H. (1946). *J. Res. nat. Bur. Stand.* **36**, 187.
64. Eubank, W. R. and Bogue, R. H. (1948). *J. Res. nat. Bur. Stand.* **40**, 225.
65. Eubank, W. R. (1950). *J. Res. nat. Bur. Stand.* **44**, 175.
66. Parker, T. W. (1954). *Proceedings of the Third International Symposium on the Chemistry of Cement, London 1952*, p. 143. Cement and Concrete Association, London.
67. Newman, E. S. and Wells, L. S. (1946). *J. Res. nat. Bur. Stand.* **36**, 137.
68. Gutt, W. Private communication.
69. Lea, F. M. and Desch, C. H. (1956). "The Chemistry of Cement and Concrete", 2nd Ed., revised by F. M. Lea. Arnold, London.
70. Bogue, R. H. (1954). *Proc. International Symposium on the Reactivity of Solids, Gothenburg, 1952*, p. 639.
71. Jander, W. and Wuhrer, J. (1938). *Zement* **27**, 73.
72. Lacey, W. N. and Woods, H. (1929). *Industr. Engng Chem.* (*Industr. Ed.*) **21**, 1124.
73. Lacey, W. N. and Shirley, H. E. (1932). *Industr. Engng Chem.* (*Industr. Ed.*) **24**, 332.
74. Toropov, N. A. and Luginina, I. G. (1953). *Tsement, Moscow* **19**, No. 1, 4.
75. Toropov, N. A. and Luginina, I. G. (1953). *Tsement, Moscow* **19**, No. 2, 17.
76. Heilmann, T. (1962). *Chemistry of Cement, Proceedings of the Fourth International Symposium, Washington 1960*, p. 87. National Bureau of Standards Monograph 43. U.S. Department of Commerce.
77. Suzukawa, Y. and Sasaki, T. (1962). *Chemistry of Cement, Proceedings of the Fourth International Symposium, Washington 1960*, p. 83. National Bureau of Standards Monograph 43. U.S. Department of Commerce.
78. Lea, F. M. and Parker, T. W. (1935). D.S.I.R. Building Research Technical Paper No. 16. HMSO, London.
79. Dahl, L. A. (1938–1939). *Rock Prod.* **41–42**.
80. Lerch, W. and Taylor, W. C. (1937). *Concrete, N.Y.* **45**, 199, 217.
81. Lerch, W. (1938). *J. Res. nat. Bur. Stand.* **20**, 77.
82. Parker, T. W. (1939). *J. Soc. chem. Ind. Lond.* **58**, 203.
83. Lerch, W. and Brownmiller, L. T. (1937). *J. Res. nat. Bur. Stand.* **18**, 609.
84. Yung, V. N., Butt, Yu. M. and Tunastov, V. V. (1957). *Trud. khim-tekh. Inst. Mendeleeva* **24**, 25.
85. McMurdie, H. F. (1941). *J. Res. nat. Bur. Stand.* **27**, 499.
86. Insley, H. (1936, 1940). *J. Res. nat. Bur. Stand.* **17**, 353; **25**, 295.
87. Nurse, R. W., Midgley, H. G. and Welch, J. H. (1961). D.S.I.R. Building Research Station Note A95.

88. Parker, T. W. and Nurse, R. W. (1939). *J. Soc. chem. Ind. Lond.* **58**, 255.
89. Tavasci, B. (1939). *Chim. e Industr.* **21**, 329.
90. Insley, H. and Frechette, van D. (1955). "Microscopy of Ceramics and Cements". Academic Press, New York.
91. Insley, H., Flint, E. P., Newman, E. S. and Swanson, J. A. (1938). *J. Res. nat. Bur. Stand.* **21**, 355.
92. Insley, H. (1954). *Proceedings of the Third International Symposium on the Chemistry of Cement, London 1952*, p. 172. Cement and Concrete Association, London.
93. Jander, W. and Hoffmann, E. (1933). *Angew. Chem.* **46, 76.**
94. Midgley, C. M. (1952). *J. appl. Phys.* **3**, 9, 277; see also Nurse, R. W. (1954). *Proceedings of the Third International Symposium on the Chemistry of Cement, London 1952*, p. 75. Cement and Concrete Association, London.
95. Fratini, N. and Turriziani, R. (1956). *Ric. Sci.* **26**, 2747.
96. Royak, S. M., Nagerova, E. I. and Kornienko, G. G. (1952). *Trud. vsesoyuz. nauch.-issled. Inst. Tsement* No. 5, 58.
97. Takashima, S. (1958). Ann. Rep. 12th Gen. Meeting Jap. Cement Engng Assoc., p. 12.
98. Guttmann, A. and Gille, F. (1931). *Zement* **20**, 144.
99. Midgley, H. G. and Smith, J. J. Private communication.
100. Bogue, R. H. and Brownmiller, L. T. (1930). *Amer. J. Sci.* **20**, 241.
101. Yannaquis, N. (1955). *Rev. Matér. Constr.* **480**, 213.
102. Newkirk, T. F. and Thwaite, R. D. (1958). *J. Res. nat. Bur. Stand.* **61**, 233.
103. Smith, D. K. (1962). *Acta cryst.* **15**, 1146.
104. Toropov, N. A. (1962). *Chemistry of Cement, Proceedings of the Fourth International Symposium, Washington 1960*, p. 113. National Bureau of Standards Monograph 43. U.S. Department of Commerce.
105. Bertaut, E. F., Blum, P. and Sagnières, A. (1959). *Acta cryst.* **12**, 149.
106. Midgley, H. G. and Fletcher, K. E. (1961). D.S.I.R. Building Research Station Note A94. HMSO, London.
107. Mather, K. (1962). *Chemistry of Cement, Proceedings of the Fourth International Symposium, Washington 1960*, p. 34. National Bureau of Standards Monograph 43. U.S. Department of Commerce.
108. Hunt, C. M. Doctoral dissertation, Maryland University (1959).
109. Lehmann, H. and Dutz, H. (1962). *Chemistry of Cement, Proceedings of the Fourth International Symposium, Washington 1960*, p. 513. National Bureau of Standards Monograph 43. U.S. Department of Commerce.
110. Lazarev, A. N. (1960). *Optika i Spektroskopiya* **9** (2), 195.
111. Midgley, H. G. (1962). *Chemistry of Cement, Proceedings of the Fourth International Symposium, Washington 1960*, p. 479. National Bureau of Standards Monograph 43. U.S. Department of Commerce.
112. Roy, D. M. (1958). *J. Amer. ceram. Soc.* **41**, 293.
113. Midgley, H. G. (1962). D.S.I.R. Building Research Station Note D811.
114. Fahn, R. (1956). *TonindustrZtg u. keram. Rdsch.* **80**, 171.
115. Hall, D. M. (1957). *Brit. J. appl. Phys.* **8**, 295.
116. Wright, P. W. Private communication.
117. Satou, S. and Tagai, H. (1961). *J. Jap. ceram. Soc.* (*Ass.*) **69**, 102.
118. Rosiwal, A. (1898). *Verh. geol. ReichsAnst.* (*StAnst.*) *Wien* 143.
119. Chayes, F. (1949). *Amer. Min.* **34**, 1.
120. Chayes, F. (1956). "Petrographic Modal Analysis". New York.

121. Lyon, R. J. P., Tuddenham, W. M. and Thompson, C. S. (1959). *Econ. Geol.* **54**, 1047.
122. Alexander, L. and Klug, H. P. (1948). *Analyt. Chem.* **20**, 886.
123. Copeland, L. E. and Bragg, R. H. (1958). *Analyt. Chem.* **30**, 196.
124. Smolczyk, H. G. (1961). *Zement-Kalk-Gips* **12**, 558.
125. Kheĭker, D. M., Konstantinov, I. E. and Alekseev, V. A. (1959). *Kristallografiya* **4**, No. 1, 54.

CHAPTER 4

The Crystal Structures of the Anhydrous Compounds

J. W. Jeffery

Birkbeck College Crystallography Laboratory, Torrington Square, London, England

CONTENTS

I. Introduction

A. General

To a first approximation most inorganic substances are composed of ions, although, especially within complex ions, there may well be a considerable degree of covalent bonding giving rise to directed bonds. The principles governing the structures of ionic crystals were established by Goldschmidt [*1*], Bragg [*2*], Pauling [*3*] and others around 1930 on the basis of the crystal structures that had been determined by X-ray diffraction by that time. The most important of these principles are the radius ratio rule of Goldschmidt and the electrostatic valency rule of Pauling, discussions of which will be found in standard textbooks [*4*].

For 20 years these principles enabled a great range of mineral structures to be elucidated [*5*], but many minerals, typified by wollastonite (β-CS), could not be fitted into the general scheme. The key to their structures has been shown in recent years to be the preponderant role played by large cations (e.g. Ca^{2+}) in such substances. This is of special relevance to cement chemistry. Belov [6] has summarized the new results on the crystal chemistry of the silicates. Of particular importance is the fact that chains of SiO_4^{4-} tetrahedra have to accommodate themselves to the large cations by kinking to give repeat lengths of three or more tetrahedra, either as single or double chains, or as more highly condensed structures.

The main purpose of this chapter is to describe the crystal structures of the anhydrous compounds most important in cement chemistry; hydrated compounds are discussed in Chapters 5 and 6. Some general matters which are especially important in this field will, however, be considered first.

B. Polymorphism

1. *Characteristics*

Polymorphs are defined as substances having the same empirical composition but different crystal structures; examples are quartz and cristobalite (SiO_2), or β-C_2S and γ-C_2S. It is important to recognize that the degree of difference between polymorphs of a given substance can vary widely. At the one extreme there are such cases as α-CS (pseudowollastonite) and β-CS (wollastonite), in which the structures are very different; α-CS contains $Si_3O_9^{6-}$ rings, while β-CS contains infinite $(SiO_3^{2-})_\infty$ chains. At the other extreme are such cases as the monoclinic and triclinic forms of β-CS, which differ only in the way in which successive layers of the structure are stacked. Extreme cases of minor stacking variation of this kind have been called polytypes [*7*]. Intergrowths of different stacking variations can sometimes occur on an atomic scale and produce O–D (order–disorder) structures [*8*].

Whether or not there is any point in distinguishing between polymorphs depends partly on the purpose in hand. Where the structures differ markedly, as with α-CS and β-CS, it will almost always be necessary to make the distinction, which is in such cases comparable in importance to that between isomers in organic chemistry. Stability fields, chemical behaviour, unit cells, optical and X-ray data, infra-red absorption and almost all other properties are likely to differ markedly. At the other extreme, as with monoclinic and triclinic wollastonite, there seems little point for most chemical purposes in making a distinction, as only minor differences or none at all are likely to be detectable by the methods mentioned above. For some purposes, however, such as studies on the mechanism of crystal growth or recognition by selected area electron diffraction, knowledge of these minor differences may be very important.

2. Origin

Polymorphism occurs when two or more arrangements of atoms or ions are possible that have sufficiently similar free energies. This aspect of the matter is discussed in Chapter 18. In theory two polymorphs of a structure cannot both be equally stable under given conditions of pressure and temperature. If one assumes that in the process of crystallization the nuclei form mainly under the influence of electrostatic forces, possibilities occur that accidents of birth may lead to differing arrangements, and once started the arrangement tends to be carried on as further ions settle into place during crystal growth. There are exceptional cases in which this tendency is very weak and the two forms may then alternate on an atomic scale giving a crystal with an "order–disorder" structure.

If one crystal form is much more stable than another, then, even although both nucleate, the more stable form will grow at the expense of the less. If the energy difference is small, however, the rate of growth of both may be much faster than the transfer of material from the less to the more stable. In such a case both forms may coexist, even in geological formations; this is known to happen with monoclinic and triclinic wollastonite [*9*].

3. Inherited Structures

Sometimes a polymorph is formed which is not the stable one under the prevailing conditions because its structure is partly inherited from that of a starting material. So-called topotactic reactions of solids come into this category; an example is the formation of α-CS by dehydration of gyrolite at temperatures where β-CS is stable (Chapter 5). A similar effect probably sometimes operates, because of the existence of short-range order, even in cases of crystallization from liquids or devitrification of glass. This may be important in the cooling of cement clinker (Chapter 3).

4. Polymorphic Transformations

Polymorphic transformations may be brought about by changes in the thermal motion of the atoms as the temperature rises or falls. Such transformations can be broadly classified into the two groups of "displacive" and "reconstructive" transformations (but for a full discussion see Buerger [*10*]).

(*a*) *Displacive transformations.* As the name implies, these transformations occur through small displacements of the atoms from their positions in the original structure to produce the new form, without any alteration in the general arrangement of the chemical bonding. Often the high-temperature form is of high symmetry and on cooling can achieve a more stable structure by a slight rearrangement which destroys one or more of the symmetry elements. This can sometimes be accomplished in several ways corresponding to the equivalent symmetry directions of the high-temperature form. As cooling proceeds, there comes a point at which a nucleus can grow and the crystal changes over to lower symmetry. Other nuclei, growing in the other possible directions, produce a composite of low-temperature crystals arranged according to the symmetry of the high-temperature form. $\alpha \leftrightharpoons \beta$ Quartz, and in the case of cement chemistry the transformations $\alpha \leftrightharpoons \alpha' \leftrightharpoons \beta$-$C_2S$, (although not yet structurally fully worked out) are displacive transformations. Such transformations usually occur rapidly and at a well defined temperature; it is difficult or impossible to freeze-in the high-temperature form.

(*b*) *Reconstructive transformations.* In these transformations the two forms have quite a different chemical bonding and general arrangement, as in the case of the high-temperature pseudowollastonite, with rings of three SiO_4 tetrahedra, compared with the infinite $(SiO_3)_\infty$ chains of the low-temperature forms. Usually the transformation is sluggish and the high-temperature form can be quenched to a metastable state which can exist almost indefinitely.

C. Isomorphism and Solid Solution

1. *Definitions*

The concept of isomorphism was originally defined in terms of the external form of crystals, but it is more convenient today to regard two or more crystalline substances as isomorphous if they have closely similar crystal structures. Two extreme types of isomorphism can be recognized. On the one hand, substances can have similar structures but no possibility exists of the formation of mixed crystals, because the differences in the type of bonding are too great; an example is provided by KCl and PbS. Such pairs of substances may be said to be isostructural. At the other extreme, complete miscibility exists as, for example, with gehlenite $Ca_2Al_2SiO_7$ and åkermanite $Ca_2MgSi_2O_7$, where single crystals having all possible intermediate compositions can be prepared. Such mixed crystals can be said to be due either to (*i*) isomorphous replacement, or (*ii*) solid solution. This and the isostructural type of isomorphism merge into one another, e.g. in the plagioclase felspars.

2. *Isomorphous Replacement*

When the matter is approached from the viewpoint of crystal structure it is usually more convenient to consider the matter in terms of isomorphous replacement rather than of solid solution. There are three main kinds of isomorphous replacement.

(a) Simple substitution, either of one ion by another of the same valency, as in

$$Mg_2SiO_4 \rightarrow Fe_x^{2+}Mg_{2-x}SiO_4$$

or of two or more ions by others of the same total valency, as in the melilite series

$$Ca_2Al(AlSiO_7) \rightarrow Ca_2Mg_xAl_{1-x}(Al_{1-x}Si_{1+x}O_7)$$

In the latter case the pair of substituted ions would tend to substitute in neighbouring positions in order to achieve a local balance of charges.

(b) Substitution with addition of other ions into otherwise vacant sites. One possible substitution in C_3S is an example of this:

$$Ca_9Si_3O_{15} \rightarrow Ca_9(Si_{3-x}Al_x)Al_{x/3}O_{15}$$

(c) Subtractive, e.g.

$$FeO \rightarrow Fe^{2+}_{(1-x)}Fe^{3+}_{2/3x}O$$

All three cases can be complicated by a change in the site occupied, as, e.g.

$$Mg_4O_4 \rightarrow Mg_{(4-x)}Al_{2/3x}O_4$$

in which the Al^{3+} ions start to use tetrahedral sites instead of the octahedral sites left vacant by Mg^{2+}.

Finally, although cation substitution is more common, anionic substitution does occur, e.g. F^- for $(OH)^-$ in apatite or $(OH)_4{}^{4-}$ for $SiO_4{}^{4-}$ in garnet.

3. *Solid Solution*

When the matter is approached from a phase equilibrium viewpoint it is more usual to discuss it in terms of solid solution. The above examples would then be solid solutions between: (a) Mg_2SiO_4–Fe_2SiO_4 and C_2AS–C_2MS_2; (b) C_3S–"$C_{4.5}A$"; (c) FeO–Fe_2O_3; and the substitution with change of site, MgO–MA. This approach is formally identical to the previous one but when it is used two points must be borne in mind: (*i*) the series need not cover the complete range between end-members;

(*ii*) there is no necessity for both end-members to be real compounds, e.g. $C_{4\cdot5}A$ above. In the latter case the complete range will obviously not be covered. Much confusion has been caused by the failure to recognize this fact, e.g. C_3S-C_3A solid solution was postulated largely for this reason and a great deal of searching was done for an end-member of the ferrite series other than the hypothetical C_2A. It would seem preferable always to think in terms of the possibilities of isomorphous replacement before postulating a solid solution series, or in other words, to ask whether the series is consistent with the known facts of crystal chemistry. Where a hypothetical end-member is postulated, it should be put in inverted commas, e.g. C_3S–"$C_{4\cdot5}A$".

4. Factors Affecting (a) Extent of and (b) Continuity in a Solid Solution Series

(a) The simplest case of solid solution is that of two substances having the formulae XM and YM, where M is an atom or group of atoms which is the same in the two cases and X and Y are different atoms or groups of atoms. XM and YM have the same crystal structure, which implies that X and Y are not too different in size (or arrangement, if complex). Such a pair of substances, mixed in any proportions, will usually crystallize under suitable conditions to give crystals in which X and Y occupy their common structural position at random throughout the crystal. The properties of the solid solution will normally be a linear combination of the properties of the two end-members, and Végard's Law is the expression of this in the case of the cell parameters. Figure 1(a) shows this in graphical form. The parameters are averages over the crystal as a whole and the effect of the cell-to-cell variation on X-ray diffraction is similar to that of an increase in thermal motion.

(b) If XM and YM have different structures or if one end-member does not exist as a separate phase, solid solution may still be possible, but over a limited range, as in Fig. 1(b).

(c) If the structure of at least one of the end-members has the possibility of acquiring higher symmetry, e.g. by the rotation of tetrahedra into positions which produce a new plane of symmetry in the structure, an increase in solid solution of the other component may give rise to a gradual change in the structure as well as producing an effect due to the different sizes of X and Y. The change in cell dimensions will then be due to the combined effects of change in size and arrangement, until the position of higher symmetry is reached, after which only the size effect will be operative. This is shown in Fig. 1(c), and such an alteration of slope of the parameter/composition curve is known in the case of the ferrite solid solution series in the system C–A–F. A change in the site being occupied might produce a similar discontinuity. Such changes in site or arrangement may well be two of the factors accounting for deviations from Végard's Law.

(d) If X differs appreciably from Y in size or arrangement, there will be a strong tendency to ordering in the solid solution structure, especially when the molar ratio X/Y is a simple rational fraction. If ordering occurs a "superstructure" is formed. The X-ray pattern will be similar to that of the random arrangement, with extra, weak reflections, corresponding to

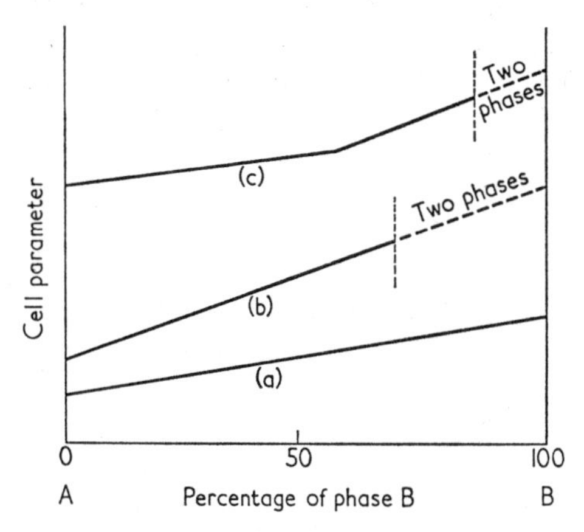

FIG. 1. Various types of parameter–composition curves for solid solutions of B in A which obey Végard's Law.

the true cell. With a 1:1 ratio the simplest possibility gives a unit cell doubled in volume, with X and Y in alternate sites. Similar ordering with a triple unit cell can occur for a 2:1 ratio. The superstructures may themselves be capable of limited solid solution with either of the end-members or each other. On the other hand, if there were two asymmetric units in the cell, ordering with a 1:1 ratio might produce no more than the loss of a symmetry element without alteration of cell size. Thus, ordering is dependent for any striking diffraction effects on accidents of symmetry and for this reason, if no other, should not necessarily be taken to indicate the formation of a separate compound. However, in some cases of ordering, e.g. monticellite ($MgCaSiO_4$) or dolomite ($MgCa(CO_3)_2$), there is a miscibility gap between the intermediate ordered arrangement and both end-members over a wide temperature range, or even at all temperatures. In such cases the ordered substance is usually considered to be a distinct compound, but where there is no such gap or a gap which exists only over a limited temperature range, its characterization as a compound is much more doubtful. In practice, if a substance can only be crystallized over a small range of compositions, it is considered for most purposes as a separate compound.

II. Crystal Structures

A. Silicates and Related Structures

1. *Pseudowollastonite* (α-*CS*)

The structure of this high-temperature form of CS has not yet been determined directly, and at least three forms have been reported which are probably different stacking modifications of the same fundamental layer structure.

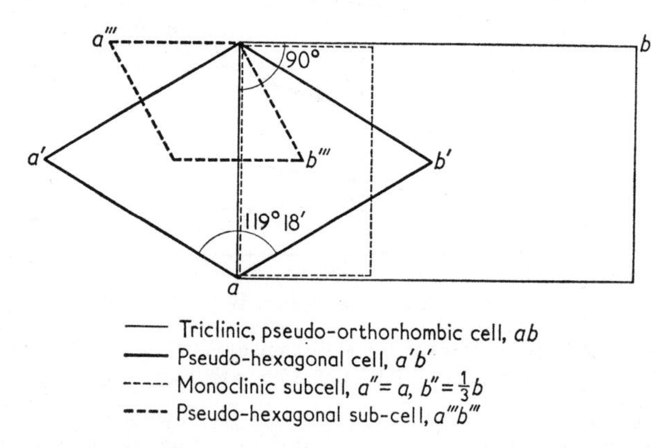

Fig. 2. The relationship between the various unit cells for the two-layer form of pseudowollanstonite; c is the same for all four cells.

The normal form has a triclinic, pseudo-orthorhombic, C face-centred cell with

$$a = 6{\cdot}90,\ b = 11{\cdot}78,\ c = 19{\cdot}65\text{Å};\quad \alpha = \gamma = 90°,\ \beta = 90°\,48'$$

which can be referred to pseudo-hexagonal axes. The pseudo-orthorhombic cell has a C face-centred monoclinic sub-cell with $b'' = \frac{1}{3}b$, which can also be referred to pseudo-hexagonal axes. The strong reflections, which show almost complete hexagonal symmetry, derive from this latter cell, although the odd layer lines in rotation photographs about c are weak, so that a pseudo-halving in the c-direction also occurs [*11*, *12*]. Figure 2 shows the relationship of the various cells.

A second form occurring in slag has been reported [*12*] which has identical strong layer lines on rotation photographs about c, but two intermediate weaker layer lines instead of one, giving a three-layer cell with a *c*-axis of approximately 30 instead of 20 Å, and typical rhombohedral row lines.

A third form has been found, also in slag, which closely resembles the first form but has a halved 001 spacing and a *c*-axis no longer perpendicular to *b* [*13*]. This is presumably a form in which the unit cell consists of a single layer. Glasser and Dent-Glasser [*14*] also refer to the polytypism of α-CS.

These forms can be explained in terms of a double sandwich layer of $Si_3O_9^{6-}$ rings between hexagonally packed Ca^{2+} ions. The three-layer cell (stacked ABCABC...) would have a *c*-axis of about 30 Å and would correspond to the pseudo-rhombohedral structure of $SrGeO_3$ [*15*], as suggested by Liebau [*16*] from the similarity of the unit cells. The two-layer cell (stacked ABAB...) would give at least pseudo-hexagonal symmetry, while the one-layer cell (stacked AAA...) would be at most monoclinic.

The larger pseudo-hexagonal cell referred to above would correspond to the whole layer, while the smaller pseudo-hexagonal cell can be explained on the basis of the Ca^{2+} positions. However, these structures require direct confirmation before they can be accepted without question.

2. *Wollastonite* (β-*CS*)

There are two closely related forms of low-temperature CS, monoclinic or parawollastonite (wollastonite-2M) and triclinic wollastonite (wollastonite-1T).

The latter form usually occurs in massive natural deposits of β-CS. The unit cells (Table I) are obviously related, and the departure of α from 90° in the triclinic form is probably not significant.

TABLE I

Crystallographic data for β-CS

Parawollastonite– monoclinic β-CS		Wollastonite– triclinic β-CS	
$a = 15{\cdot}42$		$a = 7{\cdot}94$	$\alpha = 90°\,02'$
$b = 7{\cdot}32$	$\beta = 95°\,24'$	$b = 7{\cdot}32$	$\beta = 95°\,22'$
$c = 7{\cdot}07$		$c = 7{\cdot}07$	$\gamma = 103°\,20'$
$P2_1$; $Z = 4$		$P1$; $Z = 2$	

Unit cell dimensions in Å.

Results obtained in four different laboratories [*18*, *19*] agree that infinite kinked chains of SiO_3^{2-} are present in both forms of β-CS. These chains (Fig. 3) repeat at intervals of three tetrahedra and have therefore been called "Dreierketten" [*17*]. The large size of the calcium ion prevents it from fitting into the better known pyroxene or "Zweierkette" structures as the sole metal cation, but it fits well with the 7·3 Å repeat of the Dreierketten.

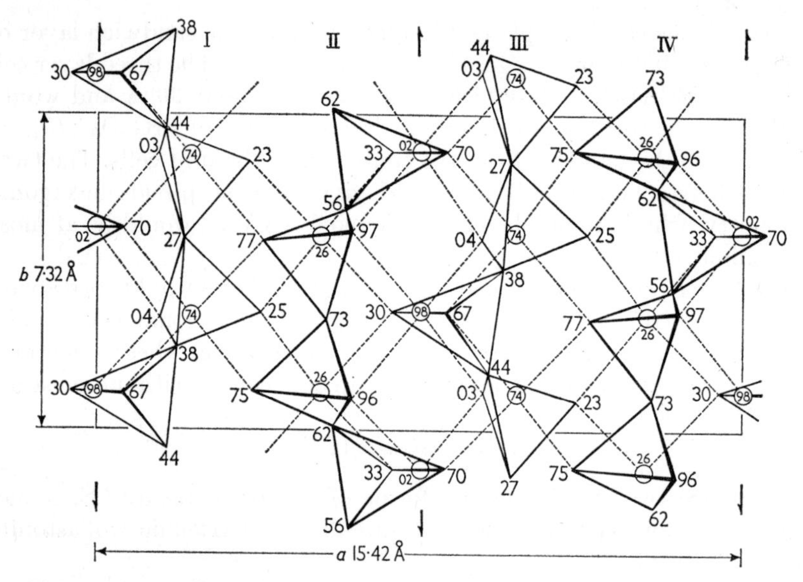

Fig. 3. The structure of parawollastonite projected on a plane perpendicular to the *c*-axis. The heights are given in hundredths of *c*. The SiO_4 groups are represented by tetrahedra. The Si atoms are not shown. The corners of the tetrahedra are the centres of O atoms. Ca atoms are shown as small circles. The origin is in the lower left-hand corner.

A detailed three-dimensional Fourier analysis of parawollastonite by Tolliday [*19*] has given atomic peaks that are clearly resolved, indicating that reliance may be placed on the results. The structure is shown in Fig. 3 projected on a plane normal to the *c*-axis. Four chains of silicon–oxygen tetrahedra cross the unit cell in the *b*-direction, with a repeat at every third tetrahedron along each chain. The chains are linked together by calcium atoms each of which is in an irregular octahedral group of six oxygen atoms. Each calcium atom has four bonds to oxygen atoms approximately in the plane of projection, which are shown by dotted lines in the figure. The other two bonds go to oxygen atoms belonging to the tetrahedra immediately above it and below it. The standard Ca—O distance is 2·38 Å, which is nearly $c/3$, and it will be noted that the difference in the heights of a calcium atom and the oxygens nearly above and below it in the projection is in each case about 0·33 (unity being added to or subtracted from the height of the oxygen atom when necessary).

Jeffery [*9*] in 1953 described the unusual diffraction effects from a crystal of wollastonite, and pointed out that they must be due to some complex type of pseudo-symmetry. The structure reported by Tolliday illustrates the nature of the peculiar relationships in atomic co-ordinates that give rise to these diffraction effects (systematic absences in the *hkl* reflections not demanded by any type of space group symmetry, and streaks parallel to a^* indicating a close intergrowth on the atomic scale of the two forms). The chains, numbered I, II, III, IV, in Fig. 3, are identical in form. They can be derived from each other by the following relations in the positions of any given atom:

I	x	y	z
II	$\frac{1}{2}-x,$	$\frac{3}{4}+y,$	$\bar{z}$
III	$\frac{1}{2}+x,$	$\frac{1}{4}+y,$	z
IV	$\bar{x},$	$\frac{1}{2}+y,$	$\bar{z}$

The structure has twofold screw axes parallel to b at the positions marked in Fig. 3† which turn I into IV and II into III. In order for there to be holohedral monoclinic symmetry, there should also be glide planes perpendicular to b, with translation $a/2$, at heights $b/4$, $3b/4$ which reflect I into III and II into IV. The symmetry would then be $P2_1/a$. Figure 3 shows, however, that this requirement is only very approximately obeyed, the departure of the atoms from the position required by the higher symmetry being quite definite. According to Tolliday [*19*], I is related to III and II to IV (and also the Ca atoms are related in pairs) by translations $\frac{1}{2}$, $\pm\frac{1}{4}$, 0, to an approximation which is very close indeed, as shown by systematic absences in the (*hkl*) spectra when k is even. If, in addition, I turned into III and II into IV by reflection as demanded by the glide plane, each chain would have to possess a transverse mirror plane which is clearly not quite the case. The structure is thus not centrosymmetrical and the chains are crystallographically of two kinds. The above relation of the co-ordinates of I and II to those of III and IV is not required by the symmetry and must be regarded as fortuitous and probably only a close approximation. The finer details of the structure perhaps still require confirmation.

Figure 4 shows the way in which Tolliday explains the relation of the monoclinic form to the triclinic form. If chain IV in Fig. 3 is translated

† Where symmetry elements are shown in the diagrams, the conventions of the International Tables for Crystallography (1952) are used.

upwards by $b/2$, its position relative to chain III becomes the same as the relation of II to I. The structure then repeats every two chains in the new a-direction, and we can outline the triclinic cells as shown in Fig. 4. This structure is also pseudo-symmetrical; the atoms are near to, but not at, positions which would give it centres of symmetry. It is therefore a rare example of a crystal with no symmetry elements.

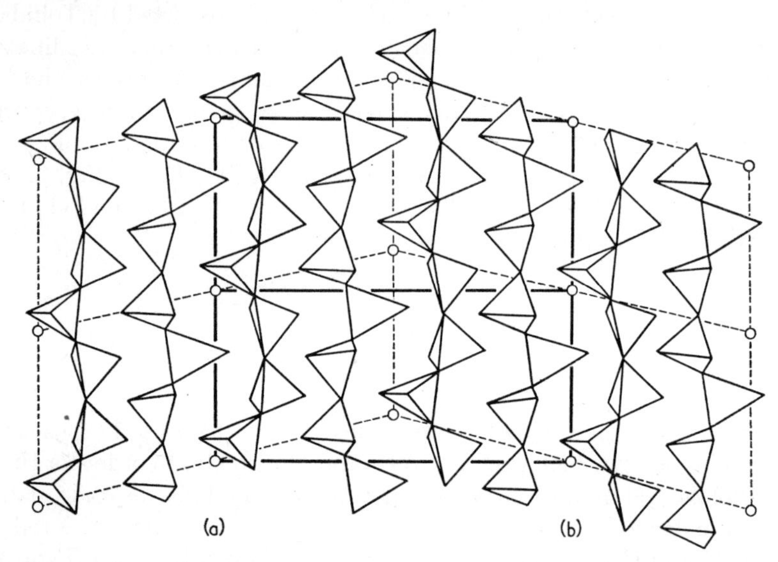

FIG. 4. The outline of the monoclinic (full lines) and triclinic (dashed lines) unit cells for β-$CaSiO_3$, showing the relationship between them and the Dreierketten.

The relationship between the monoclinic and triclinic forms explains their close intergrowth. The triclinic form can be orientated relative to the monoclinic form in the two ways (a) and (b) shown in Fig. 4, which are of course equivalent, since the monoclinic form has a 2_1 axis. It is possible to describe the monoclinic form as produced by alternate layers of the two triclinic orientations twinning every second chain.

3. *Pyroxenes*

Originally wollastonite, CS, was classed with the pyroxenes, but the large size of the Ca^{2+} ion prevents the structure from forming with SiO_3^{2-} chains of the pyroxene type. However, when half the Ca^{2+} ions are replaced by Mg^{2+} or Fe^{2+} the pyroxene structure forms. Diopside ($CaMgSi_2O_6$) and hedenbergite ($CaFeSi_2O_6$) both have a structure containing SiO_3^{2-} chains with a repeat distance of 5·25 Å in the c-direction.

The chains are linked laterally by Mg^{2+} or Fe^{2+} ions octahedrally co-ordinated by 6 O's each linked to only one Si. The Ca^{2+} ions are surrounded by 8 O's, 4 on one side, each bonded to 1 Si and 1 Mg, and 4 on the other side, each bonded to two Si's.

A complete solid solution range exists between the two end-members, and many other substitutions can occur, leading to alterations of chain stacking and symmetry. The limits of the ranges of cell size given below correspond to diopside and hedenbergite respectively:†

$$a = 9{\cdot}73\text{–}9{\cdot}85,\ b = 8{\cdot}91\text{–}9{\cdot}02,\ c = 5{\cdot}25\text{–}5{\cdot}26\ \text{Å};$$
$$\beta = 105°50'\text{–}104°20'.\quad C2/c;\quad Z = 4$$

4. *Rankinite* (C_3S_2)

This form of C_3S_2 is monoclinic. The unit cell and space group have been determined by Moody [*20*].

$$a = 10{\cdot}55,\ b = 8{\cdot}88,\ c = 7{\cdot}85\ \text{Å};\quad \beta = 120{\cdot}1°$$
$$P2_1/a;\quad Z = 4$$

The structure has not yet been determined.

5. *Kilchoanite* (C_3S_2)

The orthorhombic unit cell and probable space group have been determined from single-crystal X-ray data by Agrell and Gay [*21*].

$$a = 11{\cdot}42,\ b = 5{\cdot}09,\ c = 21{\cdot}95\ \text{Å}$$
$$Imam \text{ or } Ima2;\quad Z = 8$$

On conversion to rankinite by heating at 1000° C, a single crystal becomes a polycrystalline aggregate, thus confirming the evidence of the unit-cell sizes that no simple relationship between the structures exists. The structure of kilchoanite is also unknown.

6. *Dicalcium Silicate* (C_2S)

Of the four forms of C_2S, only two, β and γ, have known structures and even the unit-cell sizes are uncertain for the other two (α and α'). However, evidence from high-temperature powder photographs and from composite slag crystals thought to contain α- and α'-C_2S stabilized by impurities suggests that the high-temperature form, α-C_2S, is trigonal

† In this chapter data quoted without a reference are taken from Deer, W. A., Howie, R. A. and Zussman, J. "Rock Forming Minerals", Longmans, London (1963), or from the A.S.T.M. X-ray Powder Data File.

and the intermediate form, α'-, orthorhombic with a structure approximating to that of β-K_2SO_4. Nurse [*22*] has summarized the somewhat conflicting evidence for the suggested unit cells and structures of the high-temperature forms.

(*a*) *The structure of* β-C_2S. The structure of B_2O_3-stabilized (0·5%) β-C_2S was determined in 1952 by Midgley [*23*]. It is based on a monoclinic unit cell with

$$a = 5{\cdot}48,\ b = 6{\cdot}76,\ c = 9{\cdot}28\ \text{Å};\quad \beta = 94°\ 33'$$
$$P2_1/n;\quad Z = 4$$

Figure 5 shows the projection of the structure along the *b*-axis, together with the symmetry elements and heights of the atoms.

The structure is built up of isolated SiO_4 tetrahedra and Ca atoms. Four of the eight Ca atoms (Ca_1) are positioned alternately above and below SiO_4 tetrahedra in the *b*-direction so that the structure may be regarded as consisting of columns of alternating Ca atoms and tetrahedra, the columns being linked by the remaining four Ca atoms (Ca_2) which are accommodated in the holes left between the tetrahedra. The pseudo-trigonal arrangement of tetrahedra about these latter calcium atoms is illustrated in Fig. 5.

Ca_1 has a tetrahedron immediately above and below in the *b*-direction, but because of the tilt of the tetrahedron one of the base oxygens is considerably further away than the other two. There are three other tetrahedra surrounding Ca_1 at about the same level contributing three more oxygens to the inner irregular sixfold co-ordination with Ca—O distances from 2·30 to 2·75 Å. Another six oxygens, not shielded by the first six, contribute to the co-ordination at distances from 2·98 to 3·56 Å. Ca_2 has an irregular eightfold co-ordination of oxygens contributed by the six surrounding tetrahedra, with Ca—O distances from 2·36 to 2·80 Å. With reasonable assumptions about the valency contributions Pauling's rule is approximately obeyed.

The structure was based on 200 intensities for the three axial zones and gave an overall discrepancy factor of 19%. This means that the basic structure is almost certainly correct, especially as small and plausible shifts produce an orthorhombic structure of the β-K_2SO_4 type, as postulated for α'-C_2S, and rather larger but still reasonable shifts give a trigonal structure based on the unit cell shown in Fig. 5. The polysynthetic twinning is also readily understandable on the basis of the twin planes shown in Fig. 5. These would become structural symmetry planes in the α'-structure. However, if an untwinned single crystal could be produced with a known amount of stabilizer, it would be useful to confirm the structure and refine the parameters by a full three-dimensional analysis. This might also throw light on the role of such stabilizers.

(*b*) γ-C_2S. From the similarity of the powder photographs, O'Daniel and Tscheischwili [*24*] deduced that γ-C_2S and its model structure, Na_2BeF_4, were isomorphous. Since they had determined that Na_2BeF_4 had an olivine structure, it followed that the same must be true for γ-C_2S. No

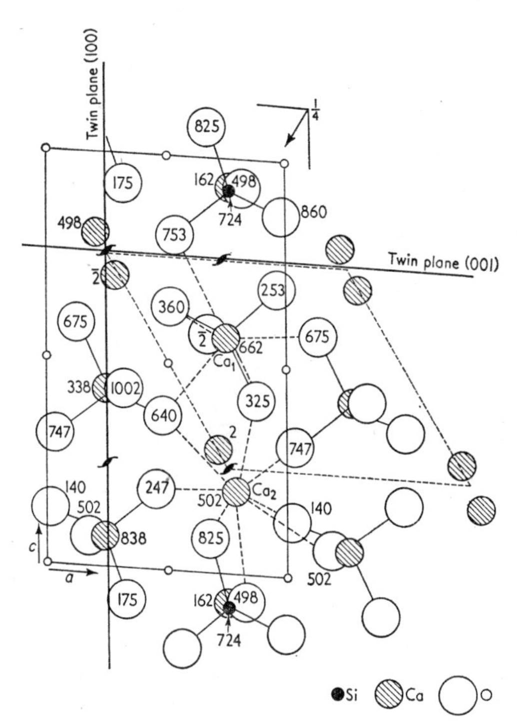

FIG. 5. The projection of the β-C_2S structure on a plane perpendicular to the *b*-axis. "Heights" in thousandths of *b* (measured downwards in the $+b$ direction). Si—O bonds: full lines; Ca—O bonds: dashed lines. Only two sets of Ca—O bonds are shown for the two distinct types of co-ordination of Ca by O.

direct confirmation of this was forthcoming, because γ-C_2S can normally only be obtained as a fine powder from the "dusting" of β-C_2S, until 1952, when Heller [*25*] reported that a somewhat imperfect single crystal of γ-C_2S was obtained on careful heating of a crystal of afwillite. Indexing of rotation photographs was consistent with the cell size and space group derived from the model structure. However, it was not until 1962 that a full structure determination from single-crystal data was accomplished by Smith, Majumdar and Ordway [26] and the olivine structure finally confirmed with the cell size and space group

$$a = 5{\cdot}06,\ b = 11{\cdot}28,\ c = 6{\cdot}78\ \text{Å}$$
$$Pbnm;\quad Z = 4$$

Figure 6 shows the projection of the structure along the a-axis. The large Ca^{2+} ions have expanded the olivine unit cell by 6% along a, 10% along b and 14% along c (qualitatively in agreement with the

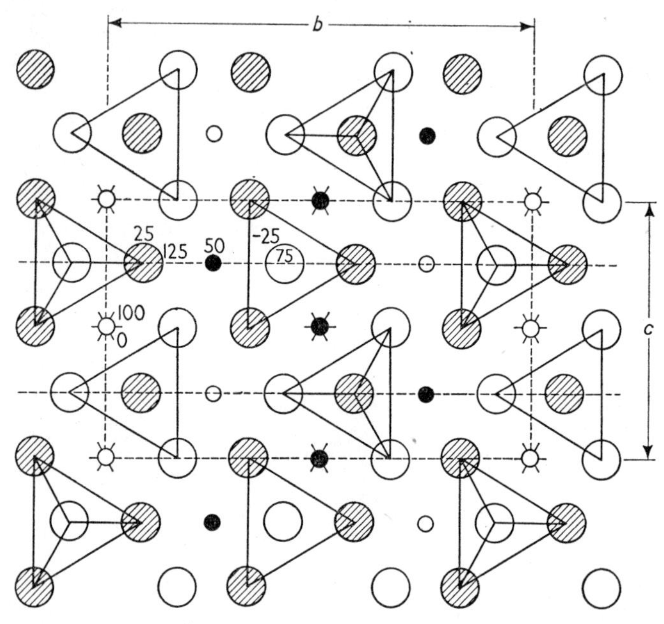

FIG. 6. The projection of the γ-C_2S structure on a plane perpendicular to the a-axis. Heights in hundredths of a. The large circles represent O atoms, shaded at $x = 25$, open at $x = 75$. They lie approximately in hexagonal close packing. The small circles, whether starred or not, represent Mg atoms, filled circles at $x = 50$, open at $x = 0$, 100. The Mg atoms are octahedrally co-ordinated by O atoms. Si atoms at the centres of the tetrahedra are not shown. The stars distinguishing certain atoms refer to monticellite. The structure is idealized.

separations of the cations in the three directions) but not altered the general regular arrangement. The atomic co-ordinates are given in Table II.

(c) *Monticellite*, CMS, also has the olivine structure, the Mg atoms at symmetry centres (starred in Fig. 6) remaining, while those on the mirror planes are replaced by Ca.

(d) *Merwinite*, C_3MS_2, is probably a distinct compound and not a solid solution of α'-C_2S, as suggested by Bredig. It is monoclinic and has cell dimensions [22]

$$a = 10{\cdot}77, b = 9{\cdot}20, c = 13{\cdot}26 \text{ Å}; \quad \beta = 91°$$

TABLE II

Atomic co-ordinates for γ-C_2S (origin at top right in Fig. 6)

Atom	Position	x	y	z
Ca (1)	4 (a)	0·000	0·000	0·000
Ca (2)	4 (c)	−0·012 ± 0·001	0·280 ± 0·001	0·250
Si	4 (c)	0·427 ± 0·001	0·099 ± 0·001	0·250
O (1)	4 (c)	−0·262 ± 0·003	0·087 ± 0·001	0·250
O (2)	4 (c)	0·302 ± 0·003	−0·042 ± 0·001	0·250
O (3)	8 (d)	0·292 ± 0·002	0·164 ± 0·001	0·060 ± 0·002

7. *Tricalcium Silicate (C_3S) and Alite*

(*a*) *The pseudo-structure.* The structures of pure C_3S and its modifications produced by small additions of Al_2O_3, Fe_2O_3, MgO and CaO are all based on a rhombohedral pseudo-structure with a hexagonal unit cell [27]

$$a = 7{\cdot}0, c = 25{\cdot}0 \text{ Å}; \quad R3m; \quad Z = 9$$

The pseudo-structure is of the type in which there are no direct links between SiO_4 tetrahedra. Three isolated tetrahedra lie on trigonal axes, held together by two triplets of calcium ions as shown in Fig. 7(b).

These are followed by three oxygen ions not attached to Si and also held together and to the topmost oxygen of the tetrahedra by triplets of calcium atoms which form almost exactly regular octahedra round the oxygens (Fig. 7(a)). The column then repeats with three more tetrahedra etc. Exactly similar columns lie on all the trigonal axes, but at different heights, so that the columns join sideways as shown in Fig. 7(a) and (b), to produce irregular octahedra of oxygen atoms surrounding the calciums and leaving holes large enough to accommodate further calcium atoms. Figure 8 is a vertical section through the long diagonal of the unit cell, showing the bottom third of the cell, and Fig. 9 a projection along the c-axis of the bottom third. The middle layer can be obtained by shifting the bottom layer up and to the left by the distance between two trigonal axes on the long diagonal and the top layer by repeating the process.

(*b*) *The true structures.* Pure C_3S is triclinic at room temperature [27].

DTA peaks at 920 and 980° C [28] and direct X-ray diffraction evidence of transitions at these temperatures [29] are consistent with the interpretation given by Jeffery [30] that the triclinic room temperature structure first changes to monoclinic at 920° C and then to trigonal symmetry at

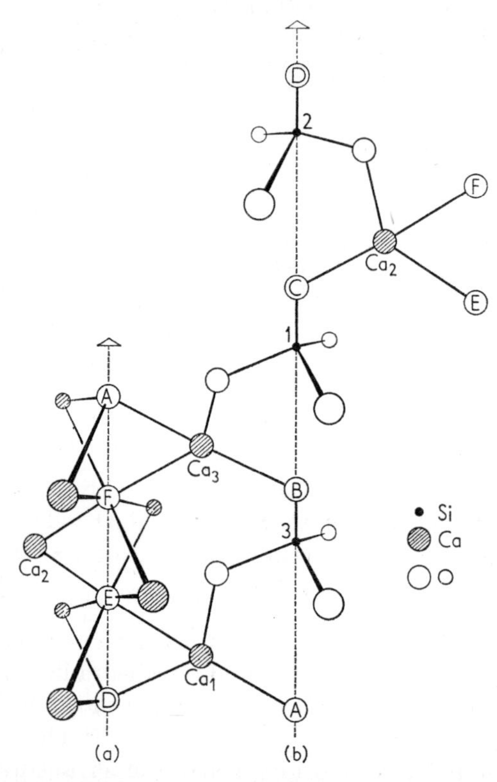

FIG. 7. Tricalcium silicate; two adjacent columns of tetrahedra and single O atoms, surrounded by Ca atoms. A complete column is obtained if (b) is placed on top of (a) (or (a) on (b)), and the labelling of the various atoms corresponds to that of Fig. 8. The Ca co-ordination is not complete, as two more O's from tetrahedra in adjacent cells come on each side of the planar fourfold co-ordination shown. Only one Ca, of the triplet required by the trigonal axis, is shown in (b), to avoid confusion.

980° C. The triclinic unit cell has not yet been determined. The cell found by Yamaguchi and Miyabe [29] does not account for a number of weak lines in the pattern and gives a calculated density which is too low [31]. It follows that the indices given by these authors are also incorrect. Extremely careful and accurate X-ray work by Yannaquis *et al.* [31] has shown that two further transitions exist in the triclinic range at 615 and

750° C, and that the transition at 980° C, although bringing the structure much nearer to trigonal symmetry, does not complete the change. They found a further transition at 1050° C which finally completed the change to the trigonal form.

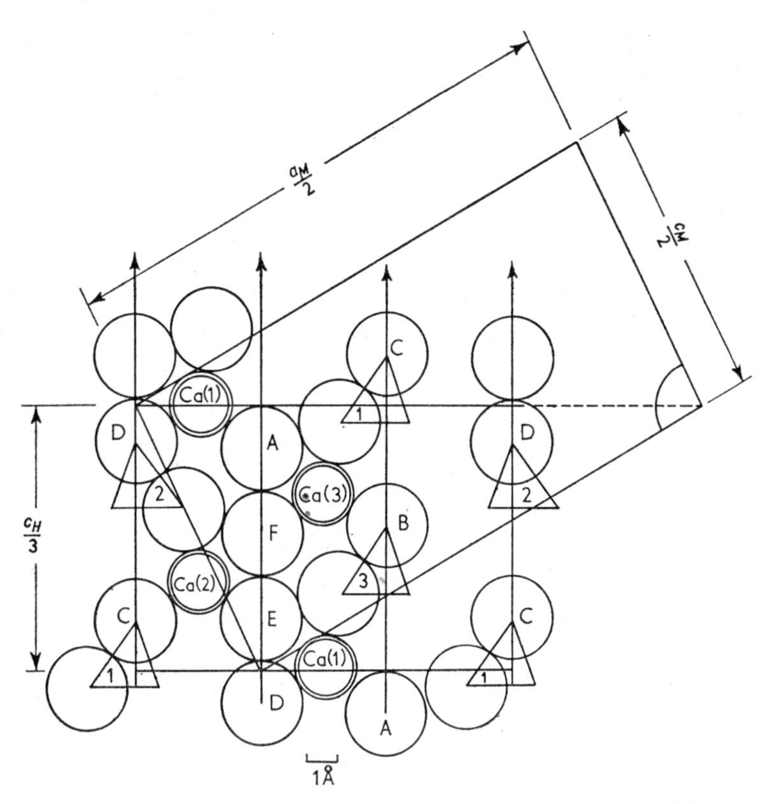

Fig. 8. Tricalcium silicate; a vertical section through the long diagonal of the hexagonal cell. The relationship of the monoclinic axes to the hexagonal cell is also shown. The various atoms are labelled, both to show the equivalences arising from the rhombohedral symmetry and to correspond with Fig. 7.

The addition of small amounts of MgO and Al_2O_3 produces a monoclinic unit cell [27] (similar to that of the monoclinic high-temperature form of pure C_3S) with

$$a = 33{\cdot}08,\ b = 7{\cdot}07,\ c = 18{\cdot}56\ \text{Å};\quad \beta = 94°\ 10'$$
$$Cm;\ \text{cell contents } 2[54CaO.16SiO_2.Al_2O_3.MgO]$$

One of the pseudo mirror planes of the rhombohedral structure becomes a true mirror plane as a result of increasing the asymmetric unit by a factor of six. The relation to the unit cell of the pseudo-structure is shown in Fig. 8. The asymmetric unit contains 18 mol. C_3S, modified by the presence of Al_2O_3 and MgO. Structurally the minimum modification of such a unit is the replacement of 1 Ca by 1 Mg. The next simplest modification would be the replacement of 2 Si by 2 Al and the addition of 1 Mg in a nearby hole to balance the charges. It would be possible for

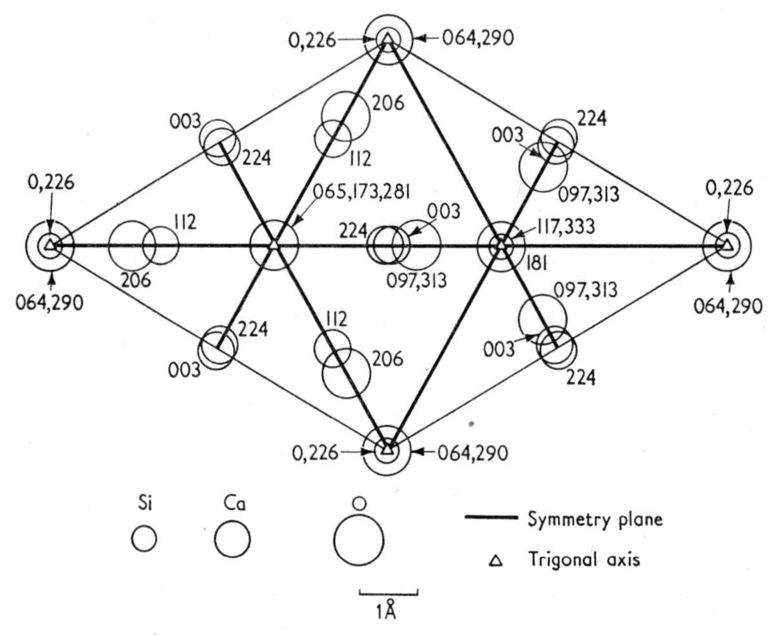

FIG. 9. Tricalcium silicate; the bottom third of the hexagonal cell projected on to a plane perpendicular to the *c*-axis. Heights in thousandths of *c*.

3 Si to be replaced by 3 Al and a further Al to be added to balance the charges. Any or all of these (and possibly other) substitutions and additions might occur simultaneously to different degrees, and this is presumably the explanation of the finding [*32*] that alites in cement vary in structure from triclinic to trigonal. An additional factor is that Fe^{3+} may replace Al^{3+} to some extent and Fe^{2+} replace Mg^{2+}.

Midgley and Fletcher [*32*] reached the following conclusions. First, substitution does not produce either vacant oxygen sites or interstitial oxygen atoms. Second, substitution arising from the addition of Al_2O_3 alone to the mix has little effect on the structural arrangement, but addition of MgO alone causes replacement of Ca by Mg and produces a structure whose powder pattern approximates to that of monoclinic alite in cement. Third, Mg-substituted alites can form solid solutions with a hypothetical end-member, "$C_{4.5}A$". Suitable additions to the mix will give, in 6 of Jeffery's assymetric units, 10 Ca replaced by Mg and 3 Si replaced by Al with a fourth, interstitial Al in a nearby hole. This gives a formula $C_{314}M_{10}S_{105}A_2$ or $(Ca, Mg)_{324}$ $[(Si, Al)_{108}O_{540}]Al$. Such a composition was found to give the best match to the X-ray diagram of monoclinic alite in clinker. However, a re-investigation of the original crystals used by Jeffery gave no reason to doubt the composition $C_{54}S_{16}AM$ or $Ca_{324}[(Si, Al)_{108}O_{540}]Mg_6$ originally arrived at, and it may well be that the amount and kind of replacement depend on the conditions of formation.

Woermann *et al.* [*67*] concluded that Mg replaces Ca to the extent of 2·0% MgO at 1550° and that this substitution affects the symmetry. They consider that Al substitution occurs as suggested by Midgley and Fletcher [*33*] up to 0·45% Al_2O_3, the intersitial Al occupying octahedral sites, but that at 0·45–1·0% Al_2O_3 replacement of Ca by Al also occurs.

Since, in all the recent work, structural changes have been followed only by X-ray powder diagrams, it is not possible to be certain of the interpretation of shifts in lines on the pattern. It may be that a sufficient number of unit cells modified by some form of substitution can "organize" the remainder of the C_3S into the monoclinic arrangement, while the effect of increasing the relative number of unmodified units would be to distort the arrangement of the modified unit and produce a change in the average unit-cell size and shape and eventually in its symmetry. This would account for the progressive change which powder photographs show as the amount of Al_2O_3 and MgO added to the alite mix is increased. The minimum quantities required (as well as the maximum possible solid solution) are still the subject of controversy and probably will only be satisfactorily determined when single crystals large enough for X-ray investigation have been produced with various amounts of substitution and addition.

A fully satisfactory X-ray analysis of the structure also awaits the production of single crystals of accurately known composition.

B. Aluminates and Related Compounds

1. CA_6

Powder photographs and cell size [*33, 34*] suggest that this compound is isomorphous with magnetoplumbite, $PbO.6Fe_2O_3$

$$a = 5{\cdot}536,\ c = 21{\cdot}825\ \text{Å};\quad P6_3/mmc;\quad Z = 2$$

The corresponding Sr and Ba aluminates and ferrites also have similar cell sizes. The structure is very similar to that of β-alumina ($Na_2O.11Al_2O_3$), 50 of the 58 atoms in the unit cell being arranged in the same way. The details of the β-alumina structure [*35*] can only be taken as "most probable" (an overall R value of 31%), and although the agreement between observed (given as s., m., w., v.w.) and calculated intensities for magnetoplumbite is said to be "perfect" [*36*], the ranges of calculated intensities covered by this description are: s 4·6–18; m 2·8–9; w 0·5–3·4; v.w. 0·2–1·8. The details of this structure therefore also require confirmation, although the cell contents are probably satisfactorily determined.

2. CA_2

Boyko and Wisnyi [*37*], in an extremely accurate study of the optical properties, showed that CA_2 is biaxial positive with a very small optic axial angle, in agreement with a number of X-ray findings that it is monoclinic. They also carried out a structure determination on this and the isomorphous strontium compound. The crystal data, in agreement with earlier work, are

$$a = 12{\cdot}89,\ b = 8{\cdot}88,\ c = 5{\cdot}45\ \text{Å};\quad \beta = 107^\circ\ 3'$$
$$C2/c;\quad Z = 4$$

The structure of the strontium analogue was determined from two projections giving R values of 15 and 22%. Using the same co-ordinates for the calcium compound, and a single zone of reflections for comparison, gave an R value of 22%. The structure clearly requires refining, but its general correctness is confirmed by Dougill [*38*]. The aluminium atoms are tetrahedrally co-ordinated by oxygen with one oxygen at the common corner of three tetrahedra. The Ca atoms are very irregularly co-ordinated with five Ca—O bonds of about 2·6 Å and four more of about 3·5 Å.

3. CF_2

The existence of a compound with this formula has been in question for over 20 years. The controversy appeared to have been closed in 1960 by the X-ray identification of a spinel-like phase with a solid solution range in this field [*39*]. The typical stoichiometric formula is $4CaO.FeO.9Fe_2O_3$. However, Chessin and Turkdogan [*40*] have prepared a compound with the formula CF_2, containing only 0·5% FeO, and have indexed a powder pattern on a very large hexagonal unit cell. This cannot be accepted with confidence until confirmed by single-crystal data.

4. CA

This structure was partially worked out by Heller [*41*], and the general character of the structure confirmed and worked out in detail by Dougill [*42*] (2400 *hkl* reflections giving $R = 10{\cdot}1\%$).

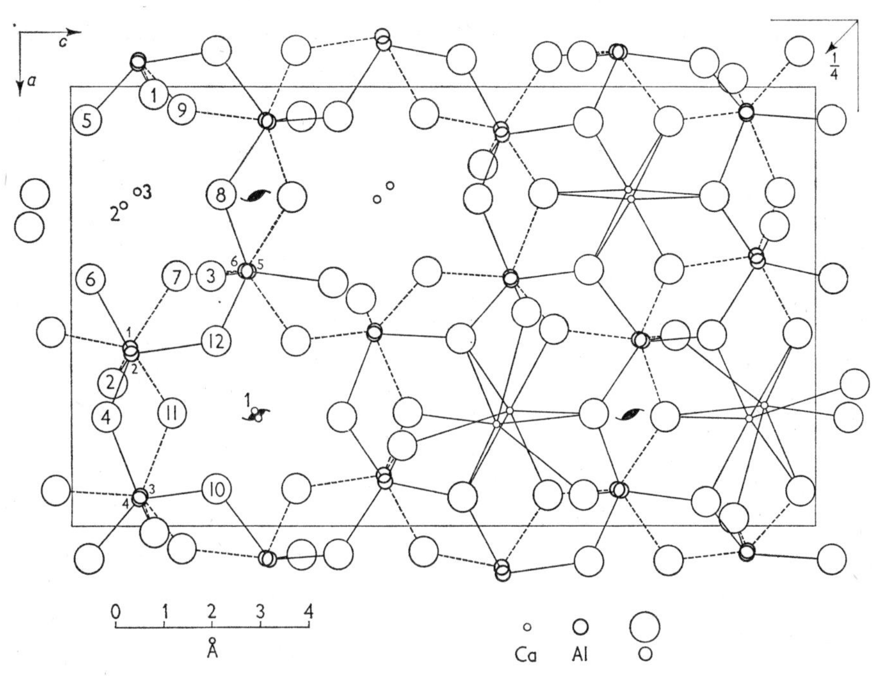

FIG. 10. The pseudo-hexagonal *b*-axis projection of CA, showing, on the left, Al—O linkages and, on the right, Al—O and Ca—O linkages. Bonds in lower layer: dashed line; bonds in upper layer: full line. One 2_1 axis has been omitted for clarity. The $+b$ direction points down.

The monoclinic unit cell has

$$a = 8{\cdot}702,\ b = 8{\cdot}102,\ c = 15{\cdot}211(\pm 0{\cdot}003\ \text{Å});\quad \beta = 90{\cdot}1°;$$
$$P2_1/\text{n};\quad Z = 12$$

The pseudo-hexagonal structure (Fig. 10) is based on that of β-tridymite with all the Si^{4+} replaced by Al^{3+} and Ca^{2+} in cavities of the framework of AlO_4 tetrahedra. The large size of the Ca^{2+} ion causes considerable distortion of the tridymite framework. Two calcium atoms (Ca_2 and Ca_3) are each surrounded by 6 oxygen atoms in an octahedral arrangement, with Ca—O distances ranging from 2·31 to 2·72 Å. The third calcium (Ca_1), near a 2_1 axis, is surrounded by 9 oxygen atoms. Two sets of

3 O's, related by the 2_1 axis, form an elongated octahedron, and are each shared by 2 Ca's to give a chain of –Ca$\xrightarrow{2{\cdot}42Å}$ 3 O's $\xrightarrow{2{\cdot}90Å}$ Ca– in the *b*-direction. The other 3 O's lie in the plane perpendicular to the chain with Ca—O distances of 2·83–3·17 Å. There are 24 Al—O distances having values between 1·68 and 1·82 Å, and the O—Al—O angles lie between 101° and 118°. A peculiar feature of the structure is that Ca_1 lies towards one end of the elongated octahedron, making normal bonds of

TABLE III

Atomic co-ordinates for $CaAl_2O_4$

	x	*y*	*z*		*x*	*y*	*z*
Ca_1	0·7412	0·5222	0·2473	O_1	0·0158	0·5316	0·1104
Ca_2	0·2686	0·4698	0·0706	O_2	0·6763	0·0229	0·0549
Ca_3	0·2392	0·0338	0·0894	O_3	0·4321	0·5261	0·1885
				O_4	0·7552	0·3565	0·0462
Al_1	0·5987	0·8294	0·0777	O_5	0·0766	0·2383	0·0230
Al_2	0·6121	0·2200	0·0801	O_6	0·4360	0·2614	0·0259
Al_3	0·9352	0·7270	0·0916	O_7	0·4330	0·8528	0·1412
Al_4	0·9393	0·3341	0·0919	O_8	0·2468	0·2440	0·2024
Al_5	0·4232	0·3276	0·2378	O_9	0·0542	0·8640	0·1495
Al_6	0·4205	0·7236	0·2332	O_{10}	0·9196	0·2441	0·1956
				O_{11}	0·7462	0·7285	0·1348
				O_{12}	0·5823	0·2259	0·1952

2·4 Å with 3 O's at one end and long bonds of 2·9 Å with the other 3. Since no other oxygens are nearer than 2·83 Å, the Ca is apparently "loose" in its co-ordination octahedron. Table III gives the atomic co-ordinates.

5. *CF*

The structure of this compound has been determined in three independent investigations [*43–45*], all of which agree within the limits of error, although the first was actually carried out on the isomorphous vanadate and the third is more completely refined. However, there is considerable variation in the cell sizes reported.

Authors	*a*	*b*	*c*
Burdese [*46*]	9·16	10·60	3·01Å
Hill, Peiser and Rait [*44*]	9·16 ± 0·03	10·67 ± 0·03	3·012 ± 0·006Å
Decker and Kasper [*45*]	9·23 ± 0·012	10·705 ± 0·014	3·024 ± 0·004Å

Pnam; $Z = 4$

All atoms are at $\frac{1}{4}$ and $\frac{3}{4}$ in z. Fe is almost regularly octahedrally co-ordinated (Fe—O range, 1·98–2·09 Å). The Ca is at the centre of a triangular prism of O's (axis parallel to c) with four Ca—O bonds to the corners of a prism face of 2·37–2·38 Å and two bonds to the opposite corners of 2·51 Å. Three other O's in the same xy-plane as Ca are at distances 2·53, 2·58 and 3·41 Å.

6. The Garnet Group

(a) Grossular, C_3AS_3, cubic.

$$a = 11{\cdot}851 \text{ Å}; \quad Ia3d: \quad Z = 8$$

The structure is composed of independent SiO_4^{4-} tetrahedra, linked to AlO_6 octahedra by corners with eight co-ordinated Ca^{2+} ions in the interstices.

(b) Pyrope, M_3AS_3. $a = 11{\cdot}459$ Å. Ca of grossular replaced by Mg.

(c) Extensive isomorphous replacement takes place in the garnet group, Fe^{2+} and Mn^{2+} replacing Mg^{2+} or Ca^{2+}, and Fe^{3+}, Ti^{3+} and Cr^{3+} replacing Al^{3+}.

7. The Melilite Group

These minerals form a solid solution series between gehlenite, C_2AS, and åkermanite, C_2MS_2, but Fe^{2+} can substitute for Mg^{2+}, Fe^{3+} for Al^{3+} and $Na^{+}+Al^{3+}$ for $Ca^{2+}+Mg^{2+}$, so that natural melilites are very variable.

The tetragonal unit cell has

$$a = 7{\cdot}69\text{–}7{\cdot}84, \; c = 5{\cdot}08\text{–}5{\cdot}01 \text{ Å}$$
$$P\bar{4}2_1m; \quad Z = 2$$

The end-values of the ranges for axial lengths are those of gehlenite and åkermanite respectively.

The structure [*52, 53*] is shown in Fig. 11. In the case of åkermanite, Mg atoms occupy the tetrahedra, T_1, and Si atoms occupy T_2 and T_3. In gehlenite the ordering of Si and 2 Al in the tetrahedra is not known and the way in which substitution proceeds also awaits sufficiently accurate structure determinations of known intermediate compositions. It is possible that the ordering changes with time.

8. C_5A_3

Orthorhombic crystals of this composition, apparently corresponding to the "unstable 5:3 phase" originally reported by Shepherd, Rankin and Wright [*47*], have been investigated by Aruja [*48*].

$$a = 10{\cdot}975, \; b = 11{\cdot}250, \; c = 10{\cdot}284 \text{ Å (probable error } \pm 0{\cdot}005 \text{ Å)}$$
$$C222_1; \quad Z = 4$$

The structure was not determined, but similarities to gehlenite (C_2AS) were pointed out which might provide a starting point for a structure analysis.

9. *Compounds Probably Related to C_5A_3*

Welch [*49*], in investigating the formation of "magnesian pleochroite", which occurs in high-alumina cement, failed to confirm the formula, C_6A_4MS, proposed by Parker [*50*], and suggested that the substance

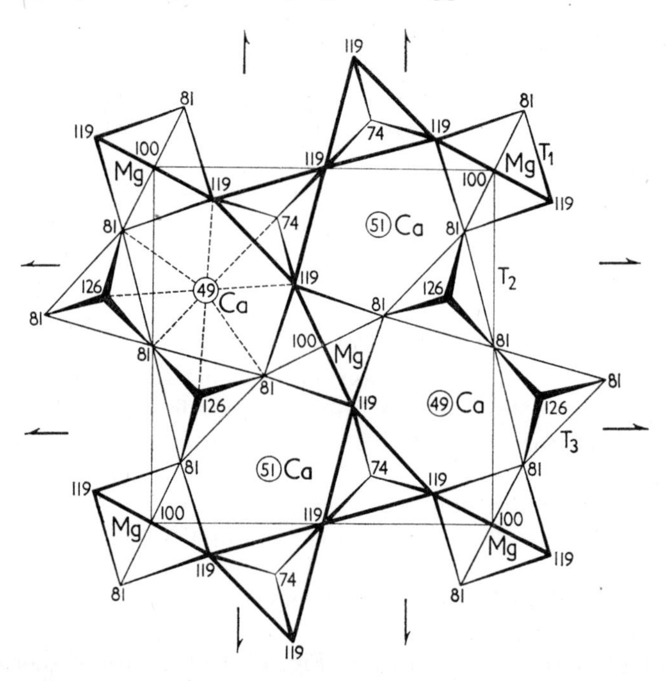

FIG. 11. The melilite structure projected on a plane perpendicular to the c-axis. Heights in hundredths of c. $\bar{4}$ axes through Mg atoms; twofold axes through mid-points of sides, m-planes joining mid-points of adjacent sides. The O atoms at the corners of tetrahedra and the (Si,Al) at the centres are not shown.

might be a solid solution of a slightly different composition. However, in the course of this work two new compounds were discovered with compositions approximating to $C_{25}A_{17}M_8$ and C_7A_5M. Subsequent single-crystal X-ray investigations by Nurse, Welch and Majumdar [*51*] the results of which are summarized in Table IV, have shown that neither of these formulae can be reconciled with the crystal data and that "magnesian pleochroite" is probably a separate compound but with a formula differing from C_6A_4MS. The true formula of "$C_{25}A_{17}M_8$" is almost certainly C_3A_2M. All these compounds, from the relation between the orthorhombic unit cells given in Table IV, appear to be structurally related between themselves, to C_5A_3 and possibly to gehlenite, C_2AS, as well.

TABLE IV

Compounds probably related to C_5A_3

	a (Å)	b (Å)	c (Å)	S.G.
C_5A_3	11·25 = 2 × 5·62	10·98	10·28 = 2 × 5·14	—[1]
"$C_{25}A_{17}M_8$"	16·77 = 3 × 5·59	10·72	5·13	*P*bma or *P*b2_1a
"C_7A_5M"	44·34 = 8 × 5·54	10·76	5·13	—
Magnesian pleochroite, "C_6A_4MS"	27·70 = 5 × 5·54	10·78	5·13	*P*2_12_12
Gehlenite, C_2AS	10·90 = 2 × 5·45	10·90	5·10	*P*42_1m[2]

(1) Original *a* and *b* interchanged.
(2) Axes correspond to the C-centred cell with axes at 45° to those of the primitive cell.

The further elucidation of the formulae of the new ternary and quaternary compounds probably requires a structure determination of C_5A_3, as a basis for deducing the atomic arrangements in the related compounds. The structure of gehlenite found by Raaz [*52*] and confirmed by Smith [*53*] does not provide any obvious basis for the other structures, and it may be that the similarity in cell size is fortuitous.

10. $C_{12}A_7$

Büssem and Eitel [*54*] determined the cubic unit cell and space group.

$$a = 11{\cdot}98\ \text{Å};\quad I\bar{4}3d;\quad Z = 2$$

They postulated a structure in which two oxygen atoms were statistically distributed over the cell. Jeevaratnam, Dent-Glasser and Glasser [*55*] confirmed the space group and cell size, and obtained an *R* value of 25% for 56 *hk*0 reflections, using Büssem and Eitel's atomic positions. This indicates that the postulated structure may be a satisfactory approximation as a starting point for a full analysis, but this can only be tested by the progress of the proposed refinement of the structure [*55*]. It has recently been found that $C_{12}A_7$ can take up water [*55–57*]. The earlier work (described in Chapter 2) produced a number of discrepancies, but further study by Nurse, Welch and Majumdar [*51*] has given the following results. $C_{12}A_7$ which has been formed under "anhydrous" conditions gains about 1·25% of its weight of water on heating slowly to 950° C in moist air. This corresponds almost exactly to 1 mol. of water to 1 formula wt. of $C_{12}A_7$. Between 950 and 1350° C most of this water is lost reversibly, but there are indications that even at the melting point (1391° C) a very small amount remains. Indeed, a small amount of water may be an essential mineralizer in the formation of the compound. Its rate of growth is so rapid that any melt (or glass) in which it nucleates converts in the first place to the maximum possible amount of $C_{12}A_7$, even though other compounds are more stable under equilibrium conditions. Crystals formed and cooled under anhydrous conditions do not hydrate readily in moist air at room temperature, so that hydrous and anhydrous crystals can both be examined by X-rays under normal conditions. No differences have yet been found in the relative intensities of the lines of the diffraction pattern, but the cell size of the anhydrous material is about 0·1% larger than that of the hydrous material. The hydrous material also has the larger refractive index (see Chapter 2).

Since the hydrated form is more stable at room temperature, it may have greater perfection of atomic arrangement, giving a defect structure on dehydration. With a very accurate structure determination it may be possible to determine the role of water in this and similar cases.

11. The Ferrite Series

A complete solid solution series C_2F–"C_2A" exists up to a composition $C_2F_{0\cdot31}A_{0\cdot69}$. Bertaut, Blum and Sagnières [*58*] determined the crystal structure of C_2F.

$$a = 5{\cdot}42,\ b = 14{\cdot}77,\ c = 5{\cdot}60\ \text{Å}$$
$$Pnma; \quad Z = 4$$

The structure (Fig. 12), which is very similar to that originally proposed by Büssem, consists of layers of FeO_6 octahedra perpendicular to *b*, joined to layers of FeO_4 tetrahedra with Ca atoms in holes between the layers having very irregular ninefold co-ordination (with one long 3·3 Å bond). All joins within and between layers are by corners only. The tetrahedral layers consist of infinite chains of tetrahedra running parallel to *a* but somewhat buckled in the *ac*-plane by compression along the chain direction.

Smith [*59*] has reviewed previous crystallographic work on the solid solution series, and reported work on single crystals with seven compositions, $C_2F_{1-p}A_p$, where *p* ranged from 0 to 0·67. He showed that the known non-linear change of some spacings with composition is accompanied by a continuous change towards a more symmetrical space group, *Imma*, which is completed at about $p = 0{\cdot}33$. By comparison of changes in calculated and observed intensity with composition, the substitution of Al was shown to take place first in the tetrahedral layers, allowing the

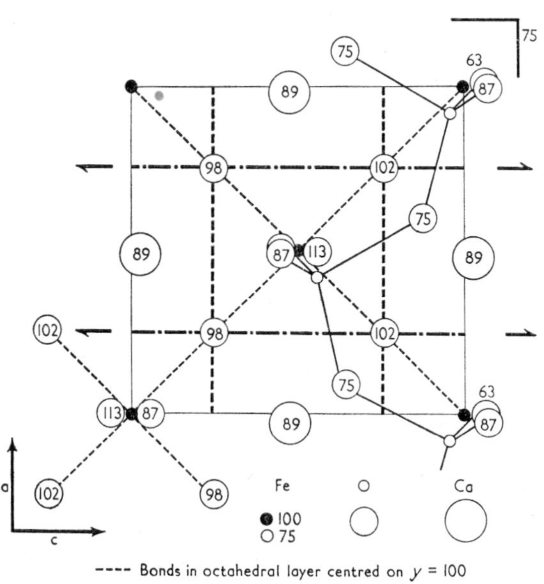

FIG. 12. Projection of the structure of C_2F on a plane perpendicular to the *b*-axis. Heights in hundredths of *b*. 2_1 axes through corners, centre and mid-points of sides of cell have been omitted. All atoms between the top of the cell and the mirror plane at $y = 75$ are shown. The mirror plane determines the positions of atoms between $y = 75$ and $y = 50$. A centre of symmetry at the body centre of the cell then gives the bottom half of the cell as the inversion of the top half. Atoms have been added to the diagram above $y = 100$, below $y = 75$ or outside the cell only as required to show typical co-ordination or bonding. Atoms on the same vertical line have been slightly staggered to render the lower one visible.

tetrahedra to contract relative to the octahedral layer and rotate into the more symmetrical position (Fig. 13). Al atoms substituting beyond $p = 0{\cdot}33$ (when the tetrahedral layer is half substituted) are distributed nearly equally between tetrahedral and octahedral sites. The composition C_4AF is not a distinct compound, since the Al atoms are substituting for about one-fourth of the octahedrally co-ordinated Fe and about three-fourths of the tetrahedrally co-ordinated Fe.

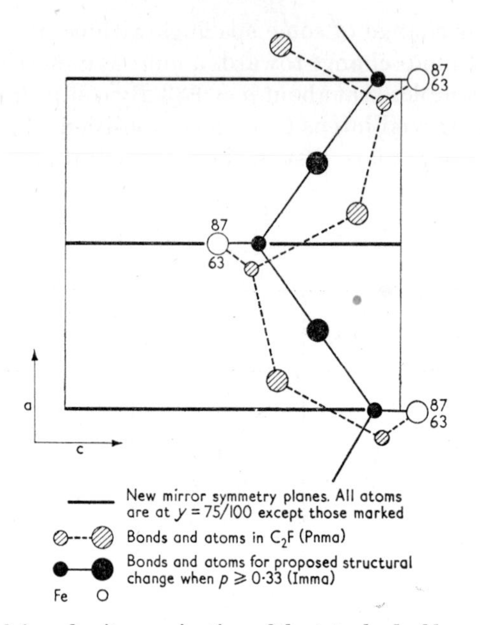

FIG. 13. Dicalcium ferrite; projection of the tetrahedral layer showing the structural change proposed to account for the higher symmetry when $p \geqslant 0{\cdot}33$. Unshaded atoms do not move appreciably. The new (Al, Fe)—O bonds are about 10% shorter than the Fe—O bonds in the CF_2 structure. The change of orientation of the tetrahedra allows the Ca atoms to move slightly on to the symmetry planes. The octahedral layer already has the higher symmetry in C_2F.

12. C_3A

The structure determination of tricalcium aluminate has still to be completed [*60*], and at present the evidence for a postulated structure or pseudo-structure is too doubtful to make discussion worth while. The cubic symmetry and space group have been determined [*61*], and accurate values of the cell size obtained [*62, 63*].

$$a = 15{\cdot}262\ \text{Å}; \quad Pa3; \quad Z = 24$$

All the strong reflections can be indexed on the basis of a cell with $a = 7{\cdot}63$ Å, indicating that the eight such units in the true cell must be similar to one another.

C. OXIDES

1. Silica, SiO_2

At atmospheric pressure silica exists in two crystalline forms, quartz and cristobalite. The presence of large foreign ions such as Ca^{2+} or Na^+

modifies the cubic packing (ABCABC...) of layers in cristobalite to produce finally hexagonal packing (ABAB...). Stacking faults both random and regular (giving polytypes) are common. When sufficient foreign ions are present to produce mainly hexagonal stacking, tridymite is produced [*64*].

Quartz and cristobalite and the main structure of tridymite are all composed of SiO_4^{4-} tetrahedra joined to other tetrahedra at all four corners. All three have high- and low-temperature forms, the transformation being "displacive". The transformation from the high-temperature cristobalite to quartz is "reconstructive", which explains why quenched cristobalite can exist indefinitely at room temperature and even have a displacive transformation at a temperature right outside its stability range. The transition temperature is variable (200–275° C), owing to variable amounts of stacking faults arising from impurities.

α-Quartz: $a = 4{\cdot}913$, $c = 5{\cdot}405$ Å; $P3_121$ or $P3_221$; $Z = 3$
β-Quartz (625° C): $a = 5{\cdot}002$, $c = 5{\cdot}454$ Å; $P6_222$ or $P6_422$; $Z = 3$
α-Cristobalite: $a = 4{\cdot}971$, $c = 6{\cdot}918$ Å; $P4_12_12$ or $P4_32_12$; $Z = 4$
β-Cristobalite (500° C): $a = 7{\cdot}18$ Å; $P2_13$; $Z = 8$

2. Corundum (α-Al_2O_3) and Hematite (α-Fe_2O_3)

These structures are isomorphous, with the following properties.

Corundum: $a = 4{\cdot}758$, $c = 12{\cdot}991$ Å; $R\bar{3}c$; $Z = 6$
Hematite: $a = 5{\cdot}028$, $c = 13{\cdot}74$ Å; $R\bar{3}c$; $Z = 6$

The rhombohedral structures are based on hexagonal close packing of oxygen ions, with two-thirds of the octahedral holes between layers filled by Al or Fe, such that pairs of AlO_6 or FeO_6 groups share an octahedral face, perpendicular to *c*.

3. γ-Al_2O_3 and γ-Fe_2O_3

γ-Al_2O_3 has been assigned a cubic, defect spinel structure with $21\frac{1}{3}$ metal atoms arranged at random in the 16 octahedral and 8 tetrahedral holes of the cubic close packed oxygen layers.

$a = 7{\cdot}90$ Å; $Fd3m$; $Z = 10\frac{2}{3}$

γ-Fe_2O_3 is near-isomorphous with γ-Al_2O_3; its pseudocell has

$a = 8{\cdot}34$ Å; $Fd3m$; $Z = 10\frac{2}{3}$

4. CaO and MgO

These have sodium chloride type structures (cubic), in which all ions are *6*-co-ordinated by ions of opposite charge, in regular octahedra.

CaO: $a = 4{\cdot}8105$ Å; $Fm3m$; $Z = 4$
MgO: $a = 4{\cdot}213$ Å; $Fm3m$; $Z = 4$

5. *Mullite*, $2Al_2O_3.SiO_2$ [$Al_{2\cdot4}Si_{0\cdot6}O_{4\cdot8}$]

$a = 7{\cdot}583 \pm 0{\cdot}002$, $b = 7{\cdot}681 \pm 0{\cdot}002$, $c = 2{\cdot}8854 \pm 0{\cdot}0005$ Å
*P*bam; $Z = 1\frac{1}{8}$ [*65*]

From the frequent appearance of diffuse scattering and superstructure effects, as well as the cell contents, this structure must be disordered, with statistical symmetry. Sadanaga, Tokonami and Takéuchi [*65*] have shown that it is a disordered phase, intermediate between the two ordered forms of $Al_2O_3.SiO_2$, sillimanite and andalusite. Columns of AlO_6 octahedra linked by edges and running parallel to *c* at the corners and centre of the cell, and cubic close packing of oxygens, are common to all three structures. The cross-links through AlO_4 and SiO_4 tetrahedra, which are ordered in different ways in sillimanite and andalusite, are completely disordered in mullite, accounting for the halving of the *c*-axis. In addition, $Si_{0\cdot4}$ of sillimanite is replaced by $Al_{0\cdot4}$ in mullite, with a consequent omission of oxygen. In the two-fifths of the cells where this substitution and omission occurs the Al moves to a nearby tetrahedral hole in the close packed oxygens which is vacant in the sillimanite structure but occupied in andalusite. This omission and rearrangement also takes place completely randomly throughout the crystal, thus giving statistical orthorhombic symmetry.

Sadanaga *et al.* [*66*] give a plausible interpretation in terms of these structures of the process of formation of mullite by heating sillimanite or andalusite.

REFERENCES

1. Goldschmidt, V. M. (1929). *Trans. Faraday Soc.* **25**, 253.
2. Bragg, W. L. (1930). *Z. Kristallogr.* **74**, 237.
3. Pauling, L. (1929). *J. Amer. chem. Soc.* **51**, 1010.
4. Wells, A. F. (1962). "Structural Inorganic Chemistry", 3rd Ed. Clarendon Press, Oxford.
5. Bragg, W. L. (1937). "Atomic Structure of Minerals". Cornell University Press, New York.
6. Belov, N. V. (1960). *Acta cryst.* **13**, 980.
7. Baumhauer, H. (1912). *Z. Kristallogr.* **50**, 33.
8. Dornberger-Schiff, K. (1956). *Acta cryst.* **9**, 593.
9. Jeffery, J. W. (1953). *Acta cryst.* **6**, 821.
10. Buerger, M. J. (1951). "Phase Transformations in Solids". Wiley, New York.
11. Jeffery, J. W. and Heller, L. (1953). *Acta cryst.* **6**, 807.
12. Smith, G. W. (1959). Private communication.
13. McGeachin, H. (1962). Private communication.
14. Glasser, F. P. and Dent-Glasser, L. S. (1961). *Z. Kristallogr.* **116**, 263.
15. Hilmer, W. (1962). *Kristallografiya* **7**, 704.

16. Liebau, F. (1957). *Acta cryst.* **10,** 790.
17. Liebau, F. (1956). *Z. phys. Chem.* **206**, 73.
18. Dornberger-Schiff, K., Liebau, F. and Thilo, E. (1955). *Acta cryst.* **8**, 752; Mamedov, Kh. S. and Belov, N. V. (1956). *Dokl. Akad. Nauk S.S.S.R.* **107,** 463; Buerger, M. J. (1956). *Proc. nat. Acad. Sci., Wash.* **42**, 113.
19. Tolliday, J. M. (1958). *Nature, Lond.* **182**, 1012; (1959). Ph.D. Thesis, London University.
20. Moody, K. M. (1952). *Miner. Mag.* **30**, 79.
21. Agrell, S. O. and Gay, P. (1961). *Nature, Lond.* **189**, 743.
22. Nurse, R. W. (1954). *Proceedings of the Third International Symposium on the Chemistry of Cement, London 1952*, p. 63. Cement and Concrete Association, London.
23. Midgley, C. M. (1952). *Acta cryst.* **5**, 307.
24. O'Daniel, H. and Tscheischwili, L. (1942). *Z. Kristallogr.* (A) **104**, 124.
25. Heller, L. (1954). *Proceedings of the Third International Symposium on the Chemistry of Cement, London 1952*, p. 241. Cement and Concrete Association, London.
26. Smith, D. K., Majumdar, A. J. and Ordway, F. *Acta cryst.* In press.
27. Jeffery, J. W. (1952). *Acta cryst.* **5**, 26.
28. Nurse, R. W., Midgley, H. G. and Welch, J. H. (1961). D.S.I.R. Building Research Station Note No. A95.
29. Yamaguchi, G. and Miyabé, H. (1960). *J. Amer. ceram. Soc.* **43**, 219.
30. Jeffery, J. W. (1954). *Proceedings of the Third International Synposium on the Chemistry of Cement, London 1952*, p. 45. Cement and Concrete Association, London.
31. Yannaquis, N., Regourd, M., Mazières, Ch. and Guinier, A. (1962). *Bull. Soc. franç. Minér.* **85**, 271.
32. Midgley, H. G. and Fletcher, K. E. (1962). D.S.I.R. Building Research Station, Note No. E 1280. HMSO, London.
33. Wallmark, S. and Westgren, A. (1937). *Ark. Kemi Min. Geol.* **12B**, No. 35.
34. Aminoff, G. (1925). *Geol. Fören. Stockh. Förh.* **47**, 266.
35. Beevers, C. A. and Ross, M. A. S. (1937). *Z. Kristallogr.* **97**, 59.
36. Adelsköld, V. (1938). *Ark. Kemi Min. Geol.* **12A**, No. 29.
37. Boyko, E. R. and Wisnyi, L. G. (1958). *Acta cryst.* **11**, 444.
38. Dougill, M. W. (1963). Private communication.
39. Burdese, A. and Borlera, M. L. (1960). *Metallurg. ital.* **52**, 710.
40. Chessin, H. and Turkdogan, E. T. (1962). *J. Amer. ceram. Soc.* **45**, 597.
41. Heller, L. (1951). Thesis, London University.
42. Dougill, M. W. (1957). *Nature, Lond.* **180**, 292; (1963). Private communication.
43. Bertaut, E. F., Blum. P. and Magnano, G. (1955). *C.R. Acad. Sci., Paris* **241**, 757.
44. Hill, P. M., Peiser, H. S. and Rait, J. R. (1956). *Acta cryst.* **9**, 981.
45. Decker, B. F. and Kasper, J. S. (1957). *Acta cryst.* **10**, 332.
46. Burdese, A., (1952). *Ric. sci.* **22**, 259.
47. Shepherd, E. S., Rankin, G. A. and Wright, F. E. (1909). *Amer. J. Sci.* **28**, 293.
48. Aruja, E. (1957). *Acta cryst.* **10**, 337.
49. Welch, J. H. (1961). *Nature, Lond.* **191**, 559.
50. Parker, T. W. (1954). *Proceedings of the Third International Symposium on the Chemistry of Cement, London 1952*, p. 487. Cement and Concrete Association, London.

51. Nurse, R. W., Welch, J. H. and Majumdar, A. J. (1963). Private communication.
52. Raaz, F. (1930). *Akad. Wiss. Wien.* **139**, 645.
53. Smith, J. V. (1953). *Amer. Min.* **38**, 643.
54. Büssem, W. and Eitel, A. (1936). *Z. Kristallogr.* **95**, 175.
55. Jeevaratnam, J., Dent-Glasser, L. S. and Glasser, F. P. (1962). *Nature, Lond.* **194**, 764.
56. Nurse, R. W. (1962). *Chemistry of Cement, Proceedings of the Fourth International Symposium, Washington 1960*, p. 307. National Bureau of Standards Monograph 43. U.S. Department of Commerce.
57. Roy, D. M. and Roy, R. (1962). *Chemistry of Cement, Proceedings of the Fourth International Symposium, Washington 1960*, p. 307. National Bureau of Standards Monograph 43. U.S. Department of Commerce.
58. Bertaut, E. F., Blum, P. and Sagnières, A. (1959). *Acta cryst.* **12**, 149.
59. Smith, D. K. (1962). *Acta cryst.* **15**, 1146.
60. Ordway, F. (1962). *Chemistry of Cement, Proceedings of the Fourth International Symposium, Washington 1960*, p. 52. National Bureau of Standards Monograph 43. U.S. Department of Commerce.
61. Ordway, F. (1954). *Proceedings of the Third International Symposium on the Chemistry of Cement, London 1952*, p. 102. Cement and Concrete Association, London.
62. Yannaquis, N. (1954). *Proceedings of the Third International Symposium on the Chemistry of Cement, London 1952*, p. 111. Cement and Concrete Association, London.
63. Swanson, H. E., Gilfrich, N. T. and Ugrinic, G. M., N.B.S. Circular 539, Vol. 5, p. 10, Washington, D.C.
64. Flörke, O. W. (1955). *Ber. dtsch. keram. Ges.* **32**, 369. Reviewed by Eitel, W. (1957). *Amer. Ceram. Soc. Bull.* **36**, 142.
65. Aramaki, S. and Roy, R. (1962). *J. Amer. ceram. Soc.* **45**, 229.
66. Sadanaga, R., Tokonami, M. and Takéuchi, Y. (1962). *Acta cryst.* **15**, 65.
67. Woermann, E., Hahn, Th. and Eysel, W. (1963). *Zement-Kalk-Gips* **16**, 370.

Part II

Chemistry of Hydrated Cement Compounds

CHAPTER 5

The Calcium Silicate Hydrates

H. F. W. TAYLOR

Department of Chemistry,
University of Aberdeen, Scotland

CONTENTS

I. Introduction

A. General Points

The calcium silicate hydrates play an essential role in the hydration reactions of most of the cements discussed in this book, including Portland cement. This chapter summarizes existing knowledge of these substances, including conditions of formation, crystal structures, and properties. It will also be convenient to consider briefly the structures and properties of $Ca(OH)_2$ and of those calcium silicates, whether hydrated or not, that contain carbonate or fluoride.

The calcium silicate hydrates include both well defined crystalline compounds, and ill-crystallized materials, which are often of somewhat indefinite composition. All are nearly insoluble in water. Reactions yielding calcium silicate hydrates below about 100° C normally give the ill-crystallized materials; the main products formed from Portland cement on hydration in pastes at room temperature are of this type. In order to obtain crystalline calcium silicate hydrates, hydrothermal conditions (reaction in the presence of water above 100° C at pressures exceeding atmospheric) are generally needed. Such conditions are used technically in the autoclave curing of cement and calcium silicate products (Chapter 11).

The more important of the calcium silicate hydrates are listed in Table I; their compositions are shown also on a triangular diagram in Fig. 1. The tobermorite group is of particular importance for cement hydration, and includes both crystalline compounds and ill-crystallized

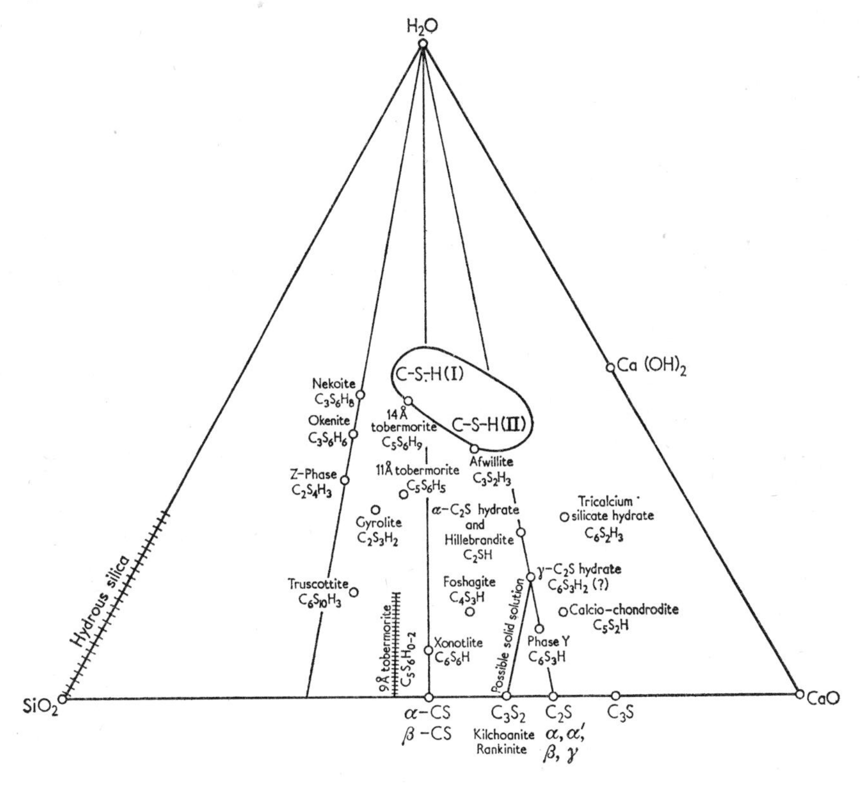

FIG. 1. Compounds in the system $CaO-SiO_2-H_2O$.

materials. The phases within this group are classified more fully in Table II (p. 182).

Reference to Fig. 1 shows that seventeen crystalline calcium silicate hydrates are known, in addition to ill-crystallized materials. There are a few additional phases, not shown in Fig. 1, whose individuality cannot be considered certain. This relatively large number of phases contrasts markedly with the simpler position existing with analogous systems containing smaller cations such as Mg^{2+} or Fe^{2+}. One reason for the

TABLE I

The calcium silicate hydrates and some related compounds (crystalline phases only)

Group	Preferred name	Composition (often only approximate) C	S	H	Probable ionic constitution	Structurally related compounds other than calcium silicate hydrates
A. Compounds structurally related to wollastonite	Nekoite	3	6	8	$Ca_3(Si_6O_{15}).8H_2O$	Wollastonite (β-$CaSiO_3$)
	Okenite	3	6	6	$Ca_3(Si_6O_{15}).6H_2O$	Scawtite ($Ca_7(Si_6O_{18})(CO_3).2H_2O$) (9)
	Xonotlite	6	6	1	$Ca_6(Si_6O_{17})(OH)_2$	Tilleyite ($Ca_5(Si_2O_7)(CO_3)_2$)
	Foshagite	4	3	1	$Ca_4(SiO_3)_3(OH)_2$	Cuspidine ($Ca_4(Si_2O_7)F_2$)
	Hillebrandite (1)	2	1	1	$Ca_2(SiO_3)(OH)_2$	
B. Tobermorite group	14 Å Tobermorite	5	6	9	$Ca_5(Si_6O_{18}H_2).8H_2O$	
	11·3 Å Tobermorite (2)	5	6	5	$Ca_5(Si_6O_{18}H_2).4H_2O$	
	9·3 Å Tobermorite (riversideite)	5	6	0–2	$Ca_5(Si_6O_{18}H_2)$ approx.	
	Other crystalline tobermorites (see Table II)					
C. Gyrolite group	Gyrolite (4)	2	3	2	$Ca_8(Si_{12}O_{30})(OH)_4.6H_2O$	Reyerite ($KCa_{14}(Si_{24}O_{60})(OH)_5.5H_2O$)
	Truscottite	6	10	3	(?)	Zeophyllite ($Ca_4Si_3O_{12}F_2H_6$)
	Z-Phase (of Assarsson) (3)	2	4	3	(?)	Pseudowollastonite (α-$CaSiO_3$)
D. Compounds structurally related to γ-Ca_2SiO_4	Calcio-chondrodite (5)	5	2	1	$Ca_5(SiO_4)_2(OH)_2$	γ-Ca_2SiO_4
	γ-C_2S hydrate (6)	2	1	<1	(?)	Kilchoanite ($Ca_3Si_2O_7$) (10)
						Chondrodite ($(Mg,Fe)_5(SiO_4)_2(OH,F)_2$)
E. Other compounds	Afwillite	3	2	3	$Ca_3(HSiO_4)_2.2H_2O$	Bultfonteinite ($Ca_2SiO_5FH_3$)
	α-C_2S hydrate (7)	2	1	1	$Ca_2(HSiO_4)(OH)$	
	Phase Y (8)	6	3	1	$Ca_6(Si_2O_7)(SiO_4)(OH)_2$	
	Tricalcium silicate hydrate	6	2	3	$Ca_6(Si_2O_7)(OH)_6$	

(1) Synonyms: C_2SH ($n = 1·60$) [*1*], $C_2SH(B)$ [*2*], C_2S β-hydrate, [*3*], dicalcium silicate hydrate (II) [*4*].
(2) Also called "tobermorite" (unqualified) [*5, 6*]: the parent substance of the group.
(3) Synonym: CS_2H_2 [*7*].
(4) Some and perhaps all specimens of the natural mineral "centrallassite" consist of gyrolite.
(5) Synonyms: Phase X [*8*].
(6) Synonyms: $C_2SH(C)$ [*2*], dicalcium silicate hydrate (III) [*4*], C_2S aq. [*1*], Phase X* [*8*], C_2S γ-hydrate [*3*]. This substance appears to be closely related to kilchoanite.
(7) Synonyms: $C_{10}S_5H_6$ [*1*], $C_2SH(A)$ [*2*], dicalcium silicate hydrate (I) [*4*], C_2S α-hydrate [*3*].
(8) Synonyms: $C_2SH(D)$ [*2*], $C_2SH_{0·5}$, [*9, 10*], $C_6S_3H_2$ [*1*].
(9) Synonym: CSH(A) [*2*], $CaO.SiO_2.H_2O$ [*1*]. The CO_3 is, however, probably essential to the structure.
(10) Synonym: Phase Z [*8*]. 1·5 C/S Hydrate [*11*] is also closely similar to kilchoanite or γ-C_2S hydrate.

Other synonyms of less importance, or for compounds within the tobermorite group, are discussed later.

greater complexity of the $CaO-SiO_2-H_2O$ system perhaps lies in the greater ionic radius and more electropositive character of calcium, which permit a number of different types of co-ordination with oxygen. Mg^{2+} and Fe^{2+}, in contrast, are nearly always octahedrally co-ordinated.

Almost without exception, the calcium silicate hydrates are not analogous in formulae or structures to Mg^{2+} or Fe^{2+} compounds. There are no known calcium analogues of such groups as the amphibole, serpentine, or mica minerals. The reason for this is again to be found in the larger radius of Ca^{2+}, which does not permit it to play more than a subsidiary role in any of these structures. An exception to this role occurs in calcio-chondrodite, which is structurally analogous to the Mg^{2+}- or Fe^{2+}-containing mineral, chondrodite; γ-Ca_2SiO_4, from which calcio-chondrodite is structurally derived, is similarly analogous to olivine ($(Mg, Fe)_2SiO_4$). The reason for these exceptions is probably that olivine and chondrodite both contain isolated SiO_4 tetrahedra, which can be separated more widely to accommodate the larger Ca^{2+} ions without causing any fundamental change in structural type. This is impossible with the other magnesium or iron silicate hydrate structures mentioned above, as all of these contain linked tetrahedra.

Even when hydrothermal conditions are used, calcium silicate hydrates made in the laboratory or in technical products are usually of small crystal size. The crystals rarely exceed 100 μ in mean dimension, and for products formed below 300° C, 10–20 μ is more usual. Most of the compounds have been found to occur also as natural minerals, and these often contain larger crystals. Parallel studies of natural specimens, for crystallographic work, and of laboratory preparations, for studying conditions of formation and other characteristics, have proved very useful. Unfortunately, none of the minerals is abundant, and some are extremely rare. They occur principally in contact zones, as at Crestmore, California [*12*], or Scawt Hill, Northern Ireland [*13*], and as inclusions in basalt, as in various localities in the north-west of Scotland [*14*].

No systematic chemical nomenclature for the calcium silicate hydrates has been devised, and it is usual to use the mineral names wherever possible. With some of the compounds, different names have been used by different authors; recommended names and synonyms are given in Table I. There are especially difficult problems as regards nomenclature in the tobermorite group and these are discussed later.

In this chapter, a phase will be described as hydrated, or as a hydrate, if it contains the elements of water, regardless of the manner in which these are combined. The existence of a true hydrate, containing molecular water, is not necessarily implied; the water may occur wholly or partly as hydroxyl.

B. Methods of Investigation

1. *Preparation and Handling*

Methods of preparation are discussed under individual compounds, and experimental techniques are more fully described in Volume 2, Chapter 25. It is important to stress that the calcium silicate hydrates, especially when moist, are attacked by atmospheric CO_2, which converts them eventually into a mixture of $CaCO_3$ and amorphous silica. Careful precautions must therefore always be taken during preparation and subsequent handling to minimize contamination. The poorly crystallized materials formed at room temperature are particularly susceptible to attack, but hydrothermal preparations are not immune. The rate of attack depends on the crystal size, and the natural minerals appear to be virtually unaffected under any normal conditions.

$CaCO_3$ formed by CO_2-attack on calcium silicate hydrates can appear either as calcite, or as vaterite. X-Ray powder data for these polymorphs are given in Volume 2, Appendix 1.

2. *Examination of Products*

The principal methods that have been used for examining preparations of calcium silicate hydrates are described in other chapters, and only a brief survey will be given here.

Chemical analysis for Ca/Si ratio is often scarcely necessary with laboratory preparations; because of the low solubilities of the calcium silicate hydrates, the Ca/Si ratio of the product formed in many methods of preparation is unlikely to differ significantly from that of the starting material. Important exceptions occur, as with preparations made at room temperature by mixing solutions of soluble silicates with ones of calcium salts, or with products obtained by hydrothermal techniques employing a high water : solids ratio. Analysis for CO_2 is always desirable as a measure of contamination. The determination of water contents is mentioned later.

Optical microscopy and X-ray powder examination are essential techniques for the identification of phases. The crystal size is often too small to allow any detailed optical study to be made, but even the mean refractive index can contribute towards identification. X-Ray single crystal studies should also be made whenever crystal size permits, since they provide a more certain means of identification, especially if several phases are present.

Weight loss curves are required if any determination of combined water is to be made. As is shown later in this chapter, a mere determination of the ignition loss after drying at 110° C or by some other arbitrarily chosen procedure is likely to give misleading results.

Other methods that have been widely used include specific gravity determinations, electron microscopy and selected area diffraction, differential thermal analysis, infra-red absorption, specific surface measurements, and chemical extraction methods for uncombined $Ca(OH)_2$ [*15–18*] or silica [*17, 19*].

Results obtained from electron microscopy and selected area electron diffraction (SED) are given mainly in Chapter 9 and Volume 2, Chapter 21, and results from infra-red absorption in Volume 2, Chapter 22. The uses of specific surface determinations and of free lime extractions are discussed mainly in Chapter 7. Optical, unit cell and X-ray powder data, as well as specific gravities, are given in Volume 2, Appendix 1.

C. Calcium Hydroxide

This is briefly considered here, mainly because its structure forms the basis from which those of many of the calcium silicate hydrates are derived.

Calcium hydroxide, or portlandite ($Ca(OH)_2$), forms hexagonal plates with (0001) cleavage. Its solubility decreases with rising temperature, and crystals several millimetres across can be obtained by raising the temperature of a saturated solution. Figure 2 gives solubility data [*20–22*] for systems at atmospheric pressure up to 100° C and under

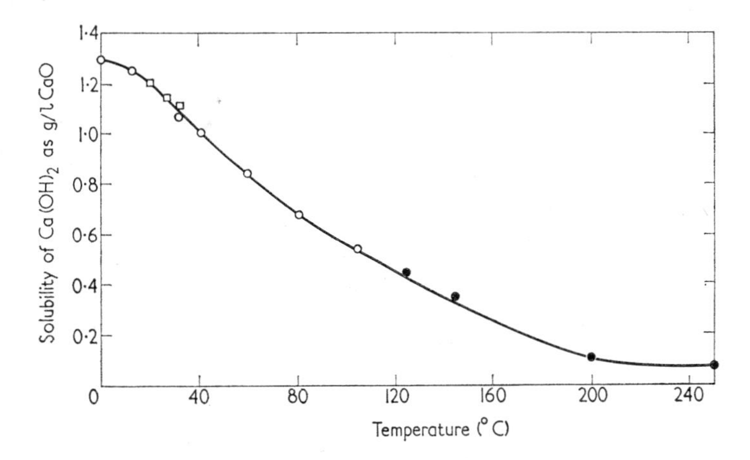

Fig. 2. The solubility of $Ca(OH)_2$. Open circles represent data of Bassett [*20*], full circles those of Peppler and Wells [*21*], and open squares those of Bates, Bower and Smith [*22*].

saturated steam conditions for higher temperatures. Data for the system CaO–H_2O at 21° C and high pressures have also been reported [*23*].

The crystal structure of $Ca(OH)_2$ (Fig. 3) has been determined from X-ray [*24*] and neutron diffraction studies [*25*]. The unit cell is hexagonal with $a = 3{\cdot}593$, $c = 4{\cdot}909$ Å [*26*]. The structure is built of layers. Each Ca^{2+} ion is octahedrally co-ordinated by oxygens, and each oxygen is tetrahedrally co-ordinated by three calciums and one hydrogen atom. The distances between the oxygen atoms in neighbouring layers are 3·333 Å; there is therefore no significant degree of hydrogen bonding and the stacking of adjacent layers is such as to allow each hydrogen atom to fit into the hollow between three oxygens of the next layer.

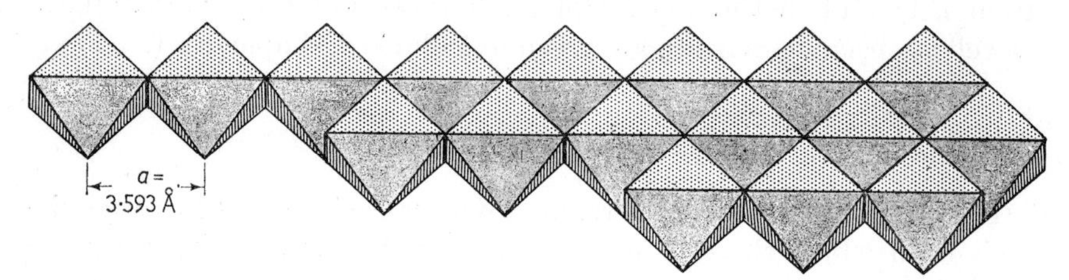

FIG. 3. Crystal structure of $Ca(OH)_2$: arrangement of linked $Ca(OH)_6$ octahedra within a single layer. Parts of three rows of octahedra are shown. The lightly dotted faces of the octahedra are lying flat, with their points away from the observer. Each octahedron has a Ca^{2+} ion at its centre and OH^- ions at each of its six apices; each OH^- ion is shared between three octahedra.

Many other metal hydroxides, including $Mg(OH)_2$ and the various polymorphs of $Al(OH)_3$ (Chapter 6) have similar structures. In the $Al(OH)_3$ polymorphs, one cation site in every three is unoccupied. These compounds differ from $Ca(OH)_2$ also in that weak hydrogen bonds are formed between the oxygen atoms of adjacent layers; this affects the way in which adjacent layers pack together.

On being heated under a water pressure of 1000 kg/cm², $Ca(OH)_2$ melts without decomposition at 835° C [*27*]. The decomposition pressure of $Ca(OH)_2$ reaches one atmosphere at 512° C [*28*]. Dehydration occurs sharply at between 360° and 410° C in a stream of air having a partial water vapour pressure of 6·5 mm, the temperature varying from sample to sample [*29*].

II. Compounds Structurally Related to Wollastonite

A. Introduction

These compounds include nekoite ($C_3S_6H_8$), okenite (CS_2H_2), xonotlite C_6S_6H), foshagite (C_4S_3H), and hillebrandite (C_2SH). All have a repeat distance of about 7·3 Å in one direction, which in varying degrees is a direction of fibrous or prismatic growth. The anhydrous compound, wollastonite (β-$CaSiO_3$), also shows these characteristics and may be regarded as having the structure from which those of the hydrated compounds of this group are derived. The structure of wollastonite was described in Chapter 4. There are infinite, metasilicate chains which are linked so as to repeat every third tetrahedron, and which are therefore called *dreierketten*; this gives rise to the 7·3 Å repeat. The chains are kinked in this particular way because it allows them to share oxygen

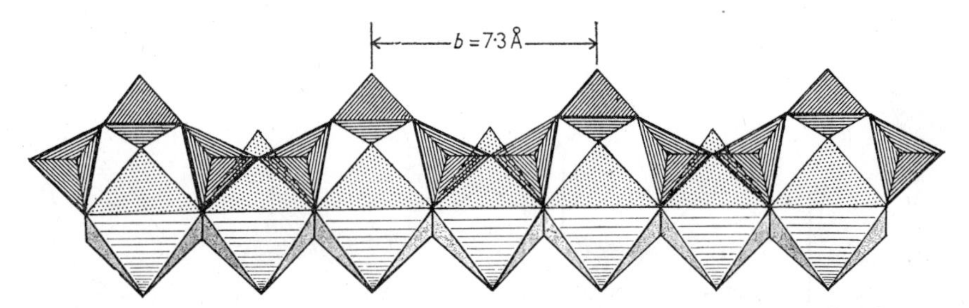

FIG. 4. A metasilicate chain (dreierkette) from the structure of wollastonite, and an associated column of CaO_6 octahedra (after Belov). The repeat distance of 7·3 Å along the *b*-axis is shown.

atoms with CaO_2 sheets somewhat similar to those existing in $Ca(OH)_2$. Figure 4 shows a chain from the wollastonite structure and its mode of attachment to a single row of CaO_6 octahedra; this row forms part of one of the calcium–oxygen sheets. In all of the compounds to be described, the chains lie parallel to the crystallographic *b*-axis.

Figure 4 shows that while the Si—O parts of the wollastonite structure repeat at intervals of 7·3 Å along *b*, the Ca—O parts repeat at intervals of ($\frac{1}{2} \times 7{\cdot}3$) or 3·65 Å in this direction. This results in a marked pseudo-halving; all X-ray reflections having odd values of *k* are weak. This effect is particularly apparent in oscillation or rotation photographs about *b*. With some specimens of wollastonite the odd layer-lines in such photographs are not only weak but also streaked, because of partial disorder in the stacking of Si—O chains, which can suffer random shifts of $b/2$ (see Chapter 4). These features of the X-ray pattern apply also in varying degrees to the hydrated compounds of the group under discussion. As would be expected, the effect is greatest with the compounds of high Ca/Si ratio (hillebrandite and foshagite), smaller with xonotlite, and least with those of low Ca/Si ratio (nekoite and okenite).

B. NEKOITE ($C_3S_6H_8$) AND OKENITE (CS_2H_2)

These are the only calcium silicate hydrates known to occur as well-defined natural minerals which have not yet been synthesized. Their high water contents suggest that synthesis is likely to be possible only at relatively low temperatures, little above or perhaps even below 100° C. Hydrothermal runs at 100–200° C with starting materials of appropriate composition seem, however, always to yield compounds of the tobermorite or gyrolite groups together with hydrous silica. There is one

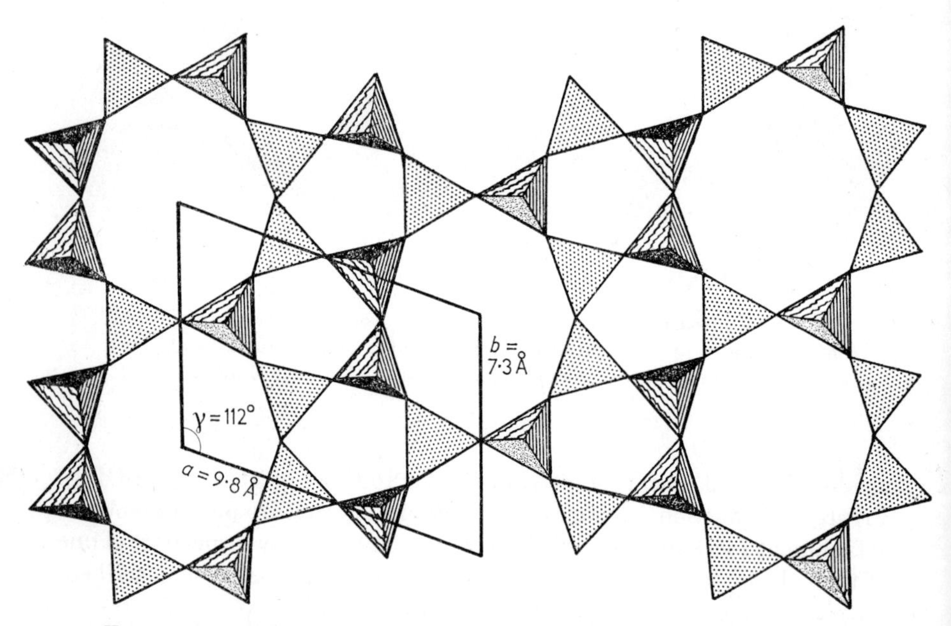

FIG. 5. Suggested arrangement of linked SiO_4 tetrahedra in a sheet of empirical composition Si_2O_5 occurring in okenite and nekoite (after Mamedov and Belov [*34*]).

recent report [*30*] of the synthesis of nekoite, but identification of the product was based only on optical examination and must be considered doubtful.

Okenite has been known as a natural mineral for more than a century; it was originally found in Greenland [*31*] and has since been found at several other localities [*32, 33*]. Nekoite has so far been found only at one locality. It was originally characterized as okenite [*12*] but it has later been shown that this is not correct [*33*]. The two minerals are none the less closely similar in structure and composition.

Both okenite and nekoite form lath-like crystals with elongation parallel to *b*. Their unit cells have been determined [*33*], but not their crystal structures. From the parameters of the unit cell, Mamedov and Belov [*34*] suggested a possible crystal structure for okenite. This was based on infinite $Si_2O_5^{2-}$ sheets of the type shown in Fig. 5, together with Ca^{2+} ions and water molecules. Structures of this type would account well for most of the available data for the two minerals, but direct confirmation is still lacking.

Early analyses indicated the formula CS_2H_2 for both minerals. There appear to be no analyses of okenite more recent than 1924 [*35*] but a new analysis of nekoite shows the true composition to be $C_3S_6H_8$ [*36*]. The water is mostly lost by 200° C; this, together with infra-red evidence (Volume 2, Chapter 22), is consistent with the view that it occurs as molecules. As the temperature is raised, a succession of changes occurs; wollastonite is formed by 800° C (p. 222).

C. Xonotlite (C_6S_6H)

This compound occurs in many localities as a natural mineral and is readily prepared hydrothermally. It was first described as a mineral by

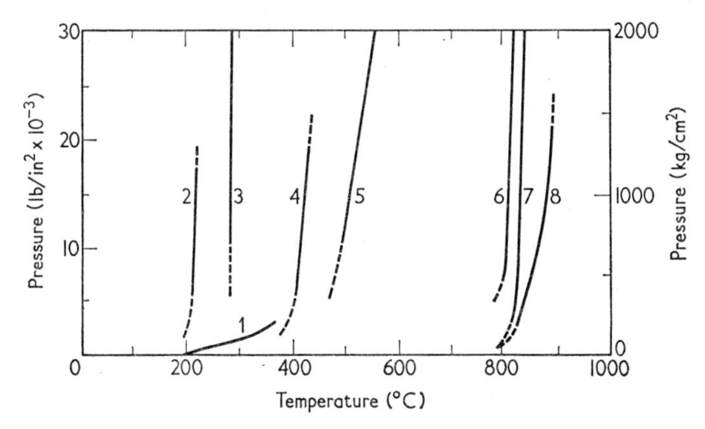

FIG. 6. *P–T* Equilibrium curves for some phase transformations in the $CaO–SiO_2–H_2O$ system (Roy [*8*]; Buckner, Roy and Roy [*41*]).

1. Vapour pressure curve of water (saturated steam curve).
2. Afwillite = kilchoanite + H_2O.
3. 11Å Tobermorite = xonotlite + H_2O + (?) truscottite.
4. Xonotlite = wollastonite + H_2O.
5. Tricalcium silicate hydrate = calcio-chondrodite + $Ca(OH)_2$ + H_2O.
6. Phase Y = α'-C_2S + H_2O.
7. Kilchoanite (assumed hydrated) = rankinite + H_2O.
8. Calcio-chondrodite = α'-C_2S + CaO + H_2O.

Rammelsberg [37]. The first well established synthesis was by Nagai [38]. Syntheses up to 1950 were reviewed by Taylor and Bessey [3]; since that time many more syntheses have been reported. Xonotlite is formed

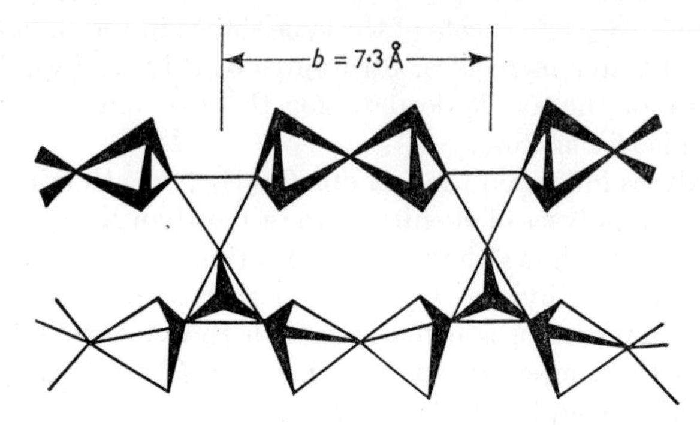

FIG. 7. A double dreierkette from the xonotlite structure (after Mamedov and Belov [43]).

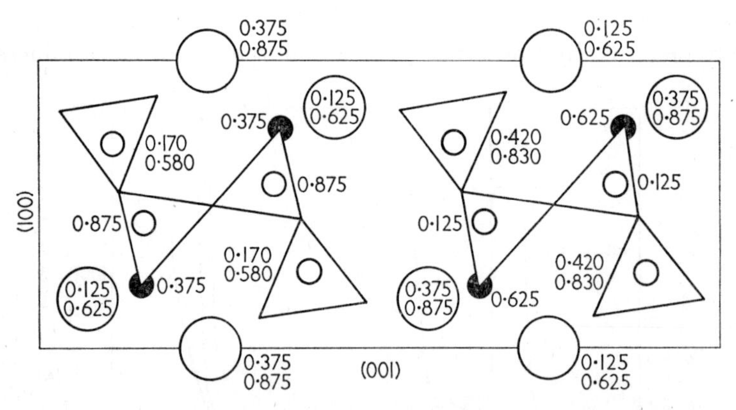

FIG. 8. The structure of xonotlite, as determined by Mamedov and Belov [43]; projection along b (the chain direction). Large open circles denote calcium, small open circles silicon, and small closed circles hydroxyl. SiO_4 tetrahedra are shown as triangles. Heights of Ca^{2+}, Si^{4+} and OH^- ions are shown as fractions of b.

reproducibly when any sufficiently reactive starting material of 1:1 Ca/Si ratio is treated hydrothermally at 150–400° C, saturated steam pressure being assumed up to the critical point (374° C). Its formation from lime–silica mixtures proceeds through the intermediate stages of C-S-H(II), C-S-H(I), and tobermorite [39, 40]. Xonotlite has also been prepared hydrothermally at pressures above that of saturated steam, and P–T curves defining its field of formation have been determined [*41*] (Fig. 6).

Xonotlite forms prismatic crystals or fibrous aggregates with elongation parallel to *b*. Synthetic crystals have a characteristic lath-like appearance in the optical or electron microscope [*1*, *42*]. The crystal structure was determined by Mamedov and Belov [*43*] who found that double dreierketten of empirical formula $(Si_6O_{17})^{12-}$ were present, together with Ca^{2+} and OH^- ions. Figure 7 shows one of these chains, and Fig. 8 shows the projection of the structure along the chain direction.

Mamedov and Belov concluded from their X-ray study that the formula of xonotlite is $Ca_6(Si_6O_{17})(OH)_2$, or C_6S_6H. Published analyses tend to show a somewhat higher water content, many of them suggesting a composition near C_5S_5H. Thermal weight-loss curves [*41*, *44*] show a step at 680–700° C which corresponds approximately to the theoretical H_2O/Si ratio of 1:6, together with additional water lost at lower temperatures. Possibly the structure can accommodate some molecular water in addition to the atoms or ions represented by the formula $Ca_6(Si_6O_{17})(OH)_2$.

Decomposition of xonotlite occurs topotactically at about 700° C, and gives wollastonite [*45*, *46*] (p. 222). The same process can be effected hydrothermally at about 400° C [*47*].

D. Foshagite (C_4S_3H)

This compound was discovered as a natural mineral at Crestmore and was originally considered to have the composition $C_5S_3H_3$ [*48*]. There are no detailed descriptions of its occurrence elsewhere in nature, though it probably also occurs in Mexico [*49*]. Its individuality as a species distinct from hillebrandite was at one time questioned, but was later established with certainty [*1*, *50*] though the original formula is incorrect [*50*].

Foshagite was first synthesized by Flint, McMurdie and Wells [*1*], by hydrothermal treatment of glasses of suitable composition at 300–350° C under saturated steam pressures. Jander and Franke [*9*] obtained a product at similar temperatures which was probably foshagite; syntheses have also been described by Gard and Taylor [*50*] at 500° C and 400 kg/cm^2 and by Roy [*8*] at 450° C and 1000–2000 kg/cm^2. The known range of formation is thus at 300–500° C and 80–2000 kg/cm^2. Various starting materials may be used; this suggests that foshagite is probably an equilibrium product, at least over a considerable part of this range.

Foshagite forms decidedly acicular or fibrous crystals. The crystal structure was determined [*51*], partly from evidence resulting from a study of the topotactic transformation to wollastonite and other products that occurs at about 700° C [*45, 50*]. It is built from Ca^{2+} and OH^- ions and single dreierketten, giving the constitutional formula $Ca_4(Si_3O_9)(OH)_2$. An alternative structure, involving double dreierketten, has been suggested [*52*], but the X-ray structure determination did not support it [*51*].

E. Hillebrandite (C_2SH)

This compound was discovered as a natural mineral at Velardeña, in Mexico, by Wright [*53*], and has since been found also at a few other localities. The first adequately documented synthesis was that of Nagai [*38*] in 1932. It has subsequently been prepared by many investigators [*1, 3, 11, 39, 42, 54–56*]. Several of the earlier workers considered that their synthetic products were not identical with the natural mineral; Taylor and Bessey [*3*] showed that this view was based on incorrect X-ray powder data for the latter, and that the synthetic and natural materials were substantially identical. Later work [*56*] showed that slight differences exist between the natural and synthetic materials. Although these have still not been explained, there is no doubt that the synthetic products are substantially the same compound as natural hillebrandite.

Hillebrandite is difficult to prepare. It has been obtained from β-C_2S, from mixtures of lime and silica, and from ill-crystallized tobermorites, by hydrothermal treatment under saturated steam pressure at temperatures usually between 150° and 250° C. It does not seem to have been obtained from γ-C_2S or from mixtures of C_3S with silica. Some investigators seem to have obtained it easily and reproducibly, but others have experienced much difficulty; in one recent, careful study [*10*] it was not obtained at all, even though the conditions were such that its formation might reasonably have been expected. There are indications that hillebrandite is an equilibrium product at 140–180° C in the presence of water and saturated steam; Heller and Taylor [*56*] obtained an apparently pure product by 166-day treatment of β-C_2S at 140° C, while Peppler [*55*] found it to be stable in six-month runs at 180° C. Roy and Harker [*57*] presented diagrams which indicate that hillebrandite can be a stable phase up to 300–350° C.

Hillebrandite appears to be formed from the starting materials mentioned above through the successive intermediate stages of C-S-H(II) or some other ill-crystallized material and α-C_2S hydrate [*11, 39, 56*]. The difficulty of preparing it is perhaps partly due to the fact that at 140° C the change from α-C_2S hydrate to hillebrandite is very slow, while at 160° C other metastable products, such as γ-C_2S hydrate, are readily formed and are not easily converted to hillebrandite. Kalousek, Logiudice and Dodson [*11*] considered that in preparations from lime and silicic acid, the physical state and reactivity of the silicic acid were important factors.

Hillebrandite forms fibrous or prismatic crystals showing the 7·3 Å repeat distance. The unit cell has been determined [*58*] but not the structure. The 7·3 Å repeat distance shows that dreierketten are almost certainly present, and if the empirical formula C_2SH is correct, the ionic constitution is perhaps $Ca_2(SiO_3)(OH)_2$. Mamedov and Belov [*59*] reasoned from the cell dimensions that double dreierketten, similar to those in xonotlite, should occur, giving $Ca_{12}(Si_6O_{17})(OH)_{14}$ or $C_{12}S_6H_7$. The water content is not at present known with enough accuracy to test these hypotheses. Dehydration occurs at 520–540° C, and gives unoriented β-C_2S; Toropov, Nikogosyan and Boikova [*60*] found that there was an intermediate, amorphous stage, and considered that this provided support for the structure suggested by Mamedov and Belov.

Kalousek, Logiudice and Dodson [*11*] concluded that hillebrandite had a variable Ca/Si ratio; they believed that foshagite, hillebrandite and tricalcium silicate hydrate might represent points on a single solid solution series. Subsequent work has shown that this is not correct, as the unit cells of the three compounds are quite different [*50*, *58*, *61*]. The possibility of a narrower range of variable composition for hillebrandite is nevertheless not excluded. As already seen, foshagite, while quite distinct in structure from hillebrandite, is quite closely related to it. Tricalcium silicate hydrate forms markedly acicular crystals, but its relationship to the group of compounds with a 7·3 Å repeat distance is uncertain; this is considered later.

III. The Tobermorite Group: Introduction and Nomenclature

A. Introduction

This group includes all the calcium silicate hydrates which are directly important in cement hydration under technical conditions used at the present time. It comprises a range of phases which vary widely in both composition and degreee of crystallinity. As a group, these are characterized by a common structural resemblance to tobermorite itself, a compound of approximate composition $C_5S_6H_5$ which was originally discovered in Scotland as a natural mineral [*5*], and which is formed in autoclaved cement products [*62*]. To avoid confusion, this compound will be referred to in future as 11·3 Å tobermorite, from the thickness of

TABLE II

Classification of the tobermorites

Primary subdivisions	Type of X-ray powder pattern	Secondary subdivisions	Composition	Appearance in electron microscope
Crystalline tobermorites	Full pattern showing many *hkl* reflections; often 40–50 lines given adequate experimental technique	14 Å Tobermorite 11·3 Å Tobermorite 9·3 Å Tobermorite 12·6 Å Tobermorite 10 Å Tobermorite	$C_5S_6H_9$ $C_5S_6H_5$ $C_5S_6H_{0-2}$ (?) (?)	Flat plates or laths, usually euhedral; rarely fibres
Semi-crystalline tobermorites	Patterns of about 6–12 lines, including mainly *hk* or *hk*0 reflections	C-S-H(I)	Ca/Si $< 1·5$	Crumpled foils
	and usually a basal reflection at 9–14 Å	C-S-H(II)	Ca/Si $\geqslant 1·5$	Usually fibres
Near-amorphous tobermorites	Weak patterns of 1–3 *hk* lines or bands (3·05, 2·8, and 1·8 Å approx.)	Tobermorite gel (predominant constituent(s)), etc.	Ca/Si probably usually $\geqslant 1·5$	Irregular platelets or foils, fibres (?)

The initials C-S-H denote "calcium silicate hydrate"; hyphens are used to show that the composition $CaO.SiO_2.H_2O$ is not necessarily indicated.

the elementary layers in the structure. The compounds or phases in the group as a whole will be described as tobermorite phases or tobermorites.

Tobermorites are formed as immediate products in most, and perhaps all, reactions in which calcium ions and silicate ions are brought together in aqueous solution. At room temperature, except under certain special conditions, they appear to persist indefinitely. They also seem to persist indefinitely under hydrothermal conditions at saturated steam pressure if the Ca/Si ratio is 0·8–1·0 and the temperature not above about 140° C.

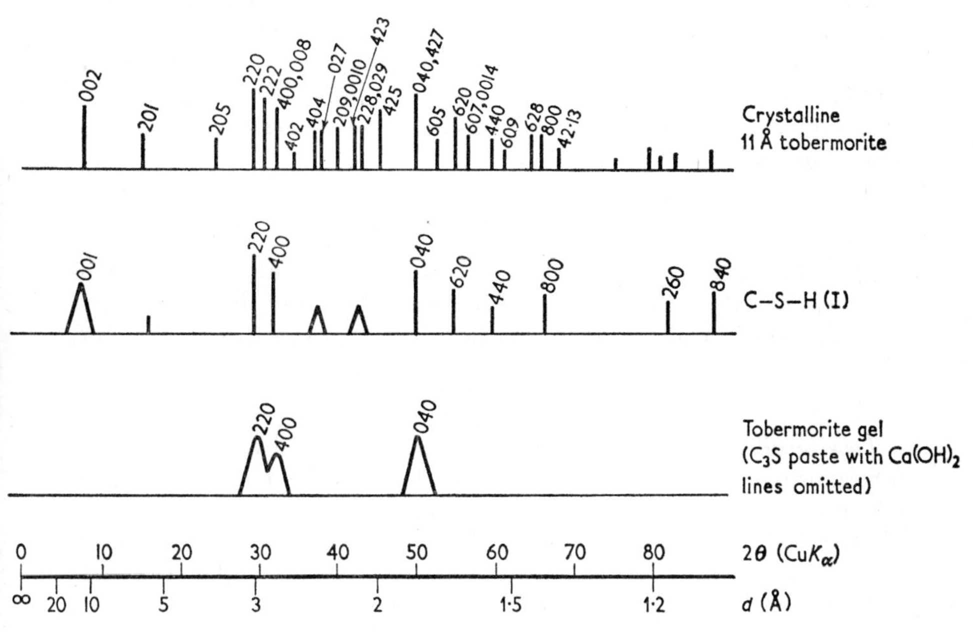

FIG. 9. Typical X-ray powder patterns for tobermorites, represented as line diagrams.

At higher temperatures, or at higher Ca/Si ratios, they are converted after varying lengths of time into calcium silicate hydrates of other types.

The tobermorites can be distinguished from calcium silicate hydrates of other groups by their X-ray powder patterns. They have layer structures, and certain reflections corresponding to repeat distances in, or nearly in, the plane of the layers are nearly always found; the most intense of these reflections have spacings of about 3·05, 2·81 and 1·82 Å. The a and b axes lie in the plane of the layers, so that these reflections have, to a first approximation, $hk0$ indices. In some tobermorites they are probably better represented as having $hk1$ or $hk2$ indices, and in near-amorphous tobermorites, which are composed of particles only a few elementary layers thick, they are better described as *hk* reflections. Crystalline tobermorites, and many semi-crystalline ones, give also a characteristic basal reflection at anything from 9 Å to 14Å; this represents the thickness of the elementary layer.

Because of the wide and often continuous variability within the tobermorite group, any system of classification is bound to be arbitrary. Table II gives a practical classification which will be used here. The primary subdivision is in terms of the degree of crystallinity, as shown by the type of X-ray powder pattern. Secondary subdivision is based on the length of the basal spacing for crystalline tobermorites, and on the Ca/Si ratio and morphology for the semi-crystalline materials. Figure 9 shows typical powder patterns for crystalline, semi-crystalline, and near amorphous tobermorites.

B. Nomenclature

This raises many problems, largely caused by the indefinite nature of many of the substances involved. The basic features of the system used here are given in Table II; some additional questions are discussed below.

1. Mineral Names

Tobermorite, as already stated, was originally used to mean 11·3 Å tobermorite and is still frequently used in this restricted sense. In mineralogy, it normally has this meaning.

Plombierite has been used as a synonym for crystalline 14 Å tobermorite [*63*], and also for a semi-crystalline material resembling C-S-H(I) [*64*]. Historical precedent supports the second usage, and the first is not recommended.

Tacharanite has been used [*65*] to designate a natural mineral which has a 12·7 Å spacing, and probably belongs to the tobermorite group. Its relationship to the material described in Table II as 12·6 Å tobermorite is uncertain.

Crestmoreite was originally used to designate oriented intergrowths of one or more crystalline tobermorites with an apatite mineral of composition $Ca_5[(P,Si,S)O_4]_3(OH,O,F)$, of a particular type found at Crestmore, California [*12*]. This usage is recommended. The use of the name to denote the tobermorite constituent(s) of these intergrowths, as opposed to the mixed material [*66*], is not recommended. The tobermorites present in the intergrowths vary; Heller and Taylor [*49*] found the 14 Å form to predominate, though all the other crystalline tobermorites were sometimes found to occur in particular fibres.

Riversideite was originally used to denote crestmoreite intergrowths rich in 9·3 Å tobermorite [*12*], but has since come to be used as a synonym for 9·3 Å tobermorite [*63*], and this use seems preferable.

2. *Synthetic Preparations*

C-S-H(I), short for calcium silicate hydrate (I) [*69*], has been called by various other names or initials, including CSH(I) [*67*], CSH(B) [*2*], and 0·8–1·33 C/S hydrate [*68*]. These are not recommended.

C-S-H(II), short for calcium silicate hydrate (II) [*69*], has been called $C_2SH(II)$ [*67*], CSH(II) [*70*], C_2SH_2 [*2*], 1·5–2·0 C/S hydrate [*68*], etc. These designations are not recommended.

Hydrate (III) [*71*] has been used to denote a form of near-amorphous tobermorite formed during reaction of C_3S with water in a ball-mill.

A need has been found for collective names. The terms "ill-crystallized tobermorite" or "C-S-H" will be used synonymously to denote any tobermorite or mixture of tobermorites belonging to the categories described in Table II as semi-crystalline or near-amorphous. Assarsson [*39, 72*] used the term "Phase B" in approximately this context, but later [*73*] used "Substance B" in a broader sense. Some workers have referred to all such materials as gels; other terms include CSH [*44*], tobermoritic material [*74*], and tobermorite-like phases [*75*].

Tobermorite gel will be used to denote the material (exclusive of $Ca(OH)_2$) which is formed in pastes of C_3S or β-C_2S at ordinary temperature, or material substantially similar to this. These pastes appear to contain more than one kind of tobermorite particle (Chapter 9), and, not only the nature of the particles themselves, but also the way in which they are associated in the solid is important. The term thus denotes not a particular kind of particle, but a solid material. C-S-H gel is a synonym; the term "tobermorite (G)" has also been used [*15*].

IV. Crystalline Tobermorites

A. 11·3 Å Tobermorite ($C_5S_6H_5$)

1. *Occurrence and Synthesis*

11·3 Å Tobermorite was discovered as a natural mineral at Tobermory, Scotland, by Heddle [*5*]. It has since been found elsewhere in Scotland [*65, 76*] and in Northern Ireland [*63*], and, as a minor constituent of crestmoreite intergrowths, in California [*49, 66*]. It has also been shown to be a major constituent of the binding material in most autoclaved cement–silica or lime–silica products [*62, 77*], and is readily synthesized also in the laboratory.

The first definite synthesis was by Flint, McMurdie and Wells in 1938 [*1*]. They assigned the formula $C_4S_5H_5$ to their product and noted that this was possibly identical with the natural mineral. The next reported syntheses were made by Heller and Taylor [*40*], and by Kalousek [*62*]. Heller and Taylor called their product "well crystallized calcium silicate hydrate (I)". Taylor [*66*] later showed that it was essentially similar to the calcium silicate hydrate constituent of the crestmoreite intergrowths from California, and Claringbull and Hey [*6*] showed that it was substantially identical with natural 11·3 Å tobermorite from the original locality. These investigations established the essential identity of the natural and synthetic materials. 11·3 Å Tobermorite has since been prepared by many other investigators [*39, 41, 42, 78–82*].

11·3 Å Tobermorite has been synthesized only under hydrothermal conditions; formation under saturated steam pressures has been reported for temperatures varying between 110° C [*40*] and 275° C [*1*]. Many different starting materials can be used, of which the more usual have been C-S-H(I) and mixtures of lime or Portland cement with either quartz or amorphous silica. When mixtures are used, reaction appears to proceed through the intermediate formation of C-S-H(II) and C-S-H(I) or of phases similar to these. The most satisfactory method of synthesis at saturated steam pressures is probably 1–3 month treatment of C-S-H(I) in aqueous suspension at 110–140° C. At these temperatures, 11·3 Å tobermorite appears to be indefinitely stable, though whether it is formed as an equilibrium product is uncertain. 11·3 Å Tobermorite can be obtained more rapidly under saturated steam pressures by treatment of lime–quartz mixtures in suspensions or pastes at 170–180° C, but it is then definitely a metastable phase which on more prolonged treatment is replaced by xonotlite. It is difficult under these conditions to obtain a pure product. 11·3 Å Tobermorite becomes unstable relative to xonotlite under saturated steam conditions at about 150° C; this transition temperature has been found, to within ± 10° C, by several investigators.

The rapid synthesis of 11·3 Å tobermorite is probably better achieved by using pressures higher than that of saturated steam; Buckner, Roy and Roy [*41*] found that at 800–1700 kg/cm², 11·3 Å tobermorite remained stable relative to xonotlite until about 285° C (Fig. 6). The treatment of C-S-H(I), or of lime–quartz mixtures of suitable composition, at about 250° C and 800–1700 kg/cm² should therefore afford a good method of synthesis.

2. *Morphology: Normal and Anomalous Varieties*

The morphology of 11·3 Å tobermorite is more fully described in Chapter 9. The crystals found in Northern Ireland are of microscopic dimensions but those from Scotland, as well as those from synthetic preparations, are clearly visible only in the electron microscope. Several varieties of 11·3 Å tobermorite can be distinguished morphologically, and it will be shown later that variations also occur in other respects. The form which has come to be regarded as the "normal" variety forms euhedral, platey crystals up to several microns across and possibly 200 Å thick [*42*, *68*, *78*, *83*]. The cleavage is (001) and the flakes usually tend to be elongated parallel to *b*. Characteristic SED patterns are readily obtained (Chapter 9). The natural material from Ballycraigy (N. Ireland) [*63*] is of this type, but at least one of the Scottish specimens is of a different type [*84*]. The conditions under which the different varieties are formed synthetically are not well established, but reaction under saturated steam pressure below 140° C seems to give the normal variety.

3. *Crystal Structure*

The unit cell was determined by McConnell [*63*] and the essential features of the crystal structure were established by Megaw and Kelsey [*85*].

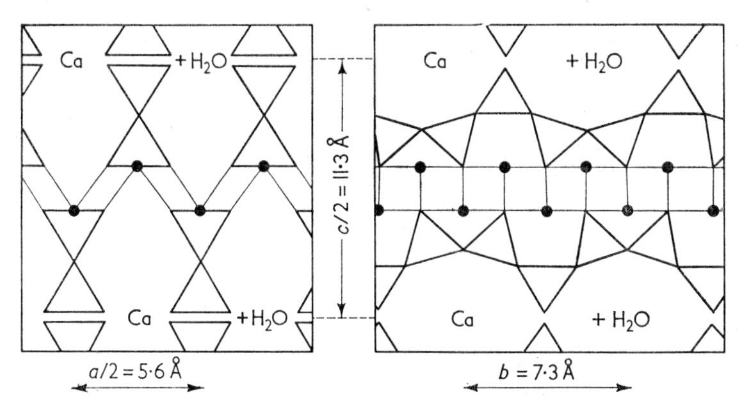

FIG. 10. The crystal structure of 11·3 Å tobermorite (idealized) based on the results of Megaw and Kelsey [*85*]. Left: projection on (010). Right: projection on (100). Small, full circles denote Ca^{2+} ions. Thick lines outline SiO_4 tetrahedra. Thin lines represent Ca—O bonds. The exact positions of the interlayer water molecules and Ca^{2+} ions are unknown. (From Taylor [*70*].)

Figures 10 and 11 show the structure in idealized form. For simplicity it is convenient to refer to a C-centred orthogonal cell with *a* 11·3, *b* 7·3, *c* 22·6 Å approx., though Megaw and Kelsey showed that 11·3 Å tobermorite does not possess orthorhombic symmetry; the true unit cell is triclinic. The structure is based on layers parallel to (001), which are not

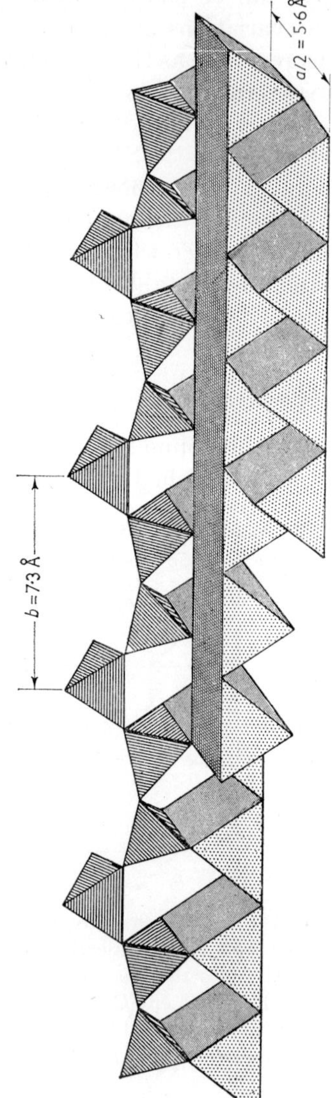

FIG. 11. Part of the Ca—O sheet from a single layer of 11·3 Å tobermorite, with a single, attached dreierkette (idealized); based on the results of Megaw and Kelsey [*85*]. Parts of three rows of CaO_6 polyhedra within the sheet are shown.

entirely unlike those existing in micas or in clay minerals such as vermiculite. Each layer is 11·3 Å thick; the doubled *c*-axis arises from the stacking of the layers. The central sheet of each layer has the empirical composition CaO_2, and may be regarded as an extremely distorted form of a $Ca(OH)_2$ sheet (this may be seen by comparing Fig. 11 with Fig. 3, p. 174). On both sides, the CaO_2 sheet is condensed, through sharing of oxygen atoms, with dreierketten, which form ribs. The composite layers thus formed have the empirical formula $Ca_4Si_6O_{18}$; they are packed together in such a way that the ribs of each layer lie immediately above those of the layer below. The remaining calcium ion in the formula unit, and four water molecules, were considered to be packed into the channels existing between the ribs, but their positions were not definitely established. The positions of the remaining two hydrogen atoms were also not established; if the features of the structure described so far are correct, these hydrogens are presumably attached to the SiO_3^{2-} chains, thus forming SiOH groups. The constitutional formula could thus be written $Ca_5(Si_6O_{18}H_2).4H_2O$. Megaw and Kelsey emphasized that their results were preliminary, and that many details of the structure remained to be settled.

Examination of Figs. 10 and 11 shows that if only the central CaO_2 sheets are considered, the structure repeats at intervals of $a/2$, $b/2$ and c, and that the resulting pseudo-cell with a 5·6, b 3·6, c 22·6 Å approx. is body centred. Because of this, those X-ray or SED reflections that can be indexed in terms of this pseudo-cell and that then have $(h+k+l)$ even are systematically strong. With very minor exceptions, only these reflections can be detected by powder methods.

For many purposes, it is convenient to refer to a structural element with a 5·6, b 7·3, c 11·3 Å; this represents the structure adequately if the stacking of successive layers is not under consideration. The structural element contains one formula unit of composition $C_5S_6H_5$ according to the picture presented above.

Mamedov and Belov [*86*] have proposed a crystal structure differing somewhat from that obtained by Megaw and Kelsey; they consider that extensive condensation between the dreierketten occurs giving anions of empirical formula $(Si_{12}O_{31})^{14-}$. At present there does not appear to be any experimental support for this structure.

4. Thermal Behaviour

When "normal" forms of 11·3 Å tobermorite are heated to about 300° C, part of the water is lost and 9·3 Å tobermorite formed. At about 800° C, wollastonite is formed topotactically [*45, 63, 87, 88*].

B. Anomalous Varieties of 11·3 Å Tobermorites

Several varieties of 11·3 Å tobermorite are known. The normal form, already described, is platey with (001) cleavage and is converted to 9·3 Å tobermorite on heating to 300° C. In single crystal X-ray or SED patterns, it gives a pattern of systematically strong reflections corresponding to the pseudo-cell mentioned above, accompanied by weaker reflections which correspond to the C-centred true cell with *a* 11·3, *b* 7·3, *c* 22·6 Å.

The natural mineral from Loch Eynort, Scotland, is anomalous in all these respects [*84*]. Its X-ray powder pattern is virtually indistinguishable from that of the normal variety, but the material is fibrous with (100) cleavage. On heating it loses water without any significant change in the basal spacing until at about 800° C conversion to wollastonite occurs. SED patterns show the normal pattern of strong reflections, but a different pattern of weak reflections. This probably means that the Si—O chains of the structure are not stacked in the same relative positions as in the normal variety.

Similar anomalies are found in some synthetic specimens, though none of these appear to be identical with the Loch Eynort material. Some preparations are seen in the electron microscope to consist of laths rather than plates [*68*, *83*] while others are distinctly fibrous [*89*, *90*]. In other cases, the 11·3 Å spacing persists up to 800° C, as with the Loch Eynort material [*44*]. Kalousek [*91*] prepared 11·3 Å tobermorite in the presence of alumina-bearing materials; he concluded that up to 5% of Al_2O_3 could enter the structure and that this caused changes in thermal behaviour, and that it also reduced the crystal size and increased the stability relative to xonotlite. The Ca/Si ratio of synthetic 11·3 Å tobermorite can apparently vary between 0·8 and 1·0 [*68*]; variations in water content have also been reported [*44*].

The reason for these variations in morphology and behaviour is unknown. They could possibly be attributed to Si—O—Si links between adjacent layers, perhaps in the manner proposed by Mamedov and Belov [*86*]. The existence of such links could account for the otherwise puzzling (100) cleavage of the Loch Eynort material, and for its ability to lose water without changing in *c*-spacing.

C. Other Crystalline Tobermorites

1. *14 Å Tobermorite* ($C_5S_6H_9$?)

Until recently this was known in nature only as a constituent of crestmoreite intergrowths [*49*, *66*] and in intergrowths with 11·3 Å tobermorite [*63*], but it has now been found pure at Crestmore [*92*]. 14 Å Tobermorite has been synthesized from lime–silica slurries at 60° C [*93*].

The structure probably closely resembles that of 11·3 Å tobermorite, with an extra layer of water molecules between adjacent layers. This would account for the larger *c*-spacing. Chemical analysis of the pure mineral from Crestmore [*94*] indicates a H_2O/Si ratio of approximately 1·5, but the interpretation is complicated because small proportions of carbonate and borate appear to be present in the structure. Dehydration to 11·3 Å tobermorite occurs at 60° C [*95*]; this is lower than the values previously reported. McConnell [*63*] found this process to be reversible with the material from Northern Ireland.

2. *9·3 Å Tobermorite* (Riversideite $C_5S_6H_{0-2}$)

This is formed when "normal" 11·3 Å tobermorite is heated in air at 300° C; it also occurs as a constituent of crestmoreite intergrowths [*49, 63, 87*]. The H_2O/Si ratio was at one time considered to be 0·5 [*87*], but subsequent work has suggested that it can vary from zero, or nearly zero, to about 0·33 [*63, 95*].

9·3 Å Tobermorite is clearly derived from the 11·3 Å form by loss of water molecules and consequent decrease in layer thickness. Megaw and Kelsey [*85*] suggested that the packing of adjacent layers differed from that in 11·3 Å tobermorite; in 9·3 Å tobermorite the ribs of one layer were likely to pack into the grooves of the next. Taylor [*88*] obtained X-ray evidence which supported this view.

When 11·3 Å tobermorite is dehydrated to the 9·3 Å form, the crystallinity deteriorates [*88*]. 9·3 Å Tobermorite has never been obtained from hydrothermal reactions using finely divided materials, though it has been found as a constituent of afwillite crystals which have been converted into pseudomorphs under hydrothermal conditions [*95*]. It should perhaps be considered a metaphase, with no true field of stability.

3. *10Å and 12·6 Å Tobermorites*

Crystalline 10 Å and 12·6 Å tobermorites occur in some specimens of crestmoreite intergrowths [*49*]; several varieties of 10 Å tobermorite have since been found pure at Crestmore [*92, 94*]. A crystalline mineral having a basal spacing of 12·7 Å and apparently belonging to the tobermorite group has also been found in Scotland and named tacharanite [*65*]; its composition is approximately CSH. The relationship of tacharanite to the 12·6 Å tobermorite from Crestmore is uncertain. None of these phases has been fully investigated.

V. Ill-crystallized Tobermorites (C-S-H)

A. Semi-crystalline Tobermorites: General Points

The term "ill-crystallized tobermorites" is used to include both semi-crystalline and near-amorphous tobermorites. The former are the normal products obtained from reactions in aqueous suspensions at room temperature, though they are also formed hydrothermally and in pastes. They have been studied intensively since the classic work of Le Chatelier, published in 1887 [*97*]; work up to 1951 has been reviewed by Steinour [*98, 99*]. As shown in Table II (p. 182) it is convenient to subdivide them into C-S-H(I) and C-S-H(II), where the initials C-S-H stand for "calcium silicate hydrate". C-S-H(I) denotes material having a Ca/Si molar ratio below 1·5 and appearing as crumpled foils under the electron microscope, while C-S-H(II) denotes material with a Ca/Si ratio of 1·5 or over and appearing usually as fibres, or as foils having a distinctly corrugated or fibrous texture. Within both categories considerable variations are possible in the Ca/Si ratio, water content, basal spacing, degree of crystallinity and other properties. Brunauer [*100*] regards both C-S-H(I) and C-S-H(II) as subgroups within the tobermorite family, and this accords with the views expressed here. For Ca/Si ratios around 1·5 classification is somewhat arbitrary but can usually be made on a basis of morphology.

It has often been assumed that all ill-crystallized calcium silicate hydrates are tobermorites, but this may not be correct (see pp. 207, 212).

B. C-S-H(I)

1. Preparation

C-S-H(I) is very easily obtained, especially at room temperatures, though it can also be made hydrothermally. The most important of the methods which have been used below 100° C [*69, 98*] are (a) reaction of lime with silica sol or gel and water and (b) reaction between a solution of a calcium salt with one of an alkali silicate. Method (a) is slow, shaking for several weeks or months being needed. Method (b) is rapid, but often gives a less well crystallized product. Ethyl orthosilicate has also been used as a starting material, especially in procedures designed to ensure that the calcium concentration in solution remains steady during precipitation [*101, 102*].

The Ca/Si ratio of C-S-H(I) made by any of the above procedures can vary over a considerable range. The lower limit is about 0·8; the upper limit is less certain, values ranging from 1·33 [*68*] to 1·5 [*69*] or even higher [*98*] having been suggested in the more recent investigations. If lime and amorphous silica are used as the starting materials, the Ca/Si ratio of the product is simply that of the starting materials, provided that shaking has been carried out for long enough to ensure completeness of reaction. In the method using a calcium salt and a soluble silicate, it is convenient to add an excess of calcium nitrate solution to the sodium silicate (Ca/Si $\approx$ 4). The Ca/Si ratio of the product is then controlled largely by the Na_2O/SiO_2 ratio in the sodium silicate solution; if this ratio is 1·0, the Ca/Si ratio of the product is also approximately 1·0. The precipitate must be washed to remove soluble salts; a $Ca(OH)_2$ solution of a concentration which causes it to be roughly in equilibrium with the precipitate may be used [*69*].

C-S-H(I) can also be made by the repeated extraction of C_3S with water at room temperature, but this method is tedious. Semi-crystalline tobermorites prepared at room temperature from C_3S or β-C_2S and having Ca/Si ratio of 1·5 or over will be regarded here as C-S-H(II).

In suspensions under hydrothermal conditions, C-S-H(I) seems to be formed usually as a short-lived intermediate product, which rapidly recrystallizes to give 11·3 Å tobermorite. Thus, Kalousek and Prebus [*68*] showed electron micrographs of typical crumpled foils of C-S-H(I) made in 2 h reactions of lime and diatomite at 175° C. C-S-H(I) can possibly also be formed in pastes, especially where adequate space is available to permit the foils to grow (Chapter 9).

2. *Equilibria*

When C-S-H(I) is placed in water or lime solutions, it gains or loses lime, depending on the concentration of the solution. A reasonably definite metastable equilibrium curve can be obtained [*68*, *98*, *99*]. Figure 12 shows some recent versions. Several weeks are usually required for attainment of equilibrium. The concentration of silica in solution is very low for Ca/Si ratios above about 1·0. For lower Ca/Si ratios it rises to a maximum of about 0·3 g SiO_2/l at Ca/Si = 0·1 [*103*].

3. *X-Ray Powder Pattern*

This was shown in Fig. 9 (p. 183). Numerous workers [*62*, *69*, *102*. etc.] have obtained data agreeing substantially with those shown there. The pattern resembles that of the crystalline tobermorites, but with only a few lines, mainly ones with $hk0$ indices and a broad basal one. If anything, the patterns are even nearer to that of tacharanite [*65*], but this resemblance may be fortuitous.

The basal spacing is broad and depends on both Ca/Si ratio and H_2O/Si

7

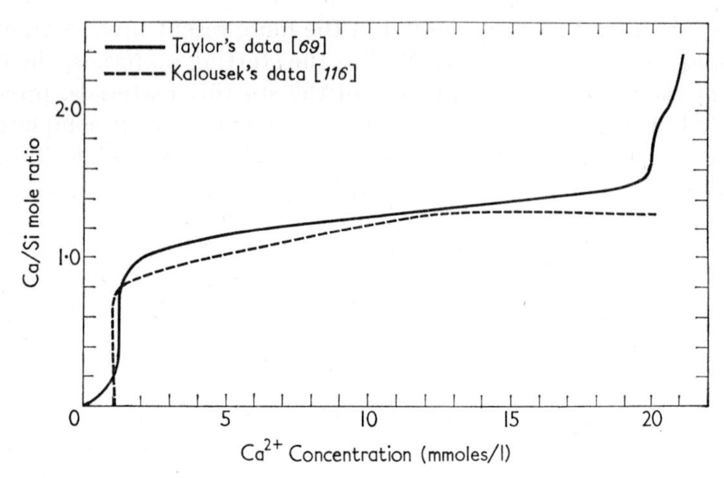

FIG. 12. Metastable equilibrium curves showing the relation between the Ca/Si mole ratio of C-S-H(I) and the concentration of Ca^{2+} in solution at room temperature.

ratio, and possibly on other factors not well understood. Samples which have been gently dried (e.g. with alcohol and ether) show a rough correlation between basal spacing and Ca/Si ratio; the basal spacing falls from 13–14 Å for Ca/Si = 0·8 to about 10 Å for Ca/Si = 1·5 [*68, 70, 102*]. Specimens with Ca/Si approximately 1·0 usually give values of 12–12·5 Å. The basal spacing decreases on heating, values of 9–11 Å being recorded for specimens heated to about 100° C and of 9–10 Å for specimens heated to about 240° C [*87, 104*].

Some specimens of C-S-H(I), especially when made by mixing solutions of a calcium salt and of a soluble silicate, show only a very weak basal reflection or none at all. Natural plombierite [*64*], which has the approximate composition $0{\cdot}8CaO.SiO_2.2H_2O$, also gives no basal reflection.

4. *Morphology and Specific Surface*

This is fully discussed in Chapter 7. The foils appear to be only a few elementary layers thick; they are often several microns across in any direction, with no obvious elongation. At high Ca/Si ratios, there is a trend towards fibrous character. Typical values for the specific surface are 135–378 m^2/g [*15*]. Assuming the density to be about 2·2 g/cm^3 (the value found for crystalline 14 Å tobermorite [*95*]) and neglecting edge effects, this corresponds to a mean foil thickness of 24–67 Å, i.e. between 2 and 6 elementary layers. This is compatible with the evidence from X-rays and electron microscopy.

5. *Dehydration and Behaviour with Lime-extracting Solvents*

Thermal weight-loss curves [*87*, *104*] show a gradual loss in weight with no sharp breaks. Specimens vary in behaviour; there is some doubt as to whether there is a correlation with Ca/Si ratio [*15*, *87*, *104*]. Typical values for H_2O/Si are 2·0 for samples dried with alcohol and ether at 20° C ,1·0 at 100° C, 0·5 at 300° C, and almost zero at 500° C.

The X-ray pattern becomes almost that of an amorphous substance at 400–600° C, only a single diffuse band at about 3·05 Å remaining. At about 800° C, wollastonite is formed [*87*]. Gard, Howison and Taylor [*83*] reported the formation of β-C_2S as an intermediate stage, but Lawrence [*105*] showed that this only occurred if the material was partly carbonated.

C-S-H(I) has been described as relatively resistant towards aceto-acetic ester and similar solvents [*68*]. Funk [*106*] reported that samples with Ca/Si = 1·0 were stable, but that those of higher Ca/Si ratio lost lime until a value of about 1·0 was reached.

6. *Structure*

The precise relationship of C-S-H(I) to 11·3 Å tobermorite has yet to be determined. Material of low Ca/Si ratio is perhaps similar to 14 Å tobermorite but consists of crystals only a few elementary layers thick. Considerable disorder probably also exists within individual layers; for example, because of the pseudo-hexagonal character, the chains may not all be parallel (Fig. 13). This might explain the absence of elongation.

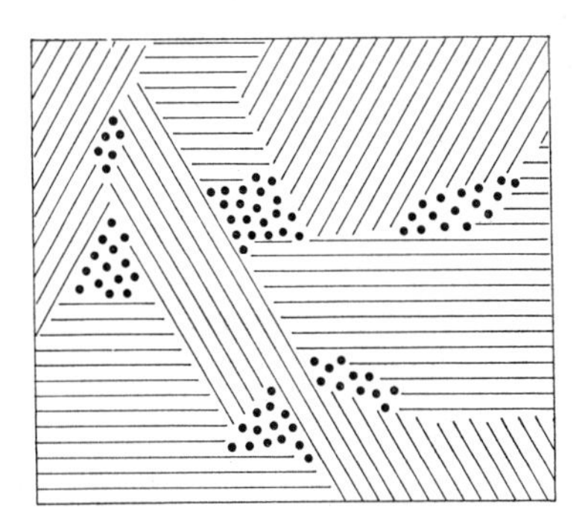

Fig. 13. C-S-H(I): diagrammatic sketch showing possible division of a single tobermorite-like sheet into domains characterized by angular displacements of the chain direction through about 60°. Dotted areas represent very disordered regions.

There has been much controversy regarding the cause of the variable Ca/Si ratio; before the structure of 11·3 Å tobermorite was known, the two main hypotheses were adsorption of $Ca(OH)_2$ on a low-lime material, and some form of solid solution [*98*]. The hypothesis of physical adsorption is now generally rejected [*68*, *104*, *107*]. The nature of the solid solution is still uncertain; the problem is closely linked with that of the structure of C-S-H(II), which is discussed later. Perhaps the most likely explanation of variable composition in C-S-H(I) is that the forms of high Ca/Si ratio are essentially very close mixtures with C-S-H(II) [*108*]. With such poorly crystallized material this does not necessarily imply that separate particles of the two substances are present, as other possibilities exist, such as (a) the random intergrowth of distinct layers of C-S-H(I) and C-S-H(II), and (b) the presence of regions of both C-S-H(I) and C-S-H(II) within each single layer, the two structures merging into each other. The materials do not seem to be mixtures of one specific type containing only two constituents, because the properties are not uniquely defined by the Ca/Si ratio; either a third constituent (such as ill-crystallized $Ca(OH)_2$ or SiO_2), can be present, or the physical form of the mixture varies, as suggested above. Perhaps all of these possibilities are realized. It must be stressed that C-S-H(I) and similar materials are not well defined crystalline substances, but very indefinite ones. With substances of this kind, distinctions between mixture, intergrowth and solid solution lose much of their significance.

C. C-S-H(II)

This term will be used to denote any semi-crystalline tobermorite with a Ca/Si ratio in the range 1·5–2·0. Material of this type is formed both at room temperatures and under hydrothermal conditions, and both in suspensions and in pastes. Some varieties of C-S-H(II) are very poorly crystallized, and there is probably no sharp dividing line between C-S-H(II) and tobermorites of the type described here as near-amorphous.

1. Preparation

The most important methods which have been used are as follows.

(a) Reaction of C_3S or β-C_2S with water in aqueous suspension at room temperature either by shaking in a bottle [*15*, *69*, *75*, *98*, *109*, *111*, *112*], or (for β-C_2S only) by treatment in a ball-mill [*109*].

(b) Hydrothermal treatment of mixtures of silica or C-S-H(I) with $Ca(OH)_2$ at 100–200° C [*39*, *62*, *68*, *72*, *83*, *112*, *113*].

(c) Reaction of β-C_2S with steam at 100° C [*75*].

(d) Reaction of calcium glycerate solution with amorphous silica [*114*, *115*].

(e) Other methods involving aqueous suspensions below 100° C. C-S-H(II) is probably in metastable equilibrium with either saturated or supersaturated $Ca(OH)_2$ solution at room temperature [*98*], and various methods can be used. Taylor [*69*] was unable to prepare it from lime and silica or from calcium nitrate and sodium silicate at room temperature, but Kalousek [*116*], Kalousek and Prebus [*68*] and Funk [*106*] have reported its preparation by methods of this type.

Fibrous tobermorites with Ca/Si ratios of 1·5 or over have also been reported in C_3S or β-C_2S pastes hydrated at room temperature or hydrothermally [*15, 75, 116, 117*]. These fibres have generally been regarded as particles of tobermorite gel, but would appear to have much in common with some forms of C-S-H(II) formed in suspensions.

Funk concluded that under most conditions (reaction in pastes, or in suspensions, or on exposure to steam) the hydration of β-C_2S proceeds topotactically; needles of C-S-H(II) are formed in a definite orientation within the reacting crystals of β-C_2S [*110, 111*]. An intermediate stage having a structure similar to that of β-C_2S but containing water and showing differences in X-ray pattern and refractive index can be recognized [*119*]. Under certain conditions (e.g. in presence of supersaturated $Ca(OH)_2$), reaction can also proceed through the solution phase; the C-S-H(II) in this case consists of undulating foils and not fibres [*106, 110*]. Topotactic conversion of α'- or β-C_2S into ill-crystallized tobermorites also occurs in nature [*64, 120*].

The morphology and other properties of the specimens of C-S-H(II) obtained by these various methods vary widely. One variety, consisting of fibre bundles, is relatively well defined.

2. *Fibre Bundles*

This variety of C-S-H(II) was first described by Taylor [*69*], who obtained it by repeated extraction of C_3S with water at room temperature. He considered it to have a Ca/Si ratio of about 2·0 and showed that it gave an X-ray powder pattern similar to but distinct from that of C-S-H(I). The basal spacing was about 10 Å. Grudemo [*102*] examined this preparation in the electron microscope and observed the characteristic morphology (Chapter 9). A closely similar product was described by Copeland and Schulz [*117*] who obtained it by shaking β-C_2S with water at room temperature; in addition to making X-ray powder studies and electron micrographs, they obtained SED patterns, which confirmed that the phase was a member of the tobermorite family. They found the Ca/Si ratio to be 1·5 for the product formed initially, but the value rose to 1·73 in the course of 9 months. Products giving a similar X-ray pattern have been obtained by reaction of calcium glycerate solutions with silica and have been found to have Ca/Si ratios of about 2 [*114, 115*]. Some of the products obtained by Funk [*75*] may also have been of this type.

The conditions under which this form of C-S-H(II) is produced are not well understood, and attempts to repeat the preparations described above have not always been successful. Thus Copeland and Schulz [*117*] were unable to obtain it from C_3S, although, as already stated, they did succeed in preparing it from β-C_2S. Most preparations of C-S-H(II) have yielded one of the less definite forms mentioned below.

3. Other Varieties

Preparations of C-S-H(II) other than those described above appear in general to have been rather poorly crystallized and there is probably no sharp distinction between them and the near-amorphous tobermorites discussed later. Their X-ray patterns seem in general to resemble those of C-S-H(I) preparations but to be less distinct; the basal spacing, if visible at all, is normally between 9·8 and 10·6 Å for gently dried material [*68, 83, 102, 116*]. Under the electron microscope most of these specimens are seen to consist of long, thin fibres, which sometimes appear to be rolled sheets [*68, 75, 83*]. One hydrothermally prepared specimen, examined by Kalousek and Prebus [*68*] appeared to consist of tubular fibres. Another, prepared from C_3S at room temperature and examined by Copeland and Schulz [*117*], consisted of sheets that were split along their length and appeared to have a corrugated structure. The undulating foils which Funk [*106*] obtained by reaction through solution were possibly of a similar character to these.

There has been considerable uncertainty about the Ca/Si ratio of these forms of C-S-H(II), though it seems definitely to lie between 1·5 and 2·0. Kalousek and Prebus [*68*] concluded that it was variable within this range. The question involves problems similar to those arising in connection with the Ca/Si ratio of tobermorite gel (Chapter 7).

Kurczyk and Schwiete [*90*] have observed what appears to be an exceptional variety of high-lime tobermorite; it was prepared by hydrating finely ground C_3S at 30° C and was reported to have a Ca/Si ratio of 1·8–1·9. Electron microscope studies showed it to consist of acicular crystals, and SED studies indicated a body-centred pseudo-cell with a basal (or 002) spacing of 13·5–14 Å.

4. Water Content, Behaviour on Dehydration, and Behaviour with Lime-extracting Solvents

As with C-S-H(I), the water content is difficult to define. Brunauer, Kantro and Copeland [*109*] concluded that a product dried over $Mg(ClO_4)_2 . 2–4H_2O$ had the composition $C_3S_2H_3$, while Funk reported H_2O/Si ratios of 1·1–1·7 for products dried over P_2O_5 [*75*]. Dehydration curves reported by Gard, Howison and Taylor [*83*] were roughly compatible with these results but, as with C-S-H(I), showed no sharp break at any temperature. About 1·0 mole of H_2O per mole of SiO_2 was retained at 120° C and about 0·5 mole at 300° C. The basal spacing remains unchanged at about 10 Å up to 100–150° C when it disappears. At 600–700° C, β-C_2S is formed.

All varieties of C-S-H(II) seem to be easily decomposed by aceto-acetic ester and similar solvents [*18*, *116*]. Hypotheses regarding the structure of C-S-H(II) are discussed later (p. 200).

D. NEAR-AMORPHOUS TOBERMORITES

These give X-ray powder patterns consisting of one to three lines or bands of approximate spacings 3·05, 2·8 and 1·8 Å (Table II and Fig. 9). Tobermorite gel, the hydrous calcium silicate which is formed in pastes of C_3S, β-C_2S, or Portland cement hydrated at room temperature, gives a pattern of this type. The morphology of tobermorite gel is considered in Chapter 9, and other properties in Chapter 7. Only a few aspects of near-amorphous tobermorites will therefore be considered here.

In addition to being formed in pastes, near-amorphous tobermorites are also formed in suspensions as initial products, which are afterwards replaced by more highly ordered material. Typical examples are the "Phase B" reported by Assarsson [*39*] and the "Hydrate III" of Kantro, Brunauer and Weise [*71*]. Near-amorphous tobermorites seem in general to have high Ca/Si ratios, similar to those of C-S-H(II) preparations; for Hydrate III and Phase B, values around 1·5 were reported. In their earlier work, Brunauer and co-workers concluded that the tobermorite gel formed in C_3S or β-C_2S pastes had Ca/Si = 1·5, but later (Chapter 7) they concluded that the value could vary, extremes of 1·39 and 1·73 being found for the relatively stable product formed when hydration approached completion. For the unstable products formed during the early stages of reaction, they found a much greater variation, from about 1·1 up to probably the Ca/Si ratio of the anhydrous starting material. The H_2O/Si ratios of near-amorphous tobermorites are similar to those found for C-S-H(II).

Near-amorphous tobermorites appear to vary in morphology. Hydrate III was shown to consist of crinkled foils [*117*]. Tobermorite gel has been reported to contain distorted platelets a few hundred ångstroms across, and much larger fibres, but there is some disagreement as to the relative proportions (Chapter 9). The fibres should perhaps be regarded as C-S-H(II) rather than as a near-amorphous tobermorite; if this view is accepted, the poorness of the X-ray pattern of the gel as a whole could only be explained by assuming the platelets to be the principal constituent. The specific surfaces reported for tobermorite gel are mostly around 250–450 m^2/g, indicating that the particles are predominantly two or three elementary layers in thickness.

E. The Structures of Lime-rich Tobermorites

If the hypothesis of physical adsorption is excluded, the high Ca/Si ratios found in C-S-H(II) and in near-amorphous tobermorites might be explained in at least two ways: the 11·3 Å tobermorite structure could be modified by incorporation of extra Ca^{2+} ions, or by removal of some of the silicon. It is also possible that both factors could operate.

Incorporation of extra Ca^{2+} is at first sight a reasonable hypothesis in view of the layer structures of tobermorites and the ease with which C-S-H(I) transfers Ca^{2+} to or from solution. However, it is most unlikely to provide the whole explanation, because the *c*-spacing normally *decreases* with rise in Ca/Si ratio. The fall in *c*-spacing suggests the contrary hypothesis, namely that high Ca/Si ratios arise from the omission of silicon.

The above remarks do not apply to the specimen examined by Kurczyk and Schwiete [*90*], which, as already mentioned, had a layer thickness of about 14 Å. Kurczyk and Schwiete suggested that the structure was that of a low-lime tobermorite interleaved with additional Ca^{2+} and OH^- ions; they suggested the formula $Ca_x[Ca_4Si_6O_{16}(OH)_2][Ca_x(OH)_2]_n$. They considered the layer thickness to be 27–28 Å and assumed that the tobermorite sheet accounted for about 17 Å of this, giving a space of 10–11 Å between individual packets. This seems to ignore the fact that the *c*-spacing of 27–28 Å must include *two* tobermorite layers; it would appear more reasonable to regard the material as being built from alternate layers of 9·3 Å tobermorite and $Ca(OH)_2$. The latter has a layer thickness of about 4·8 Å, thus giving a composite layer about 14 Å thick. The structure would then be related to that of the chlorites [*121*] in the same way as that of crystalline 11 Å tobermorite is related to that of vermiculite [*121*].

For other lime-rich tobermorites, which show a basal spacing of about 10 Å or none at all, three main hypotheses have been proposed:

(a) *Omission of "bridging" tetrahedra.* In the dreierketten of 11·3 Å tobermorite, one tetrahedron in three is not directly attached to the central CaO_2 sheet (Figs. 10 and 11). Taylor and Howison [*104*] suggested that the Si and unshared O atoms of these "bridging" tetrahedra might be omitted, leaving Si_2O_7 groups; hydrogen atoms were also omitted and extra Ca^{2+} ions added. Brunauer and Greenberg [*15*] have recently taken up this hypothesis in a modified form. This is explained in Chapter 7, p. 295. Both Taylor and Howison, and Brunauer and Greenberg, used calculations of unit cell content based on density determinations to support their hypotheses. Taylor [*70*] later expressed doubt as to whether calculations of this type can legitimately be applied to such ill-crystallized material. Even if they can, there seems to be no basis for the assumption of Braunauer and Greenberg [*15*] that the pseudo-cell with *a* 5·6, *b* 3·6 Å and one elementary layer thick contains one formula unit of $C_3S_2H_2$. A pseudo-cell of this type need not contain an integral number of formula units.

(b) *Structures based on* $H_2SiO_4^{2-}$ *groups.* The occurrence of such groups in tobermorites was suggested by Thilo [*122*] and Bernal [*67*] before the existence of dreierketten in 11·3 Å tobermorite and other calcium silicates had been established from X-ray analysis. It was indicated largely by the fact that solutions from which tobermorites are precipitated probably contain $H_2SiO_4^{2-}$ ions. The hypothesis of separate $H_2SiO_4^{2-}$ groups has since been revived, notably by Assarsson [*73*] and by Brunauer and Greenberg [*15*]. The latter investigators quoted new evidence for it, based on the behaviour of tobermorites with mineral acids.

(c) *Replacement of silicate chains by hydroxyl.* This was proposed independently by Grudemo [*118*] and by Taylor and his colleagues [*74, 83*]. The hypothesis is that whole regions of dreierketten are replaced by hydroxyl ions, which thus form part of the central CaO_2 layers. On this assumption, C-S-H(II) is roughly analogous to a clay mineral of the halloysite type; that is, the central CaO_2 sheet is condensed with dreierketten on the whole of one side, and is covered with hydroxyl on the whole of the other (Fig. 14).

This hypothesis is the only one yet proposed which explains the occurrence of high-lime tobermorites in various morphological forms. If replacement of dreierketten by hydroxyl occurred to completion on one side of each sheet, and not at all on the other (as in Fig. 14), one might expect to obtain either sheets rolled into tubes or corrugated sheets, for the same reasons as are known to apply in chrysotile (or halloysite) and in antigorite respectively [*121*]. A structure of this kind would also explain the 10 Å basal spacing [*83*]. If replacement occurred in a random way, with regions of chains and regions of hydroxyl on both sides of each CaO_2 layer, one might well expect to obtain the near-amorphous platey material. The different conditions of formation could perhaps also be explained, since better transport, made possible by the presence of excess water or by raising the temperature, would be expected to permit the formation of the more highly ordered material.

On the basis of lime-extraction experiments, Brunauer and co-workers [*15, 18, 109*] (Chapter 7) concluded that C_3S and β-C_2S pastes contain

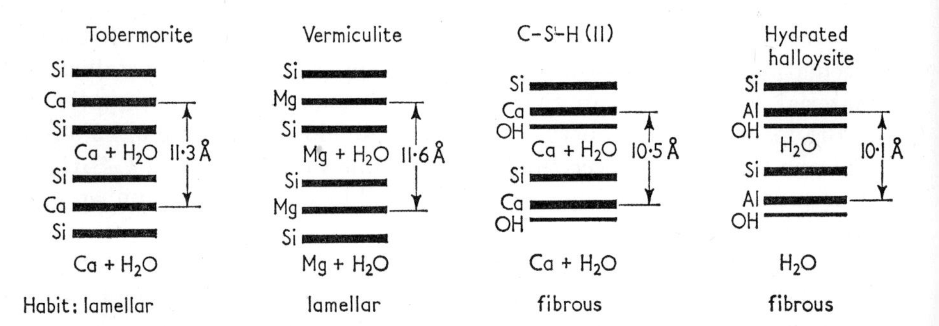

FIG. 14. Diagrammatic representation of the sequence of layers in tobermorite, vermiculite, and hydrated halloysite, and of the suggested sequence for C-S-H(II) [*83*].

not only crystalline but also amorphous $Ca(OH)_2$, which was not detectable by thermal weight-loss or X-ray methods. Buckle and Taylor [*74*] suggested that the amorphous $Ca(OH)_2$ might equally well be considered to represent lime-rich regions of the tobermorite gel. Against this, it has been shown that amorphous $Ca(OH)_2$ can be detected in cases where tobermorite gel is not involved [*71, 123*].

It is probably significant that the reflections which are given by all tobermorites are the ones that represent the repeat distances of the central CaO_2 cores of the layers rather than the Si—O parts of the structure. This suggests that in all tobermorites, including ones of high Ca/Si ratio, the CaO_2 cores have the same form as in 11·3 Å tobermorite. These cores, which are shown in Fig. 11, can be regarded as a distorted form of the CaO_2 sheets in calcium hydroxide (Fig. 3). The distortion is caused by the attachment of SiO_4 tetrahedra through sharing of edges with CaO_6 polyhedra. It is not caused by the linking with dreierketten, because if these are attached by sharing of corners, the distortion does not occur. This is the case with wollastonite (Fig. 4). The thing which distinguishes the tobermorites from wollastonite and from the hydrated compounds related to wollastonite is therefore the distortion of the CaO_2 sheet caused by the attachment of SiO_4 tetrahedra through sharing of edges. This suggests a possible synthesis of the different views of the structure of the lime-rich phases mentioned above.

Assarsson [*73*] suggested that the initial products formed in lime-silica reactions might be represented as $(Ca_{\frac{1}{2}}H_3)^{4+}(SiO_4)^{4-}$, $(CaH_2)^{4+}(SiO_4)^{4-}$, and $(Ca_2)^{4+}(SiO_4)^{4-}$. The first of these would be expected to occur at low effective Ca/Si ratios and would lead to the subsequent formation of Z-phase and gyrolite. The second would occur at moderate Ca/Si ratios and was the precursor of the tobermorite phases. The third would occur at high Ca/Si ratios and was the precursor of α-dicalcium silicate hydrate. Assarsson used the term "Substance B" to include all of these initial products.

The second of these possibilities is relevant here. The other two are further discussed elsewhere (pp. 203, 212). The process can perhaps more precisely be regarded as the reaction of a $H_2SiO_4^{2-}$ ion with a hydrated ion such as $Ca(H_2O)_5(OH)^+$ through sharing of edges and elimination of H_2O.

$$\left[\begin{array}{c} OH_2 \\ H_2O \quad | \quad OH_2 \\ Ca \\ HO \quad | \quad OH_2 \\ OH_2 \end{array}\right]^{+} + \left[\begin{array}{c} O \qquad OH \\ Si \\ O \qquad OH \end{array}\right]^{2-} \longrightarrow \left[\begin{array}{c} OH_2 \\ H_2O \quad | \quad O \qquad OH \\ Ca \qquad Si \\ HO \quad | \quad O \qquad OH \\ OH_2 \end{array}\right]^{-} + 2H_2O$$

Subsequent condensation between the resulting complex ions through formation of Ca—O—Ca links would give a CaO_2 layer having the characteristic tobermorite distortion. This layer would then serve as a matrix for the formation of dreierketten. The layer thus formed could have a very indefinite structure, in which the Ca^{2+} ions of the core were associated with anions of various kinds, including OH^-, $H_2SiO_4^{2-}$, and chains of various lengths ranging from $Si_2O_7^{6-}$ groups to complete dreierketten. Further Ca^{2+} ions and water molecules might then become adsorbed on the surfaces of the layer. The hypotheses of $H_2SiO_4^{2-}$ tetrahedra and shortened chains may thus apply to ill-crystallized tobermorites in the early stages of formation, and perhaps to near-amorphous ones at later stages, while those of halloysite-, antigorite-, or chlorite-like structures may apply to the semi-crystalline C-S-H(II).

VI. The Gyrolite Group

A. Introduction

This group comprises compounds forming hexagonal or pseudo-hexagonal crystals with a micaceous (0001) cleavage and an *a*-axis of about 9·7 Å. It includes gyrolite and truscottite, both of which crystallize well, and probably also a less definite material of composition $CS_2H_{1 \cdot 5}$ or CS_2H_2 [*7, 19, 39, 72*]. Assarsson named this Z-phase [*39, 72*]. The compounds are most easily distinguished in X-ray patterns by their longest basal spacings, which are around 22 Å for gyrolite, 19 Å for truscottite, and 15 Å for Z-phase.

The crystal structures of these compounds have not been determined, though the excellent basal cleavage suggests that infinite Si—O sheets are present. Chalmers *et al.* [*124*] suggested that these sheets may be based on Si_6O_{18} rings resembling those in beryl, but linked together into sheets of empirical composition $Si_2O_5^{2-}$ by condensation with additional tetrahedra.

B. Gyrolite ($C_2S_3H_2$)

This was discovered as a natural mineral in Skye, Scotland, by Anderson [*125*] and has since been found in many other localities. Cornu [*126*] reviewed early studies on the natural material. Gyrolite is readily synthesized; the first proven synthesis was made by Flint, McMurdie and Wells [*1*], who used glasses or gels having Ca/Si ratios of 0·5–0·66 at 150–400° C. Later work indicates that the products which they obtained at the higher temperatures in this range were almost certainly truscottite and not gyrolite. The formation of gyrolite, as distinct from truscottite or Z-phase, has since been confirmed at temperatures usually within the range 120–240° C [*19, 41, 73, 127–129*]. It seems to form, very slowly, on hydrothermal treatment at saturated steam pressure, from gels or mixtures of lime with silica at 120° C [*44*] and more rapidly at 150° C or over. Kalousek [*62*] obtained it in suspensions from lime and amorphous silica, but not from lime and quartz; Aitken and Taylor [*44*] showed that in pastes it was also formed from lime and quartz. The highest temperature at which gyrolite has been obtained is at least 240° C, but it is not stable in the presence of water at those temperatures. Harker [*128*] and Roy and Harker [*57*] found that gyrolite is unstable relative to truscottite and xonotlite under saturated steam pressure at temperatures above 220° C; in the presence of excess silica, the transition temperature is lowered to below 200° C. At about 200° C, intergrowths of gyrolite and truscottite are often obtained [*19, 57*].

These results seem to indicate that gyrolite is formed as a stable phase under saturated steam conditions over a considerable range of temperature between 120° C and about 200° C. The bottom limit may well be below 120° C. On the other hand, Peppler [*55*] considered gyrolite to be unstable relative to xonotlite and silica at 180° C.

There has been some controversy about the composition of gyrolite.

Analyses of natural specimens [*126*] mostly indicate compositions near to $C_2S_3H_2$. Mackay and Taylor [*129*] showed that this composition could be reconciled with the unit cell parameters and density; they found that three-quarters of the water was more easily lost than the remainder, and proposed the ionic constitution $Ca_4(Si_6O_{15})(OH)_2.3H_2O$. Harker [*128*] has concluded that synthetic evidence also supports this composition. In contrast, Strunz and Micheelsen [*130*] have proposed the formula $C_3S_4H_3$, while Meyer and Jaunarajs [*127*] have recently suggested the formula CS_2H_2. This latter view was based on the assumptions that the type of similarity which exists between the X-ray patterns of truscottite and gyrolite denotes identity of Ca/Si ratio, and that the Ca/Si ratio of truscottite is 0·5. Both assumptions are open to question. The formula $C_2S_3H_2$ still seems to be in the best agreement with the available evidence.

The crystal structure of gyrolite, though not known in detail, is based on a hexagonal or trigonal structural element with *a* 9·72, *c* 22·1 Å approx. [*129*]. At least two different polytypes or stacking variants are known, including a six-layer [*129*] and a one-layer [*130*] form. These may be expected to give slightly differing powder patterns, though this has not yet been adequately investigated. The reported mineral centrallassite, or at any rate some specimens of it, is identical with gyrolite [*1*, *130*].

C. Truscottite ($C_6S_{10}H_3$(?)) and Reyerite ($KCa_{14}Si_{24}O_{60}(OH)_5.5H_2O$)

These closely resemble gyrolite; even though the Ca/Si ratios may not be identical, comparisons of X-ray single crystal patterns show clearly that the structures are all very closely related [*127*, *131*]. The most certain differences lie in the lower water content and shorter *c*-axis (18·7 Å approx.) of truscottite.

There has been some uncertainty regarding the use of the names *truscottite* and *reyerite*; truscottite is the name given to a natural mineral found in Sumatra in 1914 [*132*], and reyerite is the name given to one found in Greenland in 1809 and first studied in 1906 [*133*]. Strunz and Micheelsen [*130*], and Meyer and Jaunarajs [*127*] concluded that the two minerals were identical, and if this is correct, the name reyerite clearly has priority. Recent work by Chalmers *et al.* [*124*] has shown, however, that small but distinct differences occur between the two minerals, especially in the infra-red absorption spectrum. These authors briefly studied two further natural specimens which had not previously been described, and found that one resembled truscottite, and the other, reyerite. They also studied two synthetic specimens, made from lime–silica gels, and found that both resembled truscottite. They found that the Greenland mineral contained small proportions of alkalis, and considered it possible that the potassium was essential; they concluded that the idealized formula of reyerite was $KCa_{14}Si_{24}O_{60}(OH)_5.5H_2O$. While it is probably still too soon to decide whether the two minerals should be considered distinct species, it is convenient to retain the name truscottite for materials resembling the Sumatra mineral in infra-red pattern and other respects, and to restrict the name reyerite to those resembling the Greenland mineral. Except for the two specimens studied by Chalmers *et al.*, there is little or no direct evidence as to whether the various synthetic preparations which have been described are truscottite or reyerite. On the assumption that reyerite contains essential potassium, it will be assumed here that they were truscottite.

The position is further confused by the fact that the earlier workers on synthetic materials did not distinguish truscottite (or reyerite) from gyrolite; Flint, McMurdie and Wells [*1*] concluded that all three were identical. Mackay and Taylor [*131*] showed in 1954 that truscottite differed from gyrolite in its *c*-spacing and composition.

The earliest report of the synthesis of truscottite, as distinct from gyrolite, was by Buckner, Roy and Roy [*41*], who obtained it together with other products from a hydrothermal run at 295° C and about 2000 kg/cm^2. Its preparation has since been reported by Harker [*128*], Roy and Harker [*57*], Meyer and Jaunarajs [*127*], and Funk [*19*]. It was probably also obtained by Jander and Franke [*9*] and by Funk and Thilo [*7*]. It appears to be formed as a stable product from somewhat below 200° C to over 300° C, at pressures ranging from saturated steam pressure to several thousand kg/cm^2; gels, glasses and lime–silica mixtures all seem to be satisfactory as starting materials. Temperatures above 250° C are probably advisable, since at lower temperatures the formation from gyrolite, which appears as an intermediate, is slow.

The composition of truscottite is uncertain. Analyses of the Sumatra mineral have been held to support the idealized formulae C_2S_4H [*131, 132*], $2(CaO,MgO).3SiO_2.2H_2O$ [*134*] and $Ca_{9/2}(Si—O_{18})(OH)_3$ (apparently a misprint for $Ca_{9/2}(Si_6O_{15})(OH)_3$ or C_3S_4H) [*130*]. Studies on synthetic preparations have been considered to support the formulae C_2S_4H [*19, 127*] or $C_6S_{10}H_3$ [*128*]. The evidence quoted by Meyer and Jaunarajs [*127*] regarding the Ca/Si ratio appears only to show that this lies between 0·5 and 0·67. There seems to be little evidence to support the Ca/Si ratio of 0·75 proposed by Strunz and Micheelsen [*130*]. Funk based his conclusion partly on the results of chemical extractions for the determination of uncombined silica. The formula thus seems definitely to lie in the region C_2S_4H–$C_6S_{10}H_3$. The possibility that the composition is variable, within these approximate limits, cannot be excluded.

Gyrolite, truscottite and reyerite all yield pseudowollastonite on heating at about 800° C [*19, 124, 127, 129*]; an earlier report [*131*] that truscottite behaves differently is incorrect [*124*]. The formation of pseudowollastonite rather than wollastonite, which is stable relative to pseudowollastonite at these temperatures, suggests that the hydrated compounds of this group are structurally related to pseudowollastonite.

D. Z-Phase and Related Compounds

In 1955 Funk and Thilo [*7*] described two new compounds, which they considered to have the formulae CS_2H_3 and CS_2H_2. CS_2H_3 was a near-amorphous substance, obtained by reaction of $Na(H_3SiO_4)$ and $CaCl_2$ solutions at 0° C; the first had been shown to contain $H_3SiO_4^-$ ions, and it was concluded that the precipitate had the constitution $Ca(H_3SiO_4)_2$. The X-ray powder pattern of $Ca(H_3SiO_4)_2$ was markedly different from those of semi-crystalline tobermorites, such as C-S-H(I). CS_2H_2 was obtained from $Ca(H_3SiO_4)_2$ by autoclaving at 180° C for 1–2 days. The water contents quoted above for both compounds relate to material dried over silica gel.

Shortly afterwards, Assarsson [*39, 72*] described a product of approximate composition CS_2H_2, which he obtained by autoclaving lime–silica mixtures at 180–240° C. He named it Z-phase (not the same as Roy's Phase Z or kilchoanite), and considered that it was an intermediate in the formation of gyrolite. He showed that it gave a characteristic powder spacing of 15 Å. He considered that it might be a stable product at 130–150° C; this seems unlikely, as gyrolite can be made at temperatures down to at least 120° C. If Z-phase is ever formed as a stable product, the temperature is almost certainly below 120° C.

Taylor [*135*] made a further X-ray powder study of one of Funk's and Thilo's preparations. He concluded that this was probably identical with Assarsson's Z-phase, and also that it was probably closely related to gyrolite in structure. Both Funk and Thilo and Assarsson had suggested unit cells for their products, but these cells were based only on indexing of powder data and could not be considered as established. Assarsson [*73*] did not accept the view that the two phases were identical.

In a further study, Funk [*19*] confirmed the identity of his product with Z-phase. He made new preparations by autoclaving $Ca(H_3SiO_4)_2$ at 150° C and reported DTA curves and electron micrographs, as well as better powder data. He concluded that the composition (for material dried over silica gel) was $C_2S_4H_3$. (This composition is probably not strictly comparable with the composition $C_2S_3H_2$ usually quoted for gyrolite, since the latter would probably have a lower water content if dried over silica gel.) The electron micrographs showed undulating foils resembling those of C-S-H(I), but the powder pattern was quite different from that of the latter material.

Funk [*19*] showed that further poorly crystallized products could be obtained by autoclaving $Ca(H_3SiO_4)_2$ at 120° C. These had the approximate composition CS_2H_2, but were not identical with okenite or nekoite.

VII. Compounds Structurally Related to γ-C_2S

A. Introduction

This group includes calcio-chondrodite, kilchoanite, and γ-C_2S hydrate, and also a hydrated variety of γ-C_2S. All of these phases have unit cells of which one face is closely similar in parameters to the (001) face of γ-C_2S, i.e. *a* 5·06, *b* 11·3 Å, γ 90°. Kilchoanite, or at any rate some forms of it, is anhydrous, but it is so closely related to the hydrated compounds that it is convenient to discuss it here.

B. Calcio-chondrodite (C_5S_2H)

This is not known as a natural mineral. It was first synthesized by Roy [*8*, *136*], who called it Phase X and considered it to be $C_8S_3H_3$. Buckle and Taylor [*137*] later prepared it, and showed from a single-crystal X-ray study that the cell was analogous to that of a magnesium or iron mineral of known constitution, chondrodite ($(Mg,Fe)_5(SiO_4)_2(OH,F)_2$). This showed that the formula was $Ca_5(SiO_4)_2(OH)_2$, or C_5S_2H.

Roy obtained calcio-chondrodite by hydrothermal treatment of various starting materials of appropriate Ca/Si ratio, usually at between 400° and 800° C and at pressures above 300 kg/cm^2. She concluded that it had a wide range of stability, and determined *P–T* curves defining this range relative to tricalcium silicate hydrate and α′-C_2S (Fig. 6). Buckle and Taylor obtained it, mixed with other phases, from C_3S at 600–700° C. Subsequent work [*138*] has shown that, at saturated steam pressures, calcio-chondrodite can be formed at temperatures at least as low as 250° C. A product "C_2SH(CII)" obtained by Funk [*10*] at 250–350° C appears from its X-ray powder pattern to have contained a high proportion of calcio-chondrodite.

Calcio-chondrodite forms prismatic crystals which are similar in appearance to those of olivine. They are elongated in the direction of the 5·05 Å axis and often show simple twinning across a plane in the prism zone. On heating in air, calcio-chondrodite is decomposed at 650–750° C, giving a product which appears to be a modified form of γ-C_2S [*137*].

The crystal structure of calcio-chondrodite may be inferred from analogy with that of chondrodite [*139*]. It is closely similar to that of γ-C_2S (Chapter 4) and can be described as being made up of layers of the latter of unit cell thickness alternating with layers of $Ca(OH)_2$. The *ab*-face of the unit cell is thus the same as in γ-C_2S, but the 001 spacing is raised from 6·78 Å to 8·45 Å. The formula can also be written as $2Ca_2SiO_4$. $Ca(OH)_2$, and is then seen as representing one member of a possible series of compounds having the general formula $nCa_2SiO_4.Ca(OH)_2$. With Mg^{2+}, members of this series are known for $n = 1, 2, 3$ and 4 (though only with partial replacement of F^- by OH^-). Buckle and Taylor [*137*] calculated the unit cell parameters which would be expected for Ca^{2+} compounds having $n = 1, 3$ and 4, but up to now none of these has been obtained.

C. Kilchoanite (C_3S_2)

This was first obtained synthetically by Roy [*8*, *136*], who called it Phase Z (distinct from Z-Phase of Assarsson), and was later found in Scotland as a natural mineral by Agrell and Gay [*140*]. Roy considered the formula to be C_9S_6H, but the natural mineral has the composition C_3S_2. Subsequent work by Roy *et al.* [*141*] showed that some specimens, at least, of the synthetic material also had this composition, though there were some indications of a variable composition which extended towards that of a C_2S hydrate. This is discussed further under γ-C_2S hydrate.

Roy obtained kilchoanite most satisfactorily from gels which were treated hydrothermally for runs of usually 1–3 days duration at temperatures around 700° C and pressures of 130–1300 kg/cm². She showed that at higher temperatures, rankinite was formed, and obtained a *P–T* curve for the kilchoanite–rankinite transition (Fig. 6). She also obtained it, mixed with other products, from afwillite by hydrothermal treatment little above 200° C, and obtained a *P–T* curve defining the upper stability limit of afwillite (Fig. 6). In a further study by Roy and Harker [*57*], it was concluded that kilchoanite was formed metastably, and not stably, in the region of pressure and temperature bounded by these two curves.

Heller [*45*] and Dent [*142*] found that when afwillite is heated in air at about 650° C, an unidentified product was obtained. Comparison of their data with those of Roy later showed that this product was a poorly crystallized form of kilchoanite.

The unit cell of kilchoanite has been determined by Roy *et al.* [*141*], who studied synthetic material, and independently by Agrell and Gay [*140*], who examined the natural mineral. The results of the two studies agreed; the unit cell is similar to those of γ-C_2S and calcio-chondrodite as regards the *ab*-face, but is body centred and has a *c*-axis of about 22 Å. The structural significance of this finding is still not clear, though there is almost certainly a close relationship between kilchoanite and γ-C_2S. The composition suggests, however, that pyrosilicate groups may occur, as it is difficult to see how the ionic constitution can be other than $Ca_3Si_2O_7$.

There seems little doubt that kilchoanite, when made at 600–800° C either hydrothermally or by heating of afwillite, is an anhydrous compound. Further work is needed to decide whether this is also true of the product formed at lower temperatures.

Kilchoanite is converted to rankinite on heating at 1090° C [*141*]. Flint, McMurdie and Wells [*1*] described a low-temperature polymorph of C_3S_2 which they prepared hydrothermally; they reported optical data, which do not agree with those of kilchoanite.

D. γ-C_2S Hydrate ($C_6S_3H_2$(?))

This is a product frequently formed when β- or γ-C_2S, or lime–silica mixtures of 2:1 Ca/Si ratio, are autoclaved under saturated steam pressure at 160–300° C. It was first reported by Keevil and Thorvaldson [*143*] and has since been obtained by many investigators [*1, 3, 11, 29, 56, 79, 145*, etc.]. Roy's "Phase X" [*8*] and Funk's "C_2SH(CI)" [*10*] were probably essentially similar.

The material crystallizes badly; most workers have reported only the water content, mean refractive index, and powder pattern. All of these are somewhat variable; the H_2O/Si ratio has been variously estimated at 0·26–1·0. SED studies [*144*] indicate a unit cell close to that of kilchoanite, but the weight-loss curve [*29*] and infra-red absorption spectrum [*79*] show that the compound is hydrated.

It has been suggested that γ-C_2S hydrate may have a Ca/Si ratio variable from 1·5 to 2·25 [*11*], and also that it is related to kilchoanite by the partial substitution of 4H for Si, so making the formula $C_6S_3H_2$, and that there may be a solid solution series extending from this composition to kilchoanite [*141*]; it has also been suggested that some preparations have been mixtures containing calcio-chondrodite or another member of the calcio-chondrodite series [*8, 135*]. Further work is needed to test these hypotheses.

γ-C_2S hydrate appears to form most readily from γ-C_2S; Funk was able to prepare it pure from this starting material at temperatures down to 150° C [*10*]. This can almost certainly be attributed to similarity in structures. The reaction possibly occurs without the intermediate formation of any phase of the tobermorite group.

E. A Hydrated Form of γ-C_2S

Funk [*75*] studied the hydration of β-C_2S in pastes, in suspensions, and in steam at 20–120° C. He found that if the β-C_2S was adequately stabilized (e.g. with 0·5% B_2O_3), tobermorites were formed. If, however, the β-C_2S was "inadequately stabilized", e.g. with 0·1% Na_2O, partial or complete conversion to γ-C_2S apparently occurred.

In a later study [*10*], he showed that the product was a modified form of γ-C_2S. It had the composition 1·8–2·0CaO.SiO_2.0·5–1·0H_2O after drying over P_2O_5 at 20° C, and the relative intensities of the X-ray powder lines differed slightly from those of normal preparations of γ-C_2S. It was also more reactive; on prolonged treatment at 120° C it was converted into a tobermorite, whereas normal γ-C_2S was barely affected.

VIII. Other Calcium Silicate Hydrates

These include afwillite ($C_3S_2H_3$), Phase Y (C_6S_3H), α-C_2S hydrate (C_2SH), tricalcium silicate hydrate ($C_6S_2H_3$) and several doubtful species. None of these appears to be closely related to any other anhydrous or hydrated calcium silicate, although there is a distant relationship between afwillite and γ-C_2S.

A. Afwillite ($C_3S_2H_3$)

This occurs as a natural mineral at Kimberley (S. Africa) [*146*] and elsewhere. It forms prismatic crystals, which grow to a larger size (up to a few centimetres) than those of the other calcium silicate hydrates. It is formed at low temperatures; the first reported synthesis is due to Bessey (quoted by Taylor and Bessey [*3*]), who prepared it at 98° C. It has since been obtained by ball-milling C_3S with water at room temperature [*71*, *107*] and by hydrothermal treatment of β- or γ-C_2S or of lime–silica mixtures with water at 110–160° C [*56*, *147*]. Its formation seems always to be slow. Heller and Taylor [*56*, *147*] concluded from runs of 13–151 days duration that afwillite was a stable phase in the CaO–SiO_2–H_2O system at 110–160° C. Long and McConnell [*120*] concluded from geological evidence that afwillite can exist in equilibrium with $Ca(OH)_2$ and solution at relatively low temperatures. Under saturated steam pressure, the upper stability limit is probably around 160° C [*55*, *56*, *147*] but at high water pressures afwillite is apparently stable above 200° C [*8*, *57*] (Fig. 6).

Afwillite is not easily prepared; its formation from the anhydrous calcium silicates or from lime–silica mixtures probably takes place through the intermediate formation of C-S-H(II) or similar material.

This intermediate material tends to change into other phases, especially α-C_2S hydrate, which are probably metastable relative to afwillite but are nevertheless highly persistent.

Megaw [*148*] determined the crystal structure of afwillite; it is built from Ca^{2+} ions, H_2O molecules, and $(HSiO_4)^{3-}$ tetrahedra. The constitutional formula can be written $Ca_3(HSiO_4)_2 . 2H_2O$. A system of hydrogen bonds exists among the $(HSiO_4)^{3-}$ anions and water molecules. Infra-red studies [*149*] supported the results of the X-ray structure determination. Afwillite is structurally related to bultfonteinite ($Ca_4Si_2O_{10}H_6F_2$, p. 225).

On heating in air, afwillite is dehydrated at 275–285° C; γ-C_2S is formed topotactically [*45, 150*]. The mechanism is complicated [*142*].

B. α-C_2S Hydrate (C_2SH)

This was first isolated by Thorvaldson and Shelton [*151*] from steam-cured mortars; it is not known as a natural mineral. It is readily formed, and there are many subsequent reports of its preparation at temperatures usually between 95° and 200° C and from a wide range of starting materials. These include C-S-H(II), all the polymorphs of C_2S, mixtures of lime or Portland cement with quartz or amorphous silica, and C_3S. In the last case, $Ca(OH)_2$ is also produced. α-C_2S hydrate is perhaps most reproducibly obtained pure by treatment of β-C_2S for about 14 days at 140–160° C.

Butt and Maier [*152*] and Funk [*10*] have shown that the formation of α-C_2S hydrate is much assisted by seeding. Funk found that it formed rapidly from all the polymorphs of C_2S at 100–200° C if seed crystals were present, but in their absence he could prepare it only from α-, α'- or adequately stabilized β-C_2S and then only slowly and at 120–180° C. He considered that in the absence of seed crystals, C-S-H(II) was formed as an intermediate product and changed only slowly to α-C_2S hydrate. He stressed the importance of thorough mechanical and chemical cleaning of the reaction vessel if seed crystals were to be eliminated. Lack of such cleaning may account for the wider range of formation observed by some investigators.

The formation of α-C_2S hydrate from lime–silica mixtures occurs through the intermediate formation of an ill-crystallized lime-rich phase. It has often been assumed that this is a tobermorite, probably C-S-H(II) [*70*], but it may also have a structure containing $(HSiO_4)^{3-}$ or $(SiO_4)^{4-}$ groups and thus nearer to that of α-C_2S hydrate [*73, 153*]. The crystallization of α-C_2S hydrate in autoclaved pastes of C_3S or β-C_2S is reported to be a gradual process, cryptocrystalline material being first produced [*74*].

Under favourable conditions, α-C_2S hydrate crystallizes relatively well, characteristic orthorhombic tablets being produced [*54*]. The crystal structure has been determined; it is built from Ca^{2+}, $(HSiO_4)^-$, and $(OH)^-$ ions, making the constitutional formula $Ca_2(HSiO_4)(OH)$ [*154*]. On dehydration, which occurs at about 450° C in air, unoriented β-C_2S is formed [*45*]. The possibility of variable composition has been suggested [*11*]. It is difficult to see how this could occur with well-crystallized material, but it might well happen with ill-crystallized varieties.

C. Phase Y (C_6S_3H)

This compound has not been found as a natural mineral. It was prepared by Roy [*8, 136*]. Dent Glasser, Funk, Hilmer and Taylor [*155*] showed that several products obtained by other workers were identical with it; these comprised the "$C_6S_3H_2$" of Flint, McMurdie and Wells [*1*] (called $C_2SH(D)$ by Bogue [*2*]) and products identified with it [*29, 156*], and the "$C_2SH_{0\cdot5}$" preparations of Jander and Franke [*9*] and Funk [*10*]. Thermal weight-loss curves [*29, 155, 157*] confirmed the formula C_6S_3H proposed by Roy.

Phase Y is a relatively high-temperature product, and is readily prepared; it has been obtained from α-, α'-, β- or γ-C_2S [*8, 10*], lime–quartz mixtures [*9*], α-C_2S hydrate [*1*], and gels [*8*] at 350–800° C and pressures ranging from that of saturated steam (at 350° C) [*9, 10, 29*] to 2000 kg/cm^2 (at 800° C) [*8*]. It is probably an equilibrium product over the whole of this range; Roy [*8*] reported a *P–T* curve defining its upper temperature limit of stability (Fig. 6).

Phase Y forms short, prismatic crystals, which are often twinned across a plane in the prism zone. The unit cell is known [*157*], and is not apparently related to that of any other calcium silicate hydrate, though the repeat distance along the prism axis (6·73 Å) is near that of β-C_2S. On heating in air, dehydration occurs at 640–700° C and gives unoriented β-C_2S [*157*]. The high thermal stability suggests that the water is present as ionic hydroxyl. If this is so, the constitution might be $Ca_6(SiO_4)(Si_2O_7)(OH)_2$.

D. Tricalcium Silicate Hydrate ($C_6S_2H_3$)

This has not been found as a natural mineral. Early preparations were reported by Bessey [*158, 159*] and Keevil and Thorvaldson [*143*] and Flint, McMurdie and Wells [*1*]. It has since been obtained by many other investigators. At 200–350° C under saturated steam pressures, it is formed reproducibly from either C_3S or other starting materials of appropriate Ca/Si ratio [*1, 4, 8, 61, 159*]. From C_3S, especially in pastes, it is formed also at lower temperatures; Keevil and Thorvaldson [*143*] reported its formation by the action of saturated steam on initially dry

C_3S powder at 110–374° C, Bessey [*159*] and Midgley and Chopra [*145*] obtained it from C_3S suspensions or slurries at 180° C, while Buckle and Taylor [*74*] obtained it from freshly prepared C_3S pastes at 50–300° C. If, however, C_3S or other starting materials of the same overall Ca/Si ratio are autoclaved *in suspensions* below about 180° C, α-C_2S hydrate and $Ca(OH)_2$ are produced [*1, 3, 56, 145*]. Buckle and Taylor [*74*] concluded that tricalcium silicate hydrate was not formed as an equilibrium product under saturated steam conditions below at least 170° C, but that it was formed from C_3S at lower temperatures in pastes where transport of material was difficult.

Roy [*8*] determined a *P–T* curve for the upper stability limit of tricalcium silicate hydrate, above which it is transformed to calcio-chondrodite and $Ca(OH)_2$. At 1000–2000 bars this curve lay at about 500° C (Fig. 6). Under saturated steam conditions it is difficult to prepare pure tricalcium silicate hydrate; products formed below about 250° C tend to be mixed with α-C_2S hydrate, and those formed above 250° C tend to be mixed with calcio-chondrodite [*61*]. Contamination with α-C_2S hydrate perhaps arises because of the slowness with which this phase is replaced by the more stable tricalcium silicate hydrate, while contamination with calcio-chondrodite could possibly be explained if the *P–T* curve marking the upper limit of stability of tricalcium silicate hydrate, when extended below about 400° C, roughly coincides with the saturated steam curve. Inspection of Fig. 6 suggests that this might be the case. Roy's results [*8*] suggest that pure tricalcium silicate hydrate can perhaps most reproducibly be obtained by working at 200–400° C with pressures well above those of saturated steam.

Tricalcium silicate hydrate forms long, fibrous crystals; the repeat distance along the fibre axis is 7·48 Å [*61*]. This is near that of wollastonite and tricalcium silicate hydrate is possibly structurally related to the calcium silicate hydrates containing dreierketten. The unit cell is known [*61*]. The compound dehydrates in air at 420–520° C, giving a modified form of γ-C_2S; the H_2O/Si ratio is probably 3:2, and not 2:1 as was supposed by early investigators [*61*]. The relatively high dehydration temperature suggests that the water occurs as ionic hydroxyl. This indicates the possible constitution $Ca_6(Si_2O_7)(OH)_6$, with pyrosilicate groups. Mamedov, Klebtsova and Belov [*160*] suggested a possible crystal structure, but this has not been tested experimentally.

E. Incompletely Characterized Phases

There are many reports in the literature of unexplained X-ray powder lines, and other indications of phases additional to those described here.

Two have been more fully described; a compound described as $C_6S_4H_3$ obtained from glasses and from C-S-H(II) at 150–200° C [*1, 161*], and one described as "Phase F", of possible composition $C_5S_3H_2$, formed irreproducibly in lime–quartz pastes autoclaved at 165° C [*44*]. Further work is needed before either of these can be regarded as definite compounds.

IX. Conditions of Formation and Stability

A. Introduction

A general indication of the conditions of formation of the various calcium silicate hydrates can be obtained from a plot of temperature against composition (Fig. 15); such plots have been extensively used by

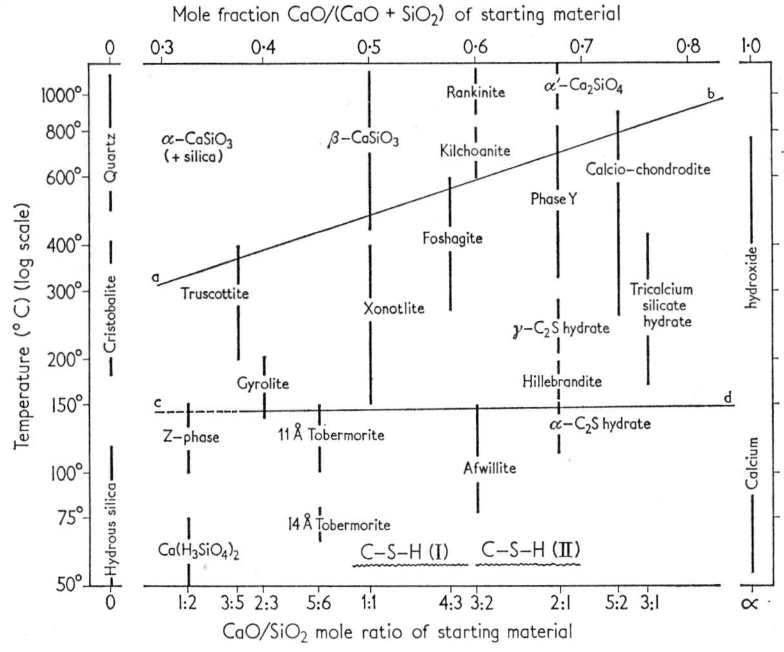

Fig. 15. Diagram showing the approximate conditions of formation of the better established anhydrous and hydrated calcium silicates under hydrothermal conditions. Vertical lines indicate Ca/Si mole ratios of compounds. The extension of each line gives an indication of the range of temperature over which the compound is formed, while displacement of the name to either side of this line indicates that the compound tends to form not only from mixtures of its own Ca/Si ratio, but also from mixtures of higher or lower Ca/Si ratio. Wavy underlining denotes variable composition. For temperatures up to 374°C, saturated steam pressure is assumed, and for higher temperature the water pressure is assumed to be around 400kg/cm². For significance of lines ab and cd, see text.

Barrer [*162*] in connection with the synthesis of aluminosilicates. Figure 15 is in no sense a phase equilibrium diagram; it merely represents the conditions under which each phase is most usually obtained, without regard to its stability. It is assumed that saturated steam conditions apply below 374° C, and that for higher temperatures the water pressure is around 400 kg/cm^2. Studies by Roy and collaborators [*8, 41, 57*] show that increasing the pressure beyond about 300 kg/cm^2 usually has only a minor effect on stability relations (Fig. 6). Reactions are also assumed to have been carried out in aqueous suspensions, using reactive starting materials such as C-S-H(I), β-C_2S, or lime–silica mixtures, and times long enough to permit the formation of *apparently* stable products (a few months below 150° C, falling to a few days at 800° C). Naturally, not all the available information can be presented on a single plot of this type. To take one example, it is not possible to show the fact that γ-C_2S hydrate is formed over a wider range if γ-C_2S is used as the starting material than from β-C_2S or other materials.

At temperatures above the line AB, anhydrous compounds are formed. Between the lines AB and CD, the usual products are hydrated compounds in which the water is largely or completely present as ionic hydroxyl, though with truscottite and perhaps also with xonotlite, some molecular water may also be present. Below the line CD, compounds are in general obtained which contain Si—OH groups; they may also contain molecular water, or ionic hydroxyl, or both.

For temperatures above CD, the degree of condensation of the silicate anion tends to fall with increase in Ca/Si ratio or in temperature; this is merely a consequence of the facts that water, if present, occurs as ionic hydroxyl, and that anhydrous compounds are produced at the higher temperatures. Below CD, the relationship is more complex, because of the occurrence of Si—OH groups.

In the light of these generalizations, the apparent formation of kilchoanite at temperatures little above 200° C (p. 209) is of interest. Either an anhydrous compound is formed at a temperature much lower than would be expected, or kilchoanite is capable of existing in a hydrated form. Further work is clearly needed.

B. Phase Equilibria

Phase equilibrium studies on the system CaO–SiO_2–H_2O fall into two main groups. Those of the first group have been concerned with the concentrations of CaO and SiO_2 existing in solutions in true or apparent equilibrium with a given solid phase. This approach has been adopted in many studies on C-S-H(I) at room temperature, the results of which have been represented by curves such as those shown in Fig. 12. It is difficult to extend such studies to higher temperatures, because the concentrations in solution become very low. Only one such study has been reported; Peppler [*55*] investigated the system at 180° C and saturated steam pressure.

The second approach was originally developed for studies on geologically important systems. Its application to calcium silicate hydrates is due especially to Roy, Harker and collaborators [*8*, *41*, *57*, *136*, *163*]. It is concerned with finding the nature of the equilibrium solid phases present for given combinations of the Ca/Si ratio, water pressure, and temperature. The results of individual studies of this type have usually been expressed in terms of *P–T* curves of the type already discussed (Fig. 6). If sufficient data are available, it is possible to establish the assemblages of phases which can coexist with each other and with the liquid (or supercritical fluid) for any desired composition at a given combination of pressure and temperature. Roy and Harker [*57*] tentatively presented this information in the form of a series of triangular diagrams for temperatures ranging from below 100° C to about 1100° C. A typical diagram, for 350° C and about 1000 kg/cm^2, is shown in Fig. 16. The compatibility triangles, bounded by thin lines, link solid phases which coexist at equilibrium under the given conditions.

In Fig. 16, lines are drawn from the compositions of the solid phases to the water apex, the fluid phase being thus regarded as pure water. The solubilities of silica and especially of lime are so low at these temperatures that a greatly enlarged scale would be needed to represent them. At temperatures above about 800° C the position is complicated by the appearance of a melt in addition to the solids and the aqueous phase.

Harker, Roy and Tuttle [*163*] studied the join Ca_2SiO_4–$Ca(OH)_2$ at temperatures up to 1110° C, under a pressure of 1000 bars. Figure 17 shows their results. The join is binary, and calcio-chondrodite melts incongruently at 955° C. A few preliminary results were also reported for the system CaO–Ca_2SiO_4–$Ca(OH)_2$, which indicated that the eutectic at which the fields of primary $Ca(OH)_2$, CaO and $Ca_5(SiO_4)_2(OH)_2$ meet is close to the CaO–$Ca(OH)_2$ side of the triangle at about 805° C. The authors reported that primary crystals formed from liquids of relatively low water content were sometimes larger and better formed than those formed in other ways and pointed out that this might provide a method of growing single crystals of hydrated phases for X-ray work. They also mentioned the possibility of employing pressures above atmospheric for the production of cement clinker.

Phase equilibrium studies in the CaO–SiO_2–H_2O system probably

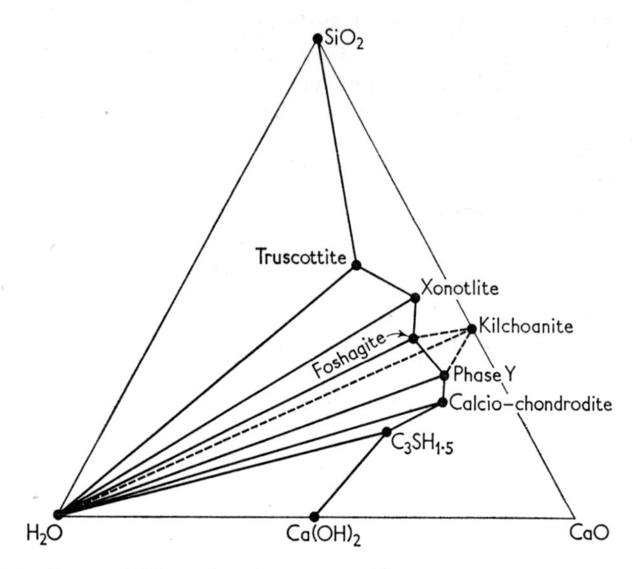

FIG. 16. Compatibility triangle representing equilibrium assemblages of solid phase in the system $CaO–SiO_2–H_2O$ at 350°C and saturated steam pressure, according to Roy and Harker [*57*]. Dashed lines indicate uncertainty.

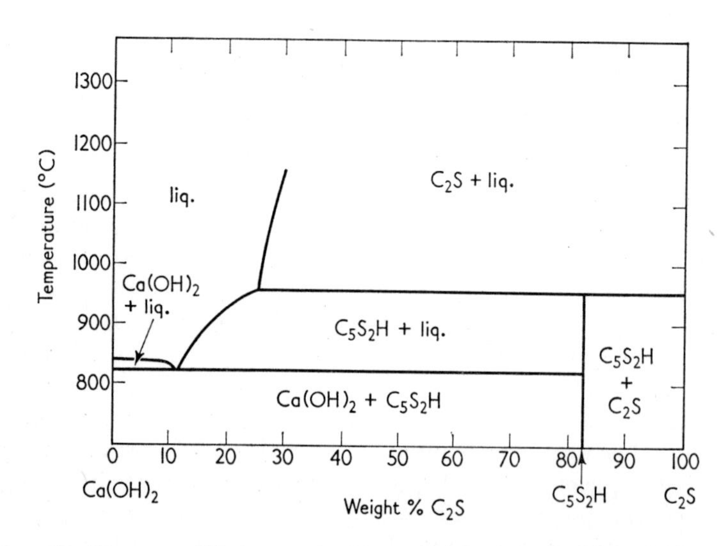

FIG. 17. Phase equilibria for the binary sub-system Ca_2SiO_4–$Ca(OH)_2$ under a pressure of 1000 bars (Harker, Roy and Tuttle [*163*]).

have their principal application above 200° C, and especially above 400° C; at these temperatures either metastable or true equilibrium is probably attained within a few days or weeks. Below 200° C, the approach to equilibrium is slow and complicated, and the products formed under given circumstances depend on many factors, such as the nature of the starting materials. Equilibrium relations, even where they can be determined, then provide only a small fraction of the useful information, and the more empirical approach represented in Fig. 15 is probably more fruitful. Even for temperatures below 200° C, however, the approach followed by Roy and collaborators is valuable in that it draws attention to the possibility of studying the entire field of pressure and temperature and not merely the line arbitrarily traced out by the saturated steam curve. This applies to non-equilibrium as well as equilibrium studies. There are no recent reports of work at pressures above that of saturated steam, but Assarsson [*112*] has reported experiments made in unsaturated or superheated steam.

C. Thermodynamics

Mchedlov-Petrosyan and collaborators [*164, 165*] have used thermodynamics to predict the phases which might be expected to form under various conditions and the energetics of the reactions involved. They used approximate methods to calculate values of ΔH_{298}, ΔF_{298}, S_{298} and C_p as a function of temperature for each of the anhydrous and hydrated compounds principally involved, and also values of ΔF_{298} and ΔH_{298} for silicate and other ions in aqueous solution. From these data they calculated free energy changes for a considerable number of reactions. These included reactions of formation of the various hydrated compounds from solutions containing Ca^{2+} and silicate ions at given Ca/Si ratios, hydration reactions of anhydrous cement compounds, and transformation reactions in which one hydrated silicate was formed from another. So far, only crystalline compounds have been considered. Figure 18 shows typical results obtained from these calculations.

In Fig. 18(a), the values of ΔF are given for the formation of various hydrated compounds from a solution in which the Ca/Si ratio is 5:6. It indicates the formation at successively higher temperatures of 14 Å, 11 Å and 9 Å tobermorites, and xonotlite. In Fig. 18(b), ΔF values are calculated for a number of real or hypothetical reactions of okenite. The theory here predicts that below 130° C okenite is unstable relative to gyrolite, and that at 130–200° C it is unstable relative to 11 Å tobermorite.

In general, the results agree with those obtained from experiment, but there are some apparent discrepancies. Thus, the theory predicts that afwillite is unstable relative to foshagite at 25–200° C. Also, some compounds, including calcio-chondrodite, α-dicalcium silicate hydrate, and tricalcium silicate hydrate, do not appear to have been considered.

The approach is interesting in that it stresses the importance of the concentrations of CaO and SiO_2 in solution in determining which phase is precipitated. This factor may partially explain the fact that different starting materials of a given Ca/Si ratio frequently yield different products. Thus, the theory predicts that gyrolite can be formed only if

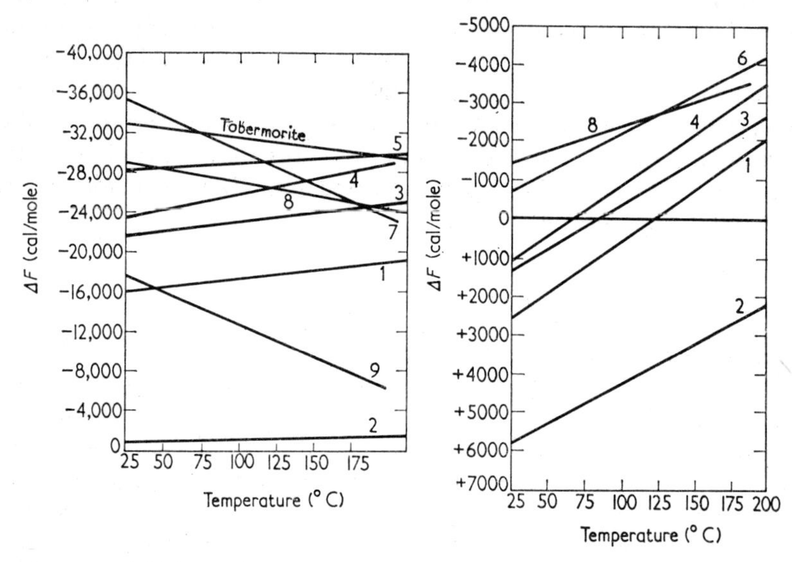

FIG. 18. Calculated free energy changes for real or hypothetical reactions in the $CaO–SiO_2–H_2O$ system. A. Formation of various compounds from a solution in which the Ca/Si ratio is 5:6. B. Transformations of okenite yielding other phases. Key to phases for both diagrams: 1, hillebrandite; 2, afwillite; 3, foshagite; 4, xonotlite; 5, 9 Å tobermorite; 6, 11 Å tobermorite; 7, 14 Å tobermorite; 8, gyrolite; 9, okenite. (After Mchedlov-Petrosyan and Babushkin [*164*].)

certain starting materials, including active silica, are used, in order to produce a sufficiently low Ca/Si ratio in the solution. This agrees with the experimental finding of Kalousek [*62*], and also with Peppler's view [*55*] that gyrolite is unstable relative to xonotlite and silica at 180° C. The theory is thus relevant to the formation of non-equilibrium as well as equilibrium products.

Mchedlov-Petrosyan and his collaborators concluded from their studies that the hydration of cements occurs by a through-solution mechanism. They also considered that hydraulic activity in the anhydrous compounds was associated with the occurrence of a low coordination number of the metal cation, though the possibility of the cation having an irregular co-ordination sphere due to the formation of additional and longer bonds to oxygen was apparently not excluded.

X. Thermal Behaviour and Topotactic Reactions

A. Weight-loss and DTA Data

The thermal behaviour of calcium silicate hydrates has been studied in three principal ways: weight-loss curves, differential thermal analysis (DTA), and X-ray or SED studies on heated material. The X-ray and SED studies have been made with powders and with single crystals. The latter method provides the greater amount of information. As shown in Chapter 25 (Volume 2), thermal weight-loss curves can be obtained using either static (or quasi-static) or dynamic techniques. The former normally give the more sharply defined decomposition temperatures; the latter are more rapid and can be compared directly with results obtained by DTA, which is necessarily a dynamic method. The decomposition temperatures obtained using dynamic methods are often about 100° C higher than those found with static methods. Table III summarizes results obtained by all of these methods. The DTA data were mostly obtained at a heating rate of about 10 deg/min.

Those calcium silicate hydrates which contain ionic hydroxyl but no SiOH nor molecular water decompose at 400–800° C under static conditions. These include xonotlite, foshagite and calcio-chondrodite, gyrolite and truscottite in their second stages of decomposition, and probably also hillebrandite, Phase Y, and tricalcium silicate hydrate. The thermal stability increases as the proportion of hydroxyl decreases; one extreme case is represented by $Ca(OH)_2$, which decomposes at about 400° C [*29*], and the other by compounds such as calcio-chondrodite or Phase Y which contain little hydroxyl and are stable to over 650° C.

Compounds containing SiOH or molecular water are less stable. The molecular water in gyrolite, truscottite, nekoite, okenite and 14 Å or 11 Å tobermorite is mostly lost by about 300° C, while afwillite and α-C_2S hydrate, which contain $HSiO_4^{3-}$ groups, are decomposed by 450° C. However, the small amount of water in 9 Å tobermorite is only lost at 300–800° C, although it probably occurs as SiOH groups.

The DTA curves in general show endothermic peaks corresponding to loss of water, irrespective of the form in which this occurs. In some cases there are additional peaks at higher temperatures corresponding to the appearance of new phases; the 850° C peak shown by afwillite probably

Table III

Thermal behaviour of the calcium silicate hydrates

Group	Compound	Static weight loss: Mainly in range (°C)	Static weight loss: References	D.T.A.: Main peak (1)	D.T.A.: References	Crystalline phase(s) formed on decomposition: Phase(s)	Whether topotactic	References
Wollastonite group	Nekoite	40–500	[*96*]	—	—	β-CS (2)	Yes	[*96*]
	Okenite	(?)	—	—	—	β-CS	Yes	[*45*]
	Xonotlite	680–700	[*41, 142*]	790–840(−)	[*41, 167, 168*]	β-CS	Yes	[*45, 46*]
	Foshagite	650–750	[*50*]	—	—	β-CS + sometimes β-C_2S	Yes	[*45, 50*]
	Hillebrandite	520–540	[*60*]	540–630(−)	[*11, 60, 167, 168*]	β-C_2S	No	[*45, 60*]
Tobermorite group	14 Å Tobermorite	55–60	[*95*]	128°(?)(−)	[*63*]	11 Å Tobermorite	Yes	[*45, 63, 87*]
	11 Å Tobermorite	100–300	[*63, 83, 84*]	250–260(−)	[*63, 78, 91*]	9 Å Tobermorite(3)	Yes	[*45, 63, 87*]
	9 Å Tobermorite	300–800	[*63, 83*]	800–850 (4)(+)	[*63, 78, 91*]	β-CS	Yes	[*45, 63, 87, 88*]
	C-S-H(I)	20–500	[*87, 104*]	830–900 (4)(+)	[*116*]	β-CS	(?)	[*105*]
	C-S-H(II)	20–500	[*83*]	Variable	[*68*]	β-C_2S	(?)	[*83*]
	Tobermorite gel	20–700	[*74*]	120°(−)	[*169*]	β-C_2S	(?)	[*166*]
Gyrolite group	Gyrolite	20–300, 600–800	[*129*]	—	—	α-CS	Yes	[*45, 129*]
	Truscottite	20–300, 700–800	[*131*]	—	—	α-CS	Yes	[*124, 127*]
γ-C_2S group	Calcio-chondrodite	650–700	[*137*]	—	—	γ-C_2S	Yes	[*137*]
	γ-C_2S hydrate	650–700	[*29*]	750°(−)	[*11*]	—	—	—
Other compounds	Afwillite	275–285	[*142, 146*]	400°(−), 850°(+)	[*142*]	γ-C_2S (5)	Yes	[*45, 142, 150*]
	α-C_2S hydrate	400–450	[*29*]	460–480° (−)	[*11*]	β-C_2S	No	[*45*]
	Phase Y	640–700	[*155, 157*]	690°(−), 980°(−)	[*155*]	β-C_2S	No	[*157*]
	Tricalcium silicate hydrate	420–550	[*61*]	—	—	γ-C_2S + CaO	(?)	[*61*]

(1) (+) denotes exotherm, (−) denotes endotherm. Only those peaks which appear to be reproducibly shown are listed.
(2) After several intermediate stages, including a xonotlite-like stage [*96*].
(3) "Normal" variety; see p. 186.
(4) This peak is very weak or absent for many (perhaps all) crystalline tobermorites, but is very strong for C-S-H(I) [*68*].
(5) Kilchoanite formed at 700°.

corresponds to the formation of kilchoanite, while the 800–900° C peak shown by C-S-H(I) corresponds to the formation of β-CS.

B. Topotactic Reactions

Reference to Table III shows that most of the calcium silicate hydrates are decomposed topotactically; that is, a single crystal of a starting material is converted into something approaching a single crystal of the product. Processes of this type occur because the crystal structures of starting material and product are related, so that complete destruction and recrystallization does not occur. Dent Glasser, Glasser and Taylor [*170*] have recently reviewed topotactic reactions in inorganic oxy-salts generally, including the dehydration reactions shown in Table III.

In those dehydrations of calcium silicate hydrates that occur at 650° C or over, the dominant tendency seems usually to be for the Ca—O parts of the structure to be preserved. This is shown particularly clearly in the dehydration of xonotlite (Fig. 19) [*46*, *171*]. The structures of both xonotlite and β-CS, which is formed on dehydration, can be considered as being composed of distorted octahedral sheets similar to those in $Ca(OH)_2$, and the observed relationship of orientations between starting material and product indicates that these sheets suffer only minor changes in the transformation. More drastic changes occur in the regions between these sheets, which contain the silicons and those oxygens not directly attached to calcium. In order to produce the single dreierketten of β-CS from the double dreierketten of xonotlite, it is necessary to postulate migrations of Si atoms or ions out of the oxygen tetrahedra in which they were originally present, into initially empty tetrahedra with which these share a common face. In addition, movements of oxygen must occur, since some of the oxygen atoms (ringed in Fig. 19) are not required in the new structure. These are not oxygens that were present as hydroxyl groups; the hydrogen atoms or ions from the latter evidently migrate separately, and the water molecules are formed later.

This mechanism implies that the SiO_4 tetrahedron is not, as has often been assumed, a stable entity at high temperatures but a relatively labile one. Studies on dehydrations of silicates containing other metallic cations, such as Mg^{2+}, together with ionic hydroxyl, lead to the same conclusion [*170*]. The processes can perhaps best be visualized using an ionic model; the Si^{4+} ion is very small and migrates relatively freely, whereas the Ca—O parts of the structure are composed entirely of large ions and are more stable.

In the calcium silicate field, two tests of the type of mechanism described for xonotlite have been made. Firstly, a similar mechanism was assumed to apply in the dehydration of foshagite; a trial crystal structure was predicted partly on this basis and was afterwards found to be correct (p. 180). Secondly, a study was made of the topotactic transformation of rhodonite ($(Mn,Ca)SiO_3$) to a manganiferous wollastonite [*172*]. The results of this study provided very strong support for the hypothesis of silicon migration.

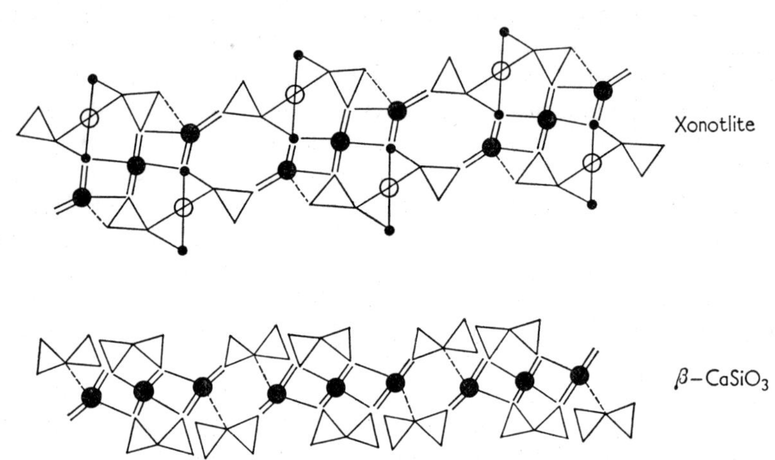

FIG. 19. Portions of the crystal structures of xonotlite and β-$CaSiO_3$, seen in projection along the chain direction in each case. The relative orientations are those found experimentally in the xonotlite–β-$CaSiO_3$ transformation. Triangles outline the projection of linked SiO_4 tetrahedra constituting the Si—O chains. Large full circles represent Ca^{2+} ions, and small full circles, OH^- ions (these coincide in projection with oxygen atoms but are not directly linked to Si, see Fig. 8). Large open circles represent oxygen atoms present in xonotlite but not in β-$CaSiO_3$ (see text).
Lines radiating from Ca atoms represent Ca—O bonds; where these are double there are twice as many in the height of the projection as where they are single. Broken lines represent Ca—O bonds that are present either in xonotlite, or in β-$CaSiO_3$, but not in both; all other Ca—O bonds occur in both structures. (After Taylor [*171*].)

Those topotactic dehydrations of calcium silicate hydrates which occur at lower temperatures seem to occur by a different mechanism. In the formation of γ-C_2S from afwillite at 275–285° C, the orientation relationship is such as to suggest that the SiO_4 tetrahedra remain intact, while migrations of Ca^{2+} and H^+ ions and water molecules occur [*142*, *150*].

In many of the topotactic dehydration processes listed in Table III, there is an apparent failure to balance the chemical equation; thus gyrolite ($C_2S_3H_2$) yields α-CS as the only detectable crystalline product.

It has often been assumed that amorphous silica is formed in such cases. While this may sometimes be the case, it appears likely from studies made by Gay and LeMaitre [*173*] on certain topotactic products formed in nature that there is often no sharp division into distinct regions or phases within the altered crystal. Considerable regions of the crystal may have a disordered structure in which only certain features of the original structure have been preserved; in some regions, which are not sharply defined, the structure approximates in varying degrees to that of a fully crystalline compound. In the case of foshagite, regions which have crystallized to β-CS are always found, but the development of β-C_2S does not always occur.

Topotactic reactions of calcium silicate hydrates occur hydrothermally. Taylor [*47*] showed that the formation of xonotlite from 11 Å tobermorite can occur in this way. Afwillite also undergoes topotactic hydrothermal reactions; under various conditions, a number of products containing dreierketten, including xonotlite, foshagite and tobermorite-like products, and also kilchoanite or a product related to it can be obtained [*96*].

XI. Calcium Silicates Containing Fluoride or Carbonate Ions

A. Fluoride Ions

Two anhydrous and two hydrated compounds are known. Cuspidine ($Ca_4Si_2O_7F_2$) occurs as a natural mineral and can also be synthesized [*156*]. Its crystal structure is known [*174*]; the $Si_2O_7^{6-}$ groups are arranged so as to resemble dreierketten with every third tetrahedron missing. Recent work [*156*] does not support earlier views that all or part of the fluorine can be replaced by hydroxyl.

A fluoride analogue of calcio-chondrodite has been prepared [*175*].

Zeophyllite ($Ca_4Si_3O_{12}F_2H_6$) is known as a natural mineral, but this has not been synthesized. It forms micaceous, pseudo-hexagonal crystals and shows some analogies to gyrolite [*176*]. The ionic constitution is probably $Ca_4(Si_3O_8)F_2(OH)_2 . 2H_2O$. Strunz and Micheelsen [*130*] showed that radiophyllite is identical with zeophyllite. Another reported mineral, foshallassite [*177*], is perhaps also identical with zeophyllite.

Bultfonteinite ($Ca_4Si_2O_{10}H_6F_2$) is also known as a mineral but has not been synthesized. It is structurally related to afwillite [*178*].

B. Carbonate Ions

The influence of carbonate ions on the anhydrous and hydrated calcium silicates is important because of the extreme difficulty of excluding atmospheric CO_2 during their preparation. In principle, there seem to be three possibilities. Firstly, the CO_3^{2-} ion may occur entirely as calcium carbonate; this normally occurs in calcium silicate preparations as either calcite or vaterite, and is readily identified by optical microscopy or X-ray powder patterns. Secondly, the CO_3^{2-} ion may enter into the structure of a calcium silicate hydrate, possibly replacing hydroxyl ions and water molecules. Gaze and Robertson [*78*] suggested that CO_3^{2-} could enter the tobermorite structure. This has been disputed [*179*], but recent infra-red work appears to confirm that it can occur in some of the compounds (Volume 2, Chapter 22). Thirdly, new phases may be formed. Three of these are known.

1. Scawtite $(Ca_7(Si_6O_{18})(CO_3).H_2O2?)$

This was originally found as a mineral. McConnell and Murdoch [*180*] have reviewed the available data. They found variations in composition, which they attributed to replacement of Si by C. They wrote the general formula $Ca_{14}(OH)_4(Si_{16-x-y}C_{4x/3}H_{4y})O_{44}$. This cannot be correct, because it implies that one Si is replaced by 4/3 of a carbon; confusion has arisen between oxidation state and co-ordination number. The formula can also be criticized in that it presupposes a structural similarity to the amphiboles; this is most unlikely for a calcium mineral, and the unit cell parameters [*181, 182*] suggest the presence of dreierketten, as in xonotlite or wollastonite, and not of zweierketten, as in the amphiboles. Further work is needed to establish the formula, but the hypothesis of replacement (in a broad sense) of C by Si appears well founded.

Flint, McMurdie and Wells [*1*] prepared a phase under saturated steam conditions at 150° C, which they considered to be $CaO.SiO_2.H_2O$. Bogue [*2*] named it CSH(A). Other investigators [*40, 41, 79, 145*] have since reported the preparation of this phase, which appears to form at temperatures intermediate between those giving 11 Å tobermorite and those giving xonotlite. Its X-ray pattern [*161*] is close to that of scawtite. Buckner, Roy and Roy [*41*] suggested that it was in fact identical with scawtite. This view appears very probable. Jeevaratnam and Taylor [*95*] found that "CSH(A)" was frequently produced in runs where inadequate precautions had been taken to exclude CO_2, but that when greater care was taken, it was never formed. This suggests that some CO_3^{2-} is essential to the structure.

2. Tilleyite $(Ca_5(Si_2O_7)(CO_3)_2)$ *and Spurrite* $(Ca_5(SiO_4)_2(CO_3))$

There are anhydrous phases which occur as natural minerals and have also been synthesized. Harker and Tuttle have made a series of studies on the anhydrous ternary system $CaO-SiO_2-CO_2$; they obtained $P_{CO_2}-T$ curves for various transformations and synthesized both tilleyite and spurrite [*183*]. McConnell [*184*] showed that both phases could also be made at lower temperatures if water at high pressure was present and CO_2 not in excess.

XII. Optical and Crystallographic Data

X-Ray powder data, unit cells, space groups, densities, and optical properties for the most important of the calcium silicate hydrates are given in Appendix I (Volume 2). These properties will not be listed for the remaining compounds as they were tabulated for most of them by Heller and Taylor [*49*]. The following important additions or corrections should, however, be noted (Table IV). Note also that the optic sign of kilchoanite is negative [*186*], and the sign of elongation of calcio-chondrodite is negative [*8*].

TABLE IV

Crystallographic data for the hydrated calcium silicates. Important additions or corrections to the compilation by Heller and Taylor [*49*]

Compound	Data	Reference
Xonotlite	Determination of (true) unit cell	[*185*]
Foshagite	Revised composition, unit cell, indexing of powder data	[*50, 51*]
9·3 Å Tobermorite	Indexing of powder data, unit cell	[*88*]
Truscottite	Revised powder data	[*124*]
Z-Phase (of Assarsson)	Data for new phase	[*19, 73*]
Kilchoanite (anhydrous)	Data for new phase	[*140*]
Calcio-chondrodite	Data for new phase	[*137*]
Phase Y	Data for new phase	[*155*]
Tricalcium silicate hydrate	Indexing of powder data, unit cell, density	[*61*]

REFERENCES

1. Flint, E. P., McMurdie, H. F. and Wells, L. S. (1938). *J. Res. nat. Bur. Stand.* **21**, 617.
2. Bogue, R. H. (1953). *Mag. Concr. Res.* **3**, 87.
3. Taylor, H. F. W. and Bessey, G. E. (1950). *Mag. Concr. Res.* **1**, 15.
4. Thorvaldson, T. (1939). *Proceedings of the Symposium on the Chemistry of Cements, Stockholm 1938*, p. 246. Ingeniörsvetenskapsakademien, Stockholm.
5. Heddle, M. F. (1880). *Miner. Mag.* **4**, 119; (1893). *Trans. geol. Soc., Glasgow* **9**, 254.

6. Claringbull, G. F. and Hey, M. H. (1952). *Miner. Mag.* **29**, 960.
7. Funk, H. and Thilo, E. (1955). *Z. anorg. Chem.* **278**, 237.
8. Roy, D. M. (1958). *Amer. Min.* **43**, 1009.
9. Jander, W. and Franke, B. (1941). *Z. anorg. Chem.* **247**, 161.
10. Funk, H. (1958). *Z. anorg. Chem.* **297**, 103.
11. Kalousek, G. L., Logiudice, J. S. and Dodson, V. H. (1954). *J. Amer. ceram. Soc.* **37**, 7.
12. Eakle, A. S. (1917). *Bull. Dept. Geol. Univ. Calif.* **10**, 327.
13. Tilley, C. E. and Harwood, H. F. (1931). *Miner. Mag.* **22**, 439.
14. Heddle, M. F. (1923). "Mineralogy of Scotland", Vol. 2, 1901, reprinted. St. Andrews University Press.
15. Brunauer, S. and Greenberg, S. A. (1962). *Chemistry of Cement, Proceedings of the Fourth International Symposium, Washington 1960*, p. 135. National Bureau of Standards Monograph 43. U.S. Department of Commerce.
16. Assarsson, G. O. and Bokström, J. M. (1953). *Analyt. Chem.* **25**, 1844.
17. Assarsson, G. O. (1954). *Zement-Kalk-Gips* **7**, 167.
18. Pressler, E. E., Brunauer, S. and Kantro, D. L. (1956). *Analyt. Chem.* **28**, 896.
19. Funk, H. (1961). *Z. anorg. Chem.* **313**, 1.
20. Bassett, H. (1934). *J. Chem. Soc.* 1270.
21. Peppler, R. B. and Wells, L. S. (1954). *J. Res. nat. Bur. Stand.* **52**, 75.
22. Bates, R. G., Bower, V. E. and Smith, E. R. (1956). *J. Res. nat. Bur. Stand.* **56**, 305.
23. Weir, C. E. (1955). *J. Res. nat. Bur. Stand.* **54**, 37.
24. Bernal, J. D. and Megaw, H. D. (1935). *Proc. roy. Soc.* **A151**, 384; Petch, H. E. and Megaw, H. D. (1954). *J. opt. Soc. Amer.* **44**, 744.
25. Busing, W. R. and Levy, H. A. (1957). *J. chem. Phys.* **26**, 563.
26. A.S.T.M. Index card 4-0733 for $Ca(OH)_2$, 1950.
27. Wyllie, P. J. and Tuttle, O. F. (1959). *J. Amer. ceram. Soc.* **42**, 448.
28. Halstead, P. E. and Moore, A. E. (1957). *J. chem. Soc.* 3873.
29. Buckle, E. R. (1959). *J. phys. Chem.* **63**, 1231.
30. Neese, H., Spangenberg, K. and Weiskirchner, W. (1957). *TonindustrZtg.* **81**, 325.
31. von Kobell, F. (1828). *Arch. gesammte Naturlehre* (*Kastner*) **14**, 333; Breithaupt, A. (1845). *Ann. Phys. Chem.* (*Poggendorff*) **64**, 170.
32. Bøggild, O. B. (1922). *Det. Kgl. Danske Videnskab. Selskab., Math.-fys. Medd.* **4**, No. 8, 1.
33. Gard, J. A. and Taylor, H. F. W. (1956). *Miner. Mag.* **31**, 5.
34. Mamedov, Kh. S. and Belov, N. V. (1958). *Dokl. Akad. Nauk S.S.S.R.* **121**, 720.
35. Christie, W. A. K. (1924). *Rec. Geol. Survey India* **56**, 199.
36. Chalmers, R. A., Nicol, A. W. and Taylor, H. F. W. (1962). *Miner. Mag.* **33**, 70.
37. Rammelsberg, C. F. (1866). *Z. dtsch. geol. Ges.* **18**, 33.
38. Nagai, S. (1932). *Z. anorg. Chem.* **206**, 177; (1932) **207**, 321; see also Büssem, W. (1939). *Proceedings of the Symposium on the Chemistry of Cements, Stockholm 1938*, p. 141. Ingeniörsvetenskapsakademien, Stockholm.
39. Assarsson, G. O. (1958). *J. phys. Chem.* **62**, 223.
40. Heller, L. and Taylor, H. F. W. (1951). *J. chem. Soc.* 2397.
41. Buckner, D. A., Roy, D. M. and Roy, R. (1960). *Amer. J. Sci.* **258**, 132.
42. Akaiwa, S. and Sudoh, G. (1956). *Semento Gijutsu Nenpo* **10**, 14.
43. Mamedov, Kh. S., and Belov, N. V. (1955). *Dokl. Akad. Nauk S.S.S.R.* **104**, 615; (1956) *Zapiskii Vsesoyuz. Mineralog. Obshchestva* **85**, 13.

44. Aitken, A. and Taylor , H. F. W. (1960). *J. appl. Chem.* **10**, 7.
45. Heller, L. (1954). *Proceedings of the Third International Symposium on the Chemistry of Cement, London 1952*, p. 237. Cement and Concrete Association, London.
46. Dent, L. S. and Taylor, H. F. W. (1956). *Acta cryst.* **9**, 1002.
47. Taylor, H. F. W. (1959). *Miner. Mag.* **32**, 110.
48. Eakle, A. S. (1925). *Amer. Min.* **10**, 97.
49. Heller, L. and Taylor, H. F. W. (1956). "Crystallographic Data for the Calcium Silicates". HMSO, London.
50. Gard, J. A. and Taylor, H. F. W. (1958). *Amer. Min.* **43**, 1.
51. Gard, J. A. and Taylor, H. F. W. (1959). *Nature, Lond.* **183**, 171; (1960). *Acta cryst.* **13**, 785.
52. Mamedov, Kh. S. and Belov, N. V. (1958). *Dokl. Akad. Nauk S.S.S.R.* **121**, 901.
53. Wright, F. E. (1908). *Amer. J. Sci.* **26**, 545.
54. Vigfusson, V. A., Bates, G. N. and Thorvaldson, T. (1934). *Canad. J. Res.* **11**, 520.
55. Peppler, R. B. (1955). *J. Res. nat. Bur. Stand.* **54**, 205.
56. Heller, L. and Taylor, H. F. W. (1952). *J. chem. Soc.* 2535.
57. Roy, D. M. and Harker, R. I. (1962). *Chemistry of Cement, Proceedings of the Fourth International Symposium, Washington 1960*, p. 196. National Bureau of Standards Monograph 43. U.S. Department of Commerce.
58. Heller, L. (1953). *Miner. Mag.* **30**, 150.
59. Mamedov, Kh. S. and Belov, N. V. (1958). *Dokl. Akad. Nauk S.S.S.R.* **123**, 741.
60. Torpov, N. A., Nikogosyan, Kh. S. and Boikova, A. I. (1959). *Zh. neorg. Khim.* **4**, 1159.
61. Buckle, E. R., Gard, J. A. and Taylor, H. F. W. (1958). *J. chem. Soc.* 1351.
62. Kalousek, G. L. (1955). *J. Amer. Concr. Inst.* **51**, 989.
63. McConnell, J. D. C. (1954). *Miner. Mag.* **30**, 293.
64. McConnell, J. D. C. (1955). *Miner. Mag.* **30**, 672.
65. Sweet, J. M. (1961). *Miner. Mag.* **32**, 745.
66. Taylor, H. F. W. (1953). *Miner. Mag.* **30**, 155.
67. Bernal, J. D. (1954). *Proceedings of the Third International Symposium on the Chemistry of Cement, London 1952*, p. 216. Cement and Concrete Association, London.
68. Kalousek, G. L. and Prebus, A. F. (1958). *J. Amer. ceram. Soc.* **41**, 124.
69. Taylor, H. F. W. (1950). *J. chem. Soc.* 3682.
70. Taylor, H. F. W. (1961). *In* "Progress in Ceramic Science", ed. by J. E. Burke, p. 89. Pergamon Press, Oxford.
71. Kantro, D. L., Brunauer, S. and Weise, C. H. (1959). *J. Colloid Sci.* **14**, 363.
72. Assarsson, G. O. (1957). *J. phys. Chem.* **61**, 473.
73. Assarsson, G. O. (1962). *Chemistry of Cement, Proceedings of the Fourth International Symposium, Washington 1960*, p. 190. National Bureau of Standards Monograph 43. U.S. Department of Commerce.
74. Buckle, E. R. and Taylor, H. F. W. (1959). *J. appl. Chem.* **9**, 163.
75. Funk, H. (1957). *Z. anorg. Chem.* **291**, 276.
76. Currie, J. (1905). *Miner. Mag.* **14**, 93.
77. Kalousek, G. L. (1954). *J. Amer. Concr. Inst.* **50**, 365.
78. Gaze, R. and Robertson, R. H. S. (1956). *Mag. Concr. Res.* **8**, 7.
79. Berkovich, T. M., Kheĭker, D. M., Gracheva, O. I., Zevin, L. S. and Kupreyeva, N. I. (1958). *Dokl. Akad. Nauk S.S.S.R.* **120**, 853.

80. Greenberg, S. (1954). *J. phys. Chem.* **58**, 362.
81. Sanders, L. D. and Smothers, W. J. (1957). *J. Amer. Concr. Inst.* **54**, 127.
82. Neese, H. (1959). *TonindustrZtg.* **83**, 124.
83. Gard, J. A., Howison, J. W. and Taylor, H. F. W. (1959). *Mag. Concr. Res.* **11**, 151.
84. Gard, J. A. and Taylor, H. F. W. (1957). *Miner. Mag.* **31**, 361.
85. Megaw, H. D. and Kelsey, C. H. (1956). *Nature, Lond.* **177**, 390; see also *Tercera reunion internacional sobre reactividad solidos, Madrid, 1956*, p. 355.
86. Mamedov, Kh. S. and Belov, N. V. (1958). *Dokl. Akad. Nauk S.S.S.R.* **123**, 163.
87. Taylor, H. F. W. (1953). *J. chem. Soc.* 163.
88. Taylor, H. F. W. (1959). *Proc. 6th National Conference on Clays and Clay Minerals, Berkeley, 1957*, p. 101.
89. Grothe, H., Schimmel, G. and zur Strassen, H. (1962). *Chemistry of Cement, Proceedings of the Fourth International Symposium, Washington 1960*, p. 194. National Bureau of Standards Monograph 43. U.S. Department of Commerce.
90. Kurczyk, H. G. and Schwiete, H. E. (1962). *Chemistry of Cement, Proceedings of the Fourth International Symposium, Washington 1960*, p. 349. National Bureau of Standards Monograph 43. U.S. Department of Commerce.
91. Kalousek, G. L. (1957). *J. Amer. ceram. Soc.* **40**, 74.
92. Murdoch, J. (1961). *Amer. Min.* **46**, 245.
93. Kalousek, G. L. and Roy, R. (1957). *J. Amer. ceram. Soc.* **40**, 236.
94. Carpenter, A. B. Private communication.
95. Jeevaratnam, J. and Taylor, H. F. W. New data.
96. Nicol, A. W. and Taylor, H. F. W. New data.
97. Le Chatelier, H. (1887). *Ann. Mines.* **11**, 345.
98. Steinour, H. H. (1947). *Chem. Rev.* **40**, 391.
99. Steinour, H. H. (1954). *Proceedings of the Third International Symposium on the Chemistry of Cement, London 1952*, p. 261. Cement and Concrete Association, London.
100. Brunauer, S. (1962). *Amer. Sci.* **50**, 210.
101. Hedin, R. (1945). *Swedish Cement and Concrete Res. Inst. Proc.* No. 3.
102. Grudemo, Å. (1955). *Swedish Cement and Concrete Res. Inst. Proc.* No. 26.
103. Flint, E. P. and Wells, L. S. (1934). *J. Res. nat. Bur. Stand.* **12**, 751.
104. Taylor, H. F. W. and Howison, J. W. (1956). *Clay Minerals Bull.* **3**, 98.
105. Lawrence, C. D. (1962). Private communication.
106. Funk, H. (1960). *Silikattechnik*, **11**, 375.
107. Brunauer, S., Copeland, L. E. and Bragg, R. H. (1956). *J. phys. Chem.* **60**, 112, 116.
108. Kalousek, G. L. (1957). Private communication.
109. Brunauer, S., Kantro, D. L. and Copeland, L. E. (1958). *J. Amer. chem. Soc.* **80**, 761.
110. Funk, H. (1962). *Chemistry of Cement, Proceedings of the Fourth International Symposium, Washington 1960*, p. 291. National Bureau of Standards Monograph 43. U.S. Department of Commerce.
111. Funk, H. (1960). *Silikattechnik* **11**, 373.
112. Assarsson, G. O. (1956). *J. phys. Chem.* **60**, 1559.
113. Assarsson, G. O. and Rydberg, E. (1956). *J. phys. Chem.* **60**, 397.
114. Toropov, N. A., Borisenko, A. I. and Shirokova, P. V. (1953). *Izv. Akad. Nauk S.S.S.R., Otd. Khim. Nauk* 65.
115. Tamás, F. D. (1960). *Silikattechnik* **11**, 378.

116. Kalousek, G. L. (1954). *Proceedings of the Third International Symposium on the Chemistry of Cement, London, 1952*, p. 296. Cement and Concrete Association, London.
117. Copeland, L. E. and Schulz, E. (1962). *J. P.C.A. Res. Dev. Labs* **4**, (1), 2.
118. Grudemo, Å. (1962). *Chemistry of Cement, Proceedings of the Fourth International Symposium, Washington 1960*, p. 615. National Bureau of Standards Monograph 43. U.S. Department of Commerce.
119. Funk, H., Schreppel, B. and Thilo, E. (1960). *Z. anorg. Chem.* **304**, 12.
120. Long, J. V. P. and McConnell, J. D. C. (1959). *Miner. Mag.* **32**, 117.
121. Brown, G. (ed.) (1961). "The X-ray Identification and Crystal Structures of Clay Minerals", 2nd Ed., p. 544. Mineralogical Society, London.
122. Thilo, E. (1950). *Abh. Dtsch. Akad. Wiss.* Berl. No. 4, 1; (1951). *Angew. Chem.* **63**, 201.
123. Takashima, S. (1956). *Semento Gijutsu Nenpo* **10**, 51.
124. Chalmers, R. A., Farmer, V. C., Harker, R. I., Kelly, S. and Taylor, H. F. W. *Miner. Mag.* In press.
125. Anderson, T. (1851). *Phil. Mag.* Ser. 4, **1**, 111.
126. Cornu, F. (1907). *S.B. Akad. Wiss. Wien.* **116**, 1213.
127. Meyer, J. W. and Jaunarajs, K. L. (1961). *Amer. Min.* **46**, 913.
128. Harker, R. I. (1960). *Geol. Soc. America 1960 Annual Meeting Program*, p. 113.
129. Mackay, A. L. and Taylor, H. F. W. (1953). *Miner. Mag.* **30**, 80.
130. Strunz, H. and Micheelsen, H. (1958). *Naturwissenschaften* **45**, 515.
131. Mackay, A. L. and Taylor, H. F. W. (1954). *Miner. Mag.* **30**, 450.
132. Hövig, P. (1914). *Jaarb. Mijnw. Ned.-Oost.-Ind.* **41**, 202.
133. Cornu, F. and Himmelbauer, A. (1906). *Miner. petrogr. Mitt.* **25**, 519.
134. Grutterink, J. A. (1925). *Verh. geol. Mijnb. Genoot. Ned. Kolon.* Geol. Ser. **8**, 197.
135. Taylor, H. F. W. (1962). *Chemistry of Cement, Proceedings of the Fourth International Symposium, Washington 1960*, p. 167. National Bureau of Standards Monograph 43. U.S. Department of Commerce.
136. Roy, D. M. (1958). *J. Amer. ceram. Soc.* **41**, 293.
137. Buckle, E. R. and Taylor, H. F. W. (1958). *Amer. Min.* **43**, 818.
138. Gard, J. A., Speakman, K. and Taylor, H. F. W. New data.
139. Taylor, W. H. and West, J. (1928). *Procl roy. Soc.* **A117**, 517; *Z. Kristallogr.* **70**, 461.
140. Agrell, S. O. and Gay, P. (1961). *Nature, Lond.* **189**, 743.
141. Roy, D. M., Gard, J. A., Nicol, A. W. and Taylor, H. F. W. (1960). *Nature, Lond.* **188**, 1187.
142. Dent, L. S. (1957). Ph.D. Thesis, Aberdeen.
143. Keevil, N. B. and Thorvaldson, T. (1936). *Canad. J. Res.* **B14**, 20.
144. Buckle, E. R. (1958). Ph.D. Thesis, Aberdeen.
145. Midgley, H. G. and Chopra, S. K. (1960). *Mag. Concr. Res.* **12**, 19.
146. Parry, J. and Wright, F. E. (1925). *Miner. Mag.* **20**, 277.
147. Heller, L. and Taylor, H. F. W. (1952). *J. chem. Soc.* 1018.
148. Megaw, H. D. (1952). *Acta cryst.* **5**, 477.
149. Petch, H. E., Sheppard, N. and Megaw, H. D. (1956). *Acta cryst.* **9**, 26.
150. Taylor, H. F. W. (1955). *Acta cryst.* **8**, 440.
151. Thorvaldson, T. and Shelton, G. R. (1929). *Canad. J. Res.* **1**, 148.
152. Butt, Yu. M. and Maier, A. A. (1957). *Trud. Mosk. Khim.-Tekh. Inst.* No. 24, 61.
153. Kalousek, G. L. (1954). *Proceedings of the Third International Symposium on the Chemistry of Cement, London 1952*, p. 334. Cement and Concrete Association, London.

154. Heller, L. (1952). *Acta cryst.* **5**, 724.
155. Dent Glasser, L. S., Funk, H., Hilmer, W. and Taylor, H. F. W. (1961). *J. appl. Chem.* **11**, 186.
156. van Valkenberg, A. and Rynders, G. F. (1958). *Amer. Min.* **43**, 1195.
157. Dent Glasser, L. S. and Roy, D. M. (1959). *Amer. Min.* **44**, 447.
158. Bessey, G. E. (1933). Rep. Building Research Board. HMSO, London, 25.
159. Bessey, G. E. (1939). *Proceedings of the Symposium on the Chemistry of Cements, Stockholm 1938*, p. 178. Ingeniörsvetenskapsakademien, Stockholm.
160. Mamedov, Kh. S., Klebtsova, R. F. and Belov, N. V. (1959). *Dokl. Akad. Nauk. S.S.S.R.* **126**, 151.
161. McMurdie, H. F. and Flint, E. P. (1943). *J. Res. nat. Bur. Stand.* **31**, 225.
162. Barrer, R. M. (1952). *Proc. International Symposium on the Reactivity of Solids, Gothenberg, 1952*, p. 373.
163. Harker, R. I., Roy, D. M. and Tuttle, O. F. (1962). *J. Amer. ceram. Soc.* **45**, 471.
164. Mchedlov-Petrosyan, O. P. and Babushkin, W. (V). I. (1962). *Chemistry of Cement, Proceedings of the Fourth International Symposium, Washington 1960*, p. 533. National Bureau of Standards Monograph 43. U.S. Department of Commerce.
165. Babushkin, V. I., Matveyev, G. M. and Mchedlov-Petrosyan, O. P. (1962). "Termodinamika Silikatov", p. 242. Gosstróyizdat, Moscow.
166. Taylor, H. F. W. New data.
167. Toropov, N. A., Nikogosyan, Kh. S. and Boikova, A. I. (1958). *Trud. Pyatogo Soveshchan. Eksp. Tekh. Miner. Petrog. Akad. Nauk S.S.S.R.* 1956, 44.
168. Serdyuchenko, D. P. (1958). *Zap. Vse. Miner. Obshch.* **87**, 31.
169. Midgley, H. E. (1962). *Chemistry of Cement, Proceedings of the Fourth International Symposium, Washington 1960*, p. 479. National Bureau of Standards Monograph 43. U.S. Department of Commerce.
170. Dent Glasser, L. S., Glasser, F. P. and Taylor, H. F. W. (1962). *Quart. Rev.* **16**, 343.
171. Taylor, H. F. W. (1960). *J. appl. Chem.* **10**, 317.
172. Dent Glasser, L. S. and Glasser, F. P. (1961). *Acta cryst.* **14**, 818.
173. Gay, P. and LeMaitre, R. W. (1961). *Amer. Min.* **46**, 92.
174. Smirnova, R. F., Rumanova, I. M. and Belov, N. V. (1955). *Zap. Vser. Miner. Obshçh.* **84**, 159.
175. Andrews, K. W. (1961). *Refractories J.* **37**, 66.
176. Chalmers, R. A., Dent, L. S. and Taylor, H. F. W. (1958). *Miner. Mag.* **31**, 726.
177. Chirvinskii, P. N. (1936). *Vernadsky Jubilee Vol. Acad. Sci. U.S.S.R.* **2**, 757.
178. Megaw, H. D. and Kelsey, C. H. (1955). *Miner. Mag.* **30**, 569.
179. Cole, W. F. and Kroone, B. (1959). *Nature, Lond.* **184**, British Association Number BA57.
180. McConnell, D. and Murdoch, J. (1958). *Amer. Min.* **43**, 498.
181. Murdoch, J. (1955). *Amer. Min.* **40**, 505.
182. McConnell, J. D. C. (1955). *Amer. Min.* **40**, 510.
183. Harker, R. I. and Tuttle, O. F. (1956). *Amer. J. Sci.* **254**, 239; Tuttle, O. F. and Harker, R. I. (1957). *Amer. J. Sci.* **255**, 226; Harker, R. I. (1959). *Amer. J. Sci.* **257**, 656.
184. McConnell, J. D. C. (1958). Mineralogical Society, London, Notice of Meeting No. 103.
185. Gard, J. A. (1962). *Proc. European Regional Conference on Electron Microscopy, Delft, 1960*, Vol. 1, p. 203.
186. Agrell, S. O. (1962). *Min. Abs.* **15**, 541.

CHAPTER 6

The Calcium Aluminate Hydrates and Related Compounds

R. TURRIZIANI

Istituto di Chimica Applicata e Metallurgia, Facoltà di Ingegneria, University of Cagliari, Italy

CONTENTS

I. Introduction

Equilibria in the $CaO–Al_2O_3–H_2O$ and $CaO–Fe_2O_3–H_2O$ ternary systems, and in the related quaternary systems containing also SO_4^{2-}, Cl^-, CO_3^{2-} or SiO_4^{4-}, must be taken into account in all discussions of the hydration reactions of the principal hydraulic cements used in building. They are particularly important in connection with the chemical resistance of these cements to natural waters. A cement can hardly be found for which these data are irrelevant. Most of the early interest centred on syntheses of calcium aluminate and ferrite hydrates containing other anions. Research on equilibria remained more or less in abeyance until about 1940; in the succeeding years it progressed steadily, particularly for the systems containing Al_2O_3. An attempt will be made here to give a reasonably accurate picture of the present state of the subject, based on the more recent and quantitative studies. More exhaustive bibliographies of early work are given in references [*1–6*].

II. The Calcium Aluminate Hydrates

A. General Comments

The calcium aluminate hydrates referred to in the literature are very numerous; the existence in systems at room temperature of compounds with all the integral CaO/Al_2O_3 ratios from 1 to 6 has been assumed. In the last decade, however, investigations on equilibria in the $CaO–Al_2O_3–H_2O$ system have reduced the number of compounds whose existence has been ascertained beyond doubt (Table I). These investigations have been helped notably by X-ray identification and by more exact control over water contents and thus of the equilibrium relationships between the hydrated solids and the vapour phase.

The methods of preparation of the calcium aluminate hydrates which have generally been used are: (a) precipitation from solutions supersaturated with respect to lime and alumina; (b) hydration of anhydrous calcium aluminates; (c) reaction of freshly prepared alumina gel with calcium hydroxide and water; (d) precipitation by mixing alkali aluminate solutions with lime-water and calcium salts; and (e) reaction between metallic aluminium and lime-water.

A supersaturated calcium aluminate solution is prepared by shaking an aqueous suspension of CA or $C_{12}A_7$, or even of alkali-free aluminous cement. The concentrations of lime and alumina in solution depend on the shaking time, temperature, grain size of the solid and liquid/solid ratio. At room temperature, using an aluminous cement that leaves no

TABLE I

The principal calcium aluminate hydrates

Composition	Morphology	Density (g/cm³)	Characteristic X-ray powder spacings (Å)[1]	Unit cell (U) or structural element (S) (Å) *a*	*b*	*c*		References to X-ray and optical data
α_1-C_4AH_{19}	Hex. plates	1·79	10·7 (B)	5·77	—	64·08	(U)	[7], [*19*], [*19a*]
α_2-C_4AH_{19}	Hex. plates	1·81	10·7 (B)	5·77	—	21·37	(U)	[7], [*19*], [*19a*]
α-C_4AH_{13}[2]	Hex. plates	2·01 (?)	8·2 (B)	5·74	—	8·2	(S)	[7], [9], [*19*], [*19a*]
β-C_4AH_{13}	Hex. plates	2·02	7·9 (B)	5·74	—	7·92	(S)	[7], [9], [*19*], [*19a*]
β-C_2AH_8	Hex. plates	1·97(8)	10·4 (B)	5·7	—	10·4	(S)	[7], [*19a*]
α_1-C_2AH_8	Hex. plates	1·95	10·7 (B)	5·7	—	10·7	(S)	[7], [*19a*]
α_2-C_2AH_8	Hex. plates	1·95	10·7 (B)	5·7	—	10·7	(S)	[7], [*19a*]
CAH_{10}	Hex. prisms (?)	—	14·2	—	(?)	—		[*25*], [*25a*]
C_3AH_6	Cubic forms	2·52	5·14, 2·30	12·576	—	—	(U)	[*31*]
$C_4A_3H_3$	Orthorhombic plates	2·71	3·61, 2·80	12·78	12·42	8·90	(U)	[*16*], [*17*], [*49a*]

(1) (B) = longest basal spacing.
(2) Perhaps not a true ternary phase; see page 241.

residue on the 170-mesh sieve and a water/cement ratio of 25 : 1, it takes 45 min shaking in a rotary agitator (15 rev/min) to obtain a solution containing 1·01 g CaO/l and 1·60 g Al_2O_3/l. Higher liquid/solid ratios (25:1–50:1) are quoted in the literature for CA; Jones and Roberts [7] found, however, that, using finely ground material, 3–8 g of CA/l are sufficient to give a solution containing 1·5–1·9 g Al_2O_3/l. The supersaturated solution is left to stand at room temperature with exclusion of CO_2. After a period ranging from a few hours to several days one or more solid phases separate, the nature of which depends on the composition and concentration of the initial solution. Jones and Roberts [7] found that an unstable form of C_4AH_{19} arises more easily, and persists longer, with the metastable solution prepared from laboratory-made CA than with that obtained from aluminous cement. Percival, Buttler and Taylor [8] emphasized that a physico-chemical investigation might be needed to define the nature of the aluminate ion in solution.

The more basic calcium aluminate hydrates are prepared by adding calcium oxide or lime water to the supersaturated solution. Examples of the preparation of individual compounds will be given later.

The method of preparation starting from anhydrous aluminates is commonly used in hydrothermal syntheses and in the study of equilibria above 100° C. At low temperatures (1–25° C) the hydrated phases may contain residues of the initial anhydrous solid, mainly because of the reduced rate of diffusion through the hydrated outer layers of the grains.

The third method of preparation is very simple; the reaction between alumina gel and calcium hydroxide is comparatively quick, and from X-ray examination the products appear to crystallize at least as well as those prepared by the first two methods. As the velocity of reaction decreases with increasing age of the gel, the latter should be used soon after preparation. Very careful washing of the gel is recommended, to remove all traces of any foreign anions, which might otherwise enter into the reaction and form quaternary aluminates.

The other two methods are seldom used and are quoted only in the older literature. The method starting from a solution of an alkali aluminate is not recommended, owing to the presence of the anion from the calcium salt and to the formation of alkali hydroxide in the reaction.

At present the calcium aluminate hydrates are generally identified using X-rays. For most of these compounds the microscopic method is less suitable, owing to the similarity in their optical and morphological properties and to the small size of the crystals. For some compounds the problem is complicated by the existence of several hydrates; careful control of the humidity of the atmosphere with which the solid is equilibrated is then required. In some cases the hydrate that exists in contact with the liquid phase differs even from the equilibrated under high relative humidity, and for this reason it is generally advisable to examine the moist solid by X-rays.

As calcium aluminate hydrates in moist air react with atmospheric CO_2, the conditioning at the desired vapour pressure is effected by placing the solid in a vacuum desiccator over a saturated salt solution. To obtain a constant vapour pressure during the process, it is necessary to maintain equilibrium between the saturated solution and the salt which is saturating it. This can be done more easily if the crystals of the salt are small and the solution is stirred at intervals. If the aluminate is dried with a very strong drying agent, such as anhydrous $CaCl_2$, P_2O_5 or $Mg(ClO_4)^2$, particular care in handling is necessary so as to avoid rehydration, which sometimes occurs rather quickly.

Thermal dehydration of the calcium aluminate hydrates has been studied using differential thermal analysis (DTA), thermogravimetric analysis and quasi-static weight loss curves. The thermal behaviour provides a method for the direct identification of the compounds in cement pastes. It can sometimes help in the determination of crystal structure, as Buttler, Dent Glasser and Taylor [*9*] showed for C_4AH_{13} and hydrocalumite. The quasi-static method appears particularly suitable for this purpose. It consists in subjecting the sample to successively higher temperatures in an atmosphere with a definite humidity, in each case until the weight has become constant to within experimental error.

With DTA it is necessary when comparing the results of different authors to remember that the form of the curves and temperatures of the peaks are to some extent affected by numerous factors, such as heating rate, weight and particle size of sample, and design of equipment (type of container, position of thermocouple, etc.). The influence of these variables is particularly felt below 200° C. These factors are fully discussed in Volume 2, Chapter 23.

Taylor and co-workers [*10, 11*] have found that the dehydration of single crystals of calcium silicate hydrates often yields single crystals or preferred orientation aggregates of the product. Such processes, called by the authors "oriented transformations", sometimes provide important data about features of the crystal structures of the solids involved. Systematic studies of this type should be of interest also for some of the calcium aluminate hydrates. New developments may also be expected from the use of infra-red absorption and nuclear magnetic resonance.

Table I includes brief crystallographic data for the principal compounds, descriptions of which now follow.

B. C_3AH_6

Tricalcium aluminate hexahydrate crystallizes in the cubic system and shows a variety of forms, which necessarily have the same refractive index and density regardless of differences in appearance. It can be prepared hydrothermally at about 150° C, either from crystalline C_3A or from glasses or even mixtures of the hydroxides of the right composition. It is also formed below 100° C when the metastable, hexagonal calcium aluminate hydrates (described later) stand in contact with aqueous solutions; they react with the solution and C_3AH_6 is formed as a stable phase. This process occurs slowly at room temperature but becomes

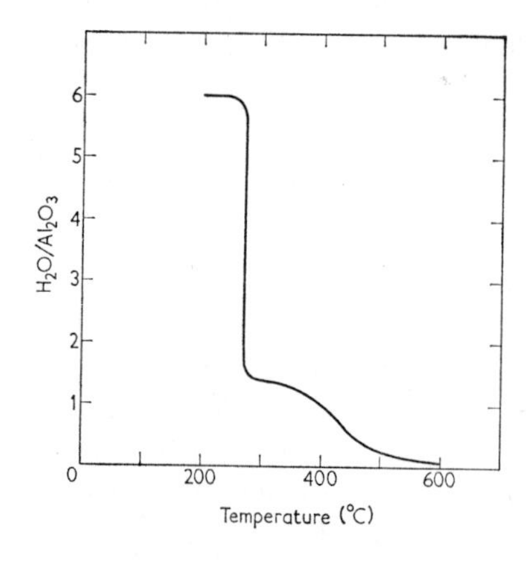

FIG. 1. Thermal weight loss curve for C_3AH_6 (Schneider and Thorvaldson [*12a*]).

increasingly rapid as the temperature is raised. Sersale [*12*] recently studied the process at 20–80° C as a function of the CaO/Al_2O_3 ratio of the mother-liquor and pH; he also studied the influence of surface-active agents. He found that the hexagonal plates change first into icositetrahedra, then into rhombododecahedra and finally into cubes. The rate of transformation increases not only with temperature but also with the CaO/Al_2O_3 ratio and pH of the solution; pH values above that of limewater were obtained by adding NaOH. At 20° C surface-active agents retard the change; this effect is reduced with increase in temperature.

Only one definite hydrate of C_3A exists. The thermaldehydration curve [*12a*] (Fig. 1) shows no appreciable loss of water below 275° C, when 4·5 molecules are liberated, leaving a solid of composition $C_3AH_{1\cdot5}$

which has refractive index 1·543. Majumdar and Roy [13] showed that $C_3AH_{1\cdot 5}$ does not exist in the CaO–Al_2O_3–H_2O system at high temperatures; they regard it as a metastable structural relict which constitutes a step in the dehydration of C_3AH_6 only at low partial pressures of water vapour. The X-ray pattern of $C_3AH_{1\cdot 5}$ closely resembles that of $C_{12}A_7$. At about 550° C the $C_3AH_{1\cdot 5}$ decomposes to give CaO and $C_{12}A_7$, which recombine around 1050° C to form C_3A. Taylor [14] suggests that the close similarity between the cell dimensions of C_3AH_6 and $C_{12}A_7$ may indicate the occurrence of an oriented transformation.

Hydrothermal dehydration of C_3AH_6 takes a different course; at 225–400° C water and $Ca(OH)_2$ are liberated and $C_4A_3H_3$ is formed. Figure 9 gives the decomposition temperature as a function of water pressure (see p. 257). At about 695° C (for a pressure of 680 kg/cm^2) the $C_4A_3H_3$ decomposes to $C_{12}A_7$, CA_2 and H_2O.

DTA curves for C_3AH_6 were reported by Kalousek, Davis and Schmertz [15], Majumdar and Roy [13] and others. At a heating rate of 8–10 deg C/min the curve shows two endothermic effects. The first of these, at 330° C is very intense and corresponds to the elimination of 4·5 H_2O; the second, at 530° C, is weaker. Working with a heating rate of 17·5 deg C/min, Turriziani and Schippa [82] found that the first endotherm occurred at a higher temperature. They also observed that the peak temperature may range from 345° C to 360° C, depending on crystal size.

Majumdar and Roy [13] also reported infra-red spectra for C_3AH_6 and $C_3AH_{1\cdot 5}$. The C_3AH_6 pattern shows the absence of the 6·2 μ absorption for so-called "free" H_2O; all the hydrogen must therefore be in hydroxyl groups. The elimination of 4·5 H_2O removes many of the main peaks.

C. $C_4A_3H_3$

Tetracalcium trialuminate trihydrate crystallizes in rectangular, orthorhombic plates with cleavage (010). It was discovered by Johnson and Thorvaldson [16], who obtained it, together with $Ca(OH)_2$, by heating C_3A or a mixture of $Ca(OH)_2$ and alumina gel at 350° C in saturated steam. Percival and Taylor [17] prepared well developed plates up to 80 μ long by treatment of the stoichiometric mixture CaO + 3CA in water under saturated steam pressure at 350° C. The probable ionic consititution is $Ca_4Al_6O_{10}(OH)_6$.

At 1° C $C_4A_3H_3$ dissolves slowly in water without forming a precipitate, but in a half-saturated lime solution it decomposes to form C_2AH_8 [18]. It decomposes on dry heating at about 720–750° C, giving first an unidentified product and afterwards $C_{12}A_7$ as the only product detectable in the X-ray powder pattern [16]. In hydrothermal runs [13] the decomposition temperature depends on the water pressure (Fig. 10; see p. 258) and the decomposition products are $C_{12}A_7$, CA_2 and water.

Majumdar and Roy [13] reported the infra-red spectrum and DTA curve. The latter is characterized by an endothermic peak at 720° C for a heating rate of 8 deg C/min.

D. C_4AH_x

The tetracalcium aluminate hydrates, first observed by Lafuma and later by Mylius, crystallize in small, thin, hexagonal plates with perfect (0001) cleavage. A compound of this group is obtained by adding CaO or lime-water to a metastable calcium aluminate solution in such a way as to combine the Al_2O_3 present as C_4AH_x and to obtain a CaO concentration under metastable equilibrium conditions of about 1·0 g/l. The time of reaction needed may range from a few days to 15–20 days, depending on whether shaking is continuous or intermittent. Buttler *et al.* [9] obtained well developed crystals at 5° C by reaction between $Ca(OH)_2$ and alumina gel and also by hydration of CA_2 in lime-water. By the last method plates some microns thick and up to 50 μ across can be obtained. Whichever method of preparation of C_4A hydrates is employed, extreme precautions are essential to exclude atmospheric CO_2, as the moist solids react very rapidly to give quaternary compounds.

According to Roberts [19], there are at least six compounds or crystalline modifications having the general formula C_4AH_x. The hydrate in metastable equilibrium with the mother-liquor at 25° C has the formula C_4AH_{19}. It occurs in two modifications: α_1-C_4AH_{19} is an unstable form which may be formed initially; it readily inverts to a second more stable form designated α_2-C_4AH_{19}. The two modifications have identical refractive indices; their X-ray patterns show only slight differences in the positions of non-basal reflections and none in those of the basal reflections. Aruja [19a] showed that they were polytypes, that is structures based on the same type of layer and differing only in the way in which successive layers are stacked.

On storage *in vacuo* at room temperature (about 18° C) and a relative humidity of 12–81%, C_4AH_{19} is converted into C_4AH_{13} [19]. At some temperature above 25° C, C_4AH_{13} may also exist in contact with aqueous solutions [20]. The question of polymorphism and polytypism in this hydrate is somewhat complicated. The early literature contains numerous references to different modifications, but much of this work was confused by the unrecognized presence of the closely related quaternary compound $C_3A.CaCO_3.11H_2O$. Roberts [19] recognized two polymorphs, which he called α- and β-C_4AH_{13}. These were distinguished by their longest basal spacings, which were 8·2 Å for the α-form and 7·9 Å for the β-form ($C_3A.CaCO_3.11H_2O$ has a basal spacing of 7·6 Å; see p. 273). Buttler *et al.* [*9*], Jones [*20*], Jones and Roberts [*7*] and Aruja [*19a*], among others, have also referred recently to these two polymorphs of C_4AH_{13}. However, Seligmann and Greening [*24*] have shown that the supposed α-form is really a second quaternary compound containing less CO_3^{2-} than does $C_3A.CaCO_3.11H_2O$, which is only formed if slight contamination by CO_2 occurs. If this is correct there is only one polymorph of C_4AH_{13}, with a basal spacing of 7·9 Å.

Even if C_4AH_{13} has only one polymorph, there is still the possibility (as with C_4AH_{19}) of different polytypes or stacking modifications. These would all necessarily have the same basal spacing of 7·9 Å, and one would expect their X-ray powder patterns to be more nearly alike than those of α- and β-C_4AH_{13}. Buttler *et al.* [*9*] obtained some indications that different polytypes of C_4AH_{13} do indeed exist, but this problem still awaits detailed examination.

In addition to the two CO_3^{2-}-containing quaternary compounds already mentioned, a third such compound closely related in structure to C_4AH_{13} is known. This is the natural mineral, hydrocalumite, which was discovered at Scawt Hill (Northern Ireland) by Tilley, Megaw and Hey [*39*]. It has the composition $Ca_{16}Al_8(OH)_{54}(H_2O)_{21}(CO_3)$ and its basal spacing is 7·86 Å. It has not yet been synthesized.

Drying of C_4AH_{19} or C_4AH_{13} over anhydrous $CaCl_2$ or solid NaOH causes dehydration to C_4AH_{11} (Table II), which, when heated to 120° C

TABLE II

Calcium aluminate hydrates: minor partial dehydration products

Composition	Conditions of dehydration	Refractive indices ω	Refractive indices ε	Density (g/cm³)	Basal spacing (Å)	References to X-ray and optical data
C_4AH_{11}	Anhydrous $CaCl_2$ or solid NaOH	1·539	1·524	2·08	7·4	[7], [*19*]
C_4AH_7	P_2O_5 or 120°C	1·555	1·544	2·28	{ 7·4 6	[*19*] [*9*]
$C_2AH_{7\cdot5}$	34% r.h.	1·520	1·505	1·98	10·6	[7], [*19*], [*19a*]
C_2AH_5	12% r.h., 102°C, $CaCl_2$ or P_2O_5	1·534	1·524	2·09	8·7	[7], [*19*], [*19a*]
C_2AH_4	120°C	1·565	1·559	2·27	7·4	[*19*]
CAH_7	45% r.h.	1·480	1·477	(?)	14·2	[*25*]

or dried over P_2O_5, gives C_4AH_7 [*19*]. In air at 81% r.h. these two latter hydrates take up water and CO_2 quickly, forming a mixture of β-C_4AH_{13} and "α-C_4AH_{13}". This can be further hydrated, giving C_4AH_{19}, by shaking with nearly saturated lime solution; in air at 99% r.h., however, this process is very slow and is still incomplete after several months [*19*].

The dehydration curve of β-C_4AH_{13} (Fig. 2) was obtained by the quasi-static method at a partial water vapour pressure of 6 mm Hg [*9*]. The loss

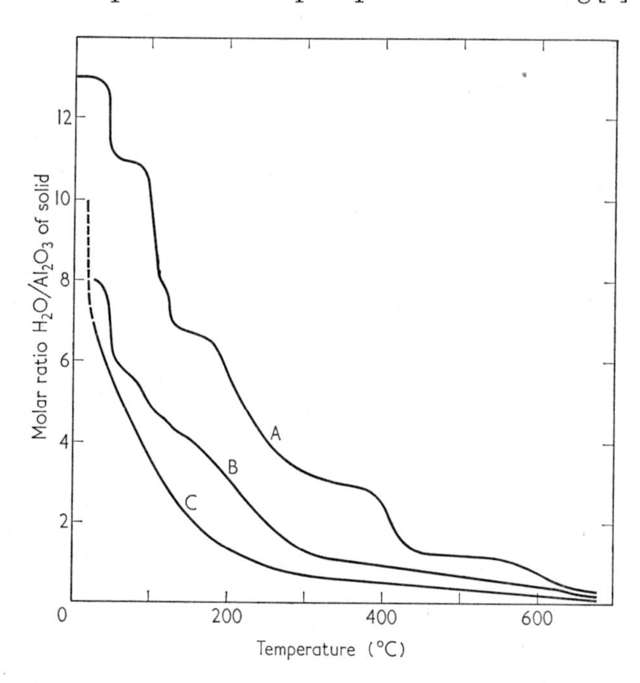

FIG. 2. Thermal weight loss curves for hydrated hexagonal calcium aluminates (quasi-static method; Buttler [*76*], Buttler *et al.* [*9*]). A, C_4AH_{13}; B, C_2AH_8; C, CAH_{10}.

up to 150° C is mainly of molecular water and the product at this temperature is essentially C_4AH_7. According to Roberts [*19*], this hydrate has a basal spacing of 7·4 Å; Buttler *et al.* [*9*], however, reported a value of about 6 Å. The 7·4 Å value apparently cannot be reconciled with the observed density at 2·28 g/cm³. Between 150° C and 300° C partial dehydroxylation occurs and $C_4A_3H_3$ and $Ca(OH)_2$ are formed; at 400° C, $Ca(OH)_2$ is decomposed to CaO; and at about 700° C the $C_4A_3H_3$ is decomposed, giving $C_{12}A_7$.

Buttler *et al.* also investigated the dehydroxylation of hydrocalumite; the process takes place in similar steps but no $C_4A_3H_3$ is formed. At

150–400° C the only crystalline phase present is $Ca(OH)_2$, which at 400° C changes into CaO. At a higher temperature $C_{12}A_7$ is formed. $C_4A_3H_3$ was also not observed in the thermal dehydration of C_3A. $CaCO_3 \cdot 11H_2O$ [*21*] (see p. 273). Its formation is possibly inhibited by the presence of CO_3^{2-} ions [*9*].

DTA curves for C_4AH_{13} have been reported by Kalousek *et al.* [*15*], Turriziani [*22*], Midgley and Rosaman [*23*] and others.

E. C_2AH_8

A dicalcium aluminate hydrate, C_2AH_8, is prepared in a similar way to C_4AH_{19}; the CaO/Al_2O_3 ratio of the initial mixture must be about 2 and the equilibrium CaO concentration about 0·45 g/l. It crystallizes in thin, hexagonal plates, which are closely similar in appearance to those of the C_4A hydrates. According to Roberts [*19*], C_2AH_8 exists in three modifications. Of these only two, designated β- and α_1-, exist in contact with aqueous solutions; the β-form is unstable with respect to the α_1-form, which is the normal one previously reported in the literature simply as C_2AH_8. From the X-ray data it appears that the structural difference between these two polymorphs is very slight. The basal spacings are 10·4 Å for β- and 10·7 Å for α_1-C_2AH_8 [*19a*]. If α_1-C_2AH_8 is subjected to successive cycles of dehydration and rehydration, it is converted into an α_2-form. The α_1- and α_2-forms are polytypes and thus have the same basal spacing [*19a*]; they resemble each other even more closely than do the α_1- and β- modifications.

Partial dehydration of any of the forms of C_2AH_8 yields successively $C_2AH_{7\cdot5}$, C_2AH_5 and C_2AH_4 (Table II). In a damp atmosphere the first two of these are reconverted into C_2AH_8, but C_2AH_4, which is obtained on heating at 120° C, is unaffected at 81% r.h. [*19*]. The thermal dehydration curve [*76*] (Fig. 2) shows no distinct steps but only slight modifications of slope below 200° C. No data concerning the decomposition products are available. Kalousek *et al.* [*15*] reported a DTA curve for C_2AH_8.

F. CAH_{16}

Monocalcium aluminate hydrate, first reported by Assarsson [*27*], is obtained at temperatures around 1° C, either from a metastable calcium aluminate solution having a CaO/Al_2O_3 molar ratio of 1·1–1·2 or by hydration of CA. In the presence of the mother-liquor the compound is stable up to 22° C, above which temperature it decomposes giving C_2AH_8.

The hydrate appears to contain $10H_2O$ when in contact with the mother-liquor and also after drying at 80% r.h. When dried at 40% r.h. it is converted into CAH_7, which has the same X-ray pattern as CAH_{10} but higher refractive indices [*25, 25a*]. Over P_2O_5 the water content is reduced to $5{\cdot}5H_2O$ and, after heating to 100–105° C, to $2{\cdot}5H_2O$. The dehydration curve [76] (Fig. 2) shows no definite steps. At 600° C the water is entirely removed and at a higher temperature CA is formed. A DTA curve and X-ray studies on heated samples have been reported [76a].

G. Doubtful Phases

The existence of calcium aluminate hydrates (Table III) other than those already described is rather doubtful and, with some of them, attempts to repeat the preparation have been unsuccessful.

TABLE III

Doubtful calcium aluminate hydrates

Composition	Morphology	Refractive indices $\omega(\gamma)$	$\epsilon(\alpha)$	Optical character	Characteristic X-ray powder spacings (Å)	References to X-ray and optical data
C_6AH_{32-33}	Needles	1·475	1·466	Uniaxial−	9·8, 5·7	[26]
C_5AH_{34}	Hex. prisms	1·487	1·480	Uniaxial−	—	[27]
C_3AH_{18-21}	Needles	1·495	1·479	Biaxial−	—	[29,] [30]
C_3AH_{10-12}	Hex. plates	1·529	1·506	Biaxial or uniaxial−	7·55–7·70	[30–33], [35], [36]

Flint and Wells [26] prepared a compound which they designated "C_6AH_{33}" by adding a solution of calcium saccharate to a metastable calcium aluminate solution; its optical properties and X-ray pattern were similar to those of ettringite ($C_3A.3CaSO_4.32H_2O$). These authors considered that the compound C_5AH_{34} detected by Assarsson [27] in aqueous extracts of CA was a hydrolysis product of C_6AH_{33}. Assarsson was unable to obtain his compound in the absence of SO_4^{2-} ions [6].

Carlson and Berman [28], and also Midgley and Rosaman [23], tried unsuccessfully to repeat the preparation of Flint and Wells. Carlson and Berman obtained a precipitate only when dissolved CO_2 was present in the saccharate solution, and the solid appeared to be the aluminate tricarbonate, $C_3A.3CaCO_3.32H_2O$.

Since a mixture of $C_3A.CaCO_3.11H_2O$ and $C_3A.3CaCO_3.32H_2O$ separates from calcium aluminate solutions when these are left in air, it is possible that "C_5AH_{34}" is in fact such a mixture. This would explain its CaO/Al_2O_3 ratio and also its reported water content if part of this were really attributable to CO_2. In conclusion, though there is no reason to deny *a priori* the existence of calcium aluminate hydrates more basic than C_4AH_x, the methods so far suggested for preparing such compounds lack experimental confirmation.

A hydrate of composition C_3AH_{18-21} was reported by Travers and Sehnoutka [*29*] and Mylius [*30*]. Travers and Sehnoutka obtained it by the reaction of potassium aluminate, calcium hydroxide and calcium nitrate solutions, Mylius by adding potassium hydroxide to a calcium aluminate solution. Drying over P_2O_5 reduced the water content to $9H_2O$ and heating at 150° C reduced it to $6H_2O$.

Among the hydrates listed in Table III, C_3AH_{10-12} is the most frequently mentioned in recent as well as older literature; it is described as forming hexagonal plates similar to those of C_4AH_{13} and C_2AH_8. Preparations have been described which are based on hydration of C_3A and on precipitation from metastable calcium aluminate solutions. X-Ray data were given by Thorvaldson, Grace and Vigfusson [*31*], Bunn and Clark [*32*] and Brocard [*33*]. The first of these patterns cannot be compared with the others, since it is confined to a narrow range of *d*-spacings which does not include the most intense, characteristic reflections that distinguish one aluminate hydrate from another. The existence of this compound was early disputed by Wells, Clarke and McMurdie [*41*]. Schippa and Turriziani [*34*] have shown that the most intense lines in the patterns given by Bunn and Clark and by Brocard are those of $C_3A.CaCO_3.11H_2O$.

Govoroff [*35*] reported the preparation of C_3AH_{12} by heating an equimolar mixture of C_4AH_{13} and C_2AH_8 at 80° C for 5 h. Roberts [*19*] repeated this experiment and obtained $C_3A.CaCO_3.11H_2O$. D'Ans and Eick [*36*] believed that they detected both C_3AH_{18-21} and C_3AH_{10-12} among the solid phases of the $CaO–Al_2O_3–H_2O$ system at 20° C by microscopic examination. Repeating their preparations, Schippa and Turriziani [*34*] found the dried precipitates to be mixtures of C_2AH_8 and "α-C_4AH_{13}". Similar results have been reported in other recent publications [*3, 7, 20, 36a*], and it might reasonably be concluded that these hydrates do not appear among the stable or metastable equilibrium solids of the $CaO–Al_2O_3–H_2O$ system at room temperature. Further investigations, based on the studies of Travers and Sehnoutka, Mylius, and Govoroff, therefore appear necessary to ascertain whether hexagonal C_3A hydrates can have their own field of existence outside the ternary aqueous equilibria.

H. Crystal Structures

The only calcium aluminate hydrate whose structure is known is C_3AH_6; the structures of all the others are largely or completely unknown. C_3AH_6 is body-centred cubic, with *a* 12·56 Å, space group *I*a3d and 8 formula units in the unit cell [*37*]. The structure is derived from that of grossularite, a garnet of formula C_3AS_3, by total replacement of SiO_4^{4-} by $4OH^-$. If the SiO_4^{4-} groups are only partly substituted, hydrogarnets are formed. These are discussed later.

The hexagonal calcium aluminate hydrates are all characterized by hexagonal or trigonal unit cells with a approximately 5·7 Å (Table I). Their morphology indicates that they have layer structures, the thickness of the elementary layer being given in each case by the characteristic longest basal spacing in the X-ray pattern. The c-axes are in general multiples of this spacing, but this is only a consequence of the way in which the layers are stacked. Buttler *et al.* [*9*] found it convenient to refer to a structural element having a 5·7 Å and c equal to the thickness of the elementary layer rather than to the unit cell. Hydrocalumite can also be referred to a structural element of this type, although its unit cell is monoclinic [*9, 39*].

The a-axis of 5·7 Å is approximately $\sqrt{3}$ times as long as that of $Ca(OH)_2$, and it is apparent that the structures of these compounds are related to those of $Ca(OH)_2$ and $Al(OH)_3$. It is generally agreed that C_4AH_{13} has the ionic constitution $2Ca(OH)_2.Al(OH)_3.3H_2O$ or $Ca_2Al(OH)_7.3H_2O$ and that C_2AH_8 can similarly be written as $2Ca(OH)_2.2Al(OH)_3.3H_2O$ or $Ca_2Al_2(OH)_{10}.3H_2O$. Brandenberger [*38*] suggested that separate layers of $Ca(OH)_2$ and $Al(OH)_3$ were present in both compounds, but this is incompatible with the thickness of the elementary layer in the case of C_4AH_{13}.

Among the individual hexagonal calcium aluminate hydrates, C_4AH_{13} and hydrocalumite have subsequently received the most attention. Tilley *et al.* [*39*], reasoning from their work on hydrocalumite, suggested tentatively that the structure of C_4AH_{13} included octahedral $Ca(OH)_2$ layers in which one Ca^{2+} ion in three was replaced by H_2O; between these layers were the Al^{3+} ions and the remaining OH^- and H_2O. The sequence of sheets within the elementary layer 7·92 Å thick was

$$(2H_2O+OH^-) : Al : (OH)_3 : (2Ca^{2+}+H_2O) : (OH)_3$$

there being thus three sheets of oxygen, present as H_2O or OH^-, in the structural element.

Bessey [*42, 43*] suggested a structure for C_2AH_8 based on the above structure for C_4AH_{13}. Later workers have questioned the substitution of H_2O for Ca^{2+} in the octahedral layer; Buttler *et al.* [*9*] considered that it was incompatible with the observed value of a for C_4AH_{13} and with the fact that this value is unchanged on dehydration to C_4AH_7. They considered it more likely that the octahedral layer contained Al^{3+}, giving the sequence

$$(3H_2O) : OH^- : (OH)_3 : (Ca_2Al) : (OH)_3$$

They suggested that hydrocalumite was derived from this structure by an ordered, partial replacement of $(2OH^- + 3H_2O)$ by CO_3^{2-}, the OH^- ions involved being those not adjacent to Ca^{2+}.

Grudemo [*40*] suggested a different modification of the structure of Tilley *et al.* [*39*], in which one Ca^{2+} position in three was vacant and the Al^{3+}, as well as some of the OH^-, lay outside the octahedral layer. He pointed out an analogy with the clay minerals, according to which the Ca^{2+} and Al^{3+} played the roles of the octahedral and tetrahedral cations, respectively. On this basis C_4A hydrates are analogous to clay minerals of the kaolinite group, while C_2A hydrates are analogous to 2:1 layer structures such as montmorillonite. A closely similar hypothesis was earlier advanced by Feitknecht and Buser [*52*] for the quaternary phases such as $C_3A.CaCl_2.xH_2O$. It must be stressed that all of these structures are purely hypothetical; in no case was a full X-ray structure determination made, though Tilley *et al.* showed that their hypothesis explained satisfactorily the relative intensities of the $00l$ reflections for hydrocalumite.

The structures of CAH_{10} and $C_4A_3H_3$ are unknown.

III. The $CaO–Al_2O_3–H_2O$ System

A. General Comments

Crystallization isotherms and solution equilibria have been examined at temperatures ranging from 1 to 1000° C. At 1–90° C solubility data have been determined by precipitating the solid phases from supersaturated calcium aluminate solutions (the supersaturation method) and by shaking initially unsaturated solutions with individual solid phases already prepared by some other method (the undersaturation method). Above 90° C, undersaturation methods, hydration of anhydrous calcium aluminates, and direct reactions starting from CaO and Al_2O_3 have been employed.

The normal way of representing the system has been to plot the concentration of lime in solution against that of alumina. Since a straight line passing through the origin is the locus of points with a given CaO/Al_2O_3 ratio, a line connecting the composition of the initial mixture with that of the equilibrium solution shows the direction of the crystallization process. At 1–90° C the equilibria are characterized by the existence of metastable phases which precipitate rapidly but transform only slowly into the stable phases. The solubility curve of a metastable phase lies above that of the stable phase.

B. Equilibria at 20–25° C

1. Introduction

Equilibria in this range have been reported by Bessey [*43*], Wells *et al.* [*41*], Jones [*44*] D'Ans and Eick [*36*], Percival and Taylor [*45*] Percival *et al.* [*8*] and Jones and Roberts [*7*]. The 25° C diagram of Jones and Roberts (Fig. 3) will be the one principally discussed here.

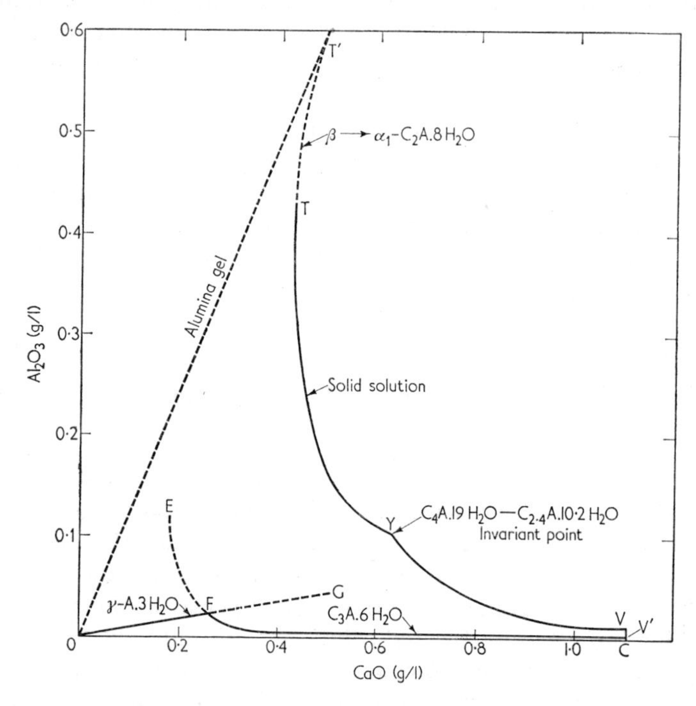

Fig. 3. Solubility relations in the system $CaO–Al_2O_3–H_2O$ at 25°C (Jones and Roberts [*7*]).

At 25° C the solid phases having true solubility curves are gibbsite or hydrargillite (AH_3), cubic tricalcium aluminate hexahydrate (C_3AH_6) and $Ca(OH)_2$. The other crystalline solid phases, viz. C_2AH_8 and C_4AH_{19}, have only metastable solubility. C_4AH_{19} is the only tetracalcium aluminate hydrate that enters into the aqueous equilibria [*7*]; C_4AH_{13} is formed only on drying it.

Hydrated alumina may have in the system several degrees of "stability" corresponding to differing crystalline forms in the sequence alumina gel (unstable)–boehmite (AH)–bayerite (AH_3)–gibbsite (AH_3; stable) The stability is higher for a better organized structure, and for a given crystalline form it increases with the crystal size. The diagram will therefore include a family of curves, representing solubilities ranging from that of a very unstable gelatinous hydrated alumina (Fig. 3, curve OT′) to that of gibbsite (curve OF).

The only stable equilibria are thus those represented by the curves OFV′C. F is the invariant point for gibbsite and C_3AH_6 with solution, and V′ is the invariant point for C_3AH_6 and $Ca(OH)_2$ with solution. C represents the solubility of $Ca(OH)_2$ (about 1·1 g CaO/l at 25° C). EF and FG are the metastable prolongations of the C_3AH_6 and gibbsite curves, respectively.

2. The Hexagonal Phases: Work of Jones and Roberts

The curve T′TYV is the crystallization isotherm of the metastable hexagonal phases C_4AH_{19} and C_2AH_8. These phases can persist for a long time in contact with aqueous solutions; they tend only slowly to be replaced by the stable C_3AH_6. The detailed significance of the curve T′TYV is not yet definitely established; the precise nature of the solid phases in metastable equilibrium with the solution, and consequently also the solubility relations of these phases, are in some respects matters for discussion.

The curve T′TYV gives the compositions of solutions in metastable equilibrium with ternary hexagonal phases, or mixtures of phases, for which the CaO/Al_2O_3 ratio varies from 2 to 4. Jones and Roberts [7] plotted the CaO/Al_2O_3 ratios of these solids against the lime and alumina concentrations in solution. The results (Fig. 4) indicated an invariant point Y at 0·63 g CaO/l and 0·105 g Al_2O_3/l, which appears also as a discontinuity on the curve T′TYV in Fig. 3. X-Ray examination of the moist solids showed the presence of C_4AH_{19} for CaO/Al_2O_3 3·8–4·0, C_4AH_{19} and C_2AH_8 or material closely resembling it (this will tentatively be called "C_2AH_8") for CaO/Al_2O_3 2·4–3·8 and "C_2AH_8" only for $CaO/Al_2O_3 < 2·4$. Solids which had been dried at 81% r.h. showed the presence of β-C_4AH_{13}, "α-C_4AH_{13}" and C_2AH_8 for CaO/Al_2O_3 2·2–3·8, the C_2AH_8 lines becoming more prominent as the ratio diminished. Dried solids with $CaO/Al_2O_3 < 2·2$ showed only C_2AH_8.

In common with other workers, Jones and Roberts found that microscopic examination of either moist or dried solids is difficult. Even when distinct reflections of two of the hexagonal hydrates are present in the X-ray pattern, optical analysis does not always provide a distinction between their separate crystals. In particular, solids with CaO/Al_2O_3 ratios below 2·4 appear under the microscope to contain only one phase.

On the basis of the above data, Jones and Roberts assumed that VY is the metastable solubility curve of C_4AH_{19} and that YT is that of a solid solution having the composition range 2·0–2·4 $CaO.Al_2O_3.8–10{\cdot}2H_2O$ (Figs. 3 and 4). Y is the invariant point C_4AH_{19}–$C_{2{\cdot}4}AH_{10{\cdot}2}$–solution. To reconcile the X-ray data for the moist and dried solids, they assumed that on drying at 81% r.h. the solid solution decomposed into "α-C_4AH_{13}", β-C_4AH_{13} and C_2AH_8. Because of the close resemblance

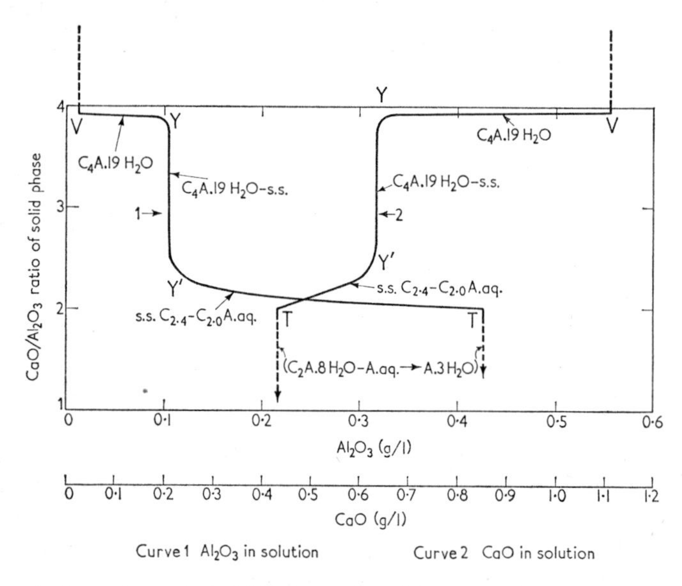

FIG. 4. Variation of CaO/Al_2O_3 ratio of hexagonal plate phase with solution composition (Jones and Roberts [7]).

between the C_4AH_{19} and C_2AH_8 patterns, the members of the solid solution series in the moist state are practically indistinguishable by X-rays from C_2AH_8. Aruja [*19a*] has concluded that the X-ray evidence by itself would be inadequate to justify the view that a solid solution existed.

TT′ represents the metastable or unstable prolongation of the solid solution curve YT, and also the locus of invariant points for the liquid with various forms of hydrated alumina and either C_2AH_8 or a member of the supposed solid solution series. The locus of these invariant points extends beyond T in the direction of Y; Jones and Roberts observed them up to a composition of 0·48 g CaO/l and 0·19 g Al_2O_3/l.

3. Earlier Work on the Hexagonal Aluminate Phases

Before the investigations of Jones and Roberts the interpretation of the metastable equilibria generally accepted was that of Wells *et al.* [*41*]. These authors found that the solubility curve of the hexagonal aluminates followed a continuous line; they found no evidence of an invariant point at Y. Their X-ray data were obtained from samples washed with alcohol and ether or dried over $CaCl_2$, and distinct lines of C_2AH_8 and "α-C_4AH_{13}" were found in the X-ray patterns of solids having CaO/Al_2O_3 ratios of 2–4. They could not differentiate the crystals of the two compounds optically, although the refractive index increased with CaO/Al_2O_3 ratio. To bridge the gap between the X-ray and microscopic data, and to reconcile the former with the solubility data in view of the phase rule, they concluded that the two compounds, because of their structural similarity, might form mixed crystals. Thus they thought that the curve T′TYV was generated by the overlapping of the individual solubility curves for C_2AH_8 and C_4AH_{13}, and on this assumption they suggested that an invariant point occurred at 0·6 g CaO/l and 0·1 g Al_2O_3/l. The presence of an invariant point was also indicated [*6*] by the fact that solids having CaO/Al_2O_3 ratios nearing 2 predominated in the initial section of the curve T′TYV, while in the final section, at high CaO concentrations, the ratios came close to 4. This behaviour was later emphasized by Wells and Carlson [*46*] and by Jones and Roberts [*7*], whose data have already been described.

For equilibria at 20° C D'Ans and Eick [*36*] confirmed the solubility relations found by Wells *et al.* [*41*]. These authors assumed the existence of distinct solubility curves for C_2AH_8 and "α-C_4AH_{13}", with an invariant point at 0·6 g CaO/l and 0·095 g Al_2O_3/l. They also reported a curve at lower concentrations of both CaO and Al_2O_3 which they attributed to β-C_4AH_{13}. D'Ans and Eick believed that some of their precipitates contained hexagonal hydrates with CaO/Al_2O_3 equal to 3, which, however, underwent rapid conversion into the other metastable aluminates. The question of the existence of hexagonal tricalcium aluminate hydrates has already been discussed.

4. $C_{2\cdot0-2\cdot4}AH_{8-10\cdot2}$: Solid Solution or Intergrowth?

It has been suggested [*47*] that a true solid solution, in which each elementary layer had a composition intermediate between those of the two end-members, would be unlikely to separate into C_2AH_8 and C_4AH_{13} as a result of so mild a treatment as drying at 81% r.h. The solubility relationships along the curve T′TYV can, however, also be interpreted without assuming the existence of a limited solid solution, by supposing instead that C_4AH_{19} and C_2AH_8 form intergrowths of a particular kind. These could not be mixed-layer structures in which the intergrowth is on a unit-cell scale, as these would not yield separately diffracting regions of C_4AH_{13} and C_2AH_8 on dehydration. At the other extreme, the individual regions of C_4AH_{19} and C_2AH_8 must be small enough to render the crystal homogeneous optically and to cause it to behave from the point of view of the phase rule as a single phase and not as two. These requirements could possibly all be satisfied if the crystal were composed of alternate groups of layers of each of the two compounds, with perhaps something over 50 elementary layers in each group.

For CaO/Al_2O_3 ratios above 3·8 or below 2·2 one would not expect to be able to observe the reflections of the minor constituent in the X-ray powder patterns of either the moist or the dry solids. In the solids with CaO/Al_2O_3 2·2–2·4 the X-ray patterns showed C_4AH_{13} and C_2AH_8 in the dried solids but only C_2AH_8, or material very similar to it, in the moist solids. This result is not unexpected if one considers the strong similarity between the C_4AH_{19} and C_2AH_8 patterns and the possible influence of the intergrowth on the stacking within the groups of layers. At CaO/Al_2O_3 ratios of 2·4–3·8 both C_2AH_8 and the appropriate C_4A hydrate were observed in both dry and moist solids, and this is also to be expected.

The above hypothesis does not modify fundamentally the interpretation of the solubility relations given by Jones and Roberts nor does it conflict with the hypothesis of Wells *et al.* A definite choice between the alternative hypotheses of true solid solution and intergrowth of groups of layers can perhaps be made only when the structures of the hexagonal aluminates are completely known.

5. CAH_{10} *in the System at 21° C*

Percival and Taylor [*45*] detected CAH_{10} in the system $CaO–Al_2O_3–H_2O$ at 21° C. This contrasts with the earlier results obtained by Bessey [*42, 43*] Wells *et al.* [*41*] and D'Ans and Eick [*36*] but is in accordance with the fact that CAH_{10} is known to be a hydration product either of CA or of aluminous cement as this temperature.

The metastable solubility curve HD for CAH_{10} (Fig. 5) lies below the curve TV, so that at 21° C C_2AH_8 in contact with solution tends to decompose into CAH_{10} and $Ca(OH)_2$. Jones and Roberts [*7*] confirmed this result and found that the transition temperature between the two phases is about 22° C. The curve HD was first determined by the undersaturation method [*45*] and later confirmed by the supersaturation method starting from very concentrated calcium aluminate solutions [*8*].

The transition temperature of 22° C and the somewhat specific precipitation conditions of CAH_{10} from supersaturation explain why previous investigators did not observe CAH_{10} in the system at 20–25° C. Jones [20]

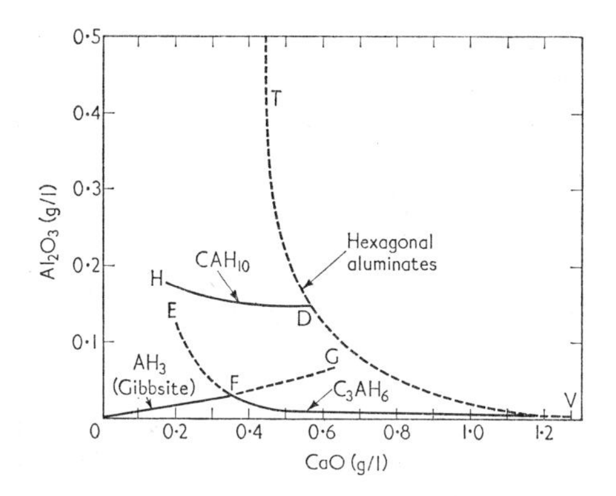

FIG. 5. Solubility relations in the system $CaO–Al_2O_3–H_2O$ at 21°C (Percival and Taylor [*45*]).

considered that the position of D was too close to Y (Fig. 3) to accord well with the probable range of compositions of the $C_{2\cdot0-2\cdot4}AH_{8-10\cdot2}$ solid solution, but Percival *et al.* [*8*] reconfirmed the position of D, thus refuting this view.

C. EQUILIBRIA AT 1–5°C

For the system at 5° C Buttler and Taylor [*48*] reported the diagram shown in Fig. 6. The solid phases were identified by X-rays after drying at 34% r.h. and their compositions were calculated by difference from the compositions of the initial mixtures and those of the final solutions. On reworking selected data of Buttler and Taylor, Jones [*20*] obtained a modified diagram (Fig. 7). He identified on the curve TYV a first invariant point Y at 0·53 g CaO/l and 0·07 g Al_2O_3/l for a supposed solid solution $C_{3\cdot5}AH_x$ with C_4AH_{19} and the liquid, and a second invariant point D at 0·38 g CaO/l and 0·13 Al_2Og_3/l for a C_3AH_x solid solution with CAH_{10} and the liquid. The existence and positions of the two invariant points were deduced by plotting the CaO/Al_2O_3 ratios of the final solid phases against the CaO and Al_2O_3 concentrations in the solution.

Buttler and Taylor [*48*] found that along the curve YD the dried solids were mixtures of C_2AH_8 and C_4AH_{13}. Assuming the equilibrium relationships as conceived by Jones to be valid, it must be inferred that the

$C_{3\cdot5-3}AH_x$ solid solution is decomposed on drying at 34% r.h. Furthermore, it must be assumed that C_2AH_8, although unstable when present as a pure phase in contact with solution at 5° C, can exist in solid solution

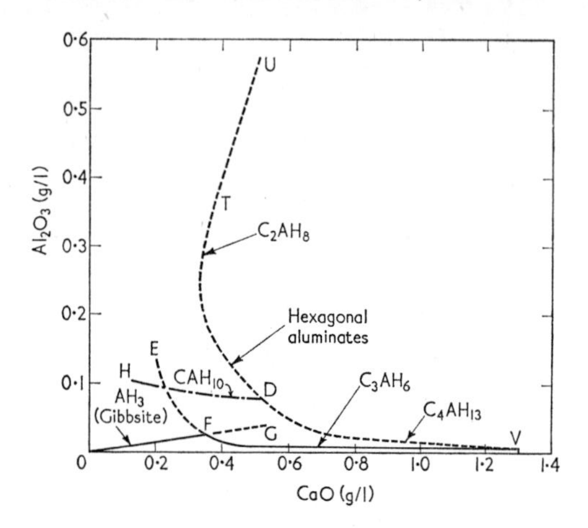

FIG. 6. Solubility relations in the system $CaO–Al_2O_3–H_2O$ at 5°C (Buttler and Taylor [*48*]).

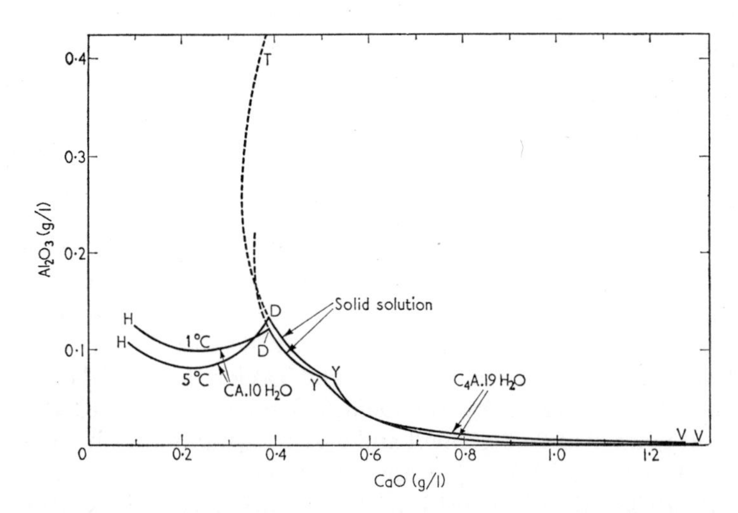

FIG. 7. System $CaO–Al_2O_3–H_2O$ at 5° and 1°C, according to Jones [*20*].

with C_4AH_{19}. Buttler and Taylor found that solids occurring in contact with solutions at or to the left of T were composed of C_2AH_8, CAH_{10}, alumina gel and sometimes also C_4AH_{13}. To reconcile this with his modified diagram Jones suggested that, in the presence of the liquid of concentration represented by T, the solid solution may decompose according to the sequence

$$C_4AH_{19} \longrightarrow C_2AH_8 \longrightarrow CAH_{10} \longrightarrow AH_3$$

though he did not exclude the possibility of direct decomposition to CAH_{10} and alumina gel. In this latter case the C_2AH_8 in the dried solids results only from decomposition of the solid solution during drying.

The diagram of Buttler and Taylor (Fig. 6) includes an invariant point at D for CAH_{10}–C_4AH_{19}–solution. Jones rejected the possibility of such an invariant point; he considered that the CAH_{10} curve could not intersect the curve TYV below Y. He attributed the low position of Buttler and Taylor's curve HD to the slowness of reaching equilibrium,

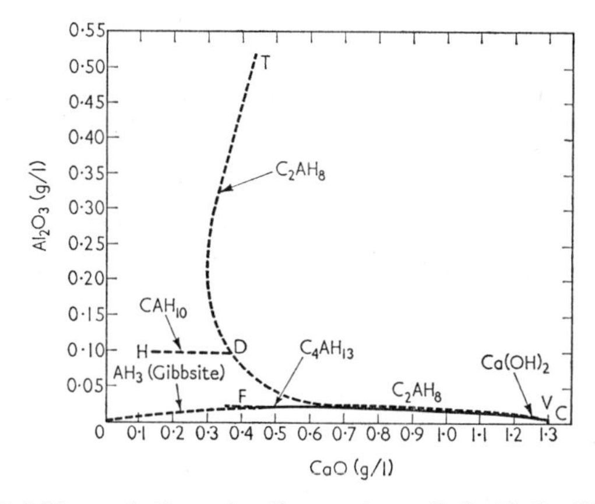

FIG. 8. Solubility relations in the system CaO–Al_2O_3–H_2O at 1°C (Carlson [*18*]).

which was characteristic of the undersaturation method which they had employed to fix the position of the intersection at D. Percival *et al.* [*8*], though recognizing that equilibria at 5° C are harder to interpret than those at 21° C, reaffirmed the position of the curve HD given by Buttler and Taylor. They maintained that the data did not justify Jones's assumption that equilibrium had not been reached.

Equilibria at 1° C (Fig. 8) were investigated by Carlson [*18*], who used X-rays to identify the solid phases in the moist state, after washing with alcohol and ether and also after drying over $CaCl_2$. Carlson used mainly supersaturation data to determine the solubility relationships since undersaturation preparations moved towards equilibrium very slowly.

On the evidence of the X-ray results for the moist solids, Carlson suggested that the solid existing along the curve TDV is C_2AH_8. At the time of his experiments the conditions of existence of the various calcium aluminate hydrates were yet to be defined, and Carlson may have confused the C_4AH_{19} pattern with that of C_2AH_8, which is very similar to it. The curve TDV clearly has the same significance as in the other diagrams.

Carlson did not give curves showing CaO/Al_2O_3 ratios of the solids plotted against solution concentrations, though Wells and Carlson [*46*] presented such a diagram. Jones [*20*] replotted the data (Fig. 7) and identified on the curve TDV a first invariant point Y at 0·5 g CaO/l and 0·075 g Al_2O_3/l between a supposed solid solution $C_{2\cdot5}AH_x$, C_4AH_{19} and the liquid, and a second invariant point D at 0·39 g CaO/l and 0·12 g Al_2O_3/l between a $C_{2\cdot1}AH_x$ solid solution, CAH_{10} and the liquid. Since C_2AH_8 is unstable at 1° C the section TD of the curve must represent an unstable extension of the solubility curve for the supposed solid solution $C_{2\cdot1}AH_x$.

Owing to the extreme slowness of dissolution, the position of OF, the solubility curve for gibbsite, in Fig. 8 is merely an estimate. The position of C, representing the solubility of $Ca(OH)_2$, was fixed from values in the literature. Carlson found that, at 1° C, C_3AH_6 appeared to be metastable, being transformed very slowly into one or more of the hexagonal hydrates.

The above data indicate that at low temperatures the solubility relationships in both metastable and stable fields are in many respects still unclarified. The possible existence and limits of composition of solid solutions or intergrowths, and their stabilities relative to the other phases, must be verified. Carlson's observation on the apparent stability of C_3AH_6 appears interesting, and further investigations to establish transition temperatures and products should be carried out.

D. Equilibria at 50–1000° C

Equilibria in this range were studied by Wells *et al.* [*41*] at 90° C, by Peppler and Wells [*49*] at 50, 120, 150, 200 and 250° C, by Majumdar and Roy [*13*] at 110–1000° C, and by Pistorius [*49a*] at 200–900° C. Peppler and Wells used polythene bottles at 50° C and stainless steel bombs at higher temperatures. The products, obtained from supersaturation and undersaturation experiments, were identified microscopically and, in doubtful cases, by X-rays. Majumdar and Roy worked with bombs of Stellite No. 25 alloy. The starting materials were crystalline preparations with various CaO/Al_2O_3 ratios, obtained by calcination of $CaCO_3$–Al_2O_3 mixtures at 1400–1500° C, or preparations made by slow firing of dried gels precipitated with ammonia from solutions containing calculated amounts of calcium and aluminium nitrates. The mixtures were placed in small gold envelopes within the bombs. The solids were identified by X-rays and microscopically. Pistorius, who worked with very high pressures, used a high-temperature "squeezer" apparatus.

The stable, hydrated phases in the 50–250° C range under saturated steam pressures comprise $Ca(OH)_2$ over the whole range; gibbsite up to 150° C and boehmite above 150° C; and C_3AH_6 below 225° C and $C_4A_3H_3$ above 225° C. The metastable hexagonal phases are present up to 90° C; at higher temperatures they are converted so rapidly into C_3AH_6 that

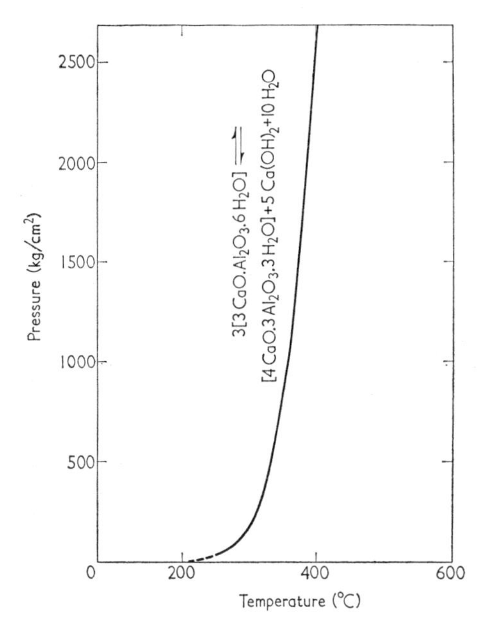

FIG. 9. Pressure–temperature curve for the dehydration of C_3AH_6 (Majumdar and Roy [*13*]).

their solubility curve cannot be determined. No information is available on the exact nature of the metastable phases above 25° C or on the mechanism of conversion to the stable phases. In the 50–250° C range the maximum solubility of the stable alumina hydrate occurs at the invariant point with C_3AH_6 and solution. The highest stable concentration of Al_2O_3 is attained at 120° C, when the value is 0·3 g/l. At 250° C the solubility is almost negligible.

The velocity of conversion of gibbsite into boehmite is still low enough at 150° C to permit the determination of solubility curves for both phases; at 200° C this is no longer possible. The solubility of C_3AH_6 reaches a maximum at about 100° C, above which it decreases rapidly. The solubility of $Ca(OH)_2$ in water decreases steadily with temperatures and at 250° C it is reduced to 0·037 g CaO/l.

Dehydration of boehmite and of $Ca(OH)_2$, and decomposition of $C_4A_3H_3$ into $C_{12}A_7$ and CA_2, occur above 250° C. At 300° C and a water pressure of 140 kg/cm² boehmite is converted into its polymorph, diaspore. Majumdar and Roy [*13*] investigated the variation of the temperatures of dehydration of $Ca(OH)_2$, and of decomposition of

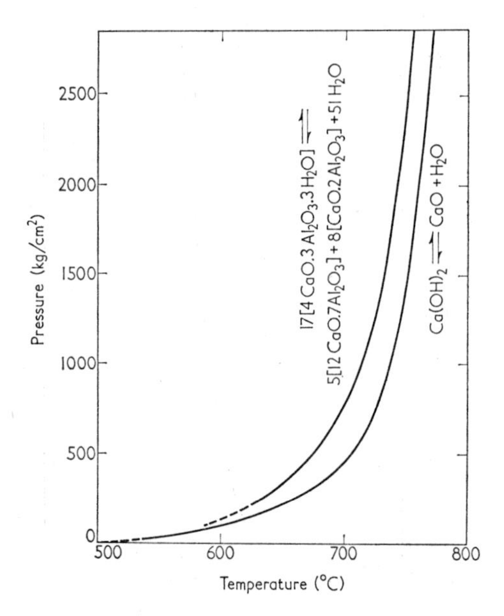

FIG. 10. Pressure–temperature curves for the dehydration of $Ca(OH)_2$ and $C_4A_3H_3$ (Majumdar and Roy [*13*]).

C_3AH_6 and $C_4A_3H_3$, as functions of water pressure. The results are shown in Figs. 9 and 10. Each curve separates the diagram into two areas; the left-hand area represents the stability field of the lower temperature phases, and the right-hand area that of the higher temperature phases.

Pistorius [*49a*] determined pressure–temperature equilibrium curves for the formation and decomposition of $C_4A_3H_3$ at water pressures of 5–50 kb. He showed that $C_4A_3H_3$ could be obtained at lower temperatures than those found by Majumdar and Roy, if starting materials of CaO/Al_2O_3 ratio 4:3 were used. His curve for the decomposition of $C_4A_3H_3$ connected smoothly with that of Majumdar and Roy, who had used lower pressures.

IV. Calcium Aluminate Hydrates Containing Other Anions

A. General Comments

At ordinary temperatures many organic and inorganic anions combine with lime, alumina and water to form slightly soluble quaternary solids. Most of these compounds belong to one or other of two series of compounds, with empirical formulae

$$C_3A.CaY_2.mH_2O \quad \text{or} \quad C_3A.CaX.mH_2O \qquad (1)$$

and

$$C_3A.3CaY_2.nH_2O \quad \text{or} \quad C_3A.3CaX.nH_2O \qquad (2)$$

where Y = a monovalent anion, X = a divalent anion, $m = 10$–12, and $n = 30$–32. The compounds of group (1) have structures closely related to those of the hexagonal calcium aluminate hydrates.

The compounds should be named by placing the name of the anion after that of the aluminate, e.g. $C_3A.CaSO_4.12H_2O$ = tetracalcium aluminate monosulphate hydrate; $C_3A.3CaSO_4.32H_2O$ = hexacalcium aluminate trisulphate hydrate. Although this nomenclature represents more correctly the relationships between the various ions, in practice it is almost completely replaced by one which puts the root of the anion name before the aluminate ending. Many years of usage have firmly established such names as sulphoaluminate, carboaluminate and chloroaluminate in place of the more correct aluminate sulphate, aluminate carbonate and aluminate chloride. With this nomenclature the compounds of the two series are distinguished by such prefixes as low-sulphate and high-sulphate; thus $C_3A.CaSO_4.12H_2O$ is called low-sulphate calcium sulphoaluminate, and $C_3A.3CaSO_4.32H_2O$ is called high-sulphate calcium sulphoaluminate. Names such as monosulphoaluminate and trisulphoaluminate are also in use.

The quaternary aluminates all appear to be hexagonal or pseudo-hexagonal; the mono- or "low" forms crystallize in thin, hexagonal plates, and the tri- or "high" forms as needles. With some ions, compounds of both series are known; with many others, however, only the mono-forms have been prepared. In general, the compounds dissolve incongruently in water.

The methods of preparation are similar to those of the calcium aluminate hydrates, viz.

(a) precipitation from calcium aluminate solutions with lime-water and calcium salt solutions;
(b) precipitation from sodium or potassium aluminate solutions by adding lime-water and calcium salt solutions;

(c) precipitation from aluminium salt solutions with lime-water and calcium salt solutions;
(d) reaction of alumina gel with lime-water and calcium salt solutions; and
(e) precipitation from a sodium aluminate solution containing also NaOH and the sodium salt of the anion, by adding calcium and sodium salt solutions of the same anion.

If the calcium salt is not sufficiently soluble the anion may be added as the sodium or, better, the ammonium salt. The overall ratios CaO/Al_2O_3 and anion/Al_2O_3 must exceed the stoichiometric ratios to avoid hydrolysis of the precipitate.

Of the numerous quaternary aluminates prepared up to now, the ones of interest to the chemistry of cement are those containing sulphate, carbonate, chloride or silicate anions. For the others the reader is referred to the original papers and past surveys [*4*, *30*, *50–53*].

The structures of the quaternary compounds are little understood. The most substantial hypothesis is that which regards the compounds of the "mono" series as closely related to the tetracalcium aluminate hydrates. Feitknecht and Buser [*54*] and, later, Buttler *et al.* [*9*] suggested that the structures are derived from that of C_4AH_{13}, the additional anions replacing hydroxyl ions not present in the octahedral calcium-containing sheet. In many cases the anions also displace water molecules; depending on the size of the anion and the number of water molecules displaced, the thickness of the structural element can be either larger or smaller than in C_4AH_{13}. Electron diffraction evidence (p. 374) confirms the close relation to C_4AH_{13}.

B. Calcium Aluminate Sulphate Hydrates

Gypsum is added to Portland cement clinker as a setting retarder. It reacts with the anhydrous aluminates to form two aluminate sulphate hydrates, the formulae of which are $C_3A.3CaSO_4.31–32H_2O$ and $C_3A.CaSO_4.xH_2O$.

1. $C_3A.3CaSO_4.31–32H_2O$

The high-sulphate compound, or trisulphate, occurs as the natural mineral, ettringite. The artificial material was first observed by Candlot and prepared by Michaëlis, who reported the formula $C_3A.3CaSO_4.30H_2O$ for the compound dried over H_2SO_4. Two methods of preparation have commonly been employed: (a) by reaction of aluminium sulphate solution with lime-water, sometimes containing gypsum and (b) by mixing a calcium aluminate solution with one of lime and gypsum.

Berman and Newman [*55*] prepared, without observing any significant differences between the products, eight samples of ettringite employing both methods and mixing the reagents according to various procedures. Since the compound dissolves incongruently in pure water, it is necessary to have excess lime and also excess gypsum in solution at the end of the reaction.

At 20° C $C_3A.3CaSO_4.31–32H_2O$ is stable at partial water vapour pressures down to 0·7 mm Hg. Intensive drying at 0·2 mm reduces the water content to about $8H_2O$. Thermal dehydration [*55*] leaves $20H_2O$ at 60° C, $8H_2O$ at 110° C, $6H_2O$ at 145° C and $2H_2O$ at 200° C. The X-ray pattern for a sample with $23H_2O$ is almost the same as that of the initial material, the only difference being an increase in the intensity of the 4·48 Å line. For a sample with $8H_2O$, lines at 7·7 and 6·1 Å appear in place of those at 9·8 and 5·6 Å. As the water content decreases to $6H_2O$ the longest spacing, at 7·7 Å, is displaced to 7·4 Å. DTA curves were reported by Kalousek *et al.* [*15*], Turriziani and Schippa [*56*], Midgley and Rosaman [*23*], Greene [*57*] and others. The curve is characterized by an endotherm at low temperatures; different authors have reported peak temperatures ranging from 110 to 150° C.

The heats of solution of the compound in different hydration states, as well as the heats of formation of $C_3A.3CaSO_4.31H_2O$ and $C_3A.3CaSO_4.32H_2O$ from C_3AH_6 and from the elements were determined by Berman and Newman [*55*]. The rate of change of the heat of solution with increase of water content in the 31–$32H_2O$ region is 2·44 kcal/mole of H_2O.

The X-ray pattern of $C_3A.3CaSO_4.31–32H_2O$ was reported by Fratini, Schippa and Turriziani [*58*], Midgley [*59*] and others. Bannister [*60*] determined the unit cell, but the structure is unknown. Taylor [*61*] suggested that this problem might be solved by using as a model the structure of the mineral, thaumasite ($(Ca_3H_2(CO_3)(SO_4)(SiO_4).13H_2O)$), recently determined by Welin [*62*], due to the dimensional analogy between the unit cells of the two minerals.

$C_3A.3CaSO_4.31–32H_2O$ dissolves in water with separation of alumina gel, $CaSO_4.2H_2O$ and $Ca(OH)_2$; the final concentrations are 35 mg CaO/l, 215 mg Al_2O_3/l and 43 mg SO_3/l. In solutions containing added lime or gypsum or both, the solubility is much lower and the decomposition is therefore largely halted. For certain concentrations of CaO and $CaSO_4$ the solubility is congruent. The compound can exist in contact with aqueous solutions of Na_2SO_4, NaCl, $NaNO_3$ or $CaCl_2$, but is decomposed by $MgSO_4$ with formation of $Mg(OH)_2$, $Al(OH)_3$ and $CaSO_4$, and by alkali carbonate solutions. D'Ans and Eick [*63*] found that with CO_2 the following reaction occurs

$$C_3A.3CaSO_4.32H_2O + 3CO_2 \longrightarrow 3CaCO_3 + Al_2O_3.xH_2O + 3CaSO_4.2H_2O + nH_2O$$

Schippa [64] confirmed this and identified the $CaCO_3$ by X-rays as aragonite.

2. $C_3A.CaSO_4.xH_2O$

This, the monosulphate or low-suphate form, was obtained by Lerch, Ashton and Bogue [65] and independently by Mylius [30]. It is prepared by adding to a metastable calcium aluminate solution a saturated lime solution containing gypsum in such concentration that the $CaSO_4/Al_2O_3$ ratio of the initial mixture is about 1. If the gypsum is in excess, ettringite is also formed; if there is not enough a supposed solid solution with C_4AH_{13} is obtained. To avoid the separation of alumina gel the overall CaO/Al_2O_3 ratio must exceed 3.

The solid phase in equilibrium with the solution exists in various modifications, depending on the temperature [66, 67]. Further modifications are produced on drying under various conditions. At 20° C the hydrate formed in equilibrium with solution contains about $16H_2O$ and its X-ray pattern is characterized by a longest basal spacing of 9·59 Å. This phase begins to dehydrate below 90% r.h., and on drying at 35% r.h. is converted into the $12H_2O$ hydrate, the basal spacing of which is 8·99 Å. At 1° C the hydrate in equilibrium with solution has a basal spacing of 10·3 Å. In the 12–17° C region mixtures of the 9·59 and 10·3 Å hydrates are precipitated in varying proportions.

At room temperature the 10·3 Å hydrate is dehydrated, giving $C_3A.CaSO_4.12H_2O$ at 90–95% r.h. Drying at 20° C and a partial water vapour pressure of 0·2 mmHg reduces the water content to about $10H_2O$ and the longest spacing to 8·26 Å. At 110° C the water content becomes $7H_2O$; dehydration over P_2O_5 reduces the water content to $8H_2O$ and the basal spacing to 8·05 Å. DTA curves for $C_3A.CaSO_4.12H_2O$ were reported by Kalousek *et al.* [15], Turriziani and Schippa [56] and Midgley and Rosaman [23]. Fratini *et al.* [58] concluded from X-ray powder data that the unit-cell dimensions were *a* 8·85, *c* 9·01 Å. Roberts [68a], taking into account the close structural resemblance that evidently exists between $C_3A.CaSO_4.xH_2O$ and C_4AH_{13}, suggested that an *a*-axis of about 5·75 Å was more likely. Electron diffraction evidence shows the true *a*-axis to be 11·4 Å with marked pseudo-halving (Chapter 9, p. 374).

Jones [44] and D'Ans and Eick [68] found that ettringite and $C_3A.CaSO_4.xH_2O$ do not form solid solutions; Turriziani and Schippa [56] confirmed this by an X-ray investigation.

Microscopic examination of the quaternary solids obtained at 25° C with CaO/Al_2O_3 ratios lower than 1 reveals the apparent existence of a single, solid hexagonal phase whose refractive index varies between those of "α-C_4AH_{13}" and $C_3A.CaSO_4.12H_2O$ (Fig. 11). Such results indicate that the two compounds form an uninterrupted series of mixed crystals. However, Turriziani and Schippa [*56*], examining by X-rays the quaternary solids with $CaSO_4/Al_2O_3$ ratios of 0–1, partially dried over $CaCl_2$, obtained patterns with distinct lines of "α-C_4AH_{13}" and of $C_3A.CaSO_4.12H_2O$. The intensity of the "α-C_4AH_{13}" lines increased with decreasing SO_4^{2-} content, although the optical properties varied as would be expected for a complete series of mixed crystals. The situation is similar to that already discussed for the metastable ternary aluminate hydrates. Before any hypothesis can be proposed, it appears necessary to have more data, obtained both for moist samples and for ones in various states of partial hydration. The solids should also have been prepared both by the supersaturation method and by reaction between C_4AH_{19} and gypsum solutions.

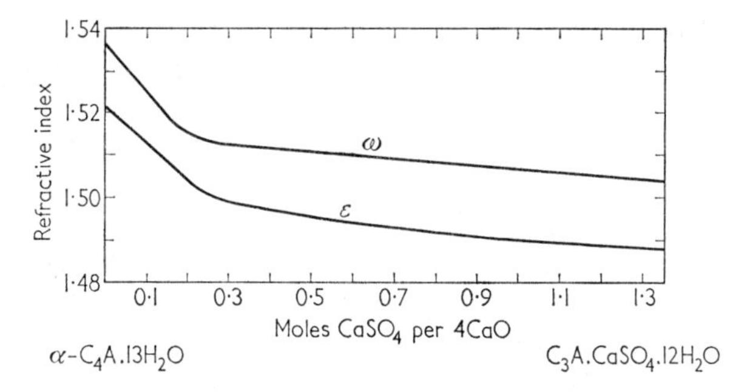

FIG. 11. Relation between refractive index and composition for the solid solution series C_4AH_{13}–$C_3A.CaSO_4.12H_2O$, according to D'Ans and Eick [*68*].

Since the literature these quaternary solids are generally called "solid solutions", this name will be used here. It is, however, implicit that in the field of these solid solutions the quaternary equilibria should be completely reviewed.

C. THE SYSTEM CaO–Al_2O_3–$CaSO_4$–H_2O

Comprehensive studies of the equilibria in this system have been reported by Jones [*44, 69*] for 25° C and by D'Ans and Eick [*68*] for 20° C. Jones represented the system as a reciprocal salt pair involving the reaction

$$3Ca(OH)_2 + Al_2(SO_4)_3 \longrightarrow Al_2O_3.xH_2O + 3CaSO_4.xH_2O$$

For the preparations he employed saturated solutions of $Ca(OH)_2$ and $CaSO_4.2H_2O$; solutions of $Al_2(SO_4)_3$; the crystalline solids $CaSO_4$.

$2H_2O$, $Ca(OH)_2$ and C_3AH_6; and distinct modifications of hydrated alumina ranging from the gel to a form with well developed crystals reported as having a structure intermediate between those of bayerite and gibbsite (bayerite crystals partially converted to gibbsite but with bayerite predominating). The solid phases were identified optically after drying at a partial water vapour pressure of 14 mmHg.

The space model developed by Jones is a square-based pyramid with H_2O at the apex. Each corner of the base represents a pure salt; the four edges meeting at the H_2O apex represent salt–water compositions and the four lateral faces represent ternary compositions of water and two salts with a common ion. The solubility relationships are reported on a plane by means of the Jänecke projection, in which any composition (for saturated solutions) in the space is projected on to the base of the pyramid at its intersection with a line connecting the H_2O apex to the composition in question. The Jänecke projection is therefore a square whose corners represent the anhydrous parts of each of the pairs of reciprocal salts. Compositions are expressed in moles per cent of components in the saturated solution. To respect the molar identity of the two sides of the reaction trimeric formulae are used at the lime and calcium sulphate corners. Thus, the C_3A composition, for example, corresponds to an equimolar mixture of $(CaO)_3$ and Al_2O_3.

In Fig. 12 the stable equilibria are represented by unbroken curves. The solid phases involved are $Ca(OH)_2$, gypsum, C_3AH_6, ettringite and gibbsite. (The alumina trihydrate actually used was the form intermediate between gibbsite and bayerite, and is designated $Al_2O_3.3H_2O$ (G) in Fig. 12; its solubility relations may be held without serious error as representing those of gibbsite.) Each solid phase is represented by a field on the diagram, over which it coexists with solution. Along the curves, two solid phases coexist with solution, and at the inter-sections of these curves are invariant points at which three solid phases coexist with solution. These invariant points are listed in Table IV.

The broken curve EH in Fig. 12 relates to the coexistence of ettringite with alumina gel. With bayerite a curve intermediate between curves EH and E_2H_2 was obtained. The ettringite and C_3AH_6 fields are thus enlarged at the expense of the hydrated alumina field as the crystallinity of the latter decreases. When the hydrated alumina occurs as gel, the ettringite field becomes very large.

The composition of the hypothetical anhydrous ettringite is situated on the $(CaO)_3$–$Al_2(SO_4)_3$ diagonal outside the ettringite field. Ettringite therefore hydrolyses in water; alumina gel separates, and the projection of the solution composition moves along the prolongation of the Al_2O_3–$C_3A.3CaSO_4$ line until the coexistence curve EH for ettringite and alumina gel is reached. The diagram shows that ettringite dissolves congruently in solutions containing lime or gypsum or both, within certain ranges of concentration. With sufficiently concentrated lime solutions it would be expected to give C_3AH_6, because the $(CaO)_3$–$C_3A.3CaSO_4$ line crosses the C_3AH_6 field. However, Jones never observed precipitation of C_3AH_6 under these conditions, and the incongruent solubility of ettringite in this zone has thus not been verified experimentally.

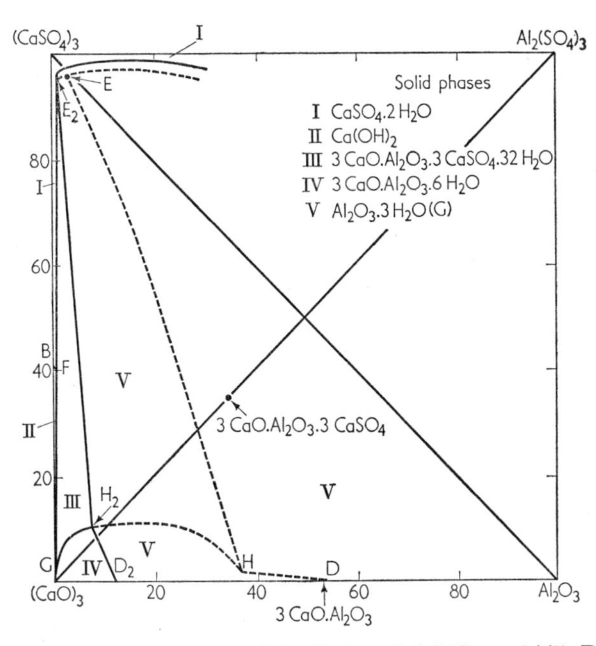

FIG. 12. System CaO–Al_2O_3–$CaSO_4$–H_2O at 25°C (Jones [*44*]). Diagram slightly distorted at the left-hand side to show the fields of primary crystallization more distinctly (Steinour [*6*]). $Al_2O_3.3H_2O(G)$ = material intermediate between bayerite and gibbsite.

In the presence of sulphate ions the ternary hexagonal plate phases do not exist as such but as solid solutions. Jones found that they had only a a metastable existence, as in the ternary system, and they therefore do not appear on Fig. 12. Figure 13 shows part of a Jänecke projection with a greatly enlarged scale for the $CaSO_4$ concentrations, for metastable equilibria involving the solid solutions and also crystalline Al_2O_3. $3H_2O(G)$. The field of the solid solutions lies close to the CaO–Al_2O_3–H_2O face of the pyramid; the $CaSO_4$ concentrations of the solutions in metastable equilibrium with them are only a few mg/l (e.g. at point R_2; see Table IV).

TABLE IV

Invariant points at 25° C (Jones)

Point (Figs. 12 and 13)	Composition of liquid phase (mg/1000 g)			Solid phases							
	CaO	$CaSO_4$	Al_2O_3	Alumina gel	Bayerite	$Al_2O_3.3H_2O(G)$	C_3AH_6	$Ca(OH)_2$	Ettringite	$C_3A.CaSO_4.H_{12}$	Gypsum
D_2	88·5	—	22·4			+	+				
E_2	37·8	2040	2·55			+			+		+
H_2	196	51	9·2			+	+		+		
R_2(1)	461	2	37			+			+	+	
F	1076	1680	6·12					+	+		+
G	1063	24·5	1·02				+	+	+		
E_1(1)	28·60	2053	1·50		+				+		+
H_1(1)	152	49	16·30		+		+		+		
E	7·6	2142	6·12	+					+		+
H	200	38	74·4	+			+		+		

(1) Not shown on Fig. 12.

H_2R_2 is the metastable prolongation of the curve E_2H_2 shown in Fig. 12. One terminal component of the solid solution series, the monosulphate $C_3A.CaSO_4.12H_2O$, coexists metastably with ettringite and solution along R_2M; another, C_4AH_{19}, occurs on the $CaO–Al_2O_3–H_2O$ face. Along $R_2S_2T_2$ the solid solution coexists with AH_3 and solution.

Jones extended this study to equilibria involving alumina gel; he regarded these as unstable as distinct from metastable. The solubility relations of the solid solution were not as simple as with crystalline AH_3. He represented the region of crystallization of the solid solution by a "purely diagrammatic sketch" (Fig. 14). Curve VT falls on the ternary face $CaO–Al_2O_3–H_2O$ and represents the solid solution; T is the unstable invariant point involving alumina gel and a solid solution of composition C_3AH_{12}, which should possibly be considered as an intergrowth of dicalcium and tetracalcium aluminate hydrates. Point V is the invariant involving $Ca(OH)_2$ and a solid solution with a CaO/Al_2O_3 ratio supposedly greater than 4 and possibly as high as 5. Along the curve RST

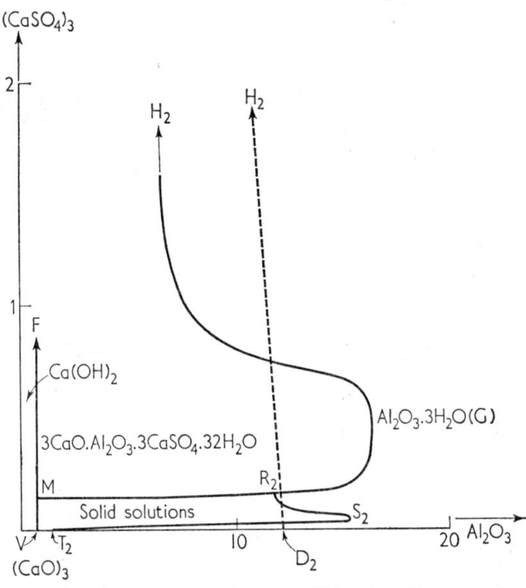

FIG. 13. Portion of the metastable crystallization diagram for the system $CaO–Al_2O_3–CaSO_4–H_2O$ at 25°C involving solid solutions (later form as given by Jones [*44*]). Detail of $(CaO)_3$ corner. $Al_2O_3.3H_2O$ (G) = material intermediate between bayerite and gibbsite.

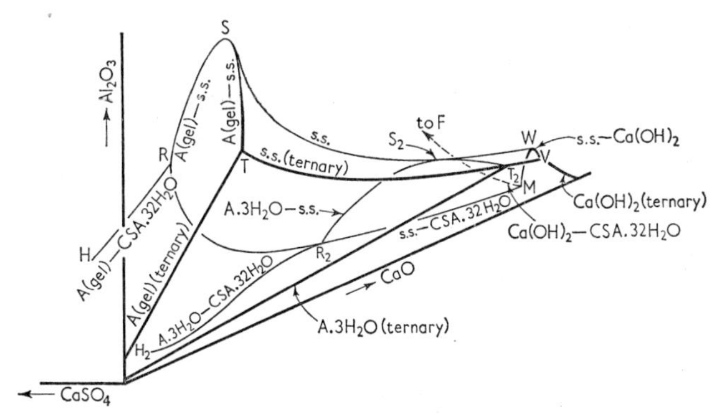

FIG. 14. Schematic diagram showing relations between various curves in equilibria involving the solid solutions (Jones [*44*]).

alumina gel coexists with a quaternary solid solution with $CaO/Al_2O_3 = 3$; along MWV Ca $(OH)_2$ coexists with the solid solution richer in lime. The aluminate monosulphate (a terminal component of the solid solution series) and ettringite coexist along RM. The limits of composition of the solid solution field are indicated schematically in Fig. 15 [6].

Between the curves VT and RM the Al_2O_3 concentrations in solution pass through a maximum which is particularly marked at low lime concentrations. The shape of the solid solution surface is such that, for certain CaO and Al_2O_3 concentrations, very small changes in $CaSO_4$ concentration cause a marked change in the composition of the solid solution. Jones considered it possible that there existed two series of solid solutions, the continuous area RSTMWV being thus interrupted; he tentatively placed the two regions within TVWS and RMWS.

D'Ans and Eick [68] represented the solubility relationships at 20°C on a diagram in which CaO, Al_2O_3 and $CaSO_4$ concentrations in the solution

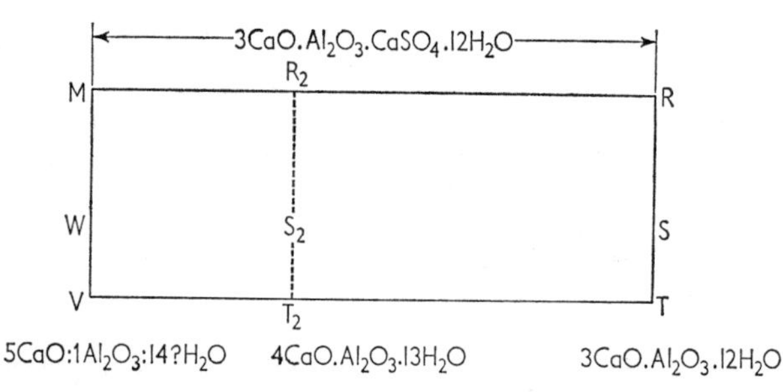

FIG. 15. Schematic diagram showing limiting compositions of the solid solution as found by Jones and represented by Steinour [6].

were plotted in mg/l on orthogonal axes (Fig. 16). Eitel [70] showed that some aspects of the equilibria can be shown more clearly if logarithmic scales are used. D'Ans and Eick used an alumina gel precipitated from an aluminate solution poor in lime and aged for 14 days. The solubility relationships involving this gel are not, therefore, strictly comparable with those found by Jones. In practice the invariant points found by D'Ans and Eick (Table V) are not very different from those found by Jones, and the alumina gel which they used was possibly intermediate between that used by Jones and bayerite, though nearer to the former. On this assumption the two sets of results can be compared. Broadly speaking, the data agree except as regards the compositions of the solid solutions and the position of their crystallization surface. D'Ans and Eick found that the compositional range of the solid solution is C_4AH_x–$C_3A.CaSO_4.xH_2O$. This conclusion was based on microscopic examination of solids precipitated from solutions supersaturated with lime and alumina, to which variable amounts of gypsum had been added. No C_2AH_8 or CAH_{10} was observed, even after a year; when the ratio Al_2O_3/

$CaSO_4$ in the initial mixture was equal to 1, hydrated alumina separated unless $CaO/Al_2O_3 \geqslant 3$. Higher $Al_2O_3/CaSO_4$ ratios required higher CaO/Al_2O_3 ratios to avoid precipitation of hydrated alumina.

According to D'Ans and Eick, the position of the primary crystallization surface for the solid solutions differs from that proposed by Jones

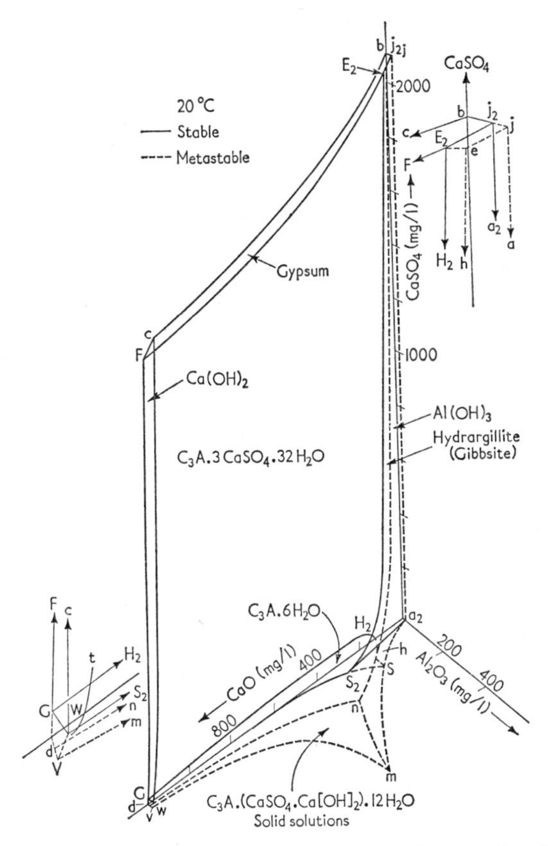

FIG. 16. Stable and metastable crystallization diagram for the system $CaO–Al_2O_3–CaSO_4–H_2O$ at 20°C (D'Ans and Eick [*36*]).

in that the Al_2O_3 concentration decreases continuously on going away from the $CaO–Al_2O_3–H_2O$ face, without passing through any maximum. On the curve vm (Fig. 16) the ternary hexagonal phase assumed by D'Ans and Eick was C_4AH_x. Point m represents the invariant with alumina gel and point v that with $Ca(OH)_2$. The coexistence of C_4AH_x with alumina gel at point m does not agree with results obtained for the ternary system at 21° C, and presupposes that small amounts of $CaSO_4$

TABLE V

Invariant points at 20° C (D'Ans and Eick)

Point (Fig. 16)	Composition of liquid phase (mg/l)			Solid phases							
	CaO	$CaSO_4$	Al_2O_3	Alumina gel(1)	Gibbsite	C_3AH_6	$Ca(OH)_2$	Ettringite	$C_3A.CaSO_4.H_{12}$	Gypsum	C_4AH_x
S_2	315	—	25		+	+					
E_2	17·7	2016	9·23		+			+		+	
H_2	159·1	35·75	16·17		+	+		+			
F	1154	1660	2·75				+	+		+	
G	1196	14·6	2·62			+	+	+			
e	8·5	2055	10·4	+				+		+	
h	195·7	11·5	64·5	+		+		+			
n	335	8	156·1	+				+	+		
d	1179	4	2·22				+	+	+		
v	1202	—	3·32				+				+
m	481·1	—	423·1	+							+
s	190	—	80	+		+					
j	—	2064	1	+						+	

(1) Aged 14 days.

suffice to direct the reaction towards the formation of mixed crystals and alumina gel rather than towards that of less basic aluminates.

The quaternary equilibria thus demand further study, especially as regards the solid solution field. A better knowledge of the metastable ternary equilibria would help to clarify many of the points about the compositions of the assumed solid solutions.

D. THE SYSTEM $CaO–Al_2O_3–CaSO_4$–Alkali–H_2O

Jones [*71, 72*] extended his study at 25° C to the equilibria of the quaternary phases in the presence of a liquid containing 1% alkali (as KOH or NaOH). This concentration was chosen as that likely to be reached in the hydration of normal Portland cements by passage into the solution of the soluble alkali in the clinker.

The preparations were kept in stainless steel containers. The hydrated alumina involved in the equilibria took the form of alumina gel. Jones used an extension of the Jänecke method of representation, in which the spatial model is a triangular prism in which only the salt proportions of the saturated solutions are represented; water contents are not shown. The three square sides of the prism are each Jänecke diagrams of reciprocal salt pairs, and two equilateral triangular ends are Jänecke diagrams of additive systems of three salts with a common ion:

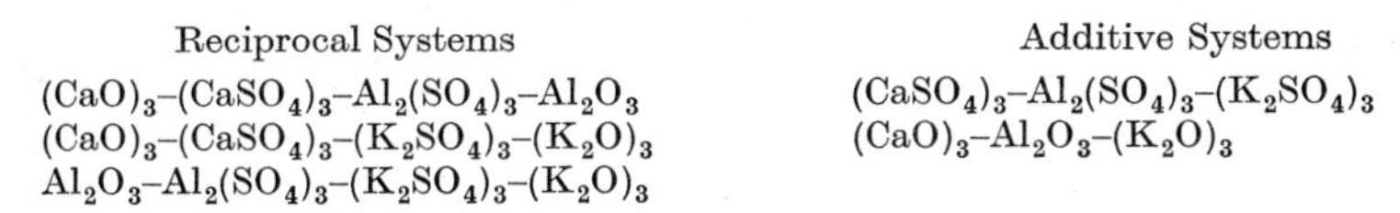

Reciprocal Systems	Additive Systems
$(CaO)_3$–$(CaSO_4)_3$–$Al_2(SO_4)_3$–Al_2O_3	$(CaSO_4)_3$–$Al_2(SO_4)_3$–$(K_2SO_4)_3$
$(CaO)_3$–$(CaSO_4)_3$–$(K_2SO_4)_3$–$(K_2O)_3$	$(CaO)_3$–Al_2O_3–$(K_2O)_3$
Al_2O_3–$Al_2(SO_4)_3$–$(K_2SO_4)_3$–$(K_2O)_3$	

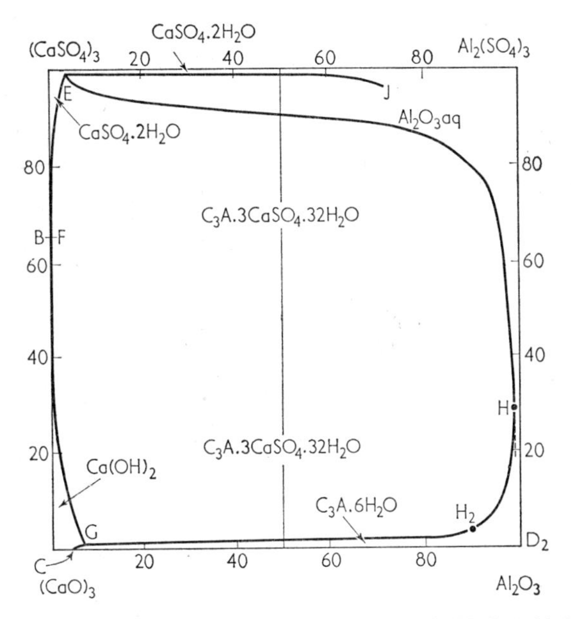

FIG. 17. Quinary system CaO–Al_2O_3–$CaSO_4$–K_2O–H_2O (1% KOH) at 25° C (Jones [*71*]). Projection on base of space figure.

Since Jones found that the solid phases take up virtually no alkali from the solution (at the most only a few tenths of 1%) and since the alkali concentration was maintained constant, the quinary system can be treated as a section of the complete system corresponding to an alkali concentration of 1%. For this reason, along any curve two solid phases coexist with solution and in any area only one solid phase, as in the quaternary system.

Figure 17 shows the projection in the $(CaO)_3$–$(CaSO_4)_3$–$Al_2(SO_4)_3$–Al_2O_3 face by radial lines from the K_2O–K_2SO_4 edge. The presence of the alkali substantially alters the liquid compositions due to the increased solubility of the Al_2O_3 and the decreased solubility of CaO, but does not change the general nature of the equilibrium relationships between the solid phases. Ettringite remains the only stable quaternary phase and, in contrast to what happens in the quaternary system, has congruent solubility. The curves were mostly determined from undersaturation. Curve GH represents stable equilibria only in the section GH_2, where H_2 is the invariant point between gibbsite, C_3AH_6, ettringite and solution. The section H_2H is the metastable extension to H, the invariant involving alumina gel, C_3AH_6 and ettringite. In contrast to the results for the quaternary system, an invariant point for alumina gel, ettringite, monosulphate and solution could not be found. Jones observed, however, that a precipitate obtained from alumina gel with KOH and $CaSO_4$ solutions had a chemical composition and optical properties approximating to those of the monosulphate. He concluded that in the quinary system, too, the hexagonal quaternary phases have metastable solubility.

Jones obtained similar results for the corresponding system with 1% NaOH. Kalousek [*73–75*] studied the equilibria $CaO–Al_2O_3–CaSO_4–Na_2O–H_2O$ at 25° C for Na_2O concentrations of 0·1–0·7 N. Unlike Jones, he determined the field of existence of the solid solution and identified various members of the series optically and with X-rays. He found that the end members are C_4AH_x and monosulphate. Kalousek carried out experiments of 18 months' duration, and in no preparations in the solid solution field was C_3AH_6 formed. In contrast to Jones, he concluded that the solid solution is a stable phase in the quinary system containing 1% or more NaOH.

E. Interconnected Solid Solution Series

Kalousek also advanced the hypothesis that the solid solution series $C_4AH_x–C_3A.CaSO_4.xH_2O$ is part of a larger series, represented by the following scheme:

$$
\begin{array}{ccc}
6CaO.Al_2O_3.32H_2O & \overset{1}{\text{———}} & 6CaO.Al_2O_3.3SO_3.32H_2O \\
\Big|\,5 & & \Big|\,2 \\
4CaO.Al_2O_3.13H_2O & \overset{4}{\text{———}} & 4CaO.Al_2O_3.SO_3.12H_2O \\
 & & \Big|\,3 \\
 & & 3CaO.Al_2O_3.12H_2O
\end{array}
$$

The existence of solid solution 4 was assumed by Jones, D'Ans and Eick, and Kalousek on the basis of optical data, but the existence of solid solution 3 was accepted only by Jones. No attempts have been made to prepare solid solutions of series 5. The results of microscopic and X-ray studies exclude the possibility of mixed crystals between the two aluminate sulphates (series 2). A recent publication of Midgley and Rosaman [23], about the possible existence of solid solution 1, is of interest. They observed that the ettringite present in hydrated Portland cement pastes differs, to an increasing extent as the paste ages, from the pure compound in X-ray pattern and DTA behaviour; the longest *d*-spacing falls and the "normalized peak temperature", T_n, increases:

	Longest *d*-spacing (Å)	T_n(°C)
Pure $C_3A.3CaSO_4.32H_2O$	9·75	130
Ettringite in 1-day-old paste	9·75	136
Ettringite in 6-month-old paste	9·63	156

T_n is the peak temperature corrected for the influence of variation in peak area, A, on the basis of the empirical relation $T = a + b \ln A$, where a and b are constants; peak temperatures are thus referred to a standard value of A. The experimental data excluded the possibility that the observed variations were caused by differences in crystal size or by solid solution with C_4AH_{13}, $C_3F.3CaSO_4.32H_2O$ or $C_3A.CaSO_4.12H_2O$, and the authors therefore considered the possibility of solid solutions of type 1. Although all their attempts to prepare C_6AH_{32} were unsuccessful, they believed they had prepared two members of this series by mixing metastable calcium aluminate solution with calcium saccharate solution (16·6 g/l CaO) containing $CaSO_4$. The DTA and X-ray data for these two preparations were similar to those found for the ettringites in set Portland cement. This conclusion, which is of interest from the point of view of the structural relationship between the compounds, also raises the question as to how such a solid solution is formed in the hydration of Portland cement.

F. Calcium Aluminate Carbonate Hydrates

Quaternary aluminates containing CO_3^{2-} have been studied mainly by Turriziani and Schippa [21], Carlson and Berman [28] and Buttler [76]. A tetracalcium aluminate monocarbonate hydrate is prepared by mixing a metastable calcium aluminate solution with $Ca(OH)_2$ and Na_2CO_3 solutions; the Al_2O_3/CO_3^{2-} ratio must be about 1. At ordinary temperatures the hydrate in equilibrium at 33% r.h. has the formula $C_3A.CaCO_3.11H_2O$. The compound crystallizes in thin, hexagonal plates with ω 1·554, ϵ 1·532 [28]. The X-ray patterns of the compound with $11H_2O$ and of the hydrate in equilibrium with the mother-liquor are practically identical. The longest basal spacing occurs at 7·6 Å.

Carlson and Berman proposed a hexagonal unit cell with *a* 8·71 *c* 7·56 Å on the basis of X-ray powder data, but in view of the close relationship with C_4AH_{13} an *a*-axis of 5·73 Å or a multiple thereof would appear more likely. Turriziani and Schippa [*21*] studied the thermal decomposition in a CO_2-free atmosphere. At 150° C the water content falls to about $6H_2O$ and the basal spacing changes to 6·3 Å. At 500° C almost all the water is lost as well as 1% of the amount of CO_2 present; the X-ray pattern shows a single, very diffuse line at 2·9–3·07 Å. At 600° C the substance is completely dehydrated and 23% of the CO_2 is lost; calcite lines appear in the X-ray pattern.

$C_3A.CaCO_3.11H_2O$ is also formed at 1–25° C, together with $CaCO_3$ and alumina gel, by exposing to air either a calcium aluminate solution or an aqueous suspension of C_2AH_8, C_4AH_{19} or C_3AH_6 [*28*]. On continued exposure it decomposes to $CaCO_3$ and alumina gel. The $CaCO_3$ is formed as calcite or more rarely as aragonite, while the alumina gel can approximate either to gibbsite or to bayerite. The formation of monocarbonate has not been observed with CAH_{10} suspensions. $C_3A.CaCO_3.11H_2O$ is also formed in pastes made from finely ground, anhydrous C_3A and $CaCO_3$ in equimolar ratio.

Hexacalcium aluminate tricarbonate hydrate [*28*] is prepared by adding to a calcium saccharate solution a calculated volume of calcium aluminate solution followed by an NH_4HCO_3 solution. A precipitate is formed, consisting of minute spherulites which are probably composed of very thin needles. After drying at 79% r.h. the substance has the composition $C_3A.3CaCO_3.32H_2O$. Its X-ray pattern, which is very similar to that of ettringite, is characterized by intense lines at 9·41, 3·80, 2·70 and 2·51 Å. On drying over $Mg(ClO_4)_2$ the compound loses $3H_2O$, and in this state has refractive indices α 1·456, γ 1·48. Carlson and Berman [*28*] tried without success to prepare the compound pure by a method similar to that used for the monocarbonate; they obtained small amounts of it, mixed with calcite or monocarbonate, by the action of atmospheric CO_2 on a dilute calcium aluminate solution. The solution was cooled to 10–20° C and air bubbled through at a moderate rate. The extent of contamination by calcite or monocarbonate was found to be significantly affected by the temperature and by the ratio of the volume of the solution to the rate of passage of air. The authors stated that it was therefore difficult to prepare more than a few milligrams of reasonably pure material. They also investigated the possible existence of solid solutions between ettringite and tricarbonate. Attempts to prepare these were unsuccessful, but their limited number renders them inconclusive.

DTA curves for monocarbonate were reported by Turriziani and

Schippa [21] and for both monocarbonates and tricarbonates by Carlson and Berman [28]. These authors also reported infra-red spectra for both hydrates.

Hydrocalumite has already been discussed (p. 241). It resembles the monocarbonate in being structurally similar to C_4AH_{13}, but contains a smaller proportion of CO_3^{2-}. "α-C_4AH_{13}" has also been discussed (p. 241).

G. Calcium Aluminate Silicate Hydrates

There are many calcium aluminate silicate hydrates belonging to the zeolite and other mineral groups, but apart from the hydrogarnets (discussed later) only one compound closely related to the calcium aluminate hydrates is definitely known. This is the so-called "gehlenite hydrate", C_2ASH_8. Strätling [77] and zur Strassen and Strätling [78] obtained it, mixed with calcium aluminate and silicate hydrates, by reaction of burnt kaolin with lime-water; by hydration of a mixture of C_3S and C_3A; and by shaking together C_3A, silica sol and lime-water. Turriziani [22] has identified it among the products of reaction of natural Italian pozzolanas with $Ca(OH)_2$. Dörr [79] prepared the pure compound by hydrating a glass of composition C_2AS in a half-saturated lime solution.

C_2ASH_8 crystallizes in weakly birefringent, hexagonal plates which have a mean refractive index of 1·512. It dissolves incongruently in water to form a solution which at room temperature contains 0·01 g SiO_2/l, 0·013 g Al_2O_3/l and 0·08 g CaO/l. According to Locher [80], it is unstable in the presence of $Ca(OH)_2$, changing to a hydrogarnet. In a saturated solution with lime and gypsum it decomposes to give ettringite, though it is stable [80] in saturated $CaSO_4$ and in 0·15 M Na_2SO_4. For short times it is unaffected by the Gille reagent [81]; it is decomposed by $MgSO_4$ solutions. Turriziani and Schippa [82] observed that at 50° C C_2ASH_8 is unstable relative to a hydrogarnet belonging to the series C_3AH_6–C_3AS_3.

X-Ray powder data were reported by Fratini and Turriziani [81] and by Schmitt [83], who derived from it the cell parameters *a* 8·85, *c* 12·66 Å and *a* 9·96, *c* 12·56 Å, respectively; however, a single crystal electron diffraction pattern reported by zur Strassen [84] indicated an *a*-axis similar to that of C_4AH_{13}. DTA curves for the product of reaction of burnt kaolin and lime-water show an endothermic peak at 220° C [85].

Schmitt [83] obtained "gehlenite hydrates" containing Fe_2O_3, which partially replaced the Al_2O_3, by hydration at ordinary temperatures in a half-saturated lime solution of quaternary glasses of compositions C_2AS–$C_2A_{0\cdot4}F_{0\cdot6}S$. In the range C_2ASH_8–$C_2A_{0\cdot7}F_{0\cdot3}SH_8$ the *a*-parameter of the postulated unit cell increased from 9·96 to 10·08 Å and the refractive index from 1·512 to 1·525. If glasses of higher iron content were used, the values of a and of the refractive index remained constant. This indicates that, in the hydrated phase, substitution is limited to 0·3 moles of Fe_2O_3.

Flint and Wells [*26*] claimed to have identified, under the microscope and by X-rays, the monosilicates and trisilicates representing the two series of quaternary aluminates, in precipitates obtained by reaction of metastable calcium aluminate solutions with calcium silicate solutions. For the compound $C_3A.CaSiO_3.xH_2O$, they reported the refractive indices ω 1·538, ϵ 1·523 and an X-ray pattern closely similar to that of "α-C_4AH_{13}". For $C_3A.3CaSiO_3.xH_2O$, they found ω 1·487, ϵ 1·479 and an X-ray pattern similar to that of ettringite. The monosilicate was obtained practically pure in only one preparation; the trisilicate was always mixed with the monosilicate and with gelatinous solids.

Turriziani [*22*] and Carlson [*86*] attempted without success to prepare the monosilicate by the method of Flint and Wells. In each case the precipitate was a mixture of tetracalcium aluminate, calcium silicate and gehlenite hydrates. Dörr [*79*] at one time claimed to have reproduced the compound, but after completing experiments of 1–2 years' duration concluded that it had not been formed. Carlson [*86*] examined X-ray films, believed to be essentially duplicates of the original ones for the actual preparations of Flint and Wells. He concluded that the patterns, which did not include the low-angle lines necessary to distinguish the hexagonal plate phases from each other, could be attributed to $Ca(OH)_2$, C_4AH_{13} and gehlenite hydrate.

Carlson and Berman [*28*] were able to examine some of Flint and Wells' original preparations, which were then 18 years old; the solids were composed of gelatinous material with a refractive index of 1·51 and acicular crystals with ω 1·495, ϵ 1·473. X-Ray patterns showed lines near to, but not identical with, those of the aluminate tricarbonate, together with lines of gehlenite hydrate and a calcium silicate hydrate. Analysis of a precipitate containing about 50% of the acicular phase gave $5{\cdot}48CaO.Al_2O_3.2{\cdot}81SiO_2.0{\cdot}73CO_2.xH_2O$. Since the optical and X-ray results excluded the presence of $CaCO_3$, the CO_2 probably occurred as the aluminate tricarbonate. The authors, taking into account the small but significant differences between the patterns of the solids examined and that of the tricarbonate, and also the low percentage of CO_2 relative to the proportion of needles, concluded that a quinary phase existed in which part of the CO_2 of the tricarbonate was replaced by SiO_2.

Mohri [*87*] claimed recently to have prepared the trisilicate by hydration at 20° C of CaO–Al_2O_3–SiO_2 glasses, but the X-ray pattern which he reported contains the strongest reflections of gehlenite hydrate [*28*].

zur Strassen and Schmitt [*99*] observed, in preparations aged for 2 years, an acicular phase which from X-ray data appeared to be the trisilicate of Flint and Wells. They did not discuss the possible relationship to the tricarbonate obtained by Carlson and Berman. In conclusion, the existence of the monosilicate is unconfirmed but that of the trisilicate, or of its solid solution with the tricarbonate, is open to discussion.

H. Calcium Aluminate Chloride Hydrates

Two calcium aluminate chloride hydrates exist, analogous to the two aluminate sulphate hydrates. $C_3A.CaCl_2.xH_2O$ is prepared by mixing at ordinary temperature a calcium aluminate solution with lime-water containing calcium chloride. The $Al_2O_3/CaCl_2$ ratio of the initial mixture must be such that the $CaCl_2$ concentration in the equilibrium solution is about 3%. The compound crystallizes as uniaxial negative, hexagonal plates, and is decomposed on adding $CaSO_4$ to the mother-liquor. The X-ray patterns of moist samples and of the material dried over anhydrous $CaCl_2$ are almost identical. The latter material contains $8H_2O$; dehydration over P_2O_5 reduces the water content to $4H_2O$.

$C_3A.CaCl_2.xH_2O$ forms a complete series of mixed crystals with hydrated tetracalcium aluminate. Turriziani and Schippa [*88*] prepared solid solutions with $CaCl_2/Al_2O_3$ ratios ranging from 0·1 to 1, starting from a metastable calcium aluminate solution. X-Ray examination of the moist precipitates indicated that the products consisted of a single solid phase. The basal spacings ranged from 8·1 Å for a $CaCl_2/Al_2O_3$ ratio of 0·22 to 7·95 Å for a $CaCl_2/Al_2O_3$ ratio of 1. Thus, on lowering the $CaCl_2/Al_2O_3$ ratio, the basal spacing trended towards the value of 8·2 Å charactersitic of "α-C_4AH_{13}". In view of the more recent findings about the C_4AH_x hydrates, further investigations on the X-ray properties of these solid solutions are needed.

$C_3A.3CaCl_2.30H_2O$ was obtained by Serb-Serbina, Savvina and Zhurina [*89*] by hydration of C_3A in 23% $CaCl_2$ at $-10°$ C. At 20° C it is unstable in lime-water or in 1·5–10% $CaCl_2$ solutions, changing to the monochloride. Serb-Serbina *et al.* reported dehydration curves for both compounds.

V. Calcium Ferrite Hydrates

Most of our knowledge of aqueous systems involving calcium ferrites is derived from the studies of Eiger [*90*], Hoffmann [*91*], Malquori and Cirilli [*92*] and Flint, McMurdie and Wells [*93*] some 20 years ago. Jones' suggestion that these systems admit of the existence of equilibria similar to those found for the calcium aluminates appears to be consistent with much of the data so far obtained, but it has failed to promote interest in the study of the calcium ferrite hydrates, perhaps because of the experimental difficulties involved.

Methods of preparation are fewer than for the aluminate hydrates because no anhydrous calcium ferrite produces supersaturated solutions which can be used to precipitate ferrite hydrates. The only direct method of preparation is the reaction between ferric hydroxide gel and lime solution. Even with freshly precipitated gels, the reaction is slow and unreacted gel remains for a long time. The only effective preparative methods are therefore ones based on precipitation: (a) from solutions of ferric salts with lime solution and (b) from solutions of ferric and calcium salts with NaOH solution. The problem with both methods is that a certain amount of the anion, depending on its concentration in the liquid, is taken up by the precipitate. To some extent this can be avoided by using very dilute solutions, but practical difficulties increase with the dilution owing to the large size of the vessels needed, the small amounts of precipitate obtained and the difficulty of obtaining adequate stirring.

The use of ferric oxide sols does not completely eliminate the presence of anions in the solution, because the sol is only stable if the liquid phase contains a certain amount of a ferric salt. Some improvement can be effected by passing the sol through an anion exchanger; ferric oxide sols stabilized with small amounts of $FeCl_3$ and having a pH of about 6 can be prepared in this way [*94*]. However, it is still necessary to verify that the concentration of stabilizer is sufficiently low to permit the preparation of calcium ferrite hydrates uncontaminated by the anion present. Ferric oxide sols are more suitable for investigations of equilibria involving quaternary ferrite hydrates (Schippa [*95*]).

At present the existence of a hexagonal C_4FH_x corresponding to C_4AH_{13} is generally accepted, but conflicting views exist about cubic C_3FH_6, which seems to be structurally unstable. The existence of less basic calcium ferrite hydrates appears uncertain; there is as yet no satisfactory evidence for the existence of C_2FH_x or CFH_x. Bogue and Lerch [*96*] and Yamauchi [*97*] were of the opinion that a C_2F hydrate was formed on hydration of C_4AF. These observations were not experimentally confirmed, and it is probable that hexagonal structures with $C/F < 4$ are unstable.

C_4FH_x was first prepared by Hoffmann [*91*] by adding a solution of $FeCl_3$ and $CaCl_2$ ($C/F = 5$) to a large volume of NaOH. Malquori and Cirilli [*92*] obtained it by gradual mixing of 20 cm^3 of $FeCl_3$ solution containing 0·04 g/cm^3 Fe_2O_3 to 20 l of lime-water. These authors stated that with a higher concentration of reactants in Hoffmann's method the monochloride $C_3F.CaCl_2.xH_2O$ is produced. Experiments performed by

Roberts at the Building Research Station and quoted by Jones [*20*] confirmed this view and indicated that, even with an initial chloride concentration of 0·05 g/l, products containing 1–5% of Cl are obtained. Optical and X-ray studies showed that these products were members of the solid solution series $C_3F.CaCl_2.xH_2O$–$C_4F.xH_2O$.

To remove any doubt as to the existence of a C_4F hydrate unstabilized by chloride, Turriziani [*98*] made X-ray examinations of preparations obtained in polythene vessels at 25–50° C from ferric oxide gels and saturated $Ca(OH)_2$ solution. Since complete reaction was never obtained, it was impossible to establish the composition of the ferrite phase. The solid in equilibrium with the mother-liquor had a basal spacing of 8·26 Å which was changed to 7·9 Å by drying at 55% r.h., and to 7·6 Å by drying over $CaCl_2$. The last two spacings are similar to those of C_4AH_{13} and C_4AH_{11}; it therefore appears reasonable to assume tentatively the existence of the analogous compounds C_4FH_{13} and C_4FH_{11}. A hydrate corresponding to C_4FH_{19} does not exist at 25° C. zur Strassen and Schmitt [*99*] also observed the formation of a C_4F hydrate in the products obtained in reaction of $Ca(OH)_2$ solution with either C_2F or ferric oxide gel in polythene vessels at room temperature. If these reactions are carried out in the presence of silica or calcium silicates, a hydrogarnet of the C_3FH_6–C_3FS_3 series is formed instead of the hexagonal ferrites.

Malquori and Cirilli [*100*] concluded from X-ray evidence that C_4FH_x and C_4AH_x form solid solutions. Chemical analyses of precipitates dried over $CaCl_2$ and KOH gave Cl contents lower than 1% of the weight of anhydrous solid.

Not all the reported preparations of C_3FH_6 can be considered to provide decisive evidence for the existence of the compound as a pure phase. Preparations made in glass vessels above room temperature unquestionably contain silica in the structure. Eiger [*90*] obtained the compound by direct reaction of the hydroxides in glass vessels at room temperature. The product contained silica which, in Eiger's opinion, was present as a calcium silicate hydrate. The investigation of Flint *et al.* [*93*] on hydrogarnets, as well as the experiments of zur Strassen and Schmitt [*99*], show that this assumption cannot be accepted; it is much more probable that part of the silica replaced hydroxyl in the C_3FH_6 structure. The stabilizing effect up to 60° C of silica can be checked by using polythene containers and up to 100° C by using polypropylene vessels.

Preparations from which silica was excluded beyond doubt were made by Burdese and Gallo [*101*] and Roberts, quoted by Jones [*20*]. Their results are not in complete agreement. Burdese and Gallo prepared a suspension by gradual addition of a solution of $FeCl_3$ to lime-water at room temperature, and then heated it at 80° C in a platinum vessel of 2 l capacity. The product, dried over $CaCl_2$, had a refractive index of 1·74 and a cubic unit cell with *a* 12·76 Å. This is the generally accepted value for pure C_3EH_6. In agreement with Malquori and Cirilli, they found the X-ray pattern and dehydration curve to be very similar to those of C_3AH_6.

On the assumption that the suspension obtained by Burdese and Gallo consisted of calcium ferrite monochloride hydrate or of its solid solution with C_4FH_x, Roberts investigated the stability of the ferrite chloride at 60° C using a polythene vessel. He found that a suspension of the compound in lime-water yielded $Fe(OH)_3$, while in N NaOH a mixture of solids was formed, among which C_3FH_6, with *a* 12·76 Å, was abundant.

Malquori and Cirilli [*100*] obtained C_3AH_6–C_3FH_6 mixed crystals by heating C_4AH_x–C_4FH_x solid solutions at 70° C in lime-water for about a month. Since their products contained less than 1% of SiO_2, they determined the refractive indices as a function only of the A/F ratio. The range of substitution studied was from 0 to 66 moles Fe_2O_3 per 100 moles R_2O_3. Flint *et al.* [*93*] tried to prepare solid solutions of this kind by precipitating ferric and aluminium chloride solutions with lime-water at 100° C in glass vessels. The products were contaminated with silica and belonged to the hydrogarnet series. When copper, iron or silver lined glass vessels were used, the same mixtures yielded products contaminated with $Fe_2O_3.xH_2O$. It therefore appears that small amounts of silica help to stabilize the C_3AH_6–C_3FH_6 mixed crystals.

VI. Calcium Ferrite Hydrates Containing Other Anions

The structural analogy between calcium ferrite and aluminate hydrates extends to the quaternary solids containing other anions. This has so far been demonstrated for the SO_4^{2-}, Cl^- and NO_3^- ions. The ferrite sulphates and chloride are of interest for the hydration of Portland cement.

McIntire and Shaw [*102*], Bogue and Lerch [*96*] and Jones [*103*] reported preparations of the trisulphate $C_3F.3CaSO_4.32H_2O$. The compound appears quite similar to ettringite in its morphological and structural properties. Malquori and Caruso [*104*] obtained the monosulphate, $C_3F.CaSO_4.xH_2O$, by reaction of a ferric sulphate solution with lime-water. This compound was subsequently prepared by Hedin [*105*] and later by Schippa [*95*], who obtained it by reaction of a ferric oxide sol (0·3 g Fe_2O_3/l and 0·07 g SO_4^{2-}/l) with lime-water containing $CaSO_4$. Schippa obtained at 18° C two hydrates; one of these contained 13–14H_2O and had a basal spacing of 10·4 Å, and the other contained 11–12H_2O and had a basal spacing of 8·98 Å. The 13–14 hydrate is stable in contact with aqueous solution or at 94% r.h.; it is changed to the 11–12 hydrate on drying at 55% r.h. X-Ray studies of the moist precipitates showed that C_4FH_x–$C_3F.CaSO_4.xH_2O$ solid solutions were not formed.

Malquori and Cirilli [*106*] found that, unlike the aluminate monosulphate, $C_3F.CaSO_4.xH_2O$ does not change to the trisulphate on contact with a solution saturated with lime and gypsum.

The monochloride, $C_3F.CaCl_2.xH_2O$ was obtained by Malquori and Caruso [*104*] by addition of $FeCl_3$ solution to lime-water containing $CaCl_2$; they prepared the mononitrate by an analogous method. Flint *et al.* reported that $C_3F.CaCl_2.xH_2O$ changes slowly to C_3FH_6 when left in contact with mother-liquor in glass vessels at room temperature. A sample analysed when 50% conversion had occurred contained 0·8% SiO_2. Another sample, placed in a lime solution (0·626 g CaO/l) for 2 weeks at 60° C changed completely into "C_3FH_6", which, however, contained 6·85% SiO_2.

The formation of solid solutions between quaternary ferrites and aluminates was observed by Malquori and Cirilli [*106*] and Cirilli [*107*] for the trisulphates and for the monochlorides, and by Roberts and Aruja [*108*] for a single member of the monosulphate series. The $C_3(F,A).3CaSO_4.32H_2O$ mixed crystals were prepared by mixing a solution containing ferric and aluminium sulphates with a saturated solution of lime and gypsum. Miscibility is incomplete; there is an upper limit of 3 to the F/A ratio. In contrast, the monochlorides appear to be completely miscible.

VII. Hydrogarnet Solid Solutions

Flint *et al.* [*93*] obtained solid solutions within the area

$$\begin{array}{ccc} C_3FH_6 & \text{———} & C_3FS_3 \ (\text{Andradite}) \\ | & & | \\ C_3AH_6 & \text{———} & C_3AS_3 \ (\text{Grossularite}) \end{array}$$

by hydration of CaO–Fe_2O_3–Al_2O_3–SiO_2 glasses in saturated steam at 200–250° C. They assumed that each end-member formed a complete series of solid solutions with the other three, the ratios of substitution being 1:1 for Al_2O_3/Fe_2O_3 and 1:2 for SiO_2/H_2O. These solid solutions are called hydrogarnets and they are cubic; the refractive index and cell parameter, and also the relative intensities of some of the low-angle X-ray powder lines, are functions of the composition. Yoder [*109*] observed that in theory the equilibrium composition of the solid solution depends on the temperature and vapour pressure.

Hydrogarnets of the C_3AH_6–C_3AS_3 series occur as natural minerals and have been prepared hydrothermally from homogeneous glasses or from mixtures of C_3A and C_3S, of C_3AH_6 with silica gel or calcium silicate hydrates, of the hydroxides, or of burnt kaolin and lime. A hydrogarnet of composition $C_3AS_{0\cdot3}H_{5\cdot4}$ was obtained at 50° C by reaction of burnt kaolin with lime-water [*85*].

zur Strassen and Schmitt [*99*] pointed out that an identification chart for the hydrogarnets can be drawn on the basis of the cell parameter and the ratio of the intensities of the 220 and 611 reflections. The intensity

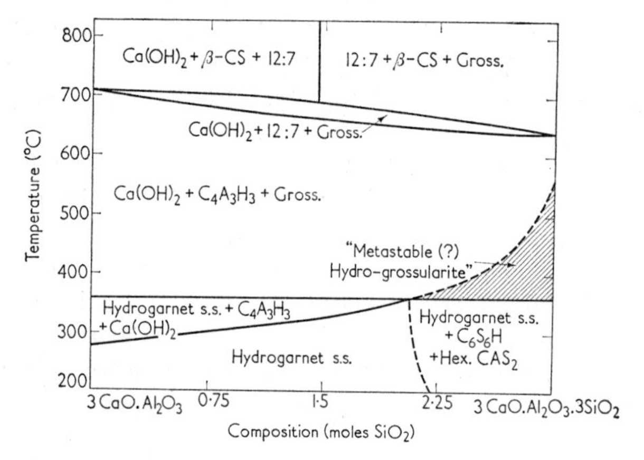

FIG. 18. Temperature–composition diagram for C_3AH_6–C_3AS_3 hydrogarnets (Roy and Roy [*111*]).

of the 220 reflections varies with composition, while that of the 611 reflection is almost independent of composition; the ratio I_{220}/I_{611} increases on moving from the grossularite corner of the field towards the C_3FH_6 corner. According to these authors [*99*], for the correct use of the method "iso-cell" and "iso-ratio" curves must be drawn from data obtained by placing glass and saturated vapour in contact with each other at the reaction temperature. In practice the usual method of heating the glass together with water in a pressure vessel until the reaction temperature is reached leads to reactions during the heating process which differ from those occurring at the final temperature and which are not reversed, so that impure products are formed.

Roy and Roy [*111*] reported a composition–temperature diagram for the series C_3AH_6–C_3AS_3 (Fig. 18), for the range of water pressures from 1000–2000 atm. In agreement with Carlson [*110*], they found that the series was incomplete at any one temperature. Two types of hydrogarnets are formed: at low temperatures a stable series C_3AH_6–$C_3AS_2H_2$ and at 300–600° C a series, which is probably metastable, ranging from C_3AS_3 to about $C_3AS_{2 \cdot 75}H_{0 \cdot 50}$. Below 360° C mixtures rich in silica yield the hexagonal polymorph of CAS_2, a calcium silicate hydrate, and the hydrogarnet $C_3AS_2H_2$. Anhydrous grossularite was prepared at about 600° C with water as catalyst.

Thermal stability in the C_3AH_6–C_3AS_2 series increases with silica content. Below 360° C the products of hydrothermal decomposition are $C_4A_3H_3$, $Ca(OH)_2$ and $C_3AS_2H_2$. Above 360° C the last phase becomes unstable, and its place is taken by grossularite or by a hydrogrossularite with a slightly larger cell parameter. At higher temperatures $C_4A_3H_3$ decomposes to $C_{12}A_7$, and at 860° C the grossularite yields wollastonite, gehlenite and anorthite.

Roy and Roy admitted the possibility of an alternative explanation of their results, namely the assumption of a complete series of mixed crystals with a thermal stability increasing with silica content. If this view is accepted, the assemblages rich in silica are metastable below 360° C and the horizontal line in Fig. 18 should be displaced from 360° C to about 550° C.

REFERENCES

1. Eitel, W. (1954). "The Physical Chemistry of the Silicates". University of Chicago Press, Chicago.
2. Bogue, R. H. (1955). "The Chemistry of Portland Cement", 2nd Ed. Reinhold, New York.
3. Lea, F. M. and Desch, C. H. (1956). "The Chemistry of Cement and Concrete", 2nd Ed. (revised by F. M. Lea). Arnold, London.
4. *Proceedings of the Symposium on the Chemistry of Cements, Stockholm 1938*, Ingeniörsvetenskapsakademien, Stockholm.
5. *Proceedings of the Third International Symposium on the Chemistry of Cement, London 1952*, Cement and Concrete Association, London.
6. Steinour, H. H. (1951). "Aqueous Cementitous Systems Containing Lime and Alumina". Bulletin 34, Portland Cement Association, Chicago.
7. Jones, F. E. and Roberts, M. H. (1959). "The System CaO–Al_2O_3–H_2O at 25° C". DSIR Building Research Station. Note E 965. HMSO, London.
8. Percival, A., Buttler, F. G. and Taylor, H. F. W. (1962). *Chemistry of Cement, Proceedings of the Fourth International Symposium, Washington 1960*, p. 277. National Bureau of Standards Monograph 43. U.S. Department of Commerce.
9. Buttler, F. G., Dent Glasser, L. S. and Taylor, H. F. W. (1959). *J. Amer. ceram. Soc.* **42**, 121.
10. Taylor, H. F. W. (1961). "Progress in Ceramic Science", edited by J. E. Burke, Vol. 1, p. 89. Pergamon Press, London.
11. Taylor, H. F. W. (1960). *J. appl. Chem.* **10**, 317.
12. Sersale, R. (1957). *Ric. Sci.* **27**, 777.

12a. Schneider, W. G. and Thorvaldson, T. (1941). *Canad. J. Res.* **B19**, 123.

13. Majumdar, A. J. and Roy. R. (1956). *J. Amer. ceram. Soc.* **39**, 434.
14. Taylor, H. F. W. (1961). "Progress in Ceramic Science", ed. by J. E. Burke, Vol. 1, p. 115. Pergamon Press, London.
15. Kalousek, G. L., Davis, C. W. and Schmertz, W. E. (1949). *J. Amer. Concr. Inst.* **45**, 693.
16. Johnson, H. and Thorvaldson, T. (1943). *Canad. J. Res.* **21B**, 236.
17. Percival, A. and Taylor, H. F. W. (1961). *Acta cryst.* **14**, 324.
18. Carlson, E. T. (1958). *J. Res. nat. Bur. Stand.* **61**, 1.
19. Roberts, M. H. (1957). *J. appl. Chem.* **7**, 543.
19a. Aruja, E. (1960, 1961). *Acta cryst.* **13**, 1018; **14**, 1213.
20. Jones, F. E. (1962). *Chemistry of Cement, Proceedings of the Fourth International Symposium, Washington 1960*, p. 205. National Bureau of Standards Monograph 43. U.S. Department of Commerce.
21. Turriziani, R. and Schippa, G. (1956). *Ric. Sci.* **26**, 2792.
22. Turriziani, R. (1954). *Ric. Sci.* **24**, 1709.
23. Midgley, H. G. and Rosaman, D. (1962). *Chemistry of Cement, Proceedings of the Fourth International Symposium, Washington 1960*, p. 259. National Bureau of Standards Monograph 43. U.S. Department of Commerce.
24. Seligmann, P. and Greening, N. R. (1962). *J. P.C.A. Res. Dev. Labs* **4**, 2. See also Roberts, M. H. (1962). *Chemistry of Cement, Proceedings of the Fourth International Symposium, Washington 1960*, p. 1033. National Bureau of Standards Monograph 43. U.S. Department of Commerce.
25. Carlson, E. T. (1957). *J. Res. nat. Bur. Stand.* **59**, No. 2, 107. P. Longuet (1954). *Proceedings of the Third International Symposium on the Chemistry of Cement, London 1952*, p. 238. Cement and Concrete Association, London.
25a. Longuet, P. (1954). *Proceedings of the Third International Symposium on the Chemistry of Cement, London 1952*, p. 238. Cement and Concrete Association, London.
26. Flint, E. P. and Wells, L. S. (1944). *J. Res. nat. Bur. Stand.* **33**, 471.
27. Assarsson, G. (1933). *Sverig. geol. Unders.* **27**, No. 379, 22.
28. Carlson, E. T. and Berman, H. A. (1960). *J. Res. nat. Bur. Stand.* **64A**, 333.
29. Travers, A. and Sehnoutka, J. (1930). *Ann. Chim.* **10**, No. 13, 253.
30. Mylius, C. R. W. (1933). *Acta Acad. åbo.* **7**, 3.
31. Thorvaldson, T., Grace, N. S. and Vigfusson, V. A. (1929). *Canad. J. Res.* **1**, 201.
32. Bunn, C. W. and Clark, L. M. (1938). *J. Soc. chem. Ind., Lond.* **57**, 399.
33. Brocard, J. (1948). *Ann. Inst. Bâtim.* No. 12.
34. Schippa, G. and Turriziani, R. (1957). *Ric. Sci.* **27**, 3654.
35. Govoroff, A. (1956). *Rev. Matér. Constr.* No. 490/491, 181.
36. D'Ans, H. and Eick H. (1953). *Zement-Kalk-Gips* **6**, 197.
36a. Rabot, R. and Mounier, M.-T. (1961). *Rev. Matér. Constr. Tech. Pub. No. 123.*
37. Brandenberger, E. (1933). *Schweiz. min. petrog. Mitt.* **13**, 569.
38. Brandenberger, E. (1936). *Schweiz. Arch. angew. Wiss.* **2**, 45.
39. Tilley, C. E., Megaw, H. D. and Hey, M. H. (1934). *Miner. Mag.* **23**, 607.
40. Grundemo, Å. (1962). *Chemistry of Cement, Proceedings of the Fourth International Symposium, Washington 1960*, p. 615. National Bureau of Standards Monograph 43. U.S. Department of Commerce.
41. Wells, L. S., Clarke, W. F. and McMurdie, H. F. (1943). *J. Res. nat. Bur. Stand.* **30**, 367.
42. Bessey, G. E. (1939). *Proceedings of the Symposium on the Chemistry of Cements, Stockholm 1938*, p. 169. Ingeniörsvetenskapsakademien, Stockholm.

43. Bessey, G. E. (1939). *Proceedings of the Symposium on the Chemistry of Cements, Stockholm 1938*, p. 178. Ingeniörsvetenskapsakademien, Stockholm.
44. Jones, F. E. (1944). *J. phys. Chem.* **48**, 311.
45. Percival, A. and Taylor, H. F. W. (1959). *J. chem. Soc.* 2629.
46. Wells, L. S. and Carlson, E. T. (1956). *J. Res. nat. Bur. Stand.* **57**, 335.
47. Taylor, H. F. W. Private communication.
48. Buttler, F. G. and Taylor, H. F. W. (1958). *J. chem. Soc.* 2103.
49. Peppler, R. B. and Wells, L. S. (1954). *J. Res. nat. Bur. Stand.* **52**, 75.
49a. Pistorius, C. W. F. T. (1962). *Amer. J. Sci.* **260**, 221.
50. Foret, J. (1935). "Récherches sur les Combinaisons entre les Sels de Calcium et les Aluminates de Calcium". Herman, Paris.
51. Feitknecht, W. (1942). *Helv. chim. acta* **25**, 106.
52. Feitknecht, W. and Buser, H. W. (1949). *Helv. chim. acta* **32**, 2298.
53. van AArdt, J. H. P. (1962). *Chemistry of Cement, Proceedings of the Fourth International Symposium, Washington 1960*, p. 835. National Bureau of Standards Monograph 43. U.S. Department of Commerce.
54. Feitknecht, W. and Buser, H. W. (1951). *Helv. chim. acta* **34**, 128.
55. Berman, H. A. and Newman, E. S. (1962). *Chemistry of Cement, Proceedings of the Fourth International Symposium, Washington 1960*, p. 247. National Bureau of Standards Monograph 43. U.S. Department of Commerce.
56. Turriziani, R. and Schippa, G. (1954). *Ric. Sci.* **24**, 2356.
57. Greene, K. T. (1962). *Chemistry of Cement, Proceedings of the Fourth International Symposium, Washington 1960*, p. 359. National Bureau of Standards Monograph 43. U.S. Department of Commerce.
58. Fratini, N., Schippa, G. and Turriziani, R. (1955). *Ric. Sci.* **25**, 57.
59. Midgley, H. G. (1957). *Mag. Conr. Res.* **9**, 17.
60. Bannister, F. A. (1936). *Miner. Mag.* **24**, 324.
61. Taylor, H. F. W. (1961). "Progress in Ceramic Science", ed. by J. E. Burke, Vol. 1, p. 111. Pergamon Press, London.
62. Welin, E. (1956). *Ark. Min.* **2**, 137.
63. D'Ans, J. and Eick, H. (1954). *Zement-Kalk-Gips* **7**, 449.
64. Schippa, G. (1961). *Ric. Sci.* **31**, 2A, 204.
65. Lerch, W., Ashton, F. W. and Bogue, R. H. (1929). *J. Res. nat. Bur. Stand.* **2**, 715.
66. Turriziani, R. and Schippa, G. (1955). *Ric. Sci.* **25**, 2894.
67. Turriziani, R. (1959). *Industr. ital. Cemento* **29**, 185, 219, 244, 276.
68. D'Ans, J. and Eick, H. (1953). *Zement-Kalk-Gips* **6**, 302.
68a. Roberts, M. H. (1962). *Chemistry of Cement, Proceedings of the Fourth International Symposium, Washington 1960*, p. 245, National Bureau of Standards Monograph 43. U.S. Department of Commerce.
69. Jones, F. E. (1939). *Trans. Faraday Soc.* **35**, 1484.
70. Eitel, W. (1957). *J. Amer. Concr. Inst.* **26**, 679.
71. Jones, F. E. (1944). *J. phys. Chem.* **48**, 356.
72. Jones, F. E. (1944). *J. phys. Chem.* **48**, 379.
73. Kalousek, G. L. (1941). Thesis, Univ. of Maryland.
74. Kalousek, G. L. (1941). *J. Amer. Concr. Inst.* **37**, 692.
75. Kalousek, G. L. (1945). *J. phys. Chem.* **49**, 405.
76. Buttler, F. G. (1958). Ph.D. Thesis, Aberdeen.
76a. Schneider, S. J. (1959). *J. Amer. ceram. Soc.* **42**, 184.
77. Strätling, W. (1938). "Die Reaktion zwischen gebranntem Kaolin und Kalk in wäseriger Lösung". Zement-Verlag, Berlin.

78. zur Strassen, H. and Strätling, W. (1940). *Z. anorg. Chem.* **245**, 267.
79. Dörr, F. H. (1954). Dissertation, Mainz; zur Strassen, H. (1958). *Zement-Kalk-Gips* **11**, 137.
80. Locher, F. W. (1962). *Chemistry of Cement, Proceedings of the Fourth International Symposium, Washington 1960*, p. 267. National Bureau of Standards Monograph 43. U.S. Department of Commerce.
81. Fratini, N. and Turriziani, R. (1954). *Ric. Sci.* **24**, 1654.
82. Turriziani, R. and Schippa, G. (1954). *Ric. Sci.* **24**, 2645.
83. Schmitt, C. H. (1962). *Chemistry of Cement, Proceedings of the Fourth International Symposium, Washington 1960*, p. 244. National Bureau of Standards Monograph 43. U.S. Department of Commerce.
84. zur Strassen, H. (1962). *Chemistry of Cement, Proceedings of the Fourth International Symposium, Washington 1960*, p. 244. National Bureau of Standards Monograph 43. U.S. Department of Commerce.
85. Turriziani, R. and Schippa, G. (1954). *Ric. Sci.* **24**, 366.
86. Carlson, E. T. (1962) *Chemistry of Cement, Proceedings of the Fourth International Symposium, Washington 1960*, p. 375. National Bureau of Standards Monograph 43. U.S. Department of Commerce.
87. Mohri, J. (1958). *Semento Gijutsu Nenpo* **12**, 43.
88. Turriziani, R. and Schippa, G. (1955). *Ric. Sci.* **25**, 3102.
89. Serb-Serbina, N. N., Savvina, Yu, A. and Zhurina, V. S. (1956). *Dokl. Akad. Nauk S.S.S.R.* **111**, 659.
90. Eiger, A. (1937). *Rev. Matér. Constr.* **161**, 187.
91. Hoffmann, H. (1936). *Zement* **25**, 113, 130, 675, 693, 711.
92. Malquori, G. and Cirilli, V. (1940). *Ric. Sci.* **11**, 316, 434.
93. Flint, E. P., McMurdie, H. F. and Wells, L. S. (1941). *J. Res. nat. Bur. Stand.* **26**, 13.
94. Nachod, F. C. (1949). "Ion Exchange Theory and Application". Academic Press, New York.
95. Schippa, G. (1958). *Ric. Sci.* **28**, 2334.
96. Bogue, R. H. and Lerch, W. (1934). *Industr. Engng Chem.* **26**, 837.
97. Yamauchi, T. (1937). *J. Jap. ceram. Soc. (Ass.)* **45**, 279.
98. Turriziani, R. Unpublished results.
99. zur Strassen, H. and Schmitt, C. H. (1962). *Chemistry of Cement, Proceedings of the Fourth International Symposium, Washington 1960*, p. 243. National Bureau of Standards Monograph 43. U.S. Department of Commerce.
100. Malquori, G. and Cirilli, V. (1943). *Ric. Sci.* **14**, 78.
101. Burdese, A. and Gallo, S. (1952). *Ann. Chim. appl., Roma* **42**, 349.
102. MacIntire, W. H. and Shaw, W. M. (1925). *Soil Sci.* **19**, 125.
103. Jones, F. E. (1945). *J. phys. Chem.* **49**, 344.
104. Malquori, G. and Caruso, E. (1938). *Int. Congr. Chem.* **2**, 713.
105. Hedin, R. (1945) *Swedish Cement and Concrete Research Institute Proc.* **3**, 155.
106. Malquori, G. and Cirilli, V. (1954). *Proceedings of the Third International Symposium on the Chemistry of Cement, London 1952*, p. 321. Cement and Concrete Association, London.
107. Cirilli, V. (1943). *Ric. Sci.* **14**, 27.
108. Roberts, M. H. and Aruja, E. Unpublished work, mentioned in reference 20.
109. Yoder, H. S. (1950). *J. Geol.* **58**, 221.
110. Carlson, E. T. (1956). *J. Res. nat. Bur. Stand.* **56**, 327.
111. Roy, D. M. and Roy, R. (1962). *Chemistry of Cement, Proceedings of the Fourth International Symposium, Washington 1960*, p. 307. National Bureau of Standards Monograph 43. U.S. Department of Commerce.

CHAPTER 7

The Hydration of Tricalcium Silicate and β-Dicalcium Silicate from 5°C to 50°C

S. BRUNAUER and D. L. KANTRO

Portland Cement Association Research and Development Laboratories, Skokie, Illinois, U.S.A.

CONTENTS

I. Introduction

Two calcium silicates, tricalcium silicate (C_3S) and β-dicalcium silicate (β-C_2S), constitute about 75% of a Portland cement by weight. In their hydration reactions both silicates produce similar calcium silicate hydrates and different amounts of calcium hydroxide. The calcium silicate hydrate, because of its similarity to the natural mineral tobermorite and because of its gel-like properties, is called tobermorite gel, and this gel is the main cementing material in hardened pastes of Portland cements, in concrete and in mortar. Tobermorite gel plays a vital role in determining the rheological properties of fresh Portland cement paste, which properties in turn determine the consistency and workability of fresh concrete. Tobermorite gel plays a dominant role in the setting and hardening of Portland cement paste and in determining the strength and dimensional stability of hardened paste and concrete. Thus, tobermorite gel is the most important constituent of concrete.

C_3S and C_2S do not appear as pure compounds in Portland cements. During the burning in the kiln, the silicates pick up and dissolve in their structures small amounts of Al_2O_3, MgO and possibly other impurities. The variety of C_3S which contains Al_2O_3 and MgO impurities is called alite [*1*], and the hydration of both C_3S and alite will be discussed in this chapter.

It is very difficult to prepare β-C_2S in a pure form, because the variety of C_2S which is stable at room temperature is γ-C_2S. However, β-C_2S can be prepared without difficulty by the use of stabilizers [*2*, *3*]. The stabilizer most frequently used in the laboratory is B_2O_3, and the hydration of C_2S reported in this chapter refers to B_2O_3-stabilized β-C_2S.

Three different methods of hydration will be considered.

(a) Hydration in the paste form, which attempts to simulate the hydration of the calcium silicates in pastes of Portland cement, in concrete and in mortar. From the point of view of its practical implications, this is the most important method of hydration, and it will be discussed in greater detail than the others. In this method the solid is mixed with a limited amount of water to form a thick slurry which sets and hardens.

(b) Hydration in a rotating polyethylene bottle with a relatively large quantity of water. This will be called "bottle" hydration.

(c) Hydration in a small steel ball mill, also with a relatively large quantity of water. This will be called ball mill hydration.

In all three methods of hydration of C_2S and in the first two methods of hydration of C_3S and alite, the calcium silicate hydrate produced is a tobermorite. As is explained in Chapter 5, the group name tobermorite has been given to calcium silicate hydrates which chemically and structurally resemble the natural mineral tobermorite. In the third method of hydration of C_3S and alite, a different calcium silicate hydrate is produced, called afwillite [*4*].

The hydration reactions will be considered from the point of view of changes in matter (stoichiometry), changes in energy (energetics) and the rates of change in the processes (kinetics).

II. Stoichiometry of the Hydration Reactions

When C_3S is mixed with a limited amount of water (e.g. 0·7 of its own weight) at room temperature and the mixture is allowed to stand, the reaction goes to completion in about a year. Examination of the hardened paste leads to the following stoichiometry of the reaction

$$2Ca_3SiO_5 + 6H_2O = Ca_3Si_2O_7.3H_2O + 3Ca(OH)_2 \tag{1}$$

C_2S reacts with water much more slowly than C_3S. The ions are densely packed in C_2S, whereas the structure of C_3S has "holes" in it [*5*], which may explain why the latter is more easily attacked by water. Also, C_3S is thermodynamically unstable with respect to C_2S plus CaO [*6*]. When C_2S is mixed with 0·7 of its own weight of water at room temperature and the mixture is allowed to stand, the reaction does not go to completion even in several years. Examination of a hardened paste more than four years old, which still contained about 15% unhydrated C_2S, has indicated the following stoichiometry

$$2Ca_2SiO_4 + 4H_2O = Ca_{3.3}Si_2O_{7.3}.3{\cdot}3H_2O + 0{\cdot}7Ca(OH)_2 \tag{2}$$

The chemical compositions of the tobermorite gel in equations (1) and (2) are somewhat different; for example, the molar CaO/SiO_2 ratio is 1·50 in equation (1) and 1·65 in equation (2). Nevertheless, the two gels are very similar; their X-ray diffraction patterns are the same, consisting of three broad lines, and electron microscopical examination reveals the same morphology. Both gels have very large specific surface areas, of the order of 300 m^2/g, but the gel coming from the hydration of C_3S has a somewhat larger area than that coming from the hydration of C_2S.

Tobermorite gel does not merely comprise the two compounds shown in equations (1) and (2). It denotes a series of hydrates of continuously varying compositions, constituting a sub-group in the tobermorite group of compounds. The composition of the gel changes in the course of hydration and also with temperature. Kantro, Brunauer and Weise [*7*] observed CaO/SiO_2 ratios ranging from 1·39 to 1·73 for stable tobermorite gels in the temperature range 5–50° C; and for unstable gels, i.e. for gels obtained in the early stages of hydration, the range was from about 1 to 3 [*8*]. Others reported even higher ratios than 1·73 for stable tobermorites [*3, 9–11*], ratios ranging up to 2·0.

The cementing properties of materials are caused by the forces residing in the surfaces of these materials. The magnitudes of these surface forces depend on the nature and the extent of the surface. Thus composition and specific surface area are two of the most important properties of tobermorite gel.

A. Distribution of Calcium in the Hydration Products of C_3S and C_2S

Both tobermorite gel and calcium hydroxide contain calcium. The calcium hydroxide content of hydrated C_3S and C_2S pastes can be determined in a number of ways: by solvent extraction methods, by X-ray quantitative analysis, by thermogravimetric analysis and by dynamic differential calorimetry.

Franke [*12*] proposed a method for the determination of uncombined CaO and $Ca(OH)_2$ in hydrated and unhydrated calcium silicates and Portland cements. A solvent mixture, consisting of acetoacetic ester and isobutyl alcohol, was refluxed with the sample, and the lime entered the solution, presumably as a calcium complex with the ester. Franke used a single extraction to obtain the free lime in the sample; however, Pressler, Brunauer and Kantro [*13*] showed that a single extraction did not remove all the free lime and that some of the lime was removed from the tobermorite gel. Utilizing the fact that the rate of removal of the uncombined lime was much greater than that of the combined lime, they developed a multiple extraction technique to correct for the decomposition of the gel. Later, Pressler *et al.* [*14*] extended this work by developing methods involving variation in the time of extraction and in the solvent to sample ratio. It was found that the result obtained on a given sample was the same, regardless of which parameter was varied.

A method of X-ray quantitative analysis for the determination of calcium hydroxide in pastes of the calcium silicates was used by Brunauer, Kantro and Copeland [*15*]. In this method, magnesium hydroxide is used as an internal standard, following the principles developed by Copeland and Bragg [*16*]. Brunauer *et al.* [*15*] found that the values obtained for the calcium hydroxide contents by X-ray quantitative analysis were invariably lower than the solvent extraction values. Various explanations for this lime discrepancy, or "missing" lime, were considered. One possibility was that the X-ray result was correct and that the extracting solvent removed some loosely bound lime from the gel. In this case the CaO/SiO_2 ratio in the gel, in the experiments which led to equation (1), was not 1·50, as is shown in that equation, but was variable within the range 1·52–1·75.

Experiments were performed in which only a part of the calcium hydroxide was extracted from the preparations. Analysis of the remaining calcium hydroxide by the X-ray method gave results in agreement with those determined by extraction. Thus it was found that the calcium hydroxide which was not measured by X-rays was more readily removed than the rest of the calcium hydroxide. This result indicated that the discrepancy between the two methods was due to calcium hydroxide not producing X-ray diffraction lines, i.e. material which was amorphous to X-rays. If this indication was correct, the CaO/SiO_2 ratio of the tobermorite gel in equation (1) was 1·51, which is, within experimental error, the value shown in the equation.

As a further argument against the hypothesis that the "missing" lime is loosely bound lime in the gel, it was shown by Kantro, Brunauer and Weise [*17*] that similar results were obtained with ball mill hydrated C_3S specimens. The calcium silicate hydrate, afwillite, produced in this method of hydration has a unique composition, and equation (1) describes this reaction exactly. Nevertheless, the X-ray quantitative analysis values for free calcium hydroxide were lower than the values predicted by equation (1), whereas the solvent extraction values were in agreement with the equation. Inasmuch as the "missing" lime here cannot be considered to be loosely bound to afwillite, it must be in a non-crystalline form.

Thus, in hydrated C_3S and C_2S preparations the calcium hydroxide appears partly in crystalline, partly in amorphous, form. The solvent extraction method can determine the total calcium hydroxide; X-ray analysis only the crystalline calcium hydroxide. The rest of the calcium is in the tobermorite gel and, in partly hydrated pastes, in the unhydrated calcium silicate residue. The latter can be conveniently determined by X-ray quantitative analysis, and so the composition of the tobermorite gel can be obtained by difference.

The results obtained by the thermogravimetric determination [*9*] and by the dynamic differential calorimetric determination [*18*] of the total calcium hydroxide are somewhat questionable, because there is no reason to expect that amorphous calcium hydroxide should have the same decomposition temperature or heat of decomposition as crystalline calcium hydroxide. Hence, the areas under the peaks obtained by these methods may not represent all of the lime liberated in the hydration reactions. The present authors have obtained thermogravimetric data indicating that the sharp break observed in the thermal weight loss curve corresponded to the free lime content determined by X-ray analysis, i.e. to the crystalline calcium hydroxide. This may be the reason why investigators using X-ray, thermogravimetric and differential calorimetric methods have reported higher CaO/SiO_2 ratios in their tobermorite gels than those using solvent extraction methods.

B. CaO/SiO_2 Ratios of Tobermorite Gels

The composition of any tobermorite can be expressed as $x CaO.SiO_2.y H_2O$, where x and y are the molar CaO/SiO_2 and H_2O/SiO_2 ratios, respectively. The H_2O/SiO_2 ratios will be discussed in the next section.

The CaO/SiO_2 ratio of a tobermorite depends on the concentration of lime in the solution which is in equilibrium with the solid. In the hydration of C_3S, lime saturation (and even supersaturation) is very quickly established in the surrounding aqueous phase, and the same happens in the hydration of C_2S, though more slowly. A number of investigators, including Bessey [*19*], Taylor [*20*], Greenberg [*21*] and Graham, Spinks and Thorvaldson [*22*], reported CaO/SiO_2 ratios of 1·5 for tobermorites in equilibrium with saturated calcium hydroxide solutions.

The same ratio is shown in equation (1) for the paste hydration of C_3S at room temperature [*15*]. However, this is not a unique ratio, as was pointed out before. As the hydration of C_2S, C_3S and alite progresses, the CaO/SiO_2 ratio changes radically; these changes will be discussed later, in the section on the kinetics of the hydration process. After a while the composition of the gel attains a stable value; at present only the CaO/SiO_2 ratios of these stable tobermorites will be discussed.

In the paste hydration of C_2S a stable ratio is reached at about 50% hydration [*7*]. The mean values of the CaO/SiO_2 ratios at 5, 25 and 50° C are 1·55, 1·58 and 1·65, respectively. There appears to be an increase in the ratio with temperature, which was noted earlier by Funk [*3*], who investigated the hydration of C_2S up to 100° C. The final ratios for C_3S pastes were found to be 1·53, 1·48 and 1·59, and for alite 1·46, 1·39 and 1·42 at 5, 25 and 50° C, respectively. In these cases there is no trend with temperature. In the calculation of the CaO/SiO_2 ratios for alite it was assumed that the small amount of aluminium present in alite became a part of the tobermorite gel structure and that it substituted for silicon just as in alite [*1*].

The CaO/SiO_2 ratio of 1·39, given above for alite, was the lowest value obtained by Kantro *et al.* [*7*] for a stable tobermorite. The highest value in the temperature range discussed in this chapter was given by a C_2S paste; it was 1·73. A C_2S paste more than five years old, maintained for a while at 80° C, gave a ratio of 1·75.

In the hydration of C_3S and C_2S with excess water in the rotating polyethylene bottle, the calcium silicate hydrates produced had CaO/SiO_2 ratios between 1·50 and 1·75 [*6*]. These hydrates, however, are not identical with the tobermorite gels produced in paste hydration. They differ in morphology (Chapter 9), but the most important difference is in the X-ray diffraction patterns. Whereas tobermorite gels have only three diffraction lines, each of which is an *hk*0 line, the hydrates obtained in bottle hydration have many diffraction lines, including the 00*l* line. These hydrates constitute another sub-group in the tobermorite group [*23*], which has been designated calcium silicate hydrate (II) or C-S-H (II).

In the ball mill hydration of C_3S the final hydration products are afwillite and calcium hydroxide [*17*]. Initially, however, a different calcium silicate hydrate forms, which later converts into afwillite. This hydrate, designated hydrate III, is very similar in many of its properties to tobermorite gel, but its X-ray diffraction pattern consists of only a single very broad line. Hydrate III may be regarded as an amorphous tobermorite gel. Its CaO/SiO_2 ratio is the same as that of afwillite, 1·50.

C. The Water of Hydration in Tobermorite Gel

In equation (1) the H_2O/SiO_2 ratio in the tobermorite gel is 1·5. Actually, this ratio was not determined experimentally, but there is evidence indicating that this is probably the correct ratio for the combined water when the paste is saturated with water. Inasmuch as the pastes are usually dried prior to study, most data on bound water contents have been obtained for at least partially desiccated pastes.

The very large surface of tobermorite gel is hydrophilic, so it adsorbs water readily. Some of the adsorbed water is held more strongly than some of the water of hydration, so that complete separation of the two types is impossible. When the material is dried, a part of each type of water is lost. However, the relative amounts of each type lost on drying depend on the extent to which the sample is dried.

Brunauer *et al.* [*15*] investigated pastes dried under two different conditions. One of these, equilibrium with the water vapour pressure over $Mg(ClO_4)_2.2H_2O$–$Mg(ClO_4)_2.4H_2O$, that is 8×10^{-3} mmHg, was designated P-drying. The other, equilibrium with the vapour pressure of ice at $-78°$ C (the sublimation temperature of solid CO_2), was called D-drying.

When a tobermorite gel with a CaO/SiO_2 ratio of 1·5 was subjected to D-drying, the resulting H_2O/SiO_2 ratio was found to be slightly greater than 1·0. The average of several samples was 1·06. It is probable that at least a part of the water in excess of 1·0 was adsorbed water. When the same materials were subjected to P-drying, the resulting H_2O/SiO_2 ratio was found to be 1·40, corresponding to tobermorite gel of the composition $Ca_3Si_2O_7.2{\cdot}8H_2O$. However, adsorption experiments showed that about 0·30 moles of water per 2 moles of silica were adsorbed on the surface; thus the corrected formula was $Ca_3Si_2O_7.2{\cdot}5H_2O$.

It is obvious that at higher vapour pressures the tobermorite contains more combined water and more adsorbed water. However, determinations of composition at these higher vapour pressures have not been performed to date. Hence, the number of water molecules in the hydrate at saturation pressure or in liquid water is somewhat conjectural. Bernal [*24*] proposed the structural formula $Ca_2[SiO_2(OH)_2]_2[Ca(OH)_2]$, indicating three molecules of water in a molecule of tobermorite, and the above experiments are consistent with Bernal's formula, except that

$CaO.H_2O$ should be written in place of $Ca(OH)_2$. After P-drying the formula is $Ca_2[SiO_2(OH)_2]_2.CaO.\frac{1}{2}H_2O$, and after D-drying the formula becomes $Ca_2[SiO_2(OH)_2]_2.CaO$.

The preceding discussion refers to a stable tobermorite gel having a CaO/SiO_2 ratio of 1·5. The variation in the CaO/SiO_2 ratios of stable gels is accompanied by a variation in the H_2O/SiO_2 ratio. Table I gives the ratios for a number of D-dried C_3S and C_2S pastes. An increase in the CaO/SiO_2 ratio above 1·5 is accompanied by a like increase in the H_2O/SiO_2 ratio above 1·0. Thus, apparently equimolar quantities of CaO and H_2O are simultaneously added to the $Ca_2[SiO_2(OH)_2]_2.CaO$ formula.

TABLE I

Compositions of tobermorite gels

Paste	CaO/SiO_2	H_2O/SiO_2 (D-dried)
C_3S	1·54	1·10
C_3S	1·56	1·14
C_3S	1·58	1·13
C_3S	1·63	1·29
C_2S	1·65	1·15
C_2S	1·63	1·09
C_2S	1·73	1·19
C_2S	1·75	1·29

D. THE SURFACE AREA OF TOBERMORITE GEL

It was stated earlier that the nature and the extent of the surface of tobermorite gel are responsible for its cementing properties. Powers and Brownyard [*25*] were the first to measure the specific surface areas of hardened Portland cement pastes, using the B.E.T. method [*26*] with water vapour as the adsorbate. They were able to demonstrate that surface area and porosity of the paste play decisive roles in determining the most important engineering properties of paste: strength, dimensional changes, permeability to water and freezing of water in the pore system. Subsequent work has shown that about 75% of the colloidal fraction of hardened Portland cement paste (the so-called "cement gel") is tobermorite gel, and at least 80% of the specific surface of hardened Portland cement paste is tobermorite gel surface.

Work on the pore size distribution of hardened pastes of calcium silicates and Portland cements is still in its very early stages; consequently, it will not be discussed in this chapter. However, much work has been done on the surfaces of these pastes. It has been established that the specific surface areas and the compositions of tobermorite gels in pastes of Portland cements are the same as in pastes of calcium silicates.

The tobermorites are layer crystals which in some ways resemble the clay minerals. The gel particles, as electron micrographs show (Chapter 9), are very thin sheets. If the average thickness of these sheets were known, the specific surface area could be calculated from crystallographic data. The crystal structures of tobermorites are discussed in detail in Chapter 5, but some aspects of the crystal structure of tobermorite gel will be described here.

Heller and Taylor [*27*] found that the unit cell of tobermorite was geometrically orthorhombic. On the basis of their work the three broad X-ray lines of tobermorite gels which, in D-dried samples, have spacings of 3·05, 2·79 and 1·82 Å, can be assigned indices of 110, 200 and 020, respectively. For simplicity, a pseudo-cell which contains a single formula unit will be described. The 020 spacing represents the distance in the *ab* projection between the successive rows of calcium atoms which occur in the central parts of the sheets. Twice this distance, 3·64 Å, is the *b* length of the orthorhombic pseudo-cell. Twice the 200 spacing, 5·59 Å, is the *a* dimension of the pseudo-cell; it is perpendicular to the *b* dimension within the layer. The area of the *ab* face of a pseudo-cell is thus 20·35 $Å^2$.

The gel gives no X-ray line corresponding to the *c* dimension, the distance between the molecular layers. Because of the absence of the *c* spacing, it was calculated from density measurements [*15*]. The density of D-dried tobermorite gel, having the formula $Ca_3Si_2O_7.2H_2O$, was found to be 2·86 g/cm^3. From the molecular weight and density the volume per formula unit is obtained. Knowing *a* and *b*, *c* can be calculated for a pseudo-cell containing one formula unit, and is thus found to be 9·3 Å. Numerous investigators found this value for certain strongly dried tobermorites which exhibit *c* spacings; it appears to be the distance of closest approach of two layers. This value agrees well with that found directly for well crystallized tobermorites that have been heated to 300° C to drive off the interlayer water (Chapter 5).

On the basis of these data one can calculate certain discrete specific surface area values for the above tobermorite gel. For sheets only a few molecular layers in thickness the edge areas can be neglected, because the two other dimensions are much greater than the thickness. If only the area of the two sides of the sheet is considered, the specific surface areas of sheets having thicknesses of one, two and three molecular layers are 755, 377 and 252 m^2/g.

The specific surface areas of tobermorite gels of other compositions can be calculated in a similar manner. Changes in the CaO/SiO_2 ratio do not alter the *a* and *b* dimensions of the unit cells of tobermorites, and changes in the H_2O/SiO_2 ratio cause only negligible changes (about 1%) in these dimensions. Kantro *et al.* [*7*] assumed that the pseudo-cell of tobermorite gel always contained one formula unit (or two silicon atoms assuming the formula $Ca_3Si_2O_7.2H_2O$), regardless of composition, and calculated specific surface areas on that basis. The areas were experimentally measured by the B.E.T. method [*26*], using water vapour as the adsorbate. Comparisons between the calculated and the experimental values for stable tobermorite gels always indicated average thicknesses between two and three molecular layers; in other words, the tobermorite gel in mature pastes of C_3S and C_2S always consisted of mixtures of two-layer and three-layer particles. In pastes of C_3S the two-layer particles predominated; in pastes of C_2S the three-layer particles.

The specific surface area of a tobermorite gel, and hence the average thickness of the particles in number of layers, has been found to depend on the CaO/SiO_2 ratio according to the equation

$$A = 865{\cdot}0 \pm 42{\cdot}9 - (354{\cdot}7 \pm 39{\cdot}9)(CaO/SiO_2) \qquad (3)$$

where A is the specific surface area in m^2/g [*7*]. This empirical equation indicates that the more lime there is in the tobermorite, the lower is the specific surface area, which in turn indicates that increase in the lime content results in a larger fraction of three-layer and a smaller fraction of two-layer sheets. It has been suggested that neighbouring crystallites of tobermorite can be cemented to each other through the adhesion of exposed calcium–oxygen sheets [*28*]. Equation (3) indicates that lime can cement the layers of tobermorite gel to each other within the crystallites.

Equation (3) also indicates that neither two-layer nor three-layer tobermorite gel has a CaO/SiO_2 ratio of 1·5; rather, the former has a lower and the latter a higher ratio. The lowest and highest ratios reported by Kantro *et al.* [*7*] are 1·39 and 1·73, respectively. The areas of these extremes are close to the calculated values for the pure two-layer and three-layer materials.

The slope of the line described by equation (3) corresponds approximately to a reduction in surface area by the area of one pseudo-cell face for two formula units of CaO added. The area contributed by a pseudo-cell surface in both two-layer and three-layer tobermorites is 20·35 $Å^2$, and the area of a molecule of water adsorbed on this surface is 11·4 $Å^2$ [*15*]; the ratio is 1·78. In the model discussed there is one silicon atom in each pseudo-cell *ab* face. If the area of the water molecule adsorbed on surface silica is the same as that adsorbed on surface lime, this ratio indicates that 0·78 formula units of CaO occurs on each pseudo-cell *ab* face. A value of 0·76 is obtained for surface lime when the areas of the adsorbed water molecules on the silica and lime are taken to be the same as those on pure hydrous amorphous silica and crystalline calcium hydroxide, that is 12·5 and 10·3 Å^2, respectively. Thus, both assumptions lead to the same surface composition. Furthermore, the model leads to the same surface composition for all tobermorite gels, regardless of the overall composition of the tobermorite gel.

The surface layer is considered to be in equilibrium with the aqueous phase in contact with it, while the inner layers of the tobermorite gel particle are not necessarily in such equilibrium. On the basis of such considerations, Kantro *et al.* [*7*] arrived at the approximate structures for two-layer and three-layer tobermorite gel particles, shown in Fig. 1. These structures refer to the stable forms of tobermorite gel.

The morphology of the particles in the tobermorites produced by hydration of C_3S and β-C_2S is discussed in Chapter 9. Copeland and Schulz [*29*] have reported on the electron microscopy of samples prepared by the present authors and their associates.

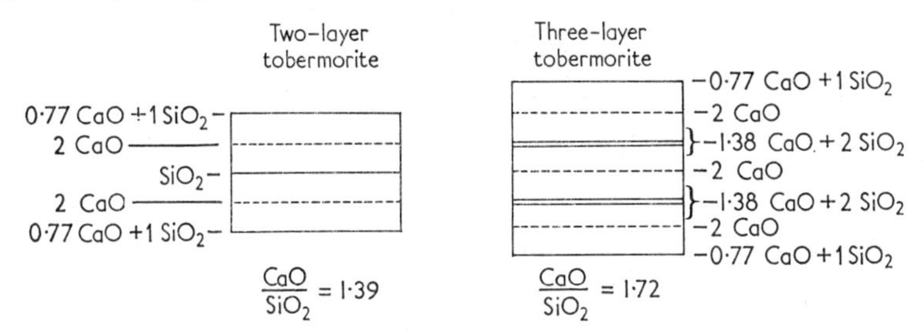

FIG. 1. Suggested structures for two-layer and three-layer tobermorite gel particles, according to Kantro *et al.* [*7*]. (Reproduced by permission from *Advances in Chemistry Series*.)

III. Energetics of the Hydration Process

A. HEATS OF HYDRATION

The reactions represented by equations (1) and (2) are accompanied by heat evolution. Because of the slowness of the reactions, Woods and Steinour [*31*] recommended the heat of solution method for the determination of the heat evolved, and all investigators since that time have been using this method. The heat of solution of C_3S and C_2S minus the heat of solution of the hydration products gives the heat of hydration.

The heat value thus obtained is a composite quantity, which may be represented by the equation

$$H_o = H_c - S + H_{H_2O} + H_{Ca(OH)_2} \tag{4}$$

where H_o is the experimentally measured overall heat of hydration, H_c is the chemical heat of hydration when a hypothetical tobermorite of negligible surface is produced in the reaction, S is the surface energy of the tobermorite gel, H_{H_2O} is the heat of adsorption of water and $H_{Ca(OH)_2}$ is the heat of adsorption of calcium hydroxide. This last term can be neglected, because the adsorption of calcium hydroxide on the tobermorite gel surface is small [*6*]. If the specific surface area of tobermorite gel were small, the last three terms on the right-hand side of equation (4) could be neglected.

The term S in equation (4) requires some elaboration. An atom, ion or molecule in the surface layer of a substance always has greater energy than in the body of that substance. This excess energy is called surface energy. All solids and liquids have a part of their energies residing in their bodies and a part residing in their surfaces. Ordinarily substances have relatively small surface areas; consequently, the surface area can be neglected. Only for colloids, such as tobermorite gel, does the surface energy constitute an appreciable fraction of the total energy.

The surface energy of a substance is usually expressed in ergs/cm^2 surface. The surface energy of tobermorite gel of composition $Ca_3Si_2O_7.2H_2O$ was found to be 386 ± 20 ergs/cm^2 [*30*]. Hence, for this tobermorite gel with a specific surface area of 300 m^2/g, the value of S is 27·7 cal/g gel. This is not a small quantity, as will be seen later.

Two important conclusions can be drawn from the surface energy work: (a) the magnitude of the surface energy of tobermorite gel clearly shows that the gel surface must contain both silica and lime; (b) the surface energies of the tobermorite gels of equations (1) and (2) appear to be the same within experimental error. The second conclusion implies that the surface compositions of the two gels are the same. Both conclusions are in line with the considerations advanced earlier for the composition of the gel surfaces.

The term H_{H_2O} in equation (4) is large, because large amounts of water are adsorbed on the surface of tobermorite gel. The heat evolved in this process adds to the heat evolved in the chemical reaction. In order to obtain the chemical heat of hydration, $H_c - S$, the adsorbed water must be removed from the paste. This can be done by D-drying. In this case, however, as was stated before, one molecule of water of hydration is also removed. Thus, the chemical heat of hydration is not determined for the reactions represented by equations (1) and (2) but for the reactions

$$2Ca_3SiO_5 + 5H_2O = Ca_3Si_2O_7.2H_2O + 3Ca(OH)_2 \quad (5)$$

$$2Ca_2SiO_4 + 3H_2O = Ca_{3\cdot3}Si_2O_{7\cdot3}.2{\cdot}3H_2O + 0{\cdot}7Ca(OH)_2 \quad (6)$$

The values so obtained are shown in the first and third rows of Table II. Similar determinations were made on P-dried pastes and corrections were made for the adsorbed water. The results are shown in the second and fourth rows of Table II. Thus, the results are in agreement within experimental error [*6*].

In the surface energy work the heat of the reaction

$$Ca_3Si_2O_7.2H_2O(s)+H_2O(l) = Ca_3Si_2O_7.3H_2O \quad (7)$$

was also determined [*30*]. Its value was −7600 cal at 23·5° C. Combining this value with the heat of reaction (5), one obtains the heat of reaction (1). The result is −42,850 cal or 93·8 cal/g C_3S. Similar calculation for the hydration of two moles of C_2S leads to the result of −10,320 cal or 30·0 cal/g C_2S. These are the results for $H_c - S$, when the tobermorite gel has a specific surface area of 300 m^2/g. For a gel with a different specific surface area correction can be made by using the surface energy of 386 $ergs/cm^2$. For a hypothetical tobermorite of negligible surface the heat of hydration of C_2S would be almost double the value given above.

Not enough data have been obtained to date for the variation of the heat of hydration with the CaO/SiO_2 ratio in the tobermorite gel. The indication is that the variation is small. Mchedlov-Petrosyan and Babushkin [*32*] calculated the heats of hydration of C_3S and C_2S from bond energy data; they obtained values of −24,500 and −6800 cal/mole, respectively. Even though they assumed that the calcium silicate

TABLE II

Heats of hydration of C_3S and C_2S

Paste	CaO/SiO_2	H_2O/SiO_2	$H_c - S$ (cal/g of silicate)
C_3S	1·54	1·10	77·5
C_3S	1·50	1·39	76·0
C_2S	1·65	1·15	7·7
C_2S	1·65	1·43	7·5

hydrate was hillebrandite with a formula $2CaO.SiO_2.1{\cdot}17H_2O$, their calculated values do not differ greatly from the experimental values reported above.

B. ENTROPIES AND FREE ENERGIES OF HYDRATION

No experimental determinations of the specific heats of tobermorite gels have been made to date; consequently, only calculated values are available for the entropy and free changes occurring in the hydration reactions. Brunauer and Greenberg [*6*] estimated that the entropy changes in the hydration of *two moles* of C_3S and C_2S are $-36{\cdot}3$ and $-19{\cdot}7$ entropy units, respectively. These values, combined with the heats of hydration given earlier, lead to free energy changes of $-32{,}000$ cal and -4400 cal, respectively. The free energy change values calculated by Mchedlov-Petrosyan and Babushkin [*32*], with hillebrandite as the hydration product, are $-18{,}700$ and -1700 cal/mole C_3S and C_2S, respectively. Again, the agreement is fair.

IV. Rates and Mechanisms of Hydration

The simplest and most convenient way to follow the course of the hydration of C_3S or C_2S is by determining the fraction or percentage hydrated at any given time. The only direct method of doing this is to determine by X-ray quantitative analysis the amount of C_3S or C_2S that was left unhydrated. A number of investigators used indirect methods to follow the reactions, such as determination of the water of hydration, the calcium hydroxide developed in the reaction, the heat evolution or the strength development. This early work was reviewed in detail by Brunauer and Greenberg [*6*]. Because the composition of the tobermorite gel changes in the course of the reaction, the percentage hydration calculated by any indirect method is unreliable.

Some investigators performed a few experiments using the direct method [*9–11*], but the only systematic use of this method was made by Kantro *et al.* [*7, 8, 17*]. They used the X-ray method to determine the fraction of C_2S or C_3S hydrated and measured the water of hydration and the calcium hydroxide produced after different times to obtain the compositions of the hydration products. In addition they measured the specific surface areas of the tobermorite gels in the hydration products. On the basis of these data they were able to deduce at least partial mechanisms for the hydration reactions.

A. Paste Hydration

The paste hydration of C_2S and C_3S was investigated from one-half hour to 400 days [*7, 8*], that of alite from 1 to 200 days [*7*]. As the hydration of alite is very similar to the hydration of C_3S, it will not be discussed here. The percentage hydration was determined by X-ray quantitative analysis.

The percentage hydration of C_2S and C_3S as a function of time, at hydration temperatures of 5, 25 and 50° C, is plotted in Figs. 2 and 3, respectively. The time axes have been made logarithmic only for the sake of convenient representation.

As Fig. 2 shows, the temperature dependence of C_2S is normal to about 70% hydration. At any given time the 50° C specimens are more hydrated

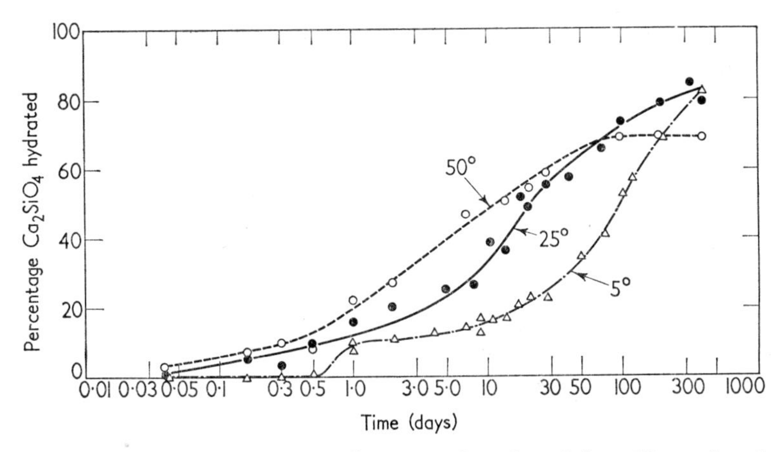

FIG. 2. Percentage hydration of β-C_2S as a function of time. (Reproduced by permission from *Journal of Physical Chemistry*.)

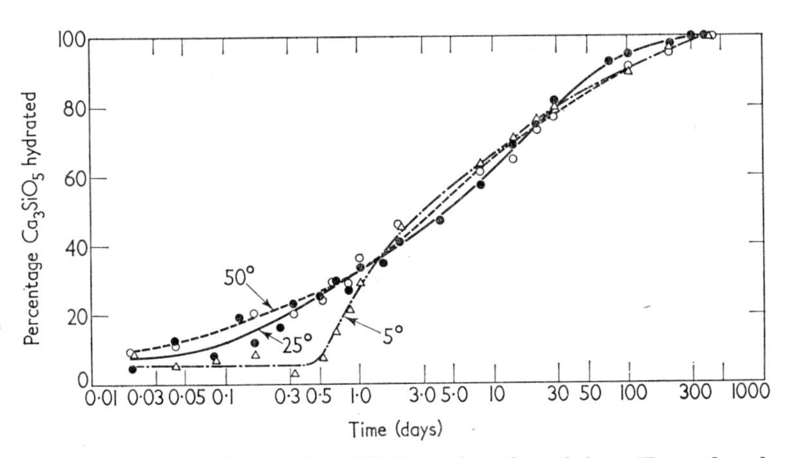

FIG. 3. Percentage hydration of C_3S as a function of time. (Reproduced by permission from *Journal of Physical Chemistry*.)

than the 25° C specimens, and these in turn are more hydrated than the 5° C specimens. After 70% hydration the curves cross each other, indicating a negative temperature dependence. As Fig. 3 shows, the hydration of C_3S has a positive temperature dependence only to about 35% hydration; after that the curves cross and recross each other. An explanation for these crossings will be given later. Schwiete and Knoblauch [33], who followed the reactions by calcium hydroxide determinations, found slight positive temperature coefficients for both silicates

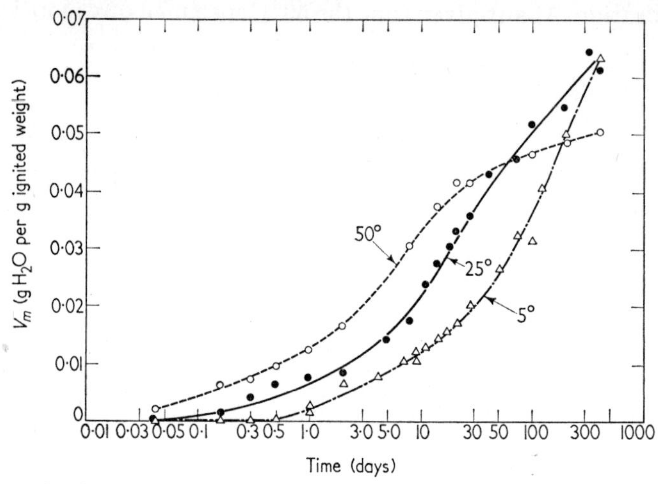

FIG. 4. Surface development in the hydration of β-C_2S. (Reproduced by permission from *Journal of Physical Chemistry*.)

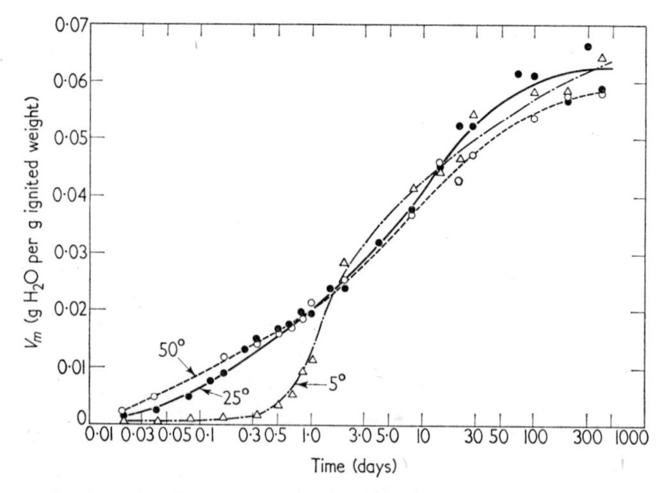

FIG. 5. Surface development in the hydration of C_3S. (Reproduced by permission from *Journal of Physical Chemistry*.)

throughout the entire range of hydration. There is an indication of a slight negative temperature coefficient in the data of van Bemst [9, 10], who used X-ray analysis.

The specific surface areas of C_3S and C_2S are negligible compared to that of tobermorite gel. Because of this, the course of hydration can be followed by measuring the surface development [7, 8]. The surface development in the hydration of C_2S and C_3S is shown in Figs. 4 and 5, respectively. The ordinate is V_m, the weight of water required to cover the entire surface with a layer one molecule thick [26]; it is directly proportional to the surface area. The data were obtained from the same pastes as the data of Figs. 2 and 3. A comparison of Fig. 2 with Fig. 4 and of Fig. 3 with Fig. 5 shows that the surface development follows closely the degree of hydration. For a simple reaction, in which the reactants have negligible surface areas in comparison with the reaction products, such a conclusion would be trivial. However, for reactions as complex as those discussed here, it is significant that through the intricate criss-crossings of the hydration curves the surface area development, by and large, faithfully follows the degree of hydration.

The curves in Figs. 2 and 4 and in Figs. 3 and 5 show similar but not identical courses. This is because the surface area depends not only on the degree of hydration but also on the composition of the tobermorite gel. This was discussed earlier for stable tobermorite gels. The composition of the gel changes in the course of hydration, and the surface area is a function of the composition for unstable tobermorite gels also.

Both the CaO/SiO_2 and H_2O/SiO_2 ratios of tobermorite gels change in the course of hydration [7, 8], but only the changes in CaO/SiO_2 ratios will be discussed here. Plots of the CaO/SiO_2 ratio as a function of the percentage hydration for C_2S and C_3S are shown in Figs. 6 and 7, respectively. Plots of the specific surface area as a function of the percentage hydration of C_2S and C_3S at 25° C are shown in Fig. 8.

The CaO/SiO_2 ratios of the gels produced in the hydration of both C_2S and C_3S are quite high initially, the curves of Figs. 6 and 7 indicating that the values at the inception of the reaction may have been the same as those of the unhydrated silicates themselves, i.e. 2·0 and 3·0. The CaO/SiO_2 ratio in the C_2S hydration passes through a minimum (Fig. 6) at the same degree of hydration as the surface area passes through a maximum (Fig. 8). The ratio in the C_3S hydration does not show a minimum (Fig. 7), but the area does show a maximum (Fig. 8).

The results have been interpreted to indicate the occurrence of three distinct stages in the course of the hydration reactions: (a) the formation of a high CaO/SiO_2, low area intermediate; (b) the conversion of this to a low CaO/SiO_2, high area intermediate; and (c) the conversion of this to the stable product.

The first step in the hydration process is the formation of a hydrated "skin" on the unhydrated substrate, which has the same CaO/SiO_2 ratio as the substrate. Such a high-lime gel coating must be unstable with respect to its surroundings and must decompose with the liberation of calcium hydroxide. This decomposition is accompanied by the splitting off of thin layers of the second intermediate. The second intermediate has a low CaO/SiO_2 ratio and, as indicated by the surface area values, consists of sheets one and two molecular layers thick. The split-off sheets take up lime from the solution, and this uptake is accompanied

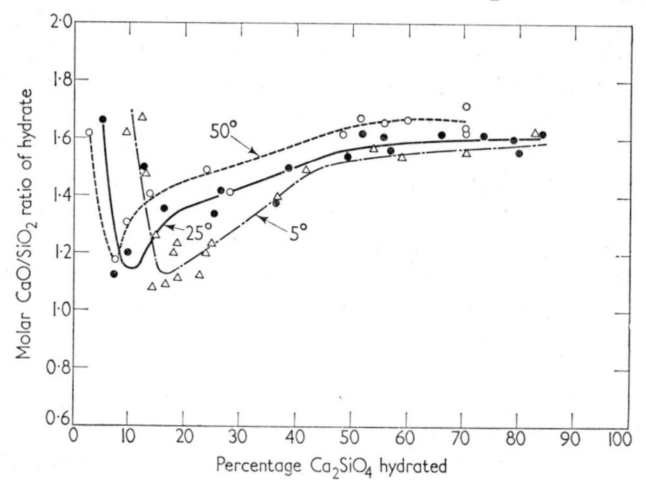

FIG. 6. Molar CaO/SiO_2 ratio of tobermorite gel as a function of the percentage hydration of β-C_2S. Reproduced by permission from *Journal of Physical Chemistry*.)

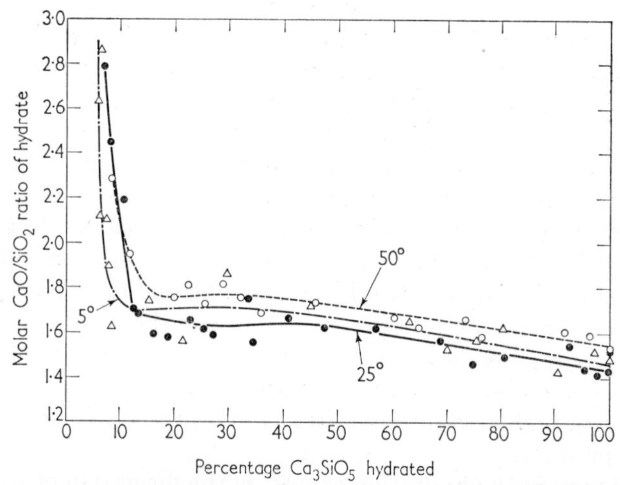

FIG. 7. Molar CaO/SiO_2 ratio of tobermorite gel as a function of the percentage hydration of C_3S. (Reproduced by permission from *Journal of Physical Chemistry*.)

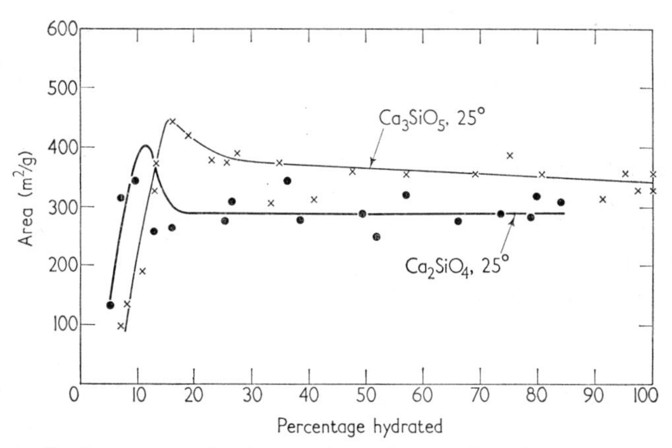

FIG. 8. Surface area of tobermorite gel as a function of percentage hydration. (Reproduced by permission from *Journal of Physical Chemistry*.)

by a condensation process resulting in the final stable two-layer and three-layer products. It was pointed out earlier that the three-layer tobermorite gel contains more lime than the two-layer material and that the extra lime acts as the binding agent. The lime taken up by the one- and two-layer intermediates presumably acts in the same way.

On the basis of certain assumptions, Kantro *et al.* [8] calculated the compositions and the thicknesses of the gel coating for their C_3S and C_2S pastes of minimum and maximum surface area. A C_3S paste hydrated for $\frac{1}{2}$ h at 5° C and a C_2S paste hydrated for 4 h at 25° C gave the minimum areas. Both pastes were about 5% hydrated. The CaO/SiO_2 ratio of the gel coating was 3·0 for the C_3S paste and 1·9 for the C_2S paste, and the thickness of the coating was about 25 molecular layers in both cases. A C_3S paste hydrated for 6 h at 25° C and a C_2S paste hydrated for 14 days at 5° C gave the maximum areas. Both pastes were 16% hydrated. The CaO/SiO_2 ratio of the gel coating was 2·9 for the C_3S paste and 1·4 for the C_2S paste. The thickness of the gel coating was about 32 layers for C_3S and about 67 layers for C_2S.

The mechanism proposed explains some heretofore unexplained experimental facts. One of these is that the specific surface areas of C_3S pastes are larger than those of C_2S pastes. Another is the anomalous behaviour of C_2S in the hydration at 50° C. As Fig. 2 shows, the hydration appears to stop when C_2S is only 70% hydrated. This results in the crossing of the hydration curves. However, crossing of the curves can occur without the stopping of the hydration process, as the curves in Fig. 3 show. A tentative explanation for this is advanced below.

In the early stages of hydration of both C_3S and C_2S the reactions have normal temperature coefficients. This indicates that the rate-determining step is a chemical reaction, such as the reaction between a silicate and a water molecule. Later, however, as Fig. 3 shows particularly well, the hydration curves at the three temperatures lie close to each other. The small temperature coefficient indicates a diffusion controlled rate-determining step. Diffusion takes over the control of rate when the gel coating is so thick that the diffusion through it is slower than the slowest step in the chemical reaction proper. Because C_2S is far less reactive than C_3S, a thicker gel coating must build up on it before diffusion becomes rate controlling. Thus, in Fig. 3 diffusion appears to take over at around 25% hydration, whereas in Fig. 2 only at around 60% hydration.

The temperature coefficient of diffusion is small, but it is still positive. Thus, diffusion by itself is not sufficient to explain the crossings of the curves. However, the rate of diffusion depends on the nature and the thickness of the gel coating. The experimental data indicate that the composition of the gel is different at different temperatures, and the thickness of the coating may also be different. Because of one or the other or both factors, the gel coating building up at higher temperatures appears to be less pervious than that building up at lower temperatures. This can explain the crossings and even the recrossings of the curves. It can explain also a complete stoppage of the reaction, as found in the 50° curve in Fig. 2.

The mechanism discussed does not account quantitatively for all features of C_3S and C_2S hydration. A complete elucidation of these complex reactions is still in the future.

B. The Ball Mill Hydration of C_3S

The stoichiometry of the hydration of C_3S in a small steel ball mill is correctly represented by equation (1). The calcium silicate hydrate that appears first is the amorphous tobermorite gel designated hydrate III [*17*], which is later converted into afwillite. The calcium hydroxide produced in the reaction is partly amorphous, partly crystalline. These results were stated earlier in this chapter.

Although hydrate III is an unstable intermediate, it takes more than a week until it completely disappears; consequently, not only its amount and composition but also several of its important properties could be determined. The amounts of unhydrated C_3S and afwillite were determined by X-ray analysis and the amounts of calcium hydroxide by the modified Franke method. These determinations, together with the water of hydration, gave the amounts and compositions of hydrate III by difference.

All of the C_3S disappeared at some time between 16 and 48 h; thus, ball mill hydration is much faster than paste hydration. (Bottle hydration is intermediate in speed between ball mill hydration and paste hydration. The kinetics of bottle hydration has not been investigated as yet.) The increase in rate was attributed to the removal of the gel coating from the surfaces of the C_3S grains by the steel balls, which resulted in the exposure of fresh anhydrous surfaces to the action of water. The rate of disappearance of the coating can be expressed by

$$-\frac{d(\text{coating})}{dt} = cA \tag{8}$$

where c is a constant depending on the grinding conditions and A is the area exposed to the steel balls. The variable area was taken to be proportional to $V^{2/3}$, where V is the volume of C_3S. The rate of disappearance of C_3S was assumed to be equal to the rate of removal of the gel coating; hence

$$-\frac{d(C_3S)}{dt} = k(C_3S)^{2/3} \tag{9}$$

where (C_3S) is the amount of unreacted C_3S in the system and the constant k includes c. The integrated form of the equation is

$$k = \frac{3[1-(C_3S)^{1/3}]}{t} \tag{10}$$

The values calculated for k were actually constant within the experimental error. Using the average value of k, 0·095 h^{-1}, the time calculated for the complete disappearance of C_3S was 31·5 h.

The agreement between the experimentally determined calcium hydroxide contents and those calculated from equation (1) implies that the stoichiometric amounts of hydrate III and calcium hydroxide are liberated as fast as C_3S disappears. This means that

$$\frac{d(\text{Hydrate III})}{dt} = k_H(C_3S)^{2/3} \tag{11}$$

and

$$\frac{d(Ca(OH)_2)}{dt} = k_c(C_3S)^{2/3}$$

It follows from equation (1) that $k_H = \frac{1}{2}k$ and $k_c = \frac{3}{2}k$.

In the initial stages of hydration, hydrate III forms faster than it disappears by conversion to afwillite, but after (C_3S) diminishes to a small value the situation is reversed. Thus, the amount of hydrate III in the system increases to a maximum and then diminishes to zero. The conversion to afwillite does not begin immediately on the appearance of hydrate III in the reaction system. After 4 h grinding in the ball mill, when C_3S is about 30% hydrated, there is still no measurable amount of afwillite present. The induction period appears to be about 6 h. The presence of the induction period probably indicates an autocatalytic mechanism, i.e. a rate dependent upon the concentrations of both afwillite and hydrate III.

REFERENCES

1. Jeffery, J. W. (1952). *Proceedings of the Third International Symposium on the Chemistry of Cement, London 1952*, p. 30. Cement and Concrete Association, London.
2. Nurse, R. W. (1952). *Proceedings of the Third International Symposium on the Chemistry of Cement, London 1952*, p. 56. Cement and Concrete Association, London.
3. Funk, H. (1957). *Z. anorg. Chem.* **291**, 276.
4. Brunauer, S., Copeland, L. E. and Bragg, R. H. (1956). *J. phys. Chem.* **60**, 116.
5. Bernal, J. D., Jeffery, J. W. and Taylor, H. F. W. (1952). *Mag. Concr. Res.* **11**, 49.
6. Braunauer, S. and Greenberg, S. A. (1962). *Chemistry of Cement, Proceedings of the Fourth International Symposium, Washington 1960*, p. 135. National Bureau of Standards Monograph 43. U.S. Department of Commerce.
7. Kantro, D. L., Brunauer, S. and Weise, C. H. (1962). *Advances in Chemistry Series* **33**, 199.
8. Kantro, D. L., Brunauer, S. and Weise, C. H. (1962). *J. phys. Chem.* **66**, 1804.
9. van Bemst, A. (1954, 1955). *Proc. 27th International Congress of Industrial Chemistry, Brussels*, Vol. 3; *Industr. chim. belge* **20**, 67.
10. van Bemst, A. (1955). *Bull. Socs. chim. belg.* **64**, 333.
11. Kurczyk, H. G. and Schwiete, H. E. (1960). *TonindustrZtg* **84** (No. 24), 585.
12. Franke, B. (1941). *Z. anorg. Chem.* **247**, 180.
13. Pressler, E. E., Brunauer, S. and Kantro, D. L. (1956). *Analyt. Chem.*, **28**, 896.
14. Pressler, E. E., Brunauer, S., Kantro, D. L. and Weise, C. H. (1961). *Analyt. Chem.* **33**, 877.
15. Brunauer, S., Kantro, D. L. and Copeland, L. E. (1958). *J. Amer. chem. Soc.* **80**, 761.
16. Copeland, L. E. and Bragg, R. H. (1958). *Analyt. Chem.* **30**, 196.
17. Kantro, D. L., Brunauer, S. and Weise, C. H. (1959). *J. Colloid Sci.* **14**, 363.
18. Schwiete, H. E. and Ziegler, G. (1958). *Ber. dtsch. keram. Ges.* **35**, 193.
19. Bessey, G. E. (1938). *Proceedings of the Symposium on the Chemistry of Cements, Stockholm 1938*, p. 178. Ingeniörsvetenskapsakademien, Stockholm.
20. Taylor, H. F. W. (1950). *J. chem. Soc.* 3682.
21. Greenberg, S. A. (1954). *J. phys. Chem.* **58**, 362.
22. Graham, W. A. G., Spinks, J. W. T. and Thorvaldson, T. (1954). *Canad. J. Chem.* **32**, 129.
23. Brunauer, S. (1962). *Amer. Scient.* **50** (March), 210.
24. Bernal, J. D. (1952). *Proceedings of the Third International Symposium on the Chemistry of Cement, London 1952*, p. 216. Cement and Concrete Association, London.

25. Powers, T. C. and Brownyard, T. L. (1947). *J. Amer. Concr. Inst.* **18**, 249; 101, 249, 469, 549, 669, 845, 933.
26. Brunauer, S., Emmett, P. H. and Teller, E. (1938). *J. Amer. chem. Soc.* **60**, 309.
27. Heller, L. and Taylor, H. F. W. (1951). *J. chem. Soc.* 2397.
28. Gard, J. A., Howison, J. W. and Taylor, H. F. W. (1959). *Mag. Concr. Res.* **11**, 151.
29. Copeland, L. E. and Schulz, E. G. (1962). *J. P.C.A. Res. Dev. Labs* **4** (No. 1), 2.
30. Brunauer, S., Kantro, D. L. and Weise, C. H. (1959). *Canad. J. Chem.* **37**, 714.
31. Woods, H. and Steinour, H. H., discussion of paper by A. S. Douglas, *J. Amer. Concr. Inst.* **3**, 195 (1931).
32. Mchedlov-Petrosyan, O. P. and Babushkin, V. I. (1962). *Chemistry of Cement, Proceedings of the Fourth International Symposium, Washington 1960*, p. 533, National Bureau of Standards Monograph 43. U.S. Department of Commerce.
33. Schwiete, H. E. and Knoblauch, H. (1959). *Research Reports of the Nordrhein-Westfalen District*, No. 748, Köln and Opladen.

Part III

Utilization of Portland Cement

CHAPTER 8

Chemistry of Hydration of Portland Cement at Ordinary Temperature

L. E. COPELAND and D. L. KANTRO

Portland Cement Association Research and Development Laboratories, Skokie, Illinois, U.S.A.

CONTENTS

I. Initial Reactions in Plastic Pastes

Portland cement clinker is a heterogeneous mixture of several "minerals". Even individual ground clinker particles are heterogeneous, for each of the principal minerals is invariably present in each particle. Small crystallites of alite (C_3S) and belite (β-C_2S) are embedded in a mixed crystalline interstitial material composed of C_3A and a solid solution of lime, alumina and ferric oxide. This solid solution has generally been regarded as tetracalciumaluminoferrite (C_4AF) but in reality can vary in composition from C_6AF_2 to C_6A_2F. In this chapter we shall refer to it simply as the "ferrite phase".

The initial reactions of Portland cement clinker with water are generally rather violent, so it is a universal custom in the manufacture of Portland cement to add a few per cent of calcium sulphate, usually as gypsum, to regulate these initial reactions. Although the hydration reactions of Portland cement are the primary concern of this chapter, the initial reactions of Portland cement clinker (no $CaSO_4$ present) with water will be discussed first.

It is not possible to discuss the initial reactions without introducing the concept of "setting". "Setting" and "hardening" are practical terms and were not meant to have theoretical significance; initial and final set are defined by methods of testing that determine the time when neat cement paste shows an arbitrarily defined resistance to a particular deformation. Although the tests are made on neat cement pastes, they were designed to have significance for the use of cement in mortar and concrete. In practice, concrete must remain plastic long enough to be properly placed and finished. "Setting" implies loss of plasticity, or workability, so the initial reactions of importance are those that produce "quick set" and those that make it possible to control time of set. In this section we shall be concerned with "flash" or "quick" set, and with "false" set. "Flash set" is a very quick set that occurs when cement insufficiently retarded is mixed with water. The chemical reactions involved liberate a large amount of heat, and the set cannot be overcome by remixing. "False set", on the other hand, is an early stiffening that sometimes amounts to initial set. Little heat is developed, and it can be overcome by remixing, or by a longer initial mixing time. More water is frequently used to minimize or avoid false set, at the expense of more desirable properties of the concrete.

A. The Clinker–Water System

Mixing of ground Portland cement clinker with water is usually accompanied by a rapid evolution of heat followed by flash set of the paste. The action is very similar to that observed when C_3A is mixed with water. The water rapidly becomes saturated with calcium hydroxide[*1*], produced by hydrolysis of C_3S, and calcium aluminate. Alkali hydroxides present in the clinker also dissolve rapidly. Under these conditions a hexagonal calcium aluminate hydrate should be precipitated, which possibly later would convert to the cubic form. Although it is difficult to establish which particular calcium aluminate hydrates are formed when only a small amount of water is used, evidence from DTA curves [*2–4*] indicates that C_4AH_{19} is first formed. However, Bogue and Lerch [*5*] found that the cubic hydrate, C_3AH_6, was formed when mixtures of the cement compounds were used under similar circumstances.

Although calcium hydroxide is formed from hydrolysis of alite, the rate of hydration of alite pastes is much too slow to produce the flash set observed in clinkers containing appreciable amounts of C_3A. Neither is the ferrite phase likely to cause false set, for clinkers in which the ratio Fe_2O_3/Al_2O_3 is greater than unity do not give flash set [*6*]. It may be that hydration of the ferrite phase coats the clinker particles with a colloidal hydration product and retards their hydration [*5*]. The use of gypsum in such clinkers accelerates the final setting.

The rate of hydration of belite is very small; consequently, it must be assumed that flash set occurs by rapid precipitation of calcium aluminate hydrates, as in the hydration of C_3A.

Alkalis in clinker seem to promote flash set [*7*]. Three possible explanations exist: (a) perhaps NC_8A_3 is more reactive than C_3A; (b) the higher solubility of the alkalis may result in a greater surface exposure of the aluminates; or (c) the alkali causes a reduction in the concentration of calcium hydroxide.

The third possibility is perhaps the most likely, for Forsén [*8*] has shown that alkali-free clinker is sometimes slow-setting even if it contains a normal C_3A content and no SO_3. Forsén attributed the long setting times to the formation of a coating of colloidal ferrite on the particles as mentioned above.

Forsén also found that the alkali-free clinkers that did flash set could almost always be retarded by the addition of 3% of quicklime (CaO). The retardation produced by lime explains the slow setting of clinkers high in C_3S, for C_3S will hydrolyse rapidly to produce a saturated solution of calcium hydroxide. Forsén reports that Assarsson first observed that a protective coating of C_4AH_{19} was formed on C_3A in saturated hydroxide solutions, and ascribed the retarding action to this protective coating.

But now the formation of calcium aluminate hydrates has been suggested as causing both flash set and retardation of set. One may well question how the same reaction can have the two different effects. The answer probably lies in the different physical states of the products, although this answer is difficult to confirm in systems containing only enough water to form a paste. A distinction between the reaction products in the two cases can be made, however.

When C_3A hydrates in water, the hydration product is an intergrown mixture of crystalline C_2AH_8 and C_4AH_{13} [*9*] or, more likely, of C_2AH_8 and C_4AH_{19} [*10*]. It follows from the mass balance for the reaction that the lime/alumina ratio (molar) for the mixture will be approximately 3. It was pointed out by Steinour [*11*] that the solubility curve for these mixed hexagonal crystals (Fig. 1 [*12*]) shows that, when the lime/alumina ratio is 3, the solution is about half saturated with calcium hydroxide and contains about 0·1 g Al_2O_3 per litre of solution. On the other hand, the lime/alumina ratio of the calcium aluminate hydrate is 4 when in saturated calcium hydroxide solution, and the solubility of Al_2O_3 is then only a few mg/l. When the solubility of alumina is so low, one might expect that particles of C_3A would become coated quickly

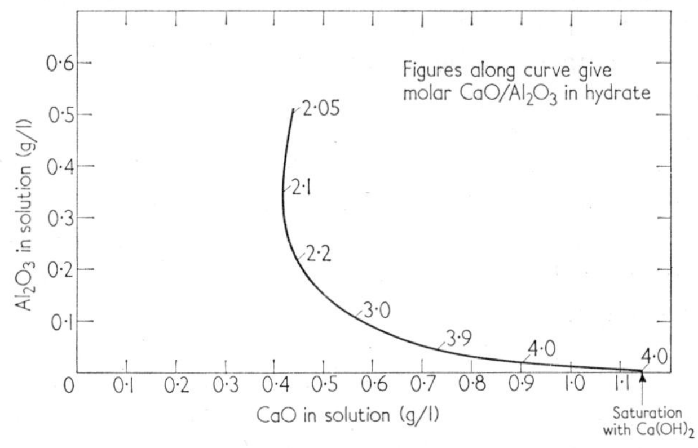

FIG. 1. Solubility curve for "hexagonal" calcium aluminate hydrates in the $CaO–Al_2O_3–H_2O$ system at 21° C (Wells, Clarke and McMurdie [*69*]).

with a film of C_4AH_{19} that would retard further hydration. If this is correct, any agent that will increase the solubility of alumina will promote flash set.

B. THE EFFECT OF GYPSUM

It was pointed out above that calcium sulphate, usually gypsum, is universally added to Portland cement clinker to control the initial reactions so as to prevent flash set. Gypsum is effective as a retarder of flash set because it reacts with calcium aluminates to form either of two practically insoluble complex compounds: calcium aluminate trisulphate hydrate and calcium aluminate monosulphate hydrate. The formulas for these compounds can be written as $3CaO.Al_2O_3.3CaSO_4.31H_2O$ and $3CaO.Al_2O_3.CaSO_4.12H_2O$, respectively. The natural mineral ettringite is calcium aluminate trisulphate hydrate.

The monosulphate enters into solid solution with tetracalcium aluminate hydrate, the formula of which may be written, for purposes of comparison, as $3\,CaO.Al_2O_3.Ca(OH)_2.12H_2O$.†

† The association of C_4AH_{13} with $C_3A.CaSO_4.12H_2O$ should perhaps be regarded as an intergrowth and not a solid solution; see Chapter 6 (Ed.).

The trisulphate forms readily in ample water, and crystallizes as needles. The monosulphate crystallizes as hexagonal plates similar in appearance to the hexagonal calcium aluminate hydrates. Both sulphates are hexagonal or pseudo-hexagonal.

When Portland cement is mixed with a relatively large amount of water on a microscope slide, the needles of the trisulphate usually form quickly. Sometimes hexagonal plates are also seen. But mixtures containing relatively large amounts of water are not good models of plastic pastes because it takes a relatively long time for the solution phase to become saturated with the soluble components. More meaningful observations are possible if solutions saturated with calcium hydroxide or with calcium hydroxide and calcium sulphate are used.

Parker [*13*] and Lerch [*14*] both found that cements that showed rapid development of crystals when mixed with water did not do so when mixed with saturated calcium hydroxide solution. With some cements no crystal formation was apparent even after 30 minutes. Evidently, the cements that show rapid crystal formation when mixed with water do not produce calcium hydroxide rapidly enough to control the initial reactions, and the retarding action in Portland cement is the result of formation of calcium hydroxide or the combined action of calcium hydroxide and calcium sulphate.

Forsén [*8*] showed that the combination of calcium hydroxide and calcium sulphate was more effective in retarding hydration of C_3A than either compound alone. He determined the rate of hydration of C_3A in water, saturated $CaSO_4$ solution, saturated $Ca(OH)_2$ solution and a solution saturated with both calcium hydroxide and calcium sulphate. His results are shown in Fig. 2, where it is apparent that calcium hydroxide alone is a better retarder than calcium sulphate alone, but that the combination of calcium hydroxide and calcium sulphate is more effective than calcium hydroxide solution. In fact, no C_3A reacted after the first minute, although the amount that reacted during the first minute was the same in calcium hydroxide solution as it was in the solution saturated with both calcium hydroxide and calcium sulphate.

Bucchi [*15*] provided evidence that a large fraction of the SO_3 in cement paste reacts in the first few minutes. He extracted the soluble SO_3 from the paste and determined the amount of calcium aluminium sulphate hydrate by difference.

Greene [*2*] used DTA to examine pastes of Portland cement mixed with water and also with saturated lime–gypsum solution. In the first case the trisulphate was found after two minutes' hydration. In the second case the trisulphate peak was "almost completely suppressed".

In neither case, however, was there any evidence that either the monosulphate or a calcium aluminate hydrate had formed.

The evidence appears to show that the trisulphate is the usual product in the early stages of hydration of Portland cements. But there is then still a question about how calcium hydroxide augments the retarding action of gypsum. In clinker, retardation by calcium hydroxide is brought about by decreasing the solubility of alumina and the formation of a protective coating of calcium aluminate hydrate. It is possible that in Portland cement a protective coating of the trisulphate is formed in

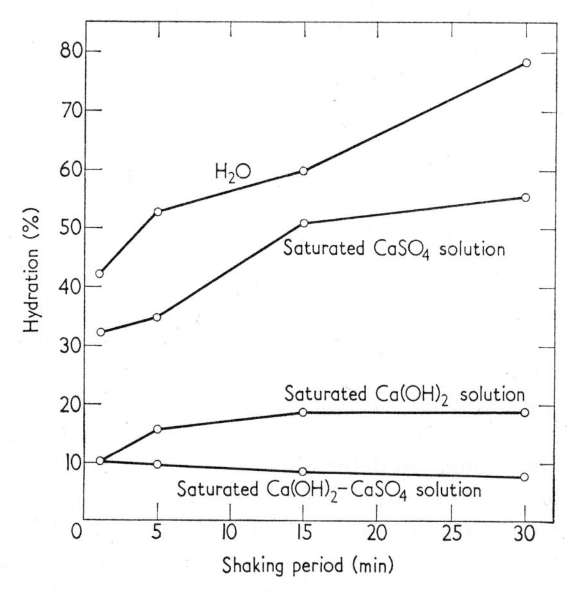

FIG. 2. Rates of hydration of tricalcium aluminate in various solutions (Forsén, Hedin).

saturated calcium hydroxide. Kalousek [*16*] apparently obtained a gel-like trisulphate in some of his experiments. Perhaps a thin coating of C_4AH_{19} is formed also [*17*], along with the trisulphate, on the surface of the C_3A.

C. Other Retarders

1. Soluble Calcium Salts

Candlot has been given credit for discovering that calcium chloride and calcium nitrate retard flash set [*8*]. Forsén [*18*] tried many other calcium salts and concluded that all those that were somewhat soluble acted to retard flash set in Portland cement clinker. Foster [19] soon afterwards reported that calcium salts in general retarded the hydration of C_3A.

Forsén showed that the set regulating calcium salts formed complex salts analogous to the calcium aluminate sulphate hydrates, and it is now well established that any inorganic calcium salt, and some organic calcium salts, can form more or less insoluble complexes with C_3A. It may thus be inferred that the retarding action of all these salts is similar to that of gypsum, and Forsén assumed that they all formed insoluble coatings that protected the C_3A.

2. *Four Classes of Retarders*

While all the soluble inorganic calcium salts are found to act like gypsum when used in moderate amounts, some of them do not act like gypsum when larger amounts are used. Calcium chloride, for example,

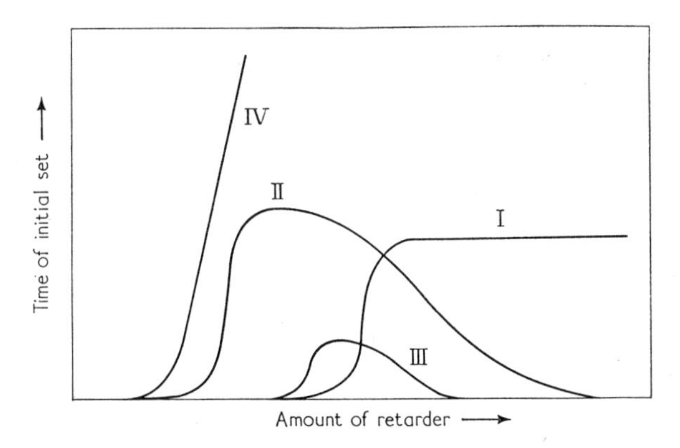

FIG. 3. Action of various retarders on clinker (Forsén). I—$CaSO_4.2H_2O$, $Ca(ClO_3)_2$, CaI_2; II—$CaCl_2$, $Ca(NO_3)_2$, $Ca(NO_2)_2$, $CaBr_2$, $CaSO_4.\frac{1}{2}H_2O$; III—Na_2CO_3, Na_2SiO_3; IV—Na_3PO_4, $Na_2B_4O_7$, Na_3AsO_4, $Ca(CH_3COO)_2$, and others.

acts as a retarder when present in low concentration, but accelerates set when used at higher concentrations. The explanation for this apparently anomalous behaviour may be that the calcium aluminate chloride hydrate is more soluble in high than in low concentrations of calcium chloride. Other inorganic salts also showed retarding action different from that of gypsum. Forsén divided retarders into four groups, depending upon their action as a function of concentration (Fig. 3).

The action of gypsum, in group 1, is to be expected because of its limited solubility. The concentration of SO_4^{2-} is independent of the amount of solid gypsum present and retardation should last as long as the gypsum lasts. It is surprising, though, that $Ca(ClO_3)_2$ and CaI_2 should be in this class; in later work Forsén shows results for CaI_2 that follow curve II rather than curve I.

Group II includes calcium chloride and compounds of a similar nature. The explanation for the decreased setting time with increasing concentration is probably the same as for calcium chloride for all compounds in this class except calcium sulphate hemihydrate. The behaviour of calcium sulphate hemihydrate will be discussed later.

Only two compounds are in group III, sodium carbonate and sodium silicate. Both compounds react with calcium hydroxide to form alkali, and Forsén attributed the little effect that they have to the precipitation of aluminum hydroxide, which would be dissolved at higher concentrations.

The compounds in group IV were called by Forsén "cement-destroying retarders". When they are used in suitable quantities they are effective retarders, but when used in larger amounts the setting period can be prolonged into what is normally the hardening period.

Probably no single mechanism can explain the action of all the substances in this class. Originally Forsén attributed the action to the precipitation of insoluble compounds as protective coatings. Later he was not so sure that this always happened and mentioned that the retarding action in some cases depended upon pH.

The agents of this class listed in Fig. 3 are those listed by Forsén in 1935 [*20*]. Many organic retarders should be added to this list. Calcium lignosulphonate belongs to group IV; in small amounts it can produce desirable effects. It is probable that most of the organic materials which Hansen [*21*] discussed as retarders for oil well cements belong to group IV. He pointed out that most of these compounds contained a H–C–OH group. Steinour [*22*], in his discussion of Hansen's paper, pointed out that there are other retarders that suggest that the active factor is simply un-ionized OH groups; calcium lignosulphonate is one of these. Both Hansen and Steinour assumed that the action of these retarders was the result of adsorption because extremely small quantities are often effective. Steinour found that some of the retarders characterized by non-ionized OH groups would retard hydration of β-C_2S in the absence of C_3A, a fact which supports the adsorption theory.

Some agents, such as sugar and borax, can seriously retard strength development even though they sometimes allow quick set. Forsén thought that these agents could allow initial solution of C_3A with hydration to precipitate a gel-like material in sufficient quantity to produce quick set, but thereafter the clinker particles would be so well coated that no further hydration could take place for hours, or sometimes days.

Salts of lead, copper and zinc are also retarders in group IV. Probably these salts hydrolyse to produce gels of the hydroxides, or hydrous oxides, that coat the clinker particles. Magnesium salts also are *retarders*, probably because magnesium hydroxide is precipitated on the clinker grains. $MgSO_4.7H_2O$ has been shown to give adequate retardation and good strength [*23*]. In some cases pH may be a factor in retardation. Films formed by weak organic acids may retard as long as the carboxyl radical is not neutralized.

D. Action of Calcium Sulphates Other than Gypsum

1. The $CaSO_4$–H_2O System

The $CaSO_4$–H_2O system has been thoroughly investigated and at least two hemihydrates and two forms of anhydrous $CaSO_4$ have been reported [*24*]. From the point of view of a cement technologist, the most important consideration is the relative solubilities of gypsum, hemihydrate (plaster of Paris) and natural anhydrite. The approximate solubility curves of these forms are shown in Fig. 4. The point of

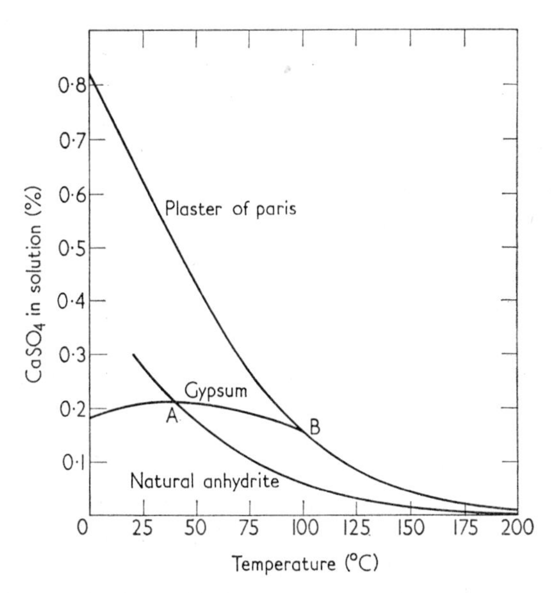

FIG. 4. Water solubility of different forms of calcium sulphate (approximate results from various investigations).

intersection of two curves gives the temperature at which the two corresponding forms are in equilibrium. At other temperatures the less soluble form is the stable form.

The curves show that above about 40° C (point A) natural anhydrite is the most stable calcium sulphate. However, gypsum in contact with water at temperatures somewhat above 40° C converts to anhydrite very slowly, if at all. Below 100° C (point B) gypsum is more stable than the hemihydrate. Saturated solutions of hemihydrate are, therefore, supersaturated with respect to gypsum, and gypsum can precipitate rapidly from these solutions.

2. *Action of Calcium Sulphate Hemihydrate*

The shape of the Type II curves of Fig. 3 shows that small amounts of hemihydrate have the same retarding action as gypsum but that as the amount of hemihydrate is increased another action is introduced, not present when gypsum is used, that shortens the time of set. It is now generally accepted that this action is the precipitation of the bladed crystals of gypsum. Under favourable conditions (unfavourable from a technical viewpoint) these crystals can form a partial, and sometimes a complete, network of particles that causes the paste to thicken, or sometimes to set.

Although many workers have studied the problem, the results of only two investigators will be summarized here. Schachtschabel [*25*] confirmed that calcium sulphate hemihydrate gave the typical Type II curve and showed further that mixing the paste for longer periods of time extended the length of the plateau. In further experiments he added 1% gypsum to cements containing 3, 4 and 6% hemihydrate. Although the cements with hemihydrate alone had setting times less than ½ h, the addition of 1% gypsum made the setting time normal for the cements with 3 and 4% hemihydrate. The initial setting time of the pastes of the cement containing 6% hemihydrate was not changed appreciably, although the time for final set was increased.

Bucchi [*15*, *26*] found that small amounts of hemihydrate gave the same setting times as equivalent small amounts of gypsum. He defined two points on the Type II curve: Q_{ci} the initial false set where the curve turns down from the plateau, and Q_{cf} the final false set.

He also used mixtures of gypsum and hemihydrate. In one set of experiments he kept the total SO_3 constant at a value that gave normal set when hemihydrate alone was used. Substitution of gypsum in small amounts (5–10% of the total sulphate) caused a markedly decreased setting time, but when more than 30–50% of the hemihydrate was replaced by gypsum the setting times were normal. In other experiments he found that when the quantity of SO_3 was greater than that corresponding to Q_{cf}, the substitution of small amounts of gypsum for hemihydrate shortened time of set to a small extent, but increasing the amount of gypsum substituted to 30% of the hemihydrate increased the time of set to the normal time. When the quantity of SO_3 corresponded to a value between Q_{ci} and Q_{cf}, small substitutions of gypsum for hemihydrate again shortened the time of set, but larger substitutions lengthened the time of set.

It is apparent from Bucchi's results that a small quantity of gypsum with the hemihydrate decreases the time of set of pastes. The gypsum apparently furnishes "seed crystals" that initiate the precipitation of gypsum from the solution of calcium sulphate supersaturated with respect to gypsum. If the number of "seed crystals" is sufficiently small, a continuous network of gypsum crystals can be formed that shortens the time of set. The results of both Bucchi and Schachtschabel indicate that if enough gypsum is present its precipitation from hemihydrate solution does not form a network, perhaps because a high degree of supersaturation of the solution cannot be achieved. These results are significant in that they provide an explanation for the phenomenon of false set which is discussed below.

3. Action of Natural Anhydrite

The solubility curves in Fig. 3 show that at normal temperature the solubility of natural anhydrite is greater than that of gypsum, and one would accordingly expect that natural anhydrite should serve as a retarder. It does so serve with many clinkers; but because it dissolves so much more slowly than gypsum it is frequently necessary to add an amount larger than the amount of gypsum required to reach the plateau of Class I retarders. With many cements it is not possible to use natura anhydrite and remain under specification (A.S.T.M.) limits for SO_3.

Control of the initial reactions requires that the retarder be capable of supplying a relatively large amount of SO_3 rapidly. After the initial reaction the rate of combination of SO_3 is relatively slow. It is reasonable, therefore, that Hansen and Hunt [*27*] found that mixtures of natural anhydrite and gypsum gave good retardation without affecting adversely other properties of the product. They concluded that, for six of the seven clinkers which they tested, natural anhydrite could be used successfully to supply from 60 and 70% of the SO_3 in the cement. Apparently the rate of solution of natural anhydrite is sufficient to supply the SO_3 required after the relatively large initial demand is supplied by the gypsum.

E. The Liquid Phase in Plastic Pastes

It is apparent from the discussion given above that within a very few minutes after cement is mixed with water the liquid phase becomes saturated, or frequently even supersaturated, with respect to $Ca(OH)_2$ and $CaSO_4.2H_2O$. Very little alumina, silica or iron are present in solution because of the extreme insolubility of their hydration products [*28, 29*].

The alkalis in cement clinker dissolve rapidly and persist as sulphate or hydroxide (produced by double decomposition with $Ca(OH)_2$). In some cases as much as 70–80% of the K_2SO_4 has been found to dissolve in the first 7 min. Na_2SO_4 dissolves more slowly, but in some cases goes into solution at almost half the rate of solution of K_2SO_4. Alkalis in other combinations go into solution less readily. After the first few minutes the rate of solution of the alkali becomes small; very little additional alkali is dissolved between 7 min and 2 h.

The liquid phase in Portland cement paste soon becomes a solution of the hydroxides and sulphates of sodium, potassium and calcium. Calcium hydroxide and gypsum are much less soluble than the other hydroxides and sulphates so the composition of the liquid phase tends to move toward the equilibrium:

$$CaSO_4.2H_2O(\text{solid}) + 2MOH(\text{aq.}) \rightleftharpoons M_2SO_4(\text{aq.}) + 2H_2O + Ca(OH)_2(\text{solid})$$

where M represents the alkalis. Because of the common ion effect, the concentrations of calcium hydroxide and calcium sulphate in solution will be depressed. Eventually the SO_4^{2-} is used up in the formation of a calcium aluminate sulphate hydrate and only the hydroxides are left.

Although the same equilibrium is approached independently of whether the alkalis enter solution as hydroxides, sulphates or even carbonates, the immediate reactions can be affected. The importance of some of these effects will become evident later.

Hansen and Pressler [*30*] have determined experimentally the equilibria corresponding to the above equation. Their data cover the range of alkali concentration encountered in fresh Portland cement pastes and are therefore useful in deciding if a paste is supersaturated with respect to gypsum [*31*].

F. False Set

Sporadically the early stiffening of plastic paste, called false set, occurs. Some of the factors that have been found to promote false set are: heating of the cement during the final grinding process, exposure to atmosphere and presence of alkali carbonate in the cement.

1. Heating of Cement During Milling

The most common cause is probably the heating of the cement during grinding. Although the calcium sulphate added to cement clinker is in the form of gypsum, it is likely that part of the gypsum is dehydrated during the grinding process to form the hemihydrate, or soluble anhydrite [*11, 32, 33*]. Accordingly, a knowledge of the stability of gypsum with respect to temperature and water vapour is important to the cement technologist. The temperature of cement during grinding will be in the range of 40–150° C.

TABLE I

Approximate relative humidity required to prevent dissociation of $CaSO_4.2H_2O$

°F	°C	Rel. humidity (%)
75	24	35
100	38	45
125	52	55
150	66	65
175	79	80
200	93	95
212	100	100
250	121	(2·5 atm vapour pressure)

From the data in Table I it would appear that the relative humidity of the atmosphere in the mill would have to exceed 50% to prevent dehydration of gypsum. Although dehydration in grinding occurs, apparently the rate of dehydration is sufficiently low that cements can be ground at 200° F (93° C) without extensive dehydration of gypsum, provided adequate cooling of the ground cement is maintained [*34*].

From the action of calcium sulphate hemihydrate one would predict that if the gypsum were markedly converted to hemihydrate, the time of set would be shortened. Lhopitallier and Stiglitz [*35*] found that supersaturation, with respect to gypsum, of the solution phase in paste precedes the development of false set. They also performed experiments, using mixtures of gypsum and hemihydrate added to siliceous sand of cement fineness. Enough calcium sulphate was used to provide 2% SO_3. Pastes were mixed for 1 min and 5 min. False set was obtained in all cases where the hemihydrate content was high and the mixing time 1 min. When the mixing time was 5 min no false set was observed except for the mixture that contained all the SO_3 as hemihydrate. Analyses of the solutions after

2 min showed supersaturation with respect to gypsum in those pastes, and in only those pastes, that showed false set.

Manabe [*36*] came to the conclusion that the dehydration of gypsum causes all cements to have a tendency to stiffen at a certain period after contact with water. If the paste is mixed until the period of stiffening is complete, there will be no false set. If the period of stiffening occurs after the paste is mixed, false set will be fully developed. If the mixing extends into the period of stiffening, the stiffening will be partially developed. The period of stiffening may occur immediately after mixing the cement with water, or it may be delayed by any agent that would tend to change the activity of the C_3A.

2. *Aeration of Cement*

Hansen [*37*] showed that two Portland cements with only slight tendencies to develop false set were made prone to false set after 24-hr exposure to laboratory air. Analyses of the liquid phase of pastes made from the cements before the exposure showed practically no supersaturation, but after exposure the supersaturation was pronounced. He assumed that before exposure the reaction of C_3A with SO_4 was immediate, but that after exposure, the C_3A was slower to react.

Swayze, quoted by Steinour [*11*] and Hansen [*38*], found that cement which contained no C_3A and no calcium sulphate became false setting after aeration. He suggested that the false set could be caused by activation of the C_3S by moisture absorption, and said that cement high in C_3S was more susceptible than cement low in C_3S.

Swayze also suggested that aeration of cements containing hemihydrate may produce a small amount of gypsum. This gypsum would serve as a seed for precipitation of the $CaSO_4$ as gypsum. If such a conversion, i.e. of a small quantity of hemihydrate to gypsum, did take place during aeration or if a small amount of gypsum was left after milling, Bucchi's experiments would indicate that false set was probable.

3. *Effect of Carbonate*

False set can also be produced by adding to the cement $\frac{1}{4}$–$\frac{1}{2}$% of alkali carbonate [*39*]. Bogue and Lerch propose that when aeration produces false set it may have done so by the formation of alkali carbonate. Manabe [*36*] describes experiments that show that aeration produced false set only when CO_2 was present in the atmosphere, and suggests that the ability of C_3A to react with SO_3 is retarded by CO_2. Support for this theory is found in work by Seligmann and Greenin g [*40*] who showed that the addition of calcite to cement paste caused the precipitation of the calcium aluminate carbonate hydrate, $3CaO.Al_2O_3.CaCO_3.xH_2O$. They suggest that a coating of $C_3A.CaCO_3.xH_2O$ on C_3A would be less reactive to SO_4^{2-} in solution.

II. Setting Reactions

A. Rates of Reactions in Plastic Pastes

1. The Rate of Heat Liberation

The determination of the rates of the chemical reactions in plastic pastes has helped in understanding the nature of the chemical reactions that are occurring and gives an excellent criterion for determining the amount of gypsum to be added to clinker in order to obtain a hardened paste with the best physical characteristics. The rate and extent of hydration of cement are easily followed by measuring the heat evolved. Various methods of measurement of heat have been used, depending upon the purpose of the investigation.

Where the primary interest is in the temperature rise of concrete in massive structures, large adiabatic calorimeters have been used. The information desired is thereby obtained directly without assumptions as to the heat capacity of the components of the concrete or the effect of the thermal history on the temperature produced.

Where the primary interest concerns the hydration reactions in the later stages, the heat of solution method, as developed by Woods, Steinour and Starke [*41*], is most useful. During this period the rate of heat evolution is too low to permit the use of either the conventional adiabatic or conduction calorimeters. Although the measurement is more laborious, the method can be made to give very precise results. Results by this method will be discussed later.

Where the primary interest concerns the early stages of the hydration reactions of cement, cement pastes in conduction calorimeters have been used [*42*]. Such devices permit the continuous and accurate recording of rate of liberation of heat from an age of about $\frac{1}{2}$ h–3 days of hydration. The technique is usually supplemented by a simple "bottle" calorimeter for measurements prior to $\frac{1}{2}$ h. The magnitude, number and time of appearance of heat liberation "peaks" that are observed during this early period, together with an identification of the predominant chemical reactions causing each peak, assist in the understanding of the various chemical reactions taking place and their effect on the properties of pastes and concrete.

(a) *The "immediate" heat of reaction* The hydration reactions of all cements undergo at least two, and sometimes three, cycles of increasing and decreasing rates [43, 44]. The first cycle is that which occurs immediately after cement is mixed with water; the rate of heat liberation increases to a very high value within the first 5 min and then decreases rapidly to a low value. It is during this cycle that the reactions discussed above occur; if the cement is not properly retarded, flash set will occur; if the cement is properly retarded, the "peak" is not quite as high as it otherwise would have been and the paste remains plastic. This initial cycle is then followed by a period of 1 or 2 h when the rate of heat liberation is relatively low and the plasticity of the paste does not change appreciably. The heat liberated during the first cycle is determined by use of the "bottle" calorimeter.

(b) *The second cycle* After a period of time that may extend from about 1 h to as much as 3 h, depending upon the composition of the cement, the rate of heat liberation starts to increase, passes through a maximum at 6–8 h after mixing and then decreases slowly. Although the time of initial set and time of final set are defined by practical tests without theoretical considerations, it so happens that in a properly retarded cement (to be defined later) the development of this second peak in the rate of heat liberation occurs at about the time of final set. The time of initial set is usually on the low side of the peak when the rate of heat liberation is increasing. The time of final set usually occurs shortly after the maximum in the rate of heat liberation. It would then appear as if the chemical reactions that are predominant during the second cycle are those mainly responsible for hardening the paste. Two reactions are dominant during this time: (1) the reaction between C_3A and $CaSO_4$ to form a calcium aluminate sulphate hydrate and (2) the hydration of alite present in the cement.

The reaction of C_3A proceeds at a slow rate as long as the solution is saturated with calcium hydroxide and gypsum. The initiation of this reaction caused the decrease in rate after the initial cycle, so one would not expect it to produce the second peak. It is true that, if the gypsum were all consumed at this time, a rapid reaction of C_3A could occur and produce a peak. It will be seen later that sometimes this happens. But the time of appearance of the second peak is not very sensitive to the amount of gypsum in the cement and the peak is still present when unreacted gypsum is present, so it appears that it is caused by the hydration of alite.

That alite can cause such a heat liberation peak can be inferred from the heat liberation curve for the hydration of C_3S both with and without gypsum [43]. Electron micrographic examination of hydrating Portland cement pastes [*45*] shows that at about the time of initial set the alite has hydrated sufficiently that laths of tobermorite gel can begin to interlock. Plate 1 of Chapter 10 shows an electron micrograph of a paste hydrated for 6 h—a little longer than the time of final set for the particular cement used. The development of tobermorite gel on cement particles which is evident in this micrograph suggests that the final set is indeed caused by interlocking of tobermorite gel particles. Thus, it is logical to associate the second peak in the heat liberation curve with the hydration of alite. The decrease in rate of heat liberation following the peak is the result of two effects: (1) The surface area of unhydrated cement particles decreases as the smaller particles become completely hydrated and the larger particles become smaller, and (2) a layer of tobermorite gel forms on the surfaces of the particles, acting the same as the protective coating of calcium aluminate sulphate hydrate formed in the earlier stages of reaction.

2. *Effect of Gypsum upon the Rate of Heat Liberation*

The effect of gypsum upon the rate of heat liberation depends upon the amount of C_3A present in the cement. In the discussion given above it was assumed that there was sufficient gypsum in the cement to retard

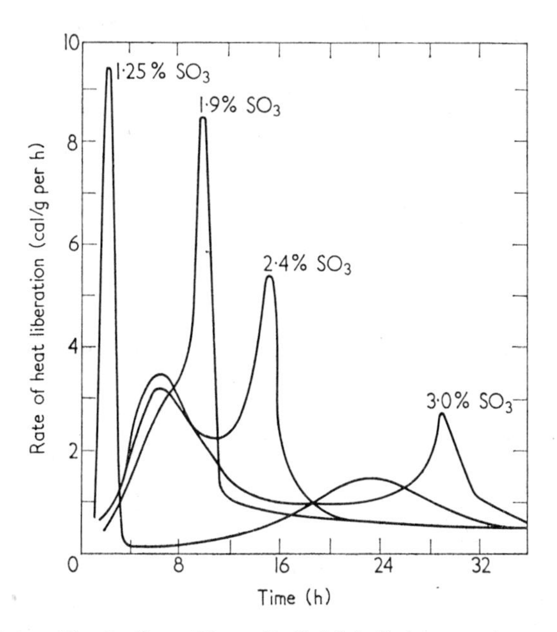

FIG. 5. Rate of hydration of low-alkali, high-C_3A cement as a function of SO_3 content.

properly the initial reactions in the paste. It is apparent from the mechanism of the retarding action that there will be a third peak in the curve if there is an appreciable quantity of unhydrated C_3A present when all of the gypsum has reacted. The curves in Fig. 5, obtained by Lerch [*44*], show that such is the case for cements high in C_3A. The peak for the immediate reaction is not shown in the figure. The time at which the SO_3 was used up is indicated by the rapid rise in heat liberation at 2, 10, 15 and 29 h, for gypsum levels corresponding to 1·25, 1·9, 2·4 and 3·0% SO_3, respectively. The peaks become progressively lower and broader as the percentage of gypsum is increased.

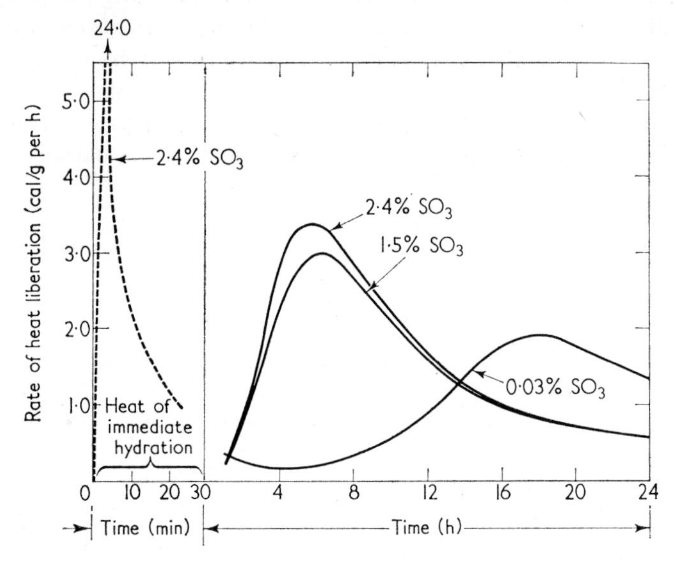

FIG. 6. Rate of hydration of low-alkali, low-C_3A cement as a function of SO_3 content (Fletcher).

It is well known that the formation of the trisulphate in mortar is accompanied by expansion of the mortar. Lerch made expansion measurements on mortar prisms made from these cements and was able to confirm that his deductions from the heat curves concerning the times of disappearance of SO_4^{2-} were correct.

In pastes made from cements low in C_3A the situation is somewhat different. The curves in Fig. 6 show that there was no evidence of a shortage of gypsum in any case and that the rate of heat liberation curves showed only two peaks. The addition of 1·5% gypsum accelerated the rate of hydration; the addition of another 1% gypsum had very little effect. Gypsum has also been found to be an accelerator for pure C_3S.

B. Optimum Gypsum

1. Gypsum Requirement for Proper Retardation

The term "properly retarded" has been used several times in the foregoing discussions although the term has not yet been defined in a precise manner.

In the discussion of Forsén's Class II retarders it was implied that any system that contained sufficient retarder to be on the plateau for time of set was "properly retarded". Such a system is properly retarded as far as control of flash set is concerned, but there are other factors too that must be considered. Many physical properties of hardened paste are affected by the quantity of gypsum that is present in the cement. Three of the most important of these are: (a) compressive strength, (b) contraction on drying and (c) delayed expansion.

(a) *Compressive strength* The compressive strengths of mortars, made from the same clinker with various amounts of gypsum added, were found to increase with increasing gypsum content, then pass through a maximum. The quantity of gypsum necessary to produce the highest strength depends upon the composition of the clinker. The effect of gypsum upon strength is pronounced at all ages, so it cannot be accounted for simply on the basis of an increased degree of hydration that occurs at shorter curing times. It seems probable that the beneficial effects are the result of an influence on the structure of the paste.

(b) *Contraction on drying* The addition of gypsum causes a decrease in contraction of mortar prisms to a minimum value, and the amount of gypsum required to produce the minimum contraction depends upon the composition of the clinker. The contractions of the different cements are very nearly equalized by the use of the proper amounts of gypsum.

(c) *Delayed expansion* The effect of excess gypsum in cement is to cause a delayed expansion of mortar or concrete. It is probably for this reason that limits have been placed on the maximum amount of SO_3 that a cement can contain. Lerch found that only a few specimens with optimum gypsum content developed larger expansions than were found with the maximum permissible (A.S.T.M.) SO_3 content when the gypsum was increased to the value giving maximum strength or minimum contraction.

(d) *Heat liberation curve* A fourth method of test for gypsum requirement is the rate of heat liberation curve discussed above. Lerch, in fact, defined the term "properly retarded" in terms of the rate of heat liberation curve: "a properly retarded cement can be considered as one which contains the minimum quantity of gypsum required to give a curve that shows two cycles of ascending and descending rates of heat liberation and that shows no appreciable change with larger additions of gypsum during the first 30 h of hydration". The quantity of gypsum determined in this way is essentially the same as the quantity of gypsum required for maximum strength and minimum shrinkage, and it avoids abnormal expansion. It can thus be called the "optimum".

2. *Factors Affecting Optimum Gypsum*

The influence of various clinker properties upon the optimum gypsum content can readily be seen by examining the rate of heat liberation curves. The specific surface of the cement and the alkali content of the clinker, as well as the C_3A content of the clinker, affect the amount of gypsum required for proper retardation.

(a) *Specific surface of cement* With clinkers high, or moderately high, in C_3A an increase in the specific surface increases the amount of gypsum required for proper retardation. The maximum rate of heat liberation becomes higher and, if sufficient gypsum is added to the finer clinker to provide proper retardation, the maximum rate of heat liberation is increased in proportion to the specific surface. With clinkers of low C_3A content no deficiency of gypsum is caused by finer grinding. The maximum rate of heat liberation is increased in proportion to the specific surface.

(b) *Alkali content* The effects produced by higher alkali contents are quite complex and not fully understood. Three characteristic differences between the action of high alkali cements and low alkali cements may be stated.

(1) At some of the lower gypsum contents the high-alkali cement may give a heat liberation curve similar to that of a properly retarded cement, but an increase in the amount of gypsum will show that it is not properly retarded. As more gypsum is added, the maximum rate of heat liberation increases, i.e. the rate of hydration of the cement increases. A third peak may form, but, as the gypsum content is increased, the curve of a properly retarded cement will be obtained.
(2) Gypsum reacts more rapidly with cements of higher alkali content.
(3) Higher gypsum contents are required for proper retardation of the higher-alkali cements.

With cements of low C_3A content there is a difference between the effect of high Na_2O and that of high K_2O. With high Na_2O and low K_2O

the rate of hydration of the cement increases with increasing gypsum content and a third peak develops in the rate of heat liberation curve. As more gypsum is added, the third peak disappears and the curve of a properly retarded cement is obtained.

With high K_2O and low Na_2O the rate of hydration of the cement increases with increasing gypsum content but the third peak does not appear. Eventually the second peak of the rate of heat liberation passes through a maximum as the optimum gypsum content is reached. The cement high in Na_2O required a larger quantity of gypsum for proper retardation than did the cement high in K_2O. It is probable that the explanation lies in the formation of NC_8A_3 and $KC_{23}S_{12}$ by the alkalies. It will be seen later that these compounds have about the same reactivities as C_3A and β-C_2S, respectively. Consequently, Na_2O from NC_8A_3 is available immediately, whereas K_2O will be made available more slowly.

III. Principal Reactions in Hardened Pastes

A. Water Content

1. States of Water in Hardened Pastes

The hydration products of Portland cement in hardened paste comprise a poorly crystallized, porous, tobermorite gel in which are embedded several more or less well crystallized hydrates and unhydrated cement particles. The water in the saturated paste exists in three states [*46*]: (1) chemically combined, (2) physically adsorbed on gel surfaces and (3) in space outside the range of surface forces. Although the water may be classified in this manner, it is not easy to determine how much water is present in each state. It is practically impossible to separate the hydration products one from the others for individual determinations of water content by chemical methods. A separation on the basis of energy, or vapour pressures, is not completely satisfactory because the energy with which water is held in physical adsorption covers such a large range that many hydrates lose their water of crystallization simultaneously with the removal of adsorbed water. Thus, there is no way to measure accurately the amount of water combined chemically with cement, although it is possible to measure precisely a quantity of water which is nearly, though not exactly, proportional to the degree of hydration of the cement.

2. Techniques for Measuring Water Content

For almost all investigations it is convenient to dry samples. The state to which the samples are dried must be chosen so that it is compatible with treatments to follow, e.g. if one is interested in determining whether ettringite is present in a paste by X-ray diffraction, the sample might be dried over saturated $CaCl_2$ solution, but it could not be dried over P_2O_5 because the ettringite would be dehydrated and none of its diffraction lines would be found.

(a) *Estimate of combined water or degree of hydration* Either of two methods are generally used to estimate the amount of water combined with cement: (1) oven drying at 100–110° C or (2) drying at room temperature over a desiccant. These procedures frequently are followed by determining the loss on ignition. The oven drying procedure is the same as is conventionally used in analytic techniques, but extreme care should be taken to protect the samples from CO_2 during the drying process. Drying at room temperature in an evacuated desiccator gives comparable results if a suitable desiccant is chosen, and provides adequate protection from carbonation. In determining the non-evaporable water [*47*] the vacuum desiccator is connected to a mechanical pump through a dry-ice trap [*47*]. The pump should be capable of maintaining a pressure less than 10 μ in the desiccator if a reasonable rate of drying is to be maintained.

(b) *Thermogravimetry* Considerably more information can be obtained from thermogravimetry. The most convenient method of measurement is by means of a thermogravimetric balance. A balance operating in a sealed or evacuated system is to be preferred to one operating in the atmosphere because of rapid carbonation of the hydrated cement. Carbonation can be eliminated by providing a CO_2-free atmosphere. The nitrogen available commercially is usually not sufficiently free of CO_2 to be used without further purification. Results can be interpreted most readily if the partial pressure of water vapour around the sample is controlled at a fixed value and the rate of temperature rise is sufficiently low to allow the sample to remain at equilibrium with the water vapour. Dehydration isobars can be obtained in this manner that are similar to those of Taylor [*48*] obtained by a more laborious procedure. His results [*49*] (Fig. 7) show a continuous and rapid loss of water below 100° C. In this region the capillary water, most of the adsorbed water and some of the water of crystallization of calcium aluminate hydrates and ettringite are lost. The continuous loss of water progresses with increasing temperature at a lower rate until almost 400° when the crystalline $Ca(OH)_2$ loses its water over a narrow temperature range. At higher temperatures there is a further slow and continuous loss in weight up to 800° C. The continuous character of the isobars, except for the step at 400° C where $Ca(OH)_2$ is dehydrated, seems to indicate that the proportion of water given up by crystalline hydrates is small or that the hydrate crystals are small and hetero-disperse.

(c) *Nuclear magnetic resonance* Broad-band nuclear-magnetic-resonance tuned to the proton frequency (NMR) has also been used to study the state of water in hardened Portland cement pastes [*50, 51*]. The reaction between cement and water causes a decrease in the intensity of the absorption peak associated with free water. Simultaneously a strong absorption near the resonance usually associated with adsorbed water together with absorption in the region associated with hydrate water appears. This latter resonance peak is strongly decreased by carbonation of the paste. Estimates of the chemically combined water from NMR agree only roughly with estimates of the combined water

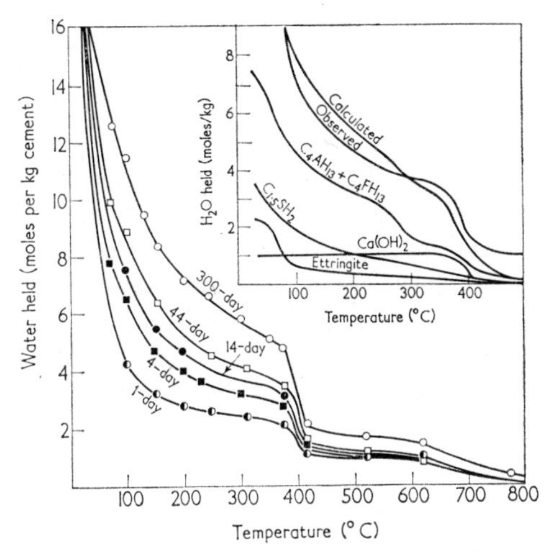

FIG. 7. Dehydration isobars for Portland cement pastes cured for various lengths of time. Inset: Observed and calculated for the 14-day specimen. CO_2-contents in moles per kg cement: 300-day 0·7, 44-day 0·6, 14-day 0·6, 7-day, 0·8, 1-day 0·1.

obtained by oven drying, but the NMR results depend upon estimates of the intensity of an absorption peak that is not at all well resolved by present techniques.

B. Hydration Products

The identification of the various hydration products of Portland cement is not always simple. The method that gives the most definite information is X-ray diffraction, and much of the discussion to follow will be based upon results of that method.

Infra-red absorption spectra have been determined for the pure compounds comprising cement and for many of the hydration products

[52, 53]. While no new conclusions regarding hydration reactions of cement have resulted as yet, evidence supporting conclusions of other work has been reported. The method promises to be a valuable tool when more information concerning infra-red absorption spectra of hydration compounds is obtained. It is possible that the hydration reactions of the individual phases in Portland cement eventually may be followed by this technique.

1. The Silicate Phases

The X-ray diffraction patterns of hydrated Portland cement pastes (Fig. 8) have certain characteristics in common with those of hydrated pastes of C_3S and C_2S [*49*, *54–56*]. In these latter pastes all the lines of $Ca(OH)_2$ as well as three diffraction lines at 3·05, 2·82 and 1·83 Å are found. These correspond to the strong 220, 400 and 040 lines of the tobermorites [*57*].

The lines of $Ca(OH)_2$ and the same three lines ascribable to calcium silicate hydrate in the pastes of the pure silicate compounds also appear in cement paste patterns. The characteristics of these lines in cement paste patterns are much the same as in the calcium silicate paste pattern. As can be seen in Fig. 8, there is a broad hump with a maximum in the vicinity of 3·05 Å. This peak is quite asymmetric, just as in the cases of C_3S and C_2S, falling off relatively sharply on the low angle side and much

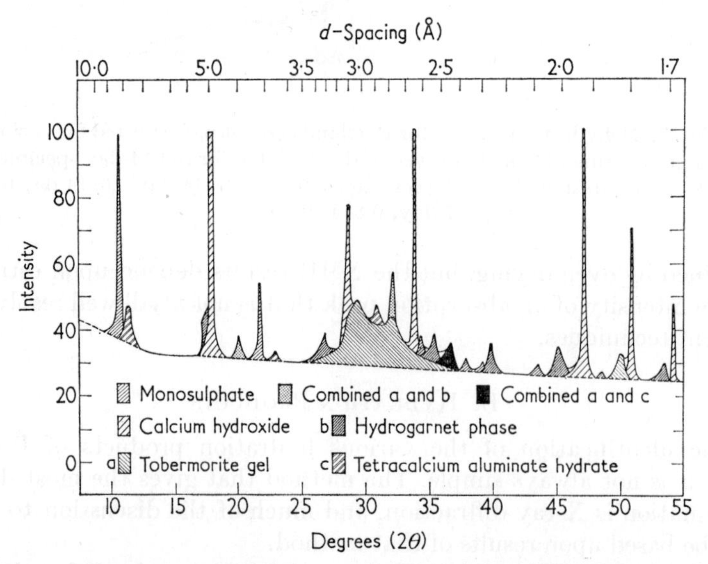

FIG. 8. Schematic X-ray diffraction pattern of normal Portland cement paste (Cu K_α radiation).

more gradually on the high angle side. In patterns obtained from saturated slices of hardened paste, in which some carbonation has occurred as a result of exposure to the atmosphere, the principal calcite line, at 3·03 Å, is superimposed on this peak.

The line found at 2·82 Å in calcium silicate hydrate patterns appears shifted to 2·78 Å in cement paste patterns. This shift may be due to the presence of lines from other hydration products at about the same spacing. These other hydration products (aluminate sulphate and aluminate hydrates) will be considered in detail in later discussion.

The 1·82 Å line is easily distinguishable. Lines of other possible hydration products may also appear at this spacing, but none would be strong enough to account for the observed intensity.

Since the same calcium silicate hydrate lines which appear in patterns of hydrated pastes of C_3S and C_2S also appear in patterns of hydrated Portland cement pastes, they tend to characterize the hydrate in cement pastes as a tobermorite gel. These X-ray results constitute a set of data too meagre in themselves, however, to give any information as to the composition of this phase.

Other information from DTA and infra-red examinations [*52*] indicates that the tobermorite gel is the same as is produced in the hydration of pure C_3S at room temperature and that it has a high C/S ratio, perhaps as high as 1·77. In addition, the DTA results of other investigators [*2*, *58*] and thermogravimetric results [*49*] confirm the conclusion that the hydrate is a tobermorite gel.

The calcium silicate hydrate which forms in Portland cement pastes probably contains small amounts of impurities, primarily alumina. It was shown by Kalousek [*59*] that aluminium can be substituted for silicon in the tobermorite lattice without the substance losing its original identity. Compositions containing as much as 4–5% Al_2O_3 were prepared hydrothermally. A similar situation may exist in Portland cement pastes during hydration at room temperature. Other substances such as alkali may also enter into the silicate structure [*60*].

2. *The Aluminate and Ferrite Phases*

As has been indicated, tobermorite gel and calcium hydroxide are not the only hydration products of Portland cement. It can be seen in Fig. 8 that many of the observed diffraction peaks are not accounted for by these substances. The most prominent sets of peaks other than those discussed previously are those of the calcium aluminate sulphate hydrates. X-Ray diffraction lines of both sulphates have been observed in patterns of hydrated pastes, sometimes separately and sometimes

Table II. *X-Ray* d-*spacings—hydrated cement wet slab and re-wetted sample data*

Specimen	C–88	C–54	C–87	C–75	C–75*	C–66–K	C–66–O	C–61–C	C–33–1	C–33–1*	C–31–1	C–32–1
Cement	15754	15754	15754	15754	15754	15754	15754	LTS–17	15622	15622	15497	15669B
Type	I	I	I	I	I	I	I	I	II	II	III	IV
Nominal w/c	0·40	0·55	0·40	0·65	0·65	0·70	0·57	0·70	0·60	0·60	0·60	0·60
Age	11 yr	11 yr	7 yr	7 yr	7 yr	11 mo	6 mo	1 yr	10 yr	10 yr	10 yr	10 yr
Spacings (Å)							9·7	9·8	9·8	9·8	9·7	9·8
		8·9		8·9	9·0	9·0		9·0		8·8	9·0	
							8·7	8·6			8·3	
	7·8	7·8	7·8	7·7	7·8	7·8	7·8	7·8	7·8	7·8	7·8	7·8
								7·4	7·4	7·4		
							5·66	5·68	5·64	5·66	5·64	5·64
	5·09	5·06	5·09	5·09		5·09	5·11	5·08	5·06		5·08	
	4·94	4·92	4·93	4·93	4·93	4·94	4·92	4·94	4·93	4·94	4·92	4·92
									4·72	4·72	4·72	4·72
				4·49	4·48			4·48				
	4·41	4·38	4·43	4·39	4·40	4·41	4·39	4·41	4·39	4·40	4·41	
		4·27	4·27		4·28		4·27		4·26		4·26	4·29
	4·00	4·00	4·03	4·02	4·02			4·02	4·04		4·02	4·02
	3·88	3·86	3·86	3·87	3·86	3·88	3·88	3·88	3·88	3·88	3·88	3·88
	3·79	3·78	3·77	3·77		3·78		3·81			3·81	
					3·67				3·67	3·67		3·69
						3·50		3·50	3·48	3·48	3·49	3·49
		3·43		3·45	3·42		3·45					
	3·35	3·35	3·34	3·35			3·34	3·36	3·35	3·38	3·35	3·36
	3·31	3·30		3·31	33·2	3·32		3·31		3·32	3·31	
												3·26
	3·12	3·11	3·12	3·12	3·12	3·12	3·12	3·12	3·12	3·12	3·12	3·12
	3·05	3·04	3·05	3·05	3·04	3·05	3·05	3·05	3·05	3·05	3·05	3·05
		2·97			2·97		2·98					
	2·88	2·88	2·88	2·88	2·89	2·89	2·88	2·89			2·89	
				2·84		2·83						2·84
	2·78	2·77	2·78	2·78	2·78	2·78	2·79	2·78	2·78	2·78	2·77	2·78
						2·69	2·69		2·67	2·69	2·70	
	2·63	2·63	2·63	2·63	2·63	2·63	2·63	2·63	2·63	2·63	2·63	2·63
	2·60	2·58	2·59	2·59	2·58	2·59	2·59	2·59	2·57	2·58	2·56	2·56
	2·53	2·52	2·54	2·53	2·54		2·54	2·54	2·54	2·53	2·53	
	2·50	2·50	2·50	2·49		2·49	2·49		2·51		2·50	2·51
	2·47	2·45	2·46	2·47	2·46		2·46	2·46	2·46		2·45	2·45
	2·43	2·42	2·43	2·43	2·42	2·43	2·43	2·43	2·42	2·44	2·43	2·41
	2·37		2·37	2·36	2·36		2·36	2·37		2·35	2·36	
		2·31	2·31									
	2·29	2·29	2·29	2·29	2·29	2·29	2·29	2·30	2·29	2·29	2·29	2·29
	2·27	2·26	2·27	2·27	2·26	2·27	2·26	2·27	2·27	2·27	2·26	
			2·19	2·19	2·20	2·19	2·20	2·20	2·21	2·21	2·21	2·21
	2·14			2·14	2·14	2·14	2·14	2·14	2·15	2·16	2·14	2·16
	2·10	2·10	2·10	2·10	2·10	2·11	2·10	2·10	2·09	2·10	2·10	2·10
	2·07	2·06	2·06		2·07	2·07		2·04	2·06	2·06	2·06	2·06
	2·01	2·01	2·02	2·01	2·01	2·01	2·02	2·01	2·01	2·01	2·01	2·01
	1·96	1·96	1·97		1·96	1·96	1·96	1·96			1·96	1·97
	1·932	1·929	1·931	1·931	1·930	1·932	1·930	1·931	1·930	1·931	1·930	1·932
	1·912	1·912	1·912	1·912			1·912		1·912		1·912	1·912
	1·881	1·875	1·879	1·879	1·875		1·879	1·881	1·879		1·875	1·879
	1·826	1·819	1·824	1·824	1·823	1·827	1·823	1·828	1·823	1·824	1·823	1·826
	1·799	1·796	1·799	1·799	1·798	1·799	1·798	1·799	1·797	1·798	1·797	1·801
	1·722	1·714	1·723	1·719	1·717	1·719	1·723	1·723	1·722	1·720	1·717	
	1·690	1·688	1·689	1·690	1·688	1·690	1·688	1·690	1·689	1·689	1·688	1·689
			1·665				1·662			1·663	1·667	1·665
	1·660	1·658	1·660	1·660	1·658	1·662	1·652	1·661	1·657	1·656	1·654	
	1·639	1·635	1·638	1·638	1·638	1·635	1·638	1·639	1·636	1·638	1·638	1·634
			1·609				1·607					

TABLE III. *X-Ray* d-*spacings—hydrated cement dried powder data*

Specimen	C–75	C–49	C–61–C	C–33–1	C–33	C–31–1	C–31	C–32–1	C–32
Cement	15754	15754	LTS-17	15622	15622	15497	15497	15669B	15669B
Type	I	I	I	II	II	III	III	IV	IV
Nominal w/c	0·65	0·65	0·70	0·60	0·60	0·60	0·60	0·60	0·60
Age	7 yr	5 yr	1 yr	10 yr	7 yr	10 yr	7 yr	10 yr	7 yr
Spacings (Å)	8·2	8·2	8·2			8·2	8·2		
	7·7	7·9	7·7	7·8	7·8	7·7			
			7·3	7·4	7·3				
	5·47	5·48	5·45	5·47	5·43	5·45		5·47	5·48
	5·06			5·06	5·06	5·07	5·06		
	4·92	4·95	4·91	4·92	4·92	4·92	4·92	4·92	4·92
	4·41	4·41	4·41	4·41	4·39	4·39	4·39	4·40	4·39
	4·09	4·10	4·07			4·08	4·08		
					4·00				
	3·89	3·92		3·86	3·89	3·86			3·92
	3·62				3·66				
	3·31	3·32	3·31	3·32	3·35	3·31	3·30		3·30
							3·21		3·18
	3·12	3·12	3·12	3·12	3·12	3·12	3·12	3·12	3·11
	3·04	3·05	3·03	3·05	3·04	3·04	3·05	3·03	3·03
	2·92					2·92			
	2·88	2·88	2·88			2·86	2·86		2·87
	2·78	2·78	2·78	2·78	2·77	2·77	2·76	2·79	2·78
		2·70	2·71	2·67	2·68				
	2·63	2·63	2·63	2·63	2·63	2·63	2·63	2·63	2·63
		2·60	2·59	2·60	2·58	2·55	2·58		
	2·53	2·53	2·53	2·53	2·53	2·53	2·53	2·53	
	2·49		2·48						
	2·46	2·46		2·46	2·45	2·46	2·47		2·45
	2·43	2·43	2·43	2·44		2·43	2·42		
	2·35	2·35	2·35			2·35	2·35		
	2·31	2·30		2·30			2·30	2·30	
	2·26	2·26	2·26	2·26	2·26	2·26	2·28		
	2·08	2·10					2·08		
		2·05		2·06	2·05				2·05
	2·01	2·01	2·01	2·01	2·01	2·01	2·00	2·01	2·02
	1·93	1·93	1·93	1·93	1·93	1·93	1·93	1·93	1·93
	1·88	1·89							1·89
	1·82	1·82	1·83	1·82	1·82	1·82	1·82	1·82	1·82
	1·80	1·80	1·80	1·80	1·80	1·80	1·80	1·80	1·80
	1·72	1·72	1·72	1·72	1·72	1·72	1·71		1·71
	1·69	1·69	1·69	1·69	1·69	1·69	1·69	1·69	1·69
	1·65	1·65	1·65	1·65	1·65	1·65	1·65	1·65	1·66
	1·64	1·64	1·64	1·64	1·64	1·64	1·64		1·64
	1·61	1·60			1·62				

simultaneously. Table II lists the X-ray diffraction *d*-spacings observed with saturated slices of several cement pastes. Table III lists *d*-spacings observed with dried ground cement pastes [*54*].

The X-ray diffraction pattern of a hydrated cement paste reflects the complexity of the mixture of hydration products. A large proportion of the diffraction lines are subject to overlapping. Hence, in identifying the lines, it is important to know what other substances may have interfering lines.

The two strongest ettringite lines, at 9·73 and 5·61 Å [*61*], occur in positions not interfered with by other possible products of cement hydration. It can be seen in Table II that, whenever the 9·73 Å line appears in the pattern of a hydrated cement, the 5·61 Å line also appears. In most cement paste patterns in which these lines appear they are weak, and the other ettringite lines are proportionately weaker and are obscured by normal background variation. However, in those patterns in which the 9·73 Å line is more intense some of the weaker lines, for example, those at 4·69 and 3·48, can be distinguished. The rather strong line occurring at 3·88 Å has the same spacing as a calcite line. However, this calcite line is weak relative to the principal calcite line at 3·03 Å, and the 3·88 Å line found in all cements showing the 9·71 and 5·61 Å lines is easily seen to be too strong to be accounted for by calcite alone.

As can be seen from Table II, ettringite lines are found in patterns of many pastes cured for long periods of time. The data for 10-year-old pastes made from cements of normal and low C_3A content indicate the presence of a significant quantity of ettringite. These observations agree with those of Turriziani [*58*], who found ettringite in pastes made from four different cement compositions and hydrated at two different water/cement ratios, at ages from 7 days to 6 years. If the cement is high in C_3A, ettringite is less likely to be found. Thus, in cement 15497 in Table II, a relatively small amount of ettringite is indicated by the X-ray diffraction pattern. In another such case, cement 15754, no ettringite is observed except in the youngest of the pastes of this cement listed in Table II, that hydrated for only 6 months. These results indicate that cement 15754 behaves in the manner described by Taylor [*49*], in that ettringite forms at early ages and then disappears. Taylor suggested that the disappearance of ettringite may be due to the formation of calcium aluminate monosulphate hydrate which then dissolves in the C_4AH_{19} phase to form a solid solution.

This suggestion of Taylor seems to be confirmed by DTA experiments of Greene [*2*] who found that ettringite is always formed as long as gypsum is present but that, if there is insufficient gypsum in the cement to react with all the C_3A to form ettringite, then either the monosulphate or a solid solution begins to form and the amount of ettringite decreases.

Considerably less success has been had in years past in identifying the monosulphate in Portland cement pastes by X-ray diffraction techniques. Turriziani [*58*] was unable to find evidence for the monosulphate in any of his specimens. He concluded that, if it does form, it does so only at a very slow rate. Diffraction patterns for the compound have been reported by Midgley [*62, 63*] and Fratini, Schippa and Turriziani [*64*]. The results are in reasonably good agreement. The principal diffraction peaks for this material are at 8·9 and 4·45 Å. The 8·9 Å line has been observed in certain of the wet slab patterns, the data for which are given in Table II. The 4·45 Å line appears not only in every one of these but also in patterns in which the 8·9 Å line does not appear. As will be seen later, a strong line from another phase also appears at 4·4 Å. The other lines of the monosulphate (or solid solution with C_4AH_{19}) are either too weak to be detectable, since the line at 8·9 Å is not very strong, or else other substances have lines appearing in the same locations, such that assignment of the diffraction maximum to a particular substance would be questionable.

The changes that occur in the X-ray diffraction patterns of the calcium aluminate sulphate hydrates as these materials are dried provides useful information for interpretation of the cement paste X-ray diffraction data. In the case of ettringite, drying results in the disappearance of the diffraction pattern. As can be seen in Table III, the 9·73 and 5·71 Å lines are absent in the patterns of dried samples of the same pastes which, when saturated, gave patterns which contained those lines.

In the case of the monosulphate the behaviour is different. The basal reflections shift to lower *d*-spacings, while the other lines show no significant change [*10*]. When the material is dried in CO_2-free air at 115° or *in vacuo* at room temperature to the $\frac{1}{2}$ μ level, the 8·9 Å line, assigned 0001 by Fratini *et al.*, contracts to 8·2 Å and the line at 4·45 Å, assigned 0002, contracts to 4·1 Å [*34*]. The 8·2 Å line appears in patterns of dried specimens of only those pastes, the wet slab patterns of which contained the 8·9 Å line. Likewise, the 4·15 Å line in patterns of dried specimens occurs only when the 8·2 Å line does. That the line at 4·4 Å does not disappear completely in patterns of dried specimens is due to the fact that another material, not affected by drying, contributes to it.

The lines of the monosulphate phase appearing in patterns of dried pastes are strong and sharp, as is illustrated by Fig. 8, indicating the presence in the paste of a significant amount of this material. This is true even for those pastes whose wet slab patterns give lines for the monosulphate phase which are very weak. The weakness of these lines in wet slab patterns may be due to carbonation of the surface of the slab.

Dried specimens of several of the pastes included in Tables II and III were brought to equilibrium with various relative humidities. No significant changes from the dried sample were observed at humidities below 28%. However, in patterns of samples bearing the monosulphate that were equilibrated at higher humidities the 8·9 Å line appeared. The 8·9 Å line showed only as a shoulder on the 8·2 Å peak at 28·8% relative humidity, was about equal in intensity with the shrinking 8·2 Å peak at 33% relative humidity, and at 42% relative humidity the 8·2 Å peak was a small shoulder on the 8·9 Å peak. At higher humidities the patterns

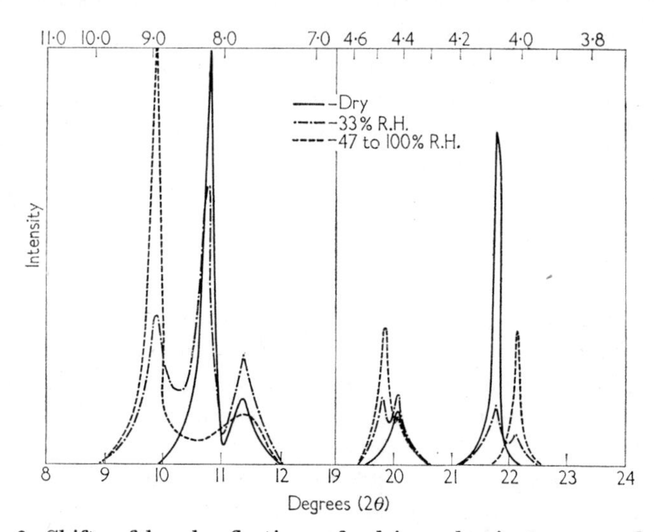

FIG. 9. Shifts of basal reflections of calcium aluminate monosulphate hydrate with changes in humidity.

appeared quite like the wet slab patterns, except that no significant amount of carbonation had occurred. As a result, the monosulphate lines were of intensities comparable to those of the lines in patterns of the dried specimens. That is, they were strong and sharp. Because of this, it was easy to distinguish the 4·45 Å monosulphate line from the line appearing at 4·40 Å due to another phase (see below). This resolution is illustrated in Fig. 9, in which are shown idealized patterns of the 9 Å and 4 Å regions of the pattern of the monosulphate in the dry form, the wet form and an intermediate condition (33% relative humidity) where both forms appear simultaneously. Furthermore, it can be seen from the results in Tables II and III and in Fig. 9 that, while the 4·1 Å line in the pattern of a dried specimen shifts to 4·45 Å in those of wet specimens, the line at 4·0 Å reappears as well when the sample is re-wetted. This corresponds to the line reported at 3·99 Å by Midgley [62, 63] and at 4·02 Å by Fratini *et al.* [64].

The ettringite lines appearing in wet slab patterns at 9·8 and 5·6 Å are absent in the dried powder patterns of these pastes. Re-wetting these pastes at humidities up to 47% has no significant effect on the diffraction pattern. Re-wetting at 100% relative humidity causes the ettringite lines to return. No significant amount of carbonation occurred with these re-wetted samples and yet the ettringite lines had intensities comparable to those in the wet slab patterns wherein considerably more carbonation occurred. Thus, if CO_2 reacts with ettringite, it does so sufficiently slowly so as not to affect the X-ray observations.

The monosulphate phase is found in pastes made from cements high in C_3A but not in those made from cements low in C_3A. On the other hand, ettringite is the only calcium aluminate sulphate hydrate found in pastes made from cements low in C_3A, while little if any of it is found in pastes of cements high in C_3A.

Many investigators in the past have reported the presence of a hexagonal tricalcium aluminate hydrate. Schippa and Turriziani [65] state that it is impossible to form such a compound under conditions such as those that exist in a hydrating cement paste and that the substance observed is actually tetracalcium aluminate carbonate hydrate. On the other hand, Midgley [52] reports the presence of C_4AH_{13} in pastes which he examined, although it appears probable, judging from his reported *d*-spacings, that he actually observed the α-C_4AH_{19} which has since been shown to be a solid solution with the composition $C_3A.\frac{1}{2}Ca(OH)_2.\frac{1}{2}CaCO_3.aq.$ [66, 67]. Greene, however, has found that C_4AH_{19} is formed in the hydration of Portland cement clinker. It is also formed if the SO_4^{2-} is low; about 2% gypsum is necessary to prevent its formation.

One of the cements low in C_3A discussed above is of especial interest in one respect, however. If the calcium aluminate monosulphate hydrate were to have formed instead of ettringite, not enough Al_2O_3 would be present in the cement to react with all the SO_3. Under these circumstances it is possible that a calcium ferrite sulphate hydrate might form. Malquori and Cirilli [68] have prepared calcium ferrite sulphate hydrates, analogous to the calcium aluminate sulphate hydrates which are found in hydrated Portland cement pastes. The X-ray diffraction patterns of these ferrite sulphate hydrates are, in general, very much like those of their aluminium analogues. As a result, the iron compounds, were they present in Portland cement pastes, would not be readily distinguishable from their aluminium counterparts. However, there is evidence that the iron-bearing phase in cement hydrates in a unique manner not involving SO_3.

The calcium aluminate and calcium ferrite hydrates have been reported as existing in hydrated cements in two different forms: cubic and hexagonal. The cubic hydrates are represented by the compounds C_3AH_6 and C_3FH_6; the hexagonal hydrates are typified by the compound C_4AH_{19}. It was shown by earlier investigators such as Wells, Clarke and McMurdie [*69*] that C_3AH_6 is the stable calcium aluminate hydrate in contact with solutions saturated with respect to calcium hydroxide. On the other hand, in none of the discussions of the stoichiometry of the hydration of Portland cement is much evidence given for the presence of the cubic tricalcium aluminate hydrate. Steinour [*70*] concluded that formation of C_4AH_{19} from clinker and water was more probable than formation of C_3AH_6. Taylor [*49*] and Turriziani [*58*] report only a hexagonal tetracalcium aluminate hydrate; they observed no cubic tricalcium aluminate hydrate in the pastes.

Recently, evidence has been presented for the existence of both types of aluminate hydrate in cement paste [*71*]. Furthermore, the cubic phase at least is most likely a solid solution of several substances.

It can be seen from the data in Tables II and III (and in Fig. 8) that a group of lines appears distinctly in patterns of all cement pastes except those of cement 15669B, which is quite low both in Al_2O_3 and in Fe_2O_3, and even in patterns of pastes of this cement there is some evidence of these lines. This group of lines appears at 5·07, 4·40, 3·32, 2·26, 2·01 and 1·72 Å. There are, in addition, four weaker lines which may be included with this group. The data in Table IV show that there is a correlation between these 10 lines and those reported for C_3AH_6 [*62, 63, 72, 73*] and for C_3FH_6. These 10 reflections are characteristic of a body-centred cubic crystal with the same space group as C_3AH_6, *I*a3d, and a cell parameter equal to 12·4 Å. If a cubic lattice of this type is present, then other lines ought to be present. The spacings of these predicted lines, calculated using a lattice parameter of 12·4 Å, are such that lines of other substances also present in the paste would interfere. When these interfering substances are removed, for example, by extraction [*54*], the resulting material gives a pattern in which lines remain at the predicted locations, as do the originally observed 10 spacings.

The value of the lattice parameter, 12·4 Å, indicates that the cubic phase is neither pure C_3AH_6 nor C_3FH_6. These compounds are end-members of the garnet–hydrogarnet solid solution series [*72, 74*], the other end-members of which are C_3AS_3 and C_3FS_3. The various hydrogarnets whose compositions fall within the solid solution range bounded by these four end-members contain silica. The cubic phase found in cement paste has a lattice parameter corresponding to a composition within this range, and is presumably a hydrogarnet.

TABLE IV

d-Spacings reported for tricalcium aluminate and ferrite hydrates and corresponding spacings in some representative hydrated cement pastes

hkl [73]	C_3AH_6 [73]	C_3AH_6 [62, 63]	C_3AH_6 [72]	$C_3AH_{3\cdot 75}$ [98]	C_3FH_6 [73]	C_3FH_6 [72]	C_3FH_2 [72]	C–75	C–33–1
211	5·16 vs	5·14 s	5·13 ms	5·03 vw	5·20 m	5·18 ms	5·07 m	5·09	5·06
220	4·47 m	4·45 m	4·45 m	4·39 vw	4·54 s	4·50 s	4·38 mw	4·39	4·39
321	3·37 m	3·37 m	3·36 mw	3·30 vw	3·40 w	3·41 mw	3·32 w	3·31	3·32
400	3·15 m	3·15 m	3·15 w	3·09 vw	3·19 s	3·19 ms	3·10 ms		
420	2·81 m	2·82 ms	2·81 ms	2·77 vw	2·85 s	2·85 ms	2·78 ms		
332			2·68 vw		2·72 vw	2·72 vw			
422	2·56 vw	2·57 w	2·55 w		2·60 m	2·61 mw	2·53 m	2·53	2·53
510, 431	2·46 w	2·47 mw	2·47 w		2·50 vw	2·51 w	2·44 vw	2·43	2·44
521	2·30 vs	2·30 vs	2·30 ms	2·26 vw	2·33 m	2·33 ms	2·27 w	2·27	2·27
440	2·22 vw	2·23 vw	2·23 w			2·26 w			
611, 532	2·04 s	2·04 s	2·04 s	2·01 vw	2·07 m	2·07 s	2·01 s	2·01	2·01
620	1·99 vw	1·99 vw	1·99 vw		1·99 w	2·02 w	1·96 w	1·96	
444	1·81 vw	1·82 vw	1·82 vw			1·84 vw	1·79 vw		
640	1·74 w	1·75 mw	1·75 ms	1·72 vw	1·77 m	1·77 m	1·72 m	1·72	1·72
633, 552, 721	1·71 w	1·71 w	1·71 w		1·73 w	1·74 vw	1·70 vw		
642	1·68 m	1·68 m	1·68 s	1·66 vw	1·70 vs	1·71 vs	1·66 vs	1·66	1·66
651, 732	1·60 vw	1·60 m			1·62 vw				
800	1·57 w	1·57 vw	1·57 w		1·59 w	1·60 w	1·55 w		
840	1·40 w	1·41 ms			1·43 m				

For additional data for C_3AH_6, see Appendix 1 (Volume 2).

Investigation of the kinetics of hydration of Portland cement [71] has resulted in indications that, as the ferrite phase hydrates, no significant change occurs in the A/F ratio of the remaining unhydrated material. Analyses of a number of cements by X-ray diffraction have led to the result that the A/F ratio of the ferrite phase is in general near unity [75]. If the hydrogarnet forming in cement pastes arises from hydration of the ferrite phase, its A/F ratio too would, in general, be near unity. A hydrogarnet with a lattice parameter 12·4 Å and an A/F ratio unity would have a composition $C_3ASH_4 . C_3FSH_4$ (see Fig. 18 in reference [74]).

There is little evidence to show either a growth or decrease of the hydrogarnet lines as a function of time. The lines appear in the pastes of cement 15754 at all ages reported in Table II. Hence, there is no indication that this hydrate would only be a hydrate of the ferrite phase after long periods of time, or that it would be the only hydration product of the ferrite phase. Whatever amount of hydrogarnet is found would appear to have formed at some early age for the most part.

Several investigators have obtained patterns of tetracalcium ferrite hydrate dried under various conditions. The low angle spacings of this material are in the same regions as those of the tetracalcium aluminate hydrate. It is not known whether the observed lines in cement paste patterns represent only the aluminate hydrate or a solid solution of the two compounds.

C. Kinetics of Hydration

1. *Methods of Study*

Probably the first kinetic studies of the hydration of Portland cement consisted in the determination of the rate of development of strength in mortars and concrete. This knowledge is essential to engineers, but its theoretical value is limited because the strength of mortar or concrete depends on many factors in addition to the extent of hydration of the cement. Although knowledge of the rate and mechanism of hydration of cement is only one step towards understanding the development of strength and many other properties of concrete, it is an important step.

Three general methods have been used in the study of kinetics of hydration of Portland cement: (1) microscopic examination of hardened neat cement pastes after known curing times, (2) observations of changes in chemical and physical properties of pastes as a function of curing time and (3) X-ray diffraction analysis of the unhydrated clinker particles in hardened pastes.

Microscopic studies of thin sections of hardened pastes give very little information about the products of hydration of cement. For the most part, only unhydrated clinker particles can be identified. These are embedded in a gel which has practically no structure visible in a light microscope. All four major phases have been found, even after 20 years' hydration. Some crystallites of a few hydration products have been found, e.g. calcium hydroxide, tetracalcium aluminate hydrate and both calcium aluminate trisulphate and monosulphate hydrates. The lack of structure in the gel has been cited to support Michaëlis' theory of the hardening of Portland cement.

Measurements of several physical and chemical properties of pastes have been made as functions of curing time. These, as well as microscopic examinations, give a sort of average rate of hydration that is of technical importance. The rate of evolution of heat has yielded a great deal of information concerning kinetics.

X-Ray diffraction is about the best technique for measuring the rates of hydration of the individual phases in Portland cement. The experimental error is large at small degrees of hydration when compared with the experimental error of standard chemical analyses but significant results can be obtained.

2. *Kinetics from Microscopic Examination*

Usually, microscopic examinations are made of polished sections or thin sections of hardened pastes. Anderegg and Hubbell [*76*], however, ground hardened pastes finely and examined the particles suspended in an oil having a refractive index of 1·67. It was thus possible to count the particles that were hydrated, for if the particle was hydrated its refractive index was lower than that of the oil and vice versa. After counting thousands of particles they determined the weight fraction of cement that was hydrated in pastes cured for various periods of time. From these results they calculated the depth of penetration of water into cement particles at each age, assuming that the shape of the original cement particles was the same as the shape of the unhydrated residues in the pastes. The results were obtained on a cement with a known particle size distribution. They noted that their A.S.T.M. Type III cement appeared to hydrate no more rapidly than their Type I cement.

Pure compounds were examined also. Pure C_2S hydrated much more slowly than pure C_3S, which hydrated more slowly than pure C_3A. Addition of 10% C_4AF caused both C_3S and C_2S to hydrate much more rapidly than either component of the mixture would hydrate when pure.

Brownmiller [*77*] estimated the rate of penetration of water into cement particles from examinations of polished sections. The results were similar to those reported by Anderegg and Hubbell. Brownmiller concluded that as hydration proceeds there is a uniform decrease in the size of the cement particles, and stated, "There is no microscopic evidence of the channelling of water into the interior of cement particles to selectively hydrate any single major constituent." However, he did observe that different phases at the boundary of a particle hydrate at different rates: C_3S hydrates more rapidly than C_2S, and some interstitial material hydrates very slowly indeed.

Rexford [78], in a microscopic examination of thin sections, confirmed the major conclusion of Brownmiller. Ward [79], however, came to different conclusions from an examination of thin sections. He observed that C_2S and C_4AF were slow to hydrate and that many cement grains were "shattered" during the hydration. He identified the fragments as C_2S. He wrote of an apparent increase in the proportion of C_2S in hardened pastes and suggested that the individual phases react separately, then at a favourable concentration a general precipitation of gel occurs.

These microscopic studies suggested a mechanism of hydration wherein the individual phases in cement do not hydrate independently of each other, and they support the theory of hardening advanced by Michaëlis. The differences observed by Ward can be reconciled with the findings of the others if it is assumed that setting is caused by his "general precipitation of gel".

The composition of the finer part of the cement may not be exactly the same as that of the coarser part of the cement. The finer fraction disappears first—probably during the first day—and thereafter the composition of the residue changes only slowly, if at all. During the initial stages of the reaction an apparent increase in the proportion of C_2S might occur, as Ward observed.

3. X-Ray Diffraction Analysis

The rates of change in physical and chemical properties of Portland cement pastes depend upon the rates of hydration of the components of cement in a complex way, and deductions about the rates of hydration of these compounds depend upon the kind of assumptions made to interpret the results. X-Ray diffraction analysis provides a method whereby rates of reaction of the individual compounds can be determined directly. The first published attempt to use X-ray analysis to estimate rates of hydration directly is that of Taylor [49]. He prepared pastes hydrated for various periods of time and separated the hydrated cement from the unhydrated particles by using a dense liquid. Upon examination of the Debye–Scherrer pattern of the fractions he estimated the weight fractions of C_3S in the unhydrated silicates, given in Table V. Simultaneously, attempts were being made to analyse cement pastes in the laboratories of the Portland Cement Association [71]. Two series of pastes were analysed: the first from a cement of average composition, the second from a cement low in C_3A and alite. The same technique for analysing

TABLE V

Weight fraction of C_3S in silicates of the unhydrated residue of cement in pastes

Days	0	1	4	7	14	44	300
$\frac{C_3S}{C_3S+\beta\text{-}C_2S}$	60	50	30	25	20	< 10	< 5

cement was used for analysing the pastes with one exception: the 511 line of silicon, instead of the 111 line, was used as the internal standard. The advantage of using this line for pastes is twofold: (1) there are no interfering lines from either the cement or its hydration products and

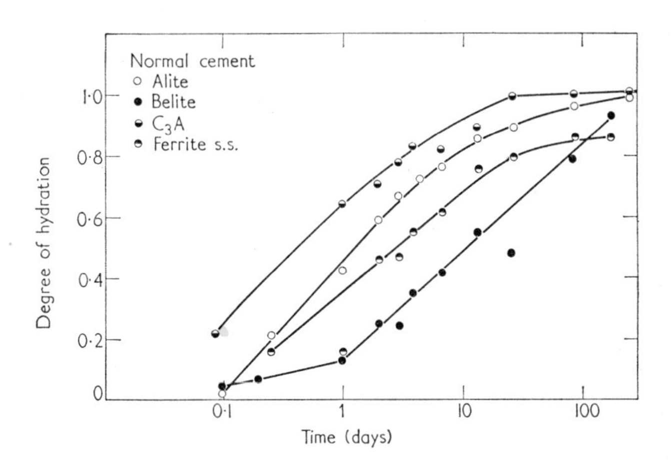

FIG. 10. Degree of hydration of a normal cement as a function of time.

(2) the strength of the line more nearly matches the strength of the cement compound lines in the pastes. The degree of hydration of each component of the cement is shown as a function of time in Figs. 10 and 11. Time is plotted on a log scale for convenience. The results show that in neither cement are the fractional rates of hydration of the components equal to

each other, and that the fractional rates of hydration of alite and the ferrite solid solution phases in the two cements are not equal. The experimental error in the determinations of rates of hydration of C_3A and belite are sufficiently large to make quantitative comparison difficult.

The order of the rates of hydration of the compounds is not the same in cements as in the pure compounds. The degree of hydration of alite increases more rapidly in the cement low in alite than in the normal cement, but the rate of hydration of alite is higher in the normal cement. A similar situation exists with belite, the actual rate of hydration of belite being greater in the cement containing the larger percentage of belite,

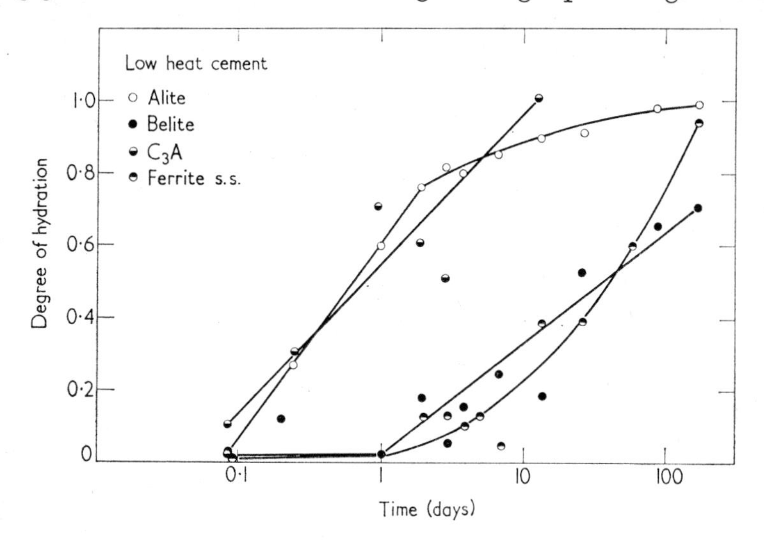

Fig. 11. Degree of hydration of a low-heat cement as a function of time.

although the degree of hydration of belite is larger in the cement containing the lesser amount of belite.

No significant conclusion concerning the comparative rates of hydration of C_3A in the two cements can be made because there is only 1% C_3A in the low-heat cement and the results are not precise. The actual rate of hydration of C_3A in the normal cement is lower than that of alite, probably because the percentage of C_3A is so low. The absolute rate of hydration of the ferrite phase, as well as the fractional rate of hydration is the lowest of all components in both cements. The position of the 141 line of the ferrite phase did not shift as hydration proceeded in either cement, indicating that the ferrite phases did not change in composition during hydration of the cement. Yamaguchi *et al.* [*80*] used X-ray diffraction analysis to study the rate of hydration of cement compounds and of Portland cement. They were able to measure degrees of hydration of the individual phases in Portland cement at curing times as short as 3 min. The order of reactivity of the compounds during the first 5 h of hydration is $C_3A > C_4AF >$ alite > belite; this is the order of reactivity of the pure compounds. They found that gypsum acts to retard the rate of hydration of C_3A but that this action needed the coexistence of $Ca(OH)_2$ from the hydration of alite. After the first day the order of reactivity was different: alite $> C_3A > C_4AF >$ belite. Differences in composition of the cement altered the rates of hydration but did not affect greatly the order of reactivity of the phases. The agreement between these two independent studies is quite satisfactory considering the complexity of the analysis.

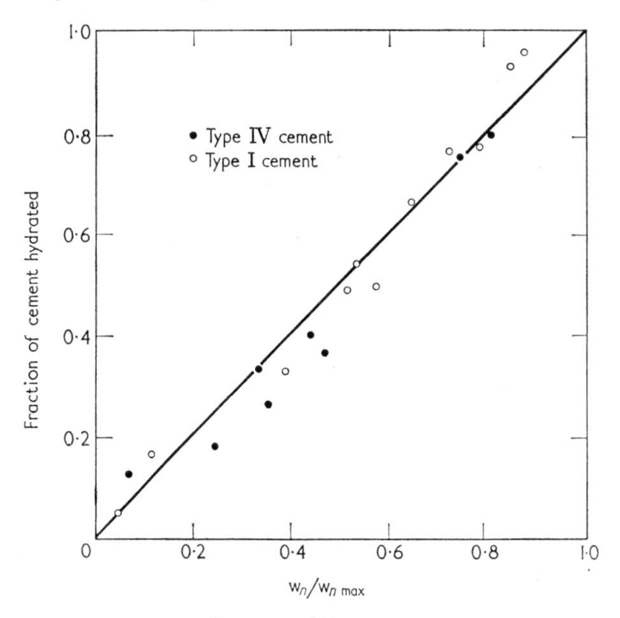

FIG. 12. Relationship between the degree of hydration of cement and ($w_n/w_{n\ max}$) in hardened pastes.

Although these X-ray analyses are not as precise as one would like, they are precise enough to show that the rate of hydration of any compound in cement is affected by the composition of the cement and that it is not the same as that of any other component in the cement. The same statement is true for fractional rates of hydration, with the possible exception of C_3A.

One may well ask, what is the significance of the term "degree of hydration" of cement or the term "maturity" of cement paste, mortar or concrete? Calculations of the heat of hydration and of the non-evaporable water content of pastes made from the X-ray analyses agree satisfactorily with the experimentally measured values. The "degree of hydration" of the cement in each paste was obtained by adding together the amounts of the four major phases that had hydrated and dividing by the original cement content of the paste. In Fig. 12 the degree of hydration so calculated is shown plotted as a function of the ratio of the non-evaporable water of the paste to the maximum non-evaporable water for the completely hydrated cement. The function is not quite linear, but the ratio $w_n/w_{n\max}$ gives a reasonably reliable estimate of the degree of hydration, particularly above 50% hydration. The corresponding ratio for heat of hydration serves equally well as an index of maturity.

4. *Kinetics from Changes in Physical and Chemical Properties*

(a) *Non-evaporable water* As yet no extensive kinetic analysis of non-evaporable water data has been made as has been done for heats of hydration. There is no need to do so since there is a linear relationship

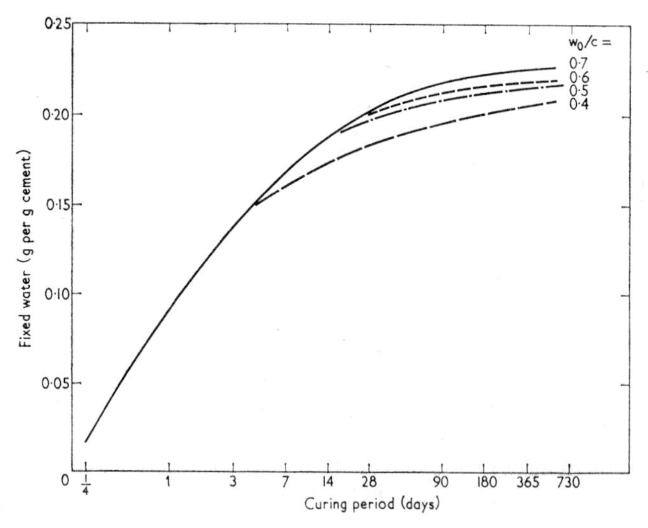

FIG. 13. Effect of w_0/c upon rate of increase of combined water in Portland cement pastes.

between non-evaporable water and heats of hydration with most cements, and, as will be seen, the heat of hydration has been examined extensively.

The effect of the original ratio of water to cement, w_0/c, can be seen by comparing non-evaporable water contents of pastes made from the same cement and cured under the same conditions. The initial rate of hydration of the cement is not appreciably affected by differences in w_0/c, but the effect of w_0/c is pronounced at an intermediate state of hydration. The smaller is w_0/c the sooner is the effect produced. In Fig. 13 it can be seen that, after several days, the smaller is w_0/c the smaller is the amount of non-evaporable water at a given age. Taplin [*81*], in more extensive measurements, found small differences in fixed water for various w_0/c from the beginning. The effect in his particular cement seem to indicate that pastes of low w_0/c hydrate more rapidly initially than pastes of higher w_0/c. Other cements show the opposite effect, so these small effects probably result from differences in composition of the cements.

The effect of availability of water after hardening is shown in Fig. 14. The upper curve shows w_n/c for a paste with $w_0/c = 0{\cdot}6$ as a function of

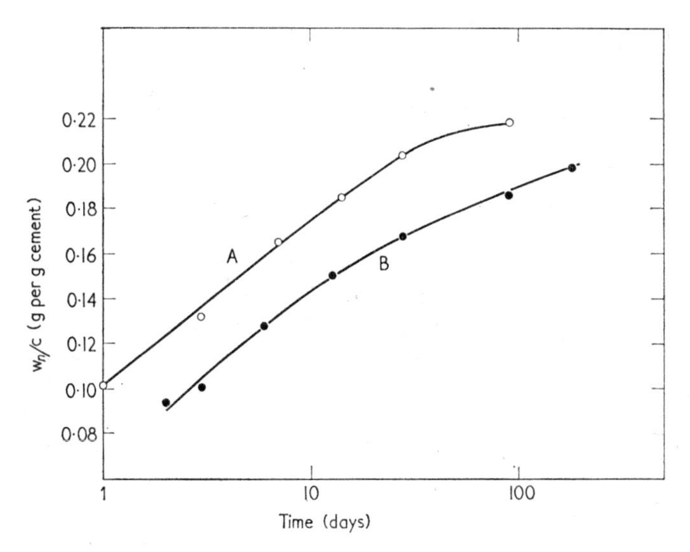

FIG. 14. Effect of self-desiccation of Portland cement pastes upon rate of hydration of the cement. A, Cured continuously moist; B, self-desiccated.

time. The paste was in contact with water at all times. The lower curve represents two samples of paste made from the same cement with the same initial water/cement ratio, but cured differently. These pastes were placed under a bell jar for 48 h after mixing. One paste was then placed in contact with water and the other was transferred to a sealed container. Both of these pastes thereafter hydrated at about the same rate as the first paste, although there was a definite retardation of hydration sometime during the first 48 h when water was not available. Taplin [*82*] reports that the availability of water had no effect at all on the rate of hydration of a cement he studied. His cement was a high alkali cement but otherwise normal in composition.

(b) *Heat of hydration* Information about the rate of hydration of cement can be obtained by determining the heat of hydration as a function of the time of cure. Woods, Steinour and Starke [*41*] were able to correlate the rate of hydration measured in this manner with the rate of development of strength in mortars. They also found a good correlation with the rate of development of strength and the composition of the cement in the mortar. Their primary motive was development of low-heat cement, and therefore it was essential to their purpose to find a relationship between heat of hydration and composition of cement. For this they employed the method of least squares, assuming that at any given age each 1% of each cement compound makes a fixed contribution to the heat evolution independent of the proportion of the compound in the cement. They were successful in accomplishing their purpose. Other workers have generally followed their example. Verbeck and his co-workers [*83, 71*] have extended the method to include effects of gypsum and glass in their computations.

By comparing the magnitudes of the calculated "contributions" of each of the components to the heat of hydration it can be inferred that the rates of hydration of the major components have the order $C_3A > C_4AF > C_3S > C_2S$. This order agrees with that obtained by comparing the rates of hydration of the pure compounds but does not agree with that deduced from X-ray diffraction analysis.

It will be instructive to examine the implications of the assumption implied by the application of the principle of least squares to heats of hydration.

Consider a series of cement pastes, each made with a cement different in composition from all the others. Let $H_j(t)$ be the heat of hydration of the paste made from 1 g of the jth cement and cured t days. Let $w_{ij}(t)$ be the weight of component i per gram of the jth cement that has hydrated t days, and h_i be the heat of hydration per gram of pure component i. The heat of hydration of the paste can now be written

$$H_j(t) = \sum_i w_{ij}(t) h_i \tag{1}$$

The w_{ij} are unknown, the h_i may be obtained from the pure components or, perhaps better, from the completely hydrated cement pastes. Let X_{ij} be the weight fraction of component i in the jth cement. In the application of the principle of least squares to heat of hydration data it is assumed that $H_j(t)$ can be written:

$$H_j(t) = \sum_i a_i(t) X \tag{2}$$

where the a_i are the "contributions" of the components i to the heat of hydration of the paste cured for t days. Equations (1) and (2) are consistent if

$$a_i(t) = \frac{w_{ij}(t)\,h_i}{X_{ij}} \tag{3}$$

The quantity $w_{ij}(t)/X_{ij}$ is the fraction of component i in the jth cement that has hydrated in t days—in other words it is the degree of hydration of component i. The assumption that the $a_i(t)$ are independent of the composition of the cement means that the degree of hydration of each component at any time, t, is independent of the composition of the cement, and the time derivative of the degree of hydration of each component, which we may call its fractional rate of hydration, is independent of the composition of the cement. It was pointed out above that the meagre amount of X-ray data available indicates that the fractional rate of hydration of any component is not independent of the composition of the cement.

Heats of hydration have also been used to calculate the rate of penetration of water into cement particles in the same manner as was done from microscopic examinations [*84*]. Similar results were obtained.

5. *Kinetics of the Overall Reaction from Heats of Hydration*

(a) *The early reactions* It will be convenient for kinetic considerations to consider only the reaction rates after hardening of the paste. The reason for this is apparent when the results of Lerch [*44*] and of Danielsson [*85*] on rate of heat liberation during the first day are considered. At this time the rate of hydration is sensitive to small differences in amounts of gypsum, alkali, $Ca(OH)_2$ and C_3A in the cement; the approximation that cement can be considered as a single component during this time is too crude to be useful.

(b) *Effect of water/cement ratio* Normally the water/cement ratio in concrete or mortar will be in the range 0·3–0·8. The nature of the hydration reactions is not affected by variation of water to cement within this range, so observed variations in heats of hydration reflect the variations in overall rate of hydration caused by the change in water/cement ratio.

The most complete investigation of the effect of water/cement ratio on heat of hydration of cement has been made by Verbeck and his co-workers. They measured heats of hydration on 20 different cements, covering a wide range of cement compositions, at three different water/cement ratios and at ages ranging from 3 days to 6½ years. Their data, averaged for specific cement types, are presented in Table VI.

It is readily seen that heat evolution is influenced significantly by the

TABLE VI

Effect of water/cement ratio on average heats of hydration of different A.S.T.M. types of cements cured at 21 °C

A.S.T.M. cement type	No. of cements averaged	Water/ cement ratio	Heat of hydration at age indicated (cal/g)						
			3 days	7 days	28 days	90 days	1 yr	6½ yr	13 yr
I	8	0·40	60·9	79·2	95·6	103·8	108·6	116·8	118·2
		0·60	65·8	87·7	107·1	114·7	120·0	123·1	
		0·80	66·3	89·4	111·6	119·5	122·2	125·0	
II	5	0·40	46·9	60·9	79·6	88·1	95·4	98·4	100·7
		0·60	49·6	61·3	83·5	94·8	102·1	104·6	
		0·80	49·3	64·3	84·7	99·2	106·8	106·1	
III	3	0·40	75·9	90·6	101·6	106·8	114·2	120·6	120·5
		0·60	86·1	103·6	119·9	124·4	127·3	127·2	
		0·80	86·7	105·0	121·0	126·3	130·0	130·6	
IV	4	0·40	40·9	50·1	65·6	74·4	80·6	85·3	87·3
		0·60	43·2	53·5	66·6	78·7	90·4	92·5	
		0·80	41·7	52·8	69·8	82·5	94·8	96·4	
All	20	0·40	55·65	70·52	86·50	94·44	100·54	106·47	107·99
		0·60	60·28	76·64	95·02	103·98	110·70	112·97	
		0·80	60·19	78·14	97·92	108·20	114·04	115·40	

water/cement ratio and that the effect is increased with increase of the heat of hydration of the cements. The least effect of water/cement ratio is seen in the low-heat cements (A.S.T.M. Type IV) at early ages. The greatest differences in heat of hydration occur at intermediate ages, the exact age for the maximum difference depending upon the composition of the cement. The maximum difference occurs latest for the cements with low heats of hydration—i.e. for those cements which hydrate more slowly. The effect is significant and should be considered when evaluating cements for mass concrete.

At the 6½-year test age, the average heat of hydration was 8% greater in pastes with 0·8 water/cement ratio than in pastes with 0·4 water/cement ratio. But at the later test ages the rate of evolution of heat was greatest in pastes with the lowest water/cement ratio; between the 1-year and 6½-year test ages, the pastes with water/cement ratio of 0·4 averaged 5·9 cal/g while those with water/cement ratio of 0·8 averaged only 1·4 cal/g. These observations indicate that the major effect of decreasing the water/cement ratio is to decrease the rate of hydration of the cement at intermediate ages rather than to limit the extent of hydration.

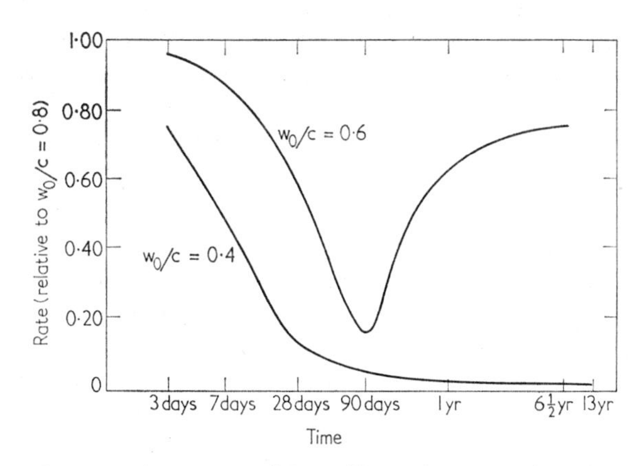

FIG. 15. Comparative rates of heat liberation as a function of time.

The effect of water/cement ratio upon the rate of heat liberation can be illustrated more accurately by comparing rates at the same degree of hydration. Such a comparison is made in Fig. 15, where the rate of heat liberation of the pastes with w_0/c equal to 0·4 and 0·6 relative to that of the paste with $w_0/c = 0·8$ at the same degree of hydration is plotted as a function of the time required for the pastes to reach that degree of hydration.

The relative rate of reaction for each of the two pastes decreases rapidly, but with the 0·6 water/cement ratio paste the relative rate commences to increase after about 90 days. At this time the cement is about 92% hydrated. In the 0·4 water/cement ratio paste the relative rate of heat liberation initially decreases more rapidly than for the 0·6 water/cement ratio paste, but its relative rate does not increase again even at 13 years, when the cement in it is almost 95% hydrated. Thus, the effect of water/cement ratio is rather complex.

A detailed analysis of such rate data has never been published, and such an analysis will not be attempted here, but the effect of water/cement ratio can be explained in a qualitative way in terms of a property of pastes similar to the gel/space ratio as defined by Powers [*86*]. This property will be designated by g and is defined as the ratio of the volume of hydrated cement to the space available for it.

At the instant of mixing cement and water, g is zero. The rate of hydration at this time is practically independent of w_0/c, as shown by the non-evaporable water curves of Fig. 13. At this time the rate of hydration should be proportional to the surface area of the cement, or approximately proportional to the $\frac{2}{3}$ power of the amount of cement present. The hydration product is primarily a colloidal gel which is deposited on the surface of the cement particles, as well as all other surfaces, as it is formed. Consequently, as soon as a continuous coating of gel is formed on the cement particles the reactants must diffuse through the gel pores before the hydration products can be precipitated, and the rate of reaction will slow down rapidly as the space available for hydration products becomes filled. The smaller is w_0/c, the more rapidly the rate of hydration will decrease.

In the present instance the decrease in rate of heat liberation resulting from filling space in the 0·4 water/cement ratio paste with gel continues throughout the 13-year period of test. After 13 years the available space is about 97% filled with hydration product. In the 0·6 water/cement ratio paste the relative rate decreases until 92% of the cement has hydrated, but the space is filled to only 74%. Apparently thereafter the decrease in surface of the unhydrated cement becomes more important than the filling of space for controlling the relative rate of heat liberation. In this paste at complete hydration the space could be filled only to 78%. The rate of hydration of the 0·8 water/cement ratio paste is affected least, for in that paste the space can be filled to only 64% at complete hydration. The effect of g upon the rate of reaction in each of the two pastes, relative to that in the paste with $w_0/c = 0{\cdot}8$, is shown in Fig. 16. The curves for the two pastes agree reasonably well in the region where the reaction is diffusion controlled.

(c) *Effect of temperature* In kinetic studies of reactions occurring in homogeneous systems (reactions occurring in a single phase) it is usually found that the rate of a reaction is proportional to a function of the composition. The constant of proportionality, called the rate constant, is a function of temperature only. An "activation energy" can be obtained from the temperature dependence of the rate constant by the use of the Arrhenius equation. The magnitude of the activation energy frequently is helpful in deciding which of two or more possible mechanisms for the reaction is most probable. Occasionally it is found

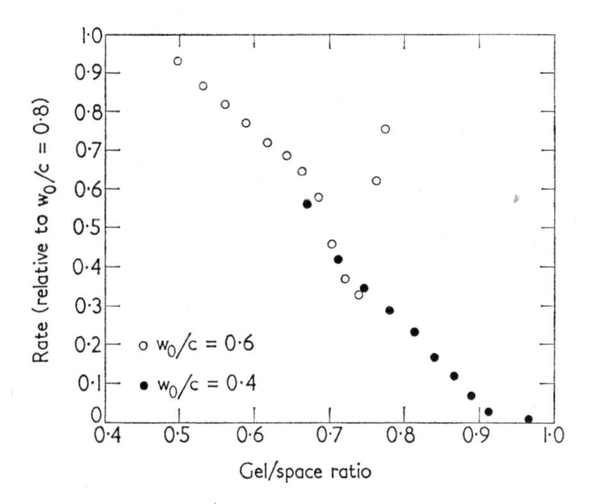

FIG. 16. Effect of gel/space ratio on comparative rates of heat liberation.

that the activation energy has more than one value, depending upon how far the reaction has progressed. In these instances it is generally true that the mechanism of the reaction changes as the reaction progresses.

In heterogeneous reactions (reactions wherein more than one phase is involved) the kinetics are much more complicated. The relationship between the rate of reaction and composition of the system is apt to be complex so that it is extremely difficult, or impossible, to obtain a rate constant. In such cases it is frequently assumed that the rate is proportional to an unknown function of the composition, and by determining the rate as a function of temperature at an arbitrarily chosen composition an "apparent activation energy" can be obtained. If this apparent activation energy is the same for all compositions, it is probable that a single mechanism for the reaction exists. If the activation energy is not constant, then either the fundamental assumption is not true (i.e. the composition and temperature variables are not separable mathematically) or the mechanism depends upon the composition.

The cement–water system is a complex heterogeneous system, and only a few experimental results showing the effect of temperature upon the rate of hydration are available. The set of published experimental results most suited for examination are those of Carlson and Forbrich [*87*], who determined the heat of hydration of a "standard" cement at 0·4 water/cement ratio. The pastes were cured at 4·4, 23·3, and 40° C for ages from 3 to 90 days. The results are presented in Table VII.

Let it be assumed that (1) the rate of hydration of cement is proportional to an unknown function of the composition, and (2) the function of composition has the same value for all temperatures at a given heat of hydration. On the basis of these assumptions an "apparent heat of activation" for the hydration of cement can be calculated by plotting

TABLE VII

Effect of curing temperature on heat of hydration

Temperature (° C)	Heat of hydration for water/cement ratio 0·40			
	3 day	7 day	28 day	90 day
4·4	29·5	43·5	78·4	88·8
23·3	52·4	72·4	83·6	90·8
40·0	72·3	80·3	86·8	93·1

the logarithm of the rate of hydration (as measured by the rate of heat liberation) as a function of the reciprocal of the absolute temperature. Satisfactory linearity of the plots is obtained at degrees of hydration greater than that corresponding to 60 cal per g cement. Plots are shown in Fig. 17. The "apparent activation energy" for the hydration of cement so calculated is not a constant, but decreases from a value of 8 to 2·3 kcal in the range of composition examined. (The composition range is probably from degrees of hydration of about 55–80%).

Either the assumptions made for the calculations are not justified or the rate determining step in the hydration process changes during the course of the reaction. The second alternative is probably true because, as was pointed out above, the rate of hydration decreases rapidly as the gel/space ratio increases, and the energy of activation for a diffusion-controlled process should be smaller than that for a process controlled by a chemical reaction.

On the other hand, the assumptions made for the calculations may not be justified for at least two reasons: the maximum strength developed in concrete is not independent of its temperature during curing [*88*]; and the value of dH/dt is reported not to be completely dependent upon the heat of hydration and temperature alone, but depends also upon the

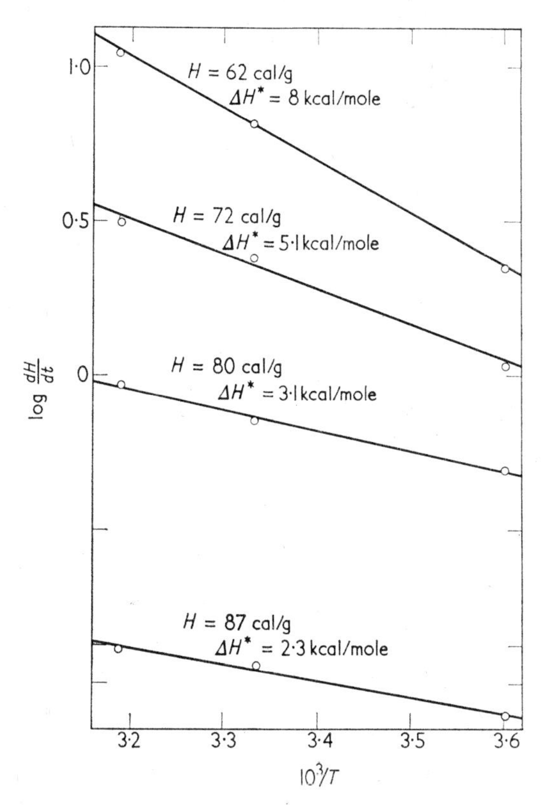

FIG. 17. Apparent activation energy for hydration of cement.

time–temperature path by which the particular state of hydration is reached [*89*]. Much more information is necessary before definite conclusions concerning the effect of temperature on rates of hydration can be reached.

D. ENERGETICS OF HYDRATION REACTIONS

1. *Effect of Intrinsic Characteristics of the Cement*

Cements differ significantly in heat evolution characteristics. One would expect that the heat of hydration of a cement should be equal to the sum of the heats of hydration of the compounds comprising the cement; it will be seen that this is essentially true.

There is a significant difference between the heats of hydration of the individual cement components, as is shown in Table VIII. The heat evolved by hydration of C_2S is lowest, 62 cal/g, while the heat of hydration of C_3A is highest of the major phases, at least 207 cal/g, with C_3S second.

TABLE VIII

Heats of complete hydration of individual compounds

Anhydrous compound	Product	Heat of hydration (cal/g anhydrous compounds)	
C_3S	C_3S+3H_2O	120	[*91*]
C_2S	C_2S+2H_2O	62	[*91*]
C_3A	$C_3A.6H_2O$	207	[*91*]
C_3A	$C_3A.6H_2O$	214	[*96*]
C_3A	$C_3A.8H_2O$	235	[*96*]
C_3A	$C_3A.10{\cdot}2H_2O$	251	[*96*]
C_3A	$C_3A.11{\cdot}6H_2O$	261	[*96*]
C_3A	$C_3A.3CaSO_4.32H_2O$	347	[*91*]
C_4AF		100	[*91*]
CaO	$Ca(OH)_2$	278·9	[*97*]

Cements with high C_3S and C_3A contents have high heats of hydration relative to those cements with high C_2S and low C_3S and C_3A contents.

It was shown above that in cement C_3A and C_3S hydrate much more rapidly than either of the other two major compounds. One should, therefore, expect to find, and one does find, a good correlation between the rate of hydration of cement and the heat of hydration of cement—those cements with high heats of hydration hydrate most rapidly.

Comparisons have been made of the observed heats of hydration of cements of different compositions at various periods of time. Various investigators [*41, 83, 90*] have applied the method of least squares in the manner described above for this purpose. The least square analysis may be performed on the basis either of calculated compound composition or of chemical oxide analysis. Least squares analyses on these two bases are equivalent since the calculated compound composition is derived by linear expressions from the oxide analysis.

The significance of the calculated coefficients at times shorter than required to hydrate the cement completely is not entirely clear; but for pastes in which the cement is completely hydrated the coefficients correspond to the heats of complete hydration of the individual compounds. This is illustrated in Table IX, where the results of Verbeck and co-workers, which have since been extended to 13 years, are compared

TABLE IX

Results of least squares analysis of heats of hydration of cements

Compound	Least squares coefficients, heat of hydration at water/cement ratio and age indicated (cal/g compound—0·40 w/c and 21° C)							Heats of complete hydration [91] (cal/g compound)
	3 days	7 days	28 days	90 days	1 yr	$6\frac{1}{2}$ yr	13 yr	
C_3S	58	53	30	104	117	117	122	120
C_2S	12	10	25	42	54	53	59	62
C_3A	212	372	329	311	279	328	324	207
C_4AF	69	118	118	98	90	111	102	100

with the heats of hydration of the pure compounds reported by Lerch and Bogue [91]. The heat coefficients obtained for C_3S, C_2S and C_4AF after 13 years agree closely with the heats of hydration of the pure compounds; but the coefficient for C_3A requires especial consideration. When pure C_3A is hydrated, C_3AH_6 is formed with the liberation of 207 cal per g C_3A. The least square coefficient for C_3A is significantly larger than this, but is approximately equal to the heat of hydration of C_3A to form ettringite. From the discussion of stoichiometry, above, it is evident that the composition of the cement influences the nature of the hydration reactions of C_3A. C_3AH_6 is rarely if ever formed, but either $C_3A.3CaSO_4.32H_2O$ or $C_3A.CaSO_4.12H_2O$ is always formed and sometimes other products as well. Because of this complicated nature of the hydration of C_3A it would indeed be surprising if the least square coefficient for C_3A did correspond precisely to the heat of hydration of C_3A to form any one of its possible hydrates.

Analyses of heat of hydration data have been based on many different models. The coefficients in Table IX were derived on the assumption:

$$H_t = a(\%C_3S) + b(\%C_2S) + c(\%C_3A) + d(\%C_4AF)$$

in which H_t = heat of hydration at the time t.

However, because of the high coefficient obtained for C_3A, other hypotheses were examined.

Formation of $C_3A.CaSO_4.12H_2O$:

$$H_t = a(\%C_3S) + b(\%C_2S) + c(\%C_3A - 1{\cdot}125\%SO_3) + d(\%C_4AF) + e(\%SO_3)$$

Formation of $C_3A.3CaSO_4.32H_2O$:

$$H_t = a(\%C_3S) + b(\%C_2S) + c(\%C_3A - 3{\cdot}375\%SO_3) + d(\%C_4AF) + e(\%SO_3)$$

The results of these analyses are shown in Table X. These assumptions did not assist in revealing further details of the nature of the hydration reactions.

TABLE X

Results of least squares analyses of heat of hydration data including the formation of calcium aluminate sulphate hydrates

Compound	Least squares coefficients (cal/g compound, 0·8 w/c—1 yr)		
	eq (1)	eq (2)	eq (3)
C_3S	124	128	125
C_2S	76	91	81
C_3A	325	396	384
SO_3		−602	672
C_4AF	95	152	121

One assumption that was instructive concerned the role of alkali in clinker on heat evolution. Newkirk [*92*] has reported that K_2O and Na_2O in Portland cement clinker forms the compounds $KC_{23}S_{12}$ and NC_8A_3.

TABLE XI

Comparison of least squares coefficients for alkali substituted compounds at early and late ages

Compound	Least squares coefficients (cal/g compound, 0·8 w/c—16 cements)		
	3 days	1 yr	$6\frac{1}{2}$ yr
C_3S	92	130	134
C_2S	22	73	76
$KC_{23}S_{12}$	57	62	74
C_3A	127	310	325
NC_8A_3	4	322	288
C_4AF	−17	88	71

Least square analysis based on calculated potential compound compositions modified by the formation of these alkali-substituted compounds produced the coefficients in Table XI. Comparison of the coefficients obtained for C_2S and $KC_{23}S_{12}$ and for C_3A and NC_8A_3, at 1 year and at

$6\frac{1}{2}$ years, suggests that the heat of hydration of the alkali-substituted compounds is nearly the same as that of the pure compounds. Comparison of the corresponding coefficients obtained at 3 days suggests that the substitution of alkali affects the rate or nature of the initial hydration reactions.

Further elucidation of the nature of the reactions by least square analyses may not be possible. Lerch [*44, 93, 94*] has shown that heat evolution is influenced, in a rather complicated way, by fineness of the cement and its content of C_3A, alkali, gypsum and glass.

The differences in rates and energies of hydration of the various compounds in Portland cement makes it possible to design Portland cements for particular purposes. If a cement is desired that will harden and develop strength quickly, it is possible to make it by increasing the proportion of C_3S and C_3A and grinding the cement finer. Frequently finer grinding alone is sufficient. If, on the other hand, a cement is desired that will develop a minimum amount of heat during hydration, then increasing the proportions of C_2S and C_4AF will provide it. These differences in rates and energies of hydration are the basis of the A.S.T.M. classifications.

2. Effect of Water

Although the external dimensions of a paste do not change appreciably as hydration of the cement proceeds, the volume of solids within the paste increases. But the volume of the hydration product formed is less than the sum of the volumes of cement and water which react to form it, so the hydration product does not fill completely the volume made available for it. If the paste is in contact with an external supply of water, more water will be drawn into the paste (the amount of water so imbibed is approximately 25% of the non-evaporable water) and hydration will proceed until either the cement is completely hydrated or the available space within the paste is completely filled.

If an external source of water is not in contact with the paste, it will be dried, or "self-desiccated", as the hydration proceeds. It is conceivable that the vapour pressure of the water may fall to such a low value that hydration is stopped before the cement is completely hydrated.

The heat of hydration of a cement, as normally measured, represents the total heat evolution that occurs. This heat is comprised of two separate parts: (1) the energy change accompanying the chemical reactions to form hydration products, and (2) the energy change associated with the simultaneous adsorption of water on the colloidal hydration product. The relative amounts of these energy changes depend to some degree upon the chemical composition of the cement, but for a completely hydrated cement approximately 80% of the heat evolved is produced by the chemical hydration reactions. The other 20% is the energy of adsorption of water on the surface of the hydration products.

It would seem that the heats of hydration of cement might be subject to a relatively large error because of the effect of adsorption, especially in those pastes made with low w_0/c and in those hydrated under conditions where the paste cannot remain saturated. Verbeck and his co-workers found that for their pastes with $w_0/c = 0{\cdot}4$ and cured 13 years the magnitude of the effect of self-desiccation amounted to 0·9 cal per g cement. The value of this correction is very close to the value calculated from the differential heat of adsorption of water on hardened cement pastes estimated from data reported by Powers and Brownyard [*95*]. In the region near saturation the differential net heat of adsorption for water is constant within experimental error and is approximately 20 cal per g H_2O. If the pastes used by Verbeck were completely hydrated in contact with water, they would have imbibed about 0·057 g water per g cement ($0{\cdot}25\ w_n/c$). The heat of solution of the paste then should be $0{\cdot}057 \times 20 = 1{\cdot}9$ cal per g cement higher than the heat of solution of a completely saturated paste. The correction would be proportionately smaller for pastes in which the cement was not completely hydrated.

Thus, the heat of hydration of cement measured by the heat of solution method can be affected by the amount of water present in two different ways: (1) the original water/cement ratio of the paste influences the rates of the hydration reactions, and indeed the extent to which the cement in the paste will hydrate; (2) the amount of water remaining in the paste at the end of the hydration period may affect the heat of solution of the paste. The first effect is the more important, the second effect being relatively small.

REFERENCES

1. Kalousek, G. L. (1946). *Proc. Amer. Soc. Test. Mater.* **46**, 1293.
2. Greene, K. T. (1962). *Chemistry of Cement, Proceedings of Fourth International Symposium, Washington 1960*, p. 359. National Bureau of Standards Monograph 43. U.S. Department of Commerce.
3. Kalousek, G. L., Davis, C. W. and Schmertz, W. E. (1949). *J. Amer. Concr. Inst.* **45**, 693.
4. Rey, M. (1957). *Silicates Ind.* **22**, 533.
5. Bogue, R. H. and Lerch, W. (1934). *Industr. Engng Chem.* (*Industr. Ed.*) **26**, 837.
6. Eckel, E. C. (1911). *Engng News* **66**, 157.
7. Schmidt, Oskar (1906). "Der Portlandzement", p. 136. Konrad Wittwer, Stuttgart.

8. Forsén, L. (1939). *Proceedings of the Symposium on the Chemistry of Cements, Stockholm 1938*, p. 298. Ingeniörsvetenskapsakademien, Stockholm.
9. Steinour, H. H. (1951). Portland Cement Association Research Dept Bull. **34**.
10. Roberts, M. H. (1957). *J. appl. Chem.* **7**, 543.
11. Steinour, H. H. (1958). Portland Cement Association Research Dept Bull. 98.
12. d'Ans, J. and Eick, H. (1953). *Zement-Kalk-Gips* **6**, 197.
13. Parker, T. W. (1939). *J. Soc. chem. Ind., Lond.* **58**, 203.
14. Lerch, W. (1941). Portland Cement Association, Unpublished Research Report.
15. Bucchi, R. (1951). *Chim. e Industr.* **33**, 685.
16. Kalousek, G. L. (1941). Thesis, University of Maryland.
17. Feichtinger, G. (1855). "Die Chemische Technologie der Mortel Materialien", p. 179. Braunschweig.
18. Forsén, L. (1930). *Zement* **19**, 1130.
19. Foster, W. D. (1932). *J. Amer. Concr. Inst.* **29**, 189.
20. Forsén, L. (1935). *Zement* **24**, 17, 33, 77, 139, 191.
21. Hansen, W. C. (1954). *Proceedings of the Third International Symposium on the Chemistry of Cement, London 1952*, p. 598. Cement and Concrete Association, London.
22. Steinour, H. H. (1954). *Proceedings of the Third International Symposium on the Chemistry of Cement, London 1952*, p. 627. Cement and Concrete Association, London.
23. Wittekindt, W. (1944). *Zement* **33**, 103.
24. Kelley, K. K., Southard, J. C. and Anderson, C. T. (1941). U.S. Bureau of Mines Technical Paper 625.
25. Schachtschabel, P. (1932). *Zement* **21**, 509, 523. Condensed account in English in *Cement & Cem. Manuf.* **6**, 54 (1933).
26. Bucchi, R. (1952). *Industr. ital. cemento* **22**, 170.
27. Hansen, W. C. and Hunt, J. O. (1949). *Bull. Amer. Soc. Test. Mater.* No. 161, 50.
28. Kuhl, H. and Wang, T. (1932). *Zement* **21**, 105, 120, 134, 165.
29. Kalousek, G. L., Jumper, C. H. and Tregoning, J. J. (1943). *J. Res. nat. Bur. Stand.* **30**, 215.
30. Hansen, W. C. and Pressler, E. E. (1947). *Industr. Engng Chem.* (*Industr. Ed.*) **39**, 1280.
31. Hansen, W. C. (1944). Portland Cement Association, Unpublished Research Report.
32. Hansen, W. C. (1962). *Chemistry of Cement, Proceedings of the Fourth International Symposium, Washington 1960*, p. 387. National Bureau of Standards Monograph 43. U.S. Department of Commerce.
33. Ahlers, G. (1962). *Chemistry of Cement, Proceedings of the Fourth International Symposium, Washington 1960*, p. 415. National Bureau of Standards Monograph 43. U.S. Department of Commerce.
34. Blanks, R. F. and Gilliland, J. L. (1951). *Proc. Amer. Concr. Inst.* **47**, 517.
35. Lhopitallier, P. and Stiglitz, P. (1950). *Research Cong. Chem. Ind., Milan*, **23**, 335.
36. Manabe, T. (1962). *Chemistry of Cement, Proceedings of the Fourth International Symposium, Washington 1960*, p. 404. National Bureau of Standards Monograph 43. U.S. Department of Commerce.
37. Hansen, W. C. (1959). Paper presented at the Third Pacific Area National Meeting of the American Society for Testing Materials, San Francisco.

38. Hansen, W. C. (1958). Paper read at A.S.T.M. meeting, June.
39. Bogue, R. H. and Lerch, W. (1933). Portland Cement Association, Unpublished Fellowship Report.
40. Seligmann, P. and Greening, N. R. (1962). *Chemistry of Cement, Proceedings of the Fourth International Symposium, Washington 1960*, p. 408. National Bureau of Standards Monograph 43. U.S. Department of Commerce.
41. Woods, H., Steinour, H. H. and Starke, H. R. (1932). *Engng News Rec.* **109**, 404; (1933). **110**, 431.
42. Carlson, R. W. (1937). *Proc. Highw. Res. Bd, Wash.* **17**, 360.
43. Forbrich, L. R. (1940). *J. Amer. Concr. Inst.* **12**, 161.
44. Lerch, W. (1946). *Proc. Amer. Soc. Test. Mater.* **46**, 1252.
45. Copeland, L. E. and Schulz, E. G. (1962). *J. P.C.A. Res. Dev. Labs* **4**, 2.
46. Powers, T. C. and Brownyard, T. L. (1946). *Proc. Amer. Concr. Inst.* **43**, 249.
47. Copeland, L. E. and Hayes, J. C. (1953). *Bull. Amer. Soc. Test. Mater.* No. 194, 70.
48. Taylor, H. F. W. (1953). *J. chem. Soc.* 163.
49. Taylor, H. F. W. (1954). *27th Congress of Industrial Chemistry, Brussels* **3**, 63.
50. Watanabe, K. and Sasaki, T. (1962). *Chemistry of Cement, Proceedings of the Fourth International Symposium, Washington 1960*, p. 491. National Bureau of Standards Monograph 43. U.S. Department of Commerce.
51. Blaine, R. L. (1962). *Chemistry of Cement, Proceedings of the Fourth International Symposium, Washington 1960*, p. 501.National Bureau of Standards Monograph 43. U.S. Department of Commerce,
52. Midgley, H. G. (1962). *Chemistry of Cement, Proceedings of the Fourth International Symposium, Washington 1960*, p. 479. National Bureau of Standards Monograph 43. U.S. Department of Commerce.
53. Lehmann, H. and Dutz, H. (1962). *Chemistry of Cement, Proceedings of the Fourth International Symposium, Washington 1960*, p. 513. National Bureau of Standards Monograph 43. U.S. Department of Commerce.
54. Kantro, D. L., Copeland, L. E. and Anderson, E. R. (1960). *Proc. Amer. Soc. Test. Mater.* **60**, 1020.
55. Brunauer, S., Copeland, L. E. and Bragg, R. H. (1956). *J. phys. Chem.* **60**, 116.
56. Brunauer, S., Kantro, D. L. and Copeland, L. E. (1958). *J. Amer. chem. Soc.* **80**, 761.
57. Heller, L. and Taylor, H. F. W. (1956). "Crystallographic Data for the Calcium Silicates", p. 36. HMSO, London.
58. Turriziani, R. (1959). *Industr. ital. Cemento* **29**, 185, 219, 244, 276.
59. Kalousek, G. L. (1957). *J. Amer. ceram. Soc.* **40**, 74.
60. Kalousek, G. L. (1944). *J. Res. nat. Bur. Stand.* **32**, 285.
61. Swanson, H. E., Gilfrich, N. T., Cook, M. I., Stinchfield, R. and Parks, P. C. (1959). "Standard X-ray Diffraction Patterns", Vol. 8, p. 3. Nat. Bur. Stand. Circular 539.
62. Midgley, H. G., quoted by Lea, F. M. and Desch, C. H. (1956). "The Chemistry of Cement and Concrete", 2nd Ed., revised by F. M. Lea, p. 604. Arnold, London.
63. Midgley, H. G. (1957). *Mag. Concr. Res.* **9**, 17.
64. Fratini, N., Schippa, G. and Turriziani, R. (1955). *Ric. sci.* **25**, 57.
65. Schippa, G. and Turriziani, R. (1957). *Ric. sci.* **27**, 3654.
66. Roberts, M. H. (1962). *Chemistry of Cement, Proceedings of the Fourth International Symposium, Washington 1960*, p. 1033. National Bureau of Standards Monograph 43. U.S. Department of Commerce.

67. Seligmann, P. and Greening, N. R. (1962). *J. P.C.A. Res. Dev. Labs* **4**, 2.
68. Malquori, G. and Cirilli, V. (1943). *Ric. sci.* **11**, 434.
69. Wells, L. S., Clarke, W. F. and McMurdie, H. F. (1943). *J. Res. nat. Bur. Stand.* **30**, 367.
70. Steinour, H. H. (1954). *Proceedings of the Third International Symposium on the Chemistry of Cement, London 1952*, p. 261. Cement and Concrete Association, London.
71. Copeland, L. E., Kantro, D. L. and Verbeck, G. (1962). *Chemistry of Cement, Proceedings of the Fourth International Symposium, Washington 1960*, p. 429. National Bureau of Standards Monograph **43**. U.S. Department of Commerce.
72. Burdese, A. and Gallo, S. (1952). *Ann. Chim.* **42**, 349.
73. Flint, E. P., McMurdie, H. F. and Wells, L. S. (1941). *J. Res. nat. Bur. Stand.* **26**, 13.
74. Jones, F. E. (1962). *Chemistry of Cement, Proceedings of the Fourth International Symposium, Washington 1960*, p. 205. National Bureau of Standards Monograph 43. U.S. Department of Commerce.
75. Kantro, D. L., Copeland, L. E. and Brunauer, S. (1962). *Chemistry of Cement, Proceedings of the Fourth International Symposium, Washington 1960*, p. 75. National Bureau of Standards Monograph 43. U.S. Department of Commerce.
76. Anderegg, F. O. and Hubbell, D. S. (1930). *Proc. Amer. Soc. Test. Mater.* **30**, 572.
77. Brownmiller, L. T. (1943). *Proc. Amer. Concr. Inst.* **39**, 193.
78. Rexford, E. P. (1943). *Proc. Amer. Concr. Inst.* **39**, 212.
79. Ward, G. W. (1944). Unpublished report of the Portland Cement Association Fellowship, Nat. Bur. Stand., Washington, D. C.
80. Yamaguchi, G., Takemoto, K., Uchikawa, H. and Takagi, S. (1962). *Chemistry of Cement, Proceedings of the Fourth International Symposium, Washington 1960*, p. 495. National Bureau of Standards Monograph 43. U.S. Department of Commerce.
81. Taplin, J. H. (1959). *Aust. J. appl. Sci.* **10**, 329.
82. Taplin, J. H. (1962). *Chemistry of Cement, Proceedings of the Fourth International Symposium, Washington 1960*, p. 465. National Bureau of Standards Monograph 43. U.S. Department of Commerce.
83. Verbeck, G. J. and Foster, C. W. (1950). *Proc. Amer. Soc. Test. Mater.* **50**, 1235.
84. Steinherz, A. R. (1958). *Rev. Matér. Constr.* No. 509, 48.
85. Danielsson, U. (1962). *Chemistry of Cement, Proceedings of the Fourth International Symposium, Washington 1960*, p. 519. National Bureau of Standards Monograph 43. U.S. Department of Commerce.
86. Powers, T. C. (1949). *Bull. Amer. Soc. Test. Mater.* No. 158, 68.
87. Carlson, R. W. and Forbrich, L. R. (1938). *Industr. Engng Chem.*, (*Anal. Ed.*) **10**, 382.
88. Bernhardt, C. J. (1956). RILEM Symposium, Winter Concreting Theory and Practice, Proc. Session BII, Danish National Institute of Building Research Special Report, Copenhagen.
89. Danielsson, U. (1956). RILEM Symp. Winter Concreting Theory and Practice, Proc. Session BII, Danish National Institute of Building Research Special Report, Copenhagen.
90. Davis, R. E., Carlson, R. W., Troxell, D. E. and Kelly, J. W. (1934). *Proc. Amer. Concr. Inst.* **30**, 485.
91. Lerch, W. and Bogue, R. H. (1934). *J. Res. nat. Bur. Stand.* **12**, 645.

92. Newkirk, T. F. (1954). *Proceedings of the Third International Symposium on the Chemistry of Cement, London 1952*, p. 151. Cement and Concrete Association, London.
93. Lerch, W. (1938). *J. Res. nat. Bur. Stand.* **20**, 78.
94. Lerch, W. (1938). *J. Res. nat. Bur. Stand.* **21**, 235.
95. Powers, T. C. and Brownyard, T. L. (1947). *Proc. Amer. Concr. Inst.* **43**, 549.
96. Thorvaldson, T., Brown, W. G. and Peaker, C. R. (1930). *J. Amer. chem. Soc.* **52**, 3927.
97. Thorvaldson, T., Brown, W. G. and Peaker, C. R. (1930). *J. Amer. chem. Soc.* **52**, 910.
98. Köberich F. (1934). Doctoral dissertation, Berlin.

CHAPTER 9

Electron Microscopy of Portland Cement Pastes

Å. Grudemo

Swedish Cement and Concrete Research Institute,
Royal Institute of Technology, Stockholm, Sweden

CONTENTS

I. Introduction

The first studies of cement hydration products using electron microscopy (EM) were reported in 1938, and selected-area electron diffraction (SED) methods were introduced about 1950. However, it is only during the last few years that EM–SED methods have been widely employed in investigations on cement hydration. Nearly all the literature reviewed below is therefore comparatively recent.

The experimental techniques of EM and ED analysis are dealt with in Chapter 21, and will not be discussed further here. It may be remarked, however, that in the preparation and examination of paste samples special techniques must sometimes be used to prevent the original gel microstructures from being damaged or changed by various operations necessary in the preparation of EM specimens. Severe grinding or ultrasonic treatment may destroy some details of the paste structures; certain suspension media may react chemically with the dispersed material; a high-intensity electron beam may, in combination with a high vacuum, cause dehydration and conversion of some crystal structures.

The hydrogel developed in a Portland cement mix of ordinary composition is, physico-chemically, an intimate but still heterogeneous mixture of several phases. The predominant phase, which is also considered to be mainly responsible for the hydraulic properties of the material, is a colloidal calcium silicate hydrate of indefinite or variable composition. This material, which can be called the C-S-H gel phase, is in certain respects structurally related to the mineral tobermorite.

The second major constituent of the cement gel is hydrated lime (CH), dissolved and reprecipitated, or otherwise separated, in the process of hydration of the calcium silicates.

The third gel phase of some importance is the group of compounds formed as products of hydration of the aluminate-bearing and ferrite-bearing clinker constituents in a solution containing excess of calcium and sulphate ions, and also reactive silica complexes. This phase, which seems in reality to be of a complex nature, will be termed the C-A-H gel phase.

With some degree of approximation the anhydrous clinker constituents can be assumed to react with water independently of other constituents present. However, various types of interactions between ions and ionic groups originating from different anhydrous compounds, or from admixtures and impurities, can be expected to occur frequently in the course of hydration and to cause disturbances in the regular formation of the pure-phase crystal lattices of solid hydrates.

II. Early Exploratory Studies and Theories

Early EM studies of the products of hydration of cement constituents were made by Eitel and his co-workers [*1*, *2*, *3*], who examined lime precipitates, C_3S in water suspension, C-S-H hydrogels and C_3A hydration products. McMurdie [*4*] studied the hexagonal plate phases formed in the hydration of C_3A, C_4AF and C_8NA_3. Sliepcevich, Gildart and Katz [*5*] examined the structures developed by Portland cement and the major cement constituents in water suspension, with or without addition of sulphate. Boutet [*6*] observed the products of hydration of some types of cements with widely varying characteristics. Several phases of different morphologic habits were observed in the two last mentioned investigations, among others one consisting of long, rather coarse fibres or splines radiating from the anhydrous particles and forming networks. This phase, which was identified as ettringite [*5*] or considered to be of variable composition [*6*], appeared essential for the formation of interparticle bonds and for the cohesive properties of the hydraulic materials. Judging from later experience, the formation of a phase of this type seems to have been caused by excessive dilution of the samples.

From these and other EM and X-ray diffraction studies Bernal [*7*] concluded that the cementing action of Portland cement and other hydraulic binders is primarily conditioned by the development of a hydrogel, composed of fibrous or needle-like particles of small dimensions, crystallizing very fast and forming a felt-like network of interlocking fibres which, in the case of Portland cement, consist mainly of a calcium silicate hydrate of some kind. A different conception of the cause of hydraulic action in cement was arrived at by Bogue [*8*], quoting data obtained in EM studies made by Swerdlow, McMurdie and Heckman [*9, 10*] on materials made by ball mill hydration or paste hydration of C_3S. In various structures observed, Bogue and his associates considered the important structural element to be extremely small (about 100Å or less) gel globules, coagulated into larger, spherulitic or disk-shaped aggregates which formed links to adjoining aggregates by intergrowth of structures in areas of contact. In transitional states of hydration various morphologic structures might occur, such as small globules alone or in clusters, thin platey or foil-like particles, and fibrous masses; and ED diagrams from these phases indicated the presence of structures related to tobermorite and hillebrandite. It was believed, however, that complete hydration yields a further dispersed and coagulated mass of very small particles having a crystal structure very similar to that of afwillite.

It has since become evident that, although each of the theories reviewed above gives a correct description of certain phenomena which may occur in particular cement hydration systems, both must be modified in essential respects to agree with later observations.

III. Observations on Lime Phases

As mentioned above, the CH precipitated from solution or formed in cement–water mixes was observed in earlier studies to assume the habit of spherulites or rounded platelets, mostly aggregated in chains. These were also observed in other studies, e.g. by Grudemo [*11*], who noticed that the ED diagrams given by most of these formations showed only one very diffuse reflection at about 3·0 Å, though some regions, in which the spherulites appeared mottled by smaller particles crystallized within them, gave ED patterns of calcite. Schimmel [*12*] established the true nature of similar precipitates as being essentially amorphous CH in which the regular process of crystallization had been disturbed by atmospheric carbon dioxide. Moderate heating in the electron beam is generally sufficient to cause dehydration and recrystallization of these particles, resulting in a mixture of calcite and calcium oxide. Upon further heating, the calcite crystals dissociate, leaving an oxide residue with crystals of extremely small dimensions. These conversion processes can readily be followed by observing the changes in the ED patterns.

It may be mentioned that in some cases a fibrous growth has been observed in CO_2-contaminated CH precipitates. A detailed account of EM examinations of various lime modifications is given by Andrievskii *et al.* [*13*].

Under CO_2-protected conditions CH is precipitated as well developed crystals assuming their natural habit of regularly hexagonal plates or prisms (Plate 1). In the interior of hardened cement paste the form of CH most readily detected by EM examination consists of large, thin plates which appear to loosen rather easily from the aggregates of gel phase in powdered samples, and which frequently show an internal pattern of irregular striations, indicating the presence of interlamellar inclusions (Plate 2). The ED patterns are somewhat variable, a common type being a single-crystal cross-grating pattern of hexagonal symmetry with $a_H = 3{\cdot}59$ Å, showing $hk0$ and sometimes also certain $hk1$ reflections. Examples of EM–ED data from such particles are given by, among others, Danielsson [*14*], Grudemo [*15, 16*] and Uchikawa and Takagi [*17*].

Apparently the larger CH crystals do not form very rapidly in the first stages of reaction in a cement paste, but grow by slow segregation and accretion processes taking place in later periods of hardening. However, part of the CH in cement paste is probably formed not by regular crystallization from solution but as one of the products of topochemical hydration processes within the anhydrous silicate crystals. The CH possibly developing in this way may well be nearly amorphous, owing to lack of space for diffusion and reorientation and to disturbances from inclusions of silica-containing groups, and it would, in addition, be very difficult to distinguish from ill crystallized C-S-H gel phases by X-ray diffraction or EM–ED observation. However, Brunauer and associates [*18, 19*] presented chemical evidence for the occurrence of amorphous lime, or of a phase very high in lime, in the products of hydration of C_3S and β-C_2S.

IV. Observations on C-A-H Gel Phases and Related Substances

Only a few EM studies on constituents of the C-A-H gel phase have been reported. Buttler, Dent Glasser and Taylor [*20*] show the typical, thin, hexagonal plates of C_4AH_{13}, or a dehydrated form of it, giving an ED spot pattern of hexagonal symmetry, corresponding to a cell with an a_H axis of about 5·7 Å. Cells of this type, with small modifications, represent the structural units of all the compounds of this group with hexagonal layer lattices. In the dehydrating conditions of the electron microscope the crystals are more or less rapidly converted to CH and a residual glass, without change in external shape. Similar habits, ED patterns and transformations upon dehydration are exhibited by the hexagonal C_2A-hydrate and by compounds in the C_2A–C_4A hydrate solid solution series.

A hexagonal plate phase of probable composition C_3AH_8 was observed by Iwai and Watanabe [*21*] as the first product of hydration of C_3A. These plates are later dissolved and reprecipitated as the cubic phase C_3AH_6 which shows crystal habits of isometric symmetry. Hexagonal plate phases occur also in the hydration of C_4AF and C_2F. The ferrite-bearing phases appear much more stable, and are at least not converted to isometric phases.

According to some available EM–ED and X-ray data [*22*], a typical course of hydration in pure C_3A pastes kept at room temperature is as follows. Small hexagonal plates are developed almost immediately upon contact with water and later grow in diameter and thickness. Hydrogarnet (C_3AH_6) starts forming after about 1 day and soon becomes a major constituent. However, even after 3 months, a considerable quantity of hexagonal plate phase remains in the paste, while it takes more than 2 weeks for the C_3A phase to disappear completely. The hexagonal plate phase gives ED patterns corresponding to the 5·7 Å unit cell; the X-ray basal spacing is about 8·0 Å, together with a weak line at about 10·3 Å, which disappears with time.

Addition of 1 mole of lightly burnt CaO per mole of C_3A in the paste retards the formation of the hexagonal plate phase during the first few hours but does not greatly influence the later course of reaction. The data indicate that, with or without the addition of lime, the hexagonal plate phase remaining after long reaction times is a C_3A-hydrate, similar to C_4AH_{13} (basal spacing about 8·0 Å) but with a relative deficit in the C/A ratio.

Addition of sulphate changes radically the course of reaction of C_3A. With a small amount of gypsum admixture (sufficient to give rapid saturation in the liquid phase) the initial formation of the hexagonal plate phase is suppressed and the aluminate trisulphate hydrate, of ettringite type, becomes the primary product of hydration. The characteristic habit of this compound is long and rather coarse rods or splines, sometimes with sharply cut-off end faces, as shown in EM studies by Stolnikov [*23*], Astreeva and Lopatnikova [*24*] and Iwai and Watanabe [*21*]. It is believed that the habits observable by EM are dehydrated pseudomorphs of the original ettringite which, from microscopic observations, is known to form long, straight rods or thin prisms, of hexagonal cross-section.

In C_3A paste with a small amount of gypsum admixture the aluminate trisulphate can be seen in EM samples as the first product of hydration and persists until the gypsum is exhausted, after about one day's hydration [*22*]. Further hydration of C_3A leads to the decomposition of the trisulphate and to the formation of hexagonal plates of the monosulphate compound or of its supposed solid solution with C_4AH_{13}. A similar trend has been observed with cement pastes, where ettringite-type formations start growing as rods radiating from the cement particles (Plates 6 and 7) but are later loosened up and decomposed with simultaneous formation of hexagonal plates of the monosulphate type and structure [*15*] (Plate 5). The dehydrated and internally collapsed pseudomorphs of ettringite crystals show characteristic striations along the fibre axis. The ED effects given by them are at most diffuse rings, indicating a non-crystalline structure.

A few per cent of calcium sulphate (usually gypsum, $CaSO_4 . 2H_2O$) is mixed with ordinary types of Portland cement clinker during grinding. Its function is to interfere with and prevent the rapid dissolution of C_3A and its reprecipitation into structures which might otherwise cause premature stiffening (so-called "quick set") of the paste. Even in types of standard cements which are low in C_3A, however, the SO_3/Al_2O_3 ratio does not much exceed unity. Thus, the equilibrium composition of the calcium aluminate sulphate hydrate formed in cement paste can be expected to be that of the monosulphate $C_3A . CaSO_4 . 10H_2O$ or of its solid solution with C_4AH_{13}. Thin plates probably of this composition are observed in small quantity in hardened pastes of greater age (some months to many years), made from standard cement with the normal content of C_3A (ASTM Type I cement, C_3A of the order of 10–15%) [*14–16*] (Plates 3 and 4). The SED single-crystal patterns given by these plates show strong reflections from a pseudo-unit cell with a'_H 5·7 Å, but in most cases there are also weaker spots indicating a true unit cell of double this size (a_H 11·4 Å) and the presence of a lattice superstructure which could be most readily interpreted as a substitution of sulphate for water and hydroxyl in every second pseudo unit cell. This substitution seems to stabilize the hexagonal lattice, for in EM observations no indications have been found of particle habits associated with the isometric hydrogarnet phase, $C_3(A,F)(S,H_2)_3$.

In old pastes made from cement of low C_3A content (ASTM Type IV, C_3A about 2–3%, $CaSO_4$ about 3%) it has not yet been possible to find any such platey crystals with the 5·7 Å hexagonal unit cell, nor have particles resembling ettringite pseudomorphs been detected with certainty in EM studies of old Portland cement pastes of any type. Possibly the quantities involved are too small to attract attention in EM surveys or the minute quantities of Al_2O_3 liberated in the disintegration of the high-sulphate compounds originally formed may be taken up by the C-S-H gel phase.

V. Observations on Tobermorite and C-S-H Gel Phases

Of the numerous crystal phases established in the C-S-H system in recent years, the mineral tobermorite and the ill-crystallized or colloidal compounds structurally related to it have attracted special attention, since they can serve as morphologic models for similar phases occurring in the hydration of cement paste.

Well crystallized tobermorite of low C/S ratio (0·8–1), either synthesized by autoclaving or obtained as a natural mineral, is shown in EM studies [*25*, *28–31*] to adopt the habit of plates, mostly flat and of considerably more than single-layer thickness, irregular or of crystalline shape, sometimes elongated in one direction. SED patterns are of the single-crystal cross-grating type, with strong pseudo-halving of the true unit cell with a 11·2, b 7·3 Å. The distribution of weak and sometimes streaky reflections indicates a high to moderate degree of stacking order over at least 10–20 elementary layers. Well crystallized tobermorite is strikingly similar to clay minerals of the vermiculite or montmorillonite type [*25*, *32*, *33*] in many respects, although it lacks their nearly hexagonal symmetry. Some typical electronmicrographs and SED patterns of tobermorite and related phases are given in Chapter 21, Volume 2.

Anomalies in morphology and modes of stacking are common. Tobermorite modifications with highly fibrous or lath-like crystals have been observed by Kurczyk and Schwiete [*26*, *27*] and by Grothe, Schimmel and zur Strassen [*34*]. Similar formations, although of higher C/S ratios and showing anomalous size and symmetry of the base of the unit cell, were observed by Akaiwa and Sudoh [*35*] and by Grudemo (cf. Plate 24). These tobermorite habits are very similar to those of xonotlite and the hydrates of the wollastonite–foshagite–hillebrandite group, which may occur in hydrothermal preparations, and care must be taken to avoid confusion with these compounds, the SED patterns of which are not always easily interpreted.

About 4–7% Al_2O_3 can be introduced into the tobermorite lattice, where it causes replacement of Si by Al, as observed by Kalousek [*37*]. The X-ray diffraction patterns do not suggest that the lattice is noticeably disturbed by this substitution, but the size of the platey crystals is reduced to about one-twentieth of the original value.

In a tobermorite of low C/S ratio, Gard, Howison and Taylor [*25*] observed aggregates of plates with parallel striations crossing each other at angles of about 60°. The SED patterns of these aggregates were composed of cross-grating spot patterns mutually rotated through about the same angle. Similar rotational superposition of sheet elements may be expected to occur, although over regions of a much smaller size, in the narrow spaces of cement paste, where the possibilities of reorientation are restricted.

Other observations [*25, 35, 36*] indicate that in synthetic preparations of higher C/S ratio (of the order of 1·2–1·5) the tobermorite plates tend to split into laths which may roll up into tubular crystals. It is not known whether the tobermorite crystal structure is preserved in this transformation or whether the latter represents a conversion to other stable crystalline compounds.

Several EM studies [*11, 16, 25, 28, 29, 36, 38–40*] have shown that the low-temperature variety of tobermorite, C-S-H (I), with a C/S ratio varying between about 0·8 and 1·5, is generally formed as a primary or intermediate product in C-S-H mixtures, even when these are processed at elevated temperatures. In its most characteristic form it develops large but very thin and flexible sheets or foils, probably of single-layer thickness, which mostly become condensed to loosely wrinkled aggregates. The ordered superposition and stacking of several sheets in parallel over large areas is probably very uncommon. The foils are sometimes so thin and transparent that only the creases are observed, giving the impression of a finely fibrous structure.

The SED patterns of C-S-H (I) aggregates or single sheets contain a variable number of *hk* cross-grating reflections from the face-centred base of the tobermorite pseudo-cell, ranging from a more or less diffuse reflection at about 3·0 Å or two reflections at 3·0 and 1·8 Å to the complete pattern containing reflections at 3·05, 2·79, 1·83, 1·66, 1·53, 1·40, 1·19, 1·11 and 1·07 Å, or spacings a few per cent lower than these, owing to contraction of the cells caused by dehydration.

In preparations of high C/S ratio (approaching 1·5) and in the presence of a saturated CH solution, C-S-H (I) particles assume truly fibrous or needle-like habits, probably formed from sheets which become rolled up or creased in one direction only [*11*]. At low C/S ratios (less than 0·8) a degeneration of the C-S-H (I) foils into more fibrous structures is also observed [*11*, *30*].

The characteristic crumpled or rolled foils of C-S-H (I) seem to appear in the hydration products of C_3S or Portland cement, wherever the solution-filled pore spaces are large enough to permit them to develop freely. Thus, they have been observed in ultrasonorated C_3S–water mixtures [*10*, *11*], after a few days' hydration of wet cement paste [*15*], in foamed asbestos–cement pastes [*39*], as recrystallization products of rehydration of fully hydrated and hardened C_3S paste after drying, grinding and suspending the powdered material in water, and in similar preparations. The structures observed by Bernard [*41*] and by Stork and Bystricky [*42*, *43*] to form on the surfaces of cement particles adjacent to larger pores in cement paste samples prepared by the "sandwich" method are probably of the same kind.

The low-temperature, high-lime compound C-S-H (II), of C/S ratio approaching 2, is probably a distinct compound, judging by its X-ray diffraction pattern and morphology, which are different from, although closely related to, those of C-S-H (I) of high C/S ratio. The original product, synthesized by Taylor and examined electron-optically by Grudemo [*11*, *38*], showed particle aggregates of characteristic habits, viz. bundles of fibres with split-up or tapering ends. The conditions required for this compound to form are not yet well defined, and some attempts to reproduce it have failed. Funk [*44*] observed a very similar phase as the product of hydration of β-C_2S in saturated steam at 100° C. Brunauer and Greenberg [*19*] reported on the formation of C-S-H (II) after long periods of hydration of β-C_2S in suspension.

VI. Observations on Calcium Silicate Pastes

Paste hydration of C_3S and β-C_2S, the pure compounds corresponding to the alite and belite constituents of cement, has been shown by various investigators to result in the formation of a C-S-H phase of C/S ratio between 1·5 and 2, together with residual lime or a phase very rich in lime. EM data on the morphology of these phases have been presented in several reports [*10*, *14*, *16*, *19*, *26*, *27*, *44–47*].

The CH phase is observed to occur as comparatively large, thin plates, giving SED single-crystal spot patterns of hexagonal symmetry, as described above. Naturally, such plates are much more common in C_3S pastes than in β -C_2S pastes.

In general, two rather distinct habits of particles are observed in the C-S-H gel phase of C_3S and β-C_2S pastes hardened at room temperature.

One of these consists of aggregated masses or flocks of small, irregular particle elements. Buckle and Taylor [*45*] observed only this C-S-H gel phase in C_3S pastes hydrated up to 5 years, and stated that it was probably composed of thin, distorted flakes of tobermoritic nature, although crypto-crystalline.

The other constituent of the C-S-H gel phase shows the habit of rather long, straight, rod-like particles, usually of the order of 1μ long and 0·1 μ across. Their EM appearance indicates that they have a tubular or reed-like internal structure, probably formed by the rolling of sheet elements. Habits resembling laths or needles are also observed in certain samples. In general, there is no great difference between particles of this type formed in C_3S or in β-C_2S pastes. However, they tend to be more slender and needle-like in C_3S pastes, more compact and columnar in β-C_2S pastes. It is evident that this phase is closely related to, although not necessarily identical with, the C-S-H (II) described in an earlier section, both with respect to habit and, most probably, to crystal structure.

There is a certain resemblance in morphology between the rod-like or tubular particles of this phase and the ettringite pseudomorphs developing in cement hydration, as described above. They are also quite similar in habit to the tricalcium silicate hydrate crystals shown in EM studies by Buckle, Gard and Taylor [*48*]. Both these similarities should be regarded as incidental.

This phase predominates in C_3S and β-C_2S pastes in advanced stages of hydration, according to Copeland and associates [*19, 47*], Kurczyk and associates [*26, 27, 46*] and Danielsson [*14*]. In similar materials studied by Grudemo [*16, 49*] the coarsely fibrous and the amorphous irregular phases were present in about equal amounts (Plate 8). In C_3S paste needle-like particles were often found to radiate from nucleation centres or from the surface of larger aggregates, whereas corresponding particles in ground β-C_2S paste mostly appeared separate (Plates 11 and 12).

C_3S crystals hydrating in paste or in suspension are, in early stages of hydration, mostly covered with radiating fibres or rolled-foil particles (Plate 10). Such formations were observed by Czernin [*50*], in preparations containing C_3S crystals in very thin paste layers studied by the carbon replica method. After only 1 h a felt-like hydration coat had developed; later, however, it ceased growing and particles of a similar type began to form in the interstitial pore space instead (Plate 13). After a few weeks the whole volume between the original anhydrous particles was filled with a hydrated mass, the texture of which appeared, however, to be granular rather than fibrous (Plate 14).

On the other hand, Funk [*51*] observed that β-C_2S particles did not develop any visible coating of hydrated particles on the surface but reacted by penetration of the liquid, which caused internal topochemical hydration and splitting up of the whole crystal into columnar particles.

Similar local reactions are likely to take place within C_3S crystals, following the development of a surface coating. Copeland and associates [*47*] succeeded in rinsing away much of the coating of the original C_3S particles by dispersing a hardened and ground C_3S paste in trichloroethane by ultrasonoration, and Grudemo [*16*] obtained a similar result by a few seconds' rapid shaking with water. This revealed an internal, hydrated structure consisting of a framework of intersecting rods or tubes (Plate 9). It would thus seem that C_3S paste hydration takes place partly by dissolution, reprecipitation and growth of hydrate crystals at or near the solid–liquid interface and partly by internal, topochemical reaction, while hydration of β-C_2S is dominated by the latter process.

The SED patterns from aggregates of the fibrous particles generally show a diffuse reflection in the 3·1–2·8 Å range, and a sharp reflection at about 1·81 Å [*16, 26*]. Isolated fibrous particles, especially in β-C_2S paste, give spot patterns consisting of six diffuse spots at about 3·0 Å, in approximately a hexagonal arrangement, and a sharp fibre reflection streak at 1·80–1·82 Å (Plate 12). These reflections correspond most nearly to the 400, 220 and $2\bar{2}0$ reflections (diffuse spots) and the 040 reflection (streak) of tobermorite. The general appearance of the SED patterns indicates that the repeat distance is well defined along the fibre (b) axis but not so well defined in the perpendicular (a) direction, probably owing to the effect of rolling. The data show that the fibrous phase is very similar to the hydrates C-S-H (I) and C-S-H (II) but do not permit a more detailed identification.

VII. Observations on Gel Phases in Hardened Cement Pastes

In a cement paste that has been cured and hardened for some time the major part of the hydrated mass consists of a phase composed of very small, cryptocrystalline particle elements and essentially similar to the C-S-H gel formed in the hydration of the pure silicates. As yet, there are no data available from systematic EM studies on possible changes in the microstructure and crystalline character of this phase with variations in, for example, the mineral composition and crystal size distribution of the cement, the water/cement ratio, the time and intensity of mixing or vibration procedures, the time and temperature of curing, or the influence of admixtures of different types.

A number of paste samples observed by Grudemo [*16*] showed in general a mixture of elements of various types. Fibrous or acicular habits were commonly found, but loosely aggregated masses of flaky or floc-like texture were predominant, at least in pastes of lower water/cement ratios (Plates 15 and 16). In general, the gel particle elements seemed finer and more irregular than those observed in pure silicate pastes. SED patterns from arbitrary sections of the gel usually contained only a diffuse ring at 2·8–3·1 Å; but, whereas fibrous structures were present in the area selected for observation, a sharp 1·81 Å reflection of weak to medium intensity could usually be detected. Occasionally aggregates of parallel oriented fibres gave patterns of diffuse spots similar to those observed in β-C_2S paste (cf. Section VI; Plate 19).

With due reservations on account of the incompleteness of the experimental EM data available the variations in character of the badly crystallized gel phases can be summed up as follows.

The fibrous phase in pastes of slow-hardening (low-heat) cement, originally containing relatively more belite, appears to have a somewhat coarser structure than the corresponding phase in standard cement paste (Plate 17). On the other hand, rapid-hardening cement paste, with a higher content of alite and more finely ground, seems to contain relatively larger quantities of a fibrous phase of finer texture (Plate 18).

The assumption has been made in some theories of cement paste reactions that the colloidal gel phases are gradually transformed into better crystallized structures. This has not been confirmed by EM observations; old pastes, cured under saturated conditions for up to 10 years, do not show a markedly higher degree of crystallinity than freshly hardened ones (Plates 20 and 21).

The effect on microstructure of variations in the water/cement ratio within the normal range (about 0·3–0·7) appears to be slight. It can be expected that the larger pore volume associated with a higher water/cement ratio would promote the development of the fibrous surface coat on at least the alite crystals, but no data from systematic EM studies on this point have yet been reported. It was observed, however, that artificial pastes of very high porosity (water/cement ratio 1·5–2) and made from finely ground standard cement, contained a high proportion of a comparatively well crystallized C-S-H gel consisting exclusively of a felt-like mass of long, slender fibres (Plates 23 and 24; cf. Plate 22).

Several investigators have used replica methods for examining cast, etched or fractured surfaces of hardened pastes [*42*, *43*, *52–57*]. The observations are generally compatible with those made by direct study of powdered samples. Large portions of the replicas show a rather featureless or finely granular substance, probably identical with the cryptocrystalline calcium silicate hydrate–calcium hydroxide matrix observed by Buckle and Taylor [*45*] and others. Better crystallized material of various habits appears embedded in this matrix (Plate 27). Larger, rounded or angular contours are most probably impressions of the broken-off surfaces of unreacted crystals of anhydrous material. Some replicas show areas with impressions of fibrous or acicular particles which have been identified tentatively as a fibrous tobermorite [*54*] but which, in some cases, might possibly be ettringite-type formations (Plate 28). Other areas, with a striated or staircase-like appearance, have been identified as segregated CH crystal plates lying parallel to each other and viewed on edge (Plate 29). These stacks are probably split up in grinding the paste, and the large, thin CH plates observed in powder samples (cf. above) may thus be the cleavage elements of larger packs.

The microstructural character of high-alumina cement pastes seems to be remarkably similar to that of Portland cement pastes, in spite of the fact that the compositions of the hydrated phases must be essentially different. In a few samples examined most of the material consisted of colloidal aggregates of rather small and thin, irregular plates (Plates 38 and 39). The SED diagrams either were weak and diffuse or showed indistinct spot patterns from an unidentified hydrate phase.

VIII. The Development of Gel Microstructures in Fresh Paste

A general observation in EM studies of the first stages of structural processes in cement pastes or thick suspensions is that the primary hydration products show particle habits of a few characteristic and recognizable types. The appearance of these formations corresponds to the period of setting in a paste. In the subsequent period of hardening and strength development, the primary structures seem to disappear gradually as the gel thickens and coagulates into a mass of a more uniform, structureless internal texture.

The evolution of structures in fresh Portland cement pastes was studied by Copeland and Schulz [*58*, *59*]. In standard cement paste (ASTM Type I) small amounts of thin platelets appeared almost instantaneously on the surfaces of the cement particles, but disappeared after about 1 h. After 2 h hydration, at about the time of initial set, acicular particles started to grow out from the surfaces of the silicate crystals and increased in size and number (cf. Plate 25). After a few more hours these particles stopped growing lengthwise and instead seemed to join up sideways with adjacent particles, forming aggregates resembling bundled fibres or striated sheets. At about 16 h the acicular particles were no longer visible. After 2 days' hydration the aggregation and densification of the creased-sheet particles had proceeded to such a stage that it became difficult to disperse the material and resolve the finer details of the hydrated structures, except along the edges of the larger aggregates. From this time on, the structures observable in the EM samples did not change much with age (cf. Plate 26).

In similar preparations of a low-heat Portland cement paste (ASTM Type IV) Copeland and Schulz [*58*] observed an analogous course of evolution of the hydrated structures, but the processes were found to proceed more slowly. In addition, a hydrated phase consisting of lath-like or rolled-tube particles, similar to those observed in C_3S or β-C_2S pastes, occurred in these samples.

IX. Effects on Microstructure of Various Admixtures and Minor Constituents

The observations just described on the course of development of paste structures were, in the main, corroborated by similar tests made with a standard cement paste of Swedish origin [*22*] (Plates 25 and 26). However, the corresponding clinker paste (i.e. without sulphate addition) did not show any signs of radial growth of acicular particles, but rather an evolution of thin plates and foils which, after a few days' hydration, were transformed into a highly dispersed mass of very small particles. In addition, separate aggregates of long, straw-like particles were observed in appreciable quantity during the period of setting (about 2–6 h after mixing; cf. Plate 31), but had disappeared after 1 day.

The effect of addition of gypsum or calcium chloride to C_3S or β-C_2S pastes was examined by Kurczyk [*26*]. In ordinary pastes, without admixtures, long and comparatively well crystallized needles of a high-lime tobermorite gel phase were formed during hardening. Addition of small amounts of gypsum or calcium chloride, however, resulted in a marked decrease in size of these crystals, accompanied by an increase in the compressive strength of the paste. With more than about 2% of admixtures the data from various analyses indicated the formation of a gel phase different from and still higher in lime than the tobermorite phase, and the compressive strength was lowered (the β-C_2S paste with 3–4% gypsum did not even set).

In other studies the influence of C_3A (about 20%) and gypsum (about 4% as SO_3) on the paste hydration of alite (containing about 2% Al_2O_3+MgO) was examined [*22*]. In the pure alite paste a surface reaction had started after 3 h, and the coat of hydration products on the alite crystals was well developed after 1 day and especially so after 2 weeks (Plate 10), whereas practically no hydrated phase except CH crystals was formed in the outer solution. The surface layer, composed of a poorly crystallized tobermorite, showed the usual forms of splinters

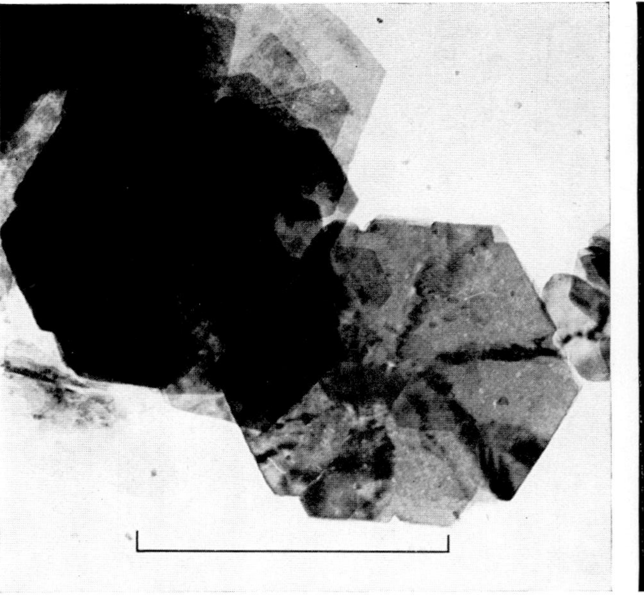

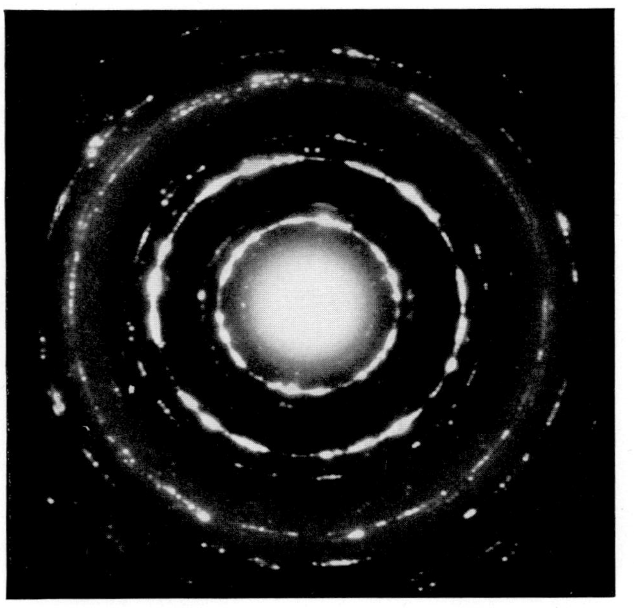

PLATE 1. Sample: cement suspension, w/c 10, rapidly stirred 10 min. EM: CH crystal plates. ED: CH spot-ring pattern (strongest reflections at 3·11, 1·80 Å).

PLATE 2. Sample: cement paste, w/c 0·65, $5\frac{1}{2}$ years old, ground and dried. EM: CH crystal plate. ED: CH $hk.0$ spots.

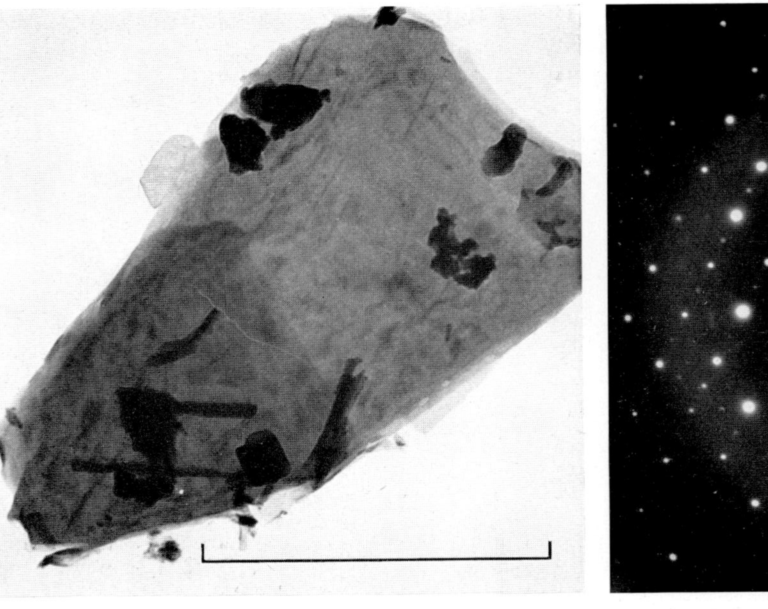

PLATE 3. Sample: cement paste, Type I cement, w/c 0·55, about 6 months old. EM: plate of C–A–H phase, probably $C_4(A, F, \bar{S})$-hydrate (hex.), with some adsorbed tobermorite material. ED: pseudo-hexagonal pattern, pseudo-cell (strong spots) with $a_H = 5{\cdot}70\text{–}5{\cdot}64$ Å, true cell of double size.

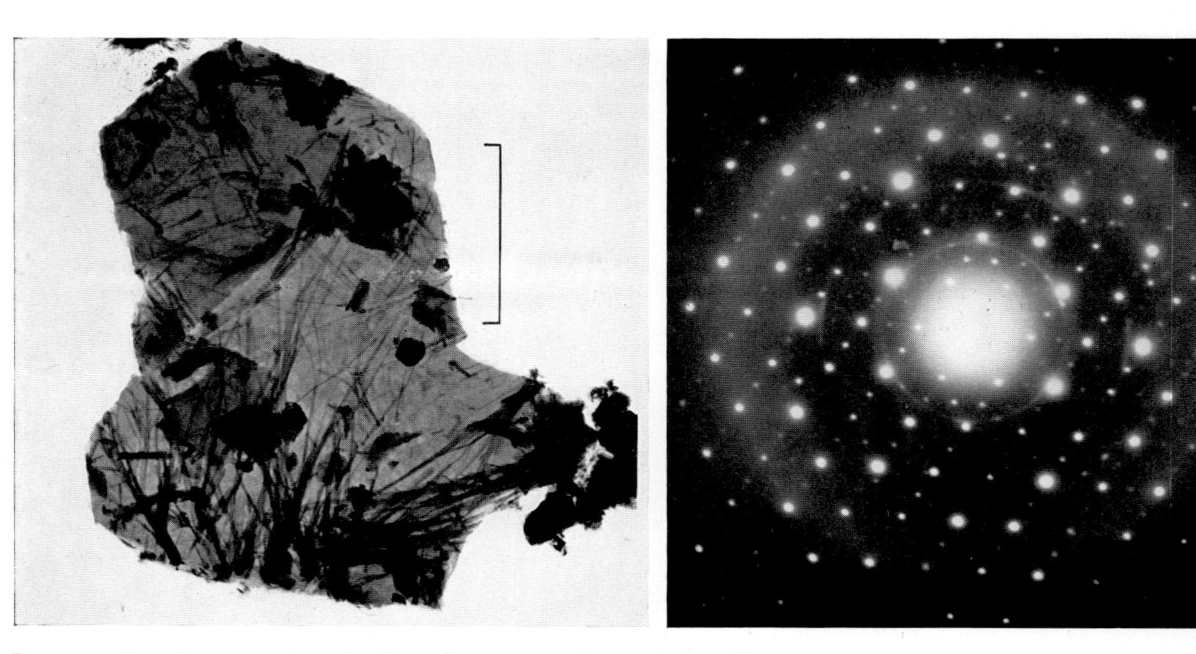

PLATE 4. Sample: cement paste, Type I cement, about 8000 cm^2/g., w/c 1·5, about $3\frac{1}{2}$ years old. EM: plate of C–A–H phase, probably $C_4(A, F, \bar{S})$-hydrate (hex.), with adsorbed fibres of tobermorite material. ED: pseudo-hexagonal pattern, pseudo-cell (strong spots) with $a_H = 5{\cdot}67$–$5{\cdot}62$ Å, true cell of double size, and tobermorite reflections (rings) at 3·03, 1·81 Å.

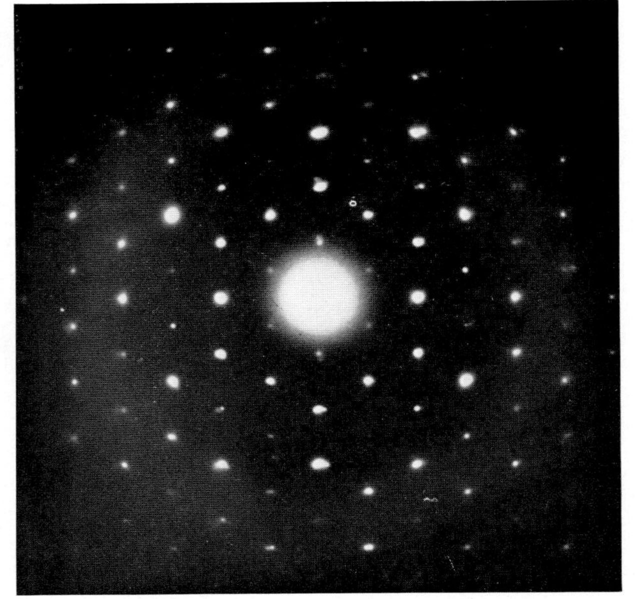

PLATE 5. Sample: cement suspension, Type I cement, w/c 4, mixed 1 day, age 4 days. EM: crystal of C–A–H phase, probably C_4(A, F, $\bar{S}$)-hydrate (hex.) with adsorbed ettringite-type rods and C–S–H (I) foils. ED: hexagonal-spot pattern, $a_H = 5{\cdot}66$ Å.

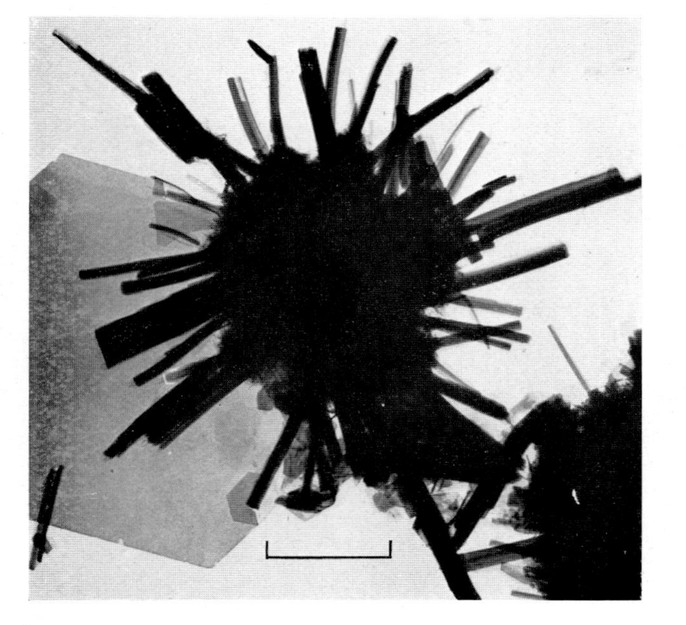

PLATE 6. Sample: as in Plate 5, immediately after mixing 1 day. EM: cement particle covered with ettringite-type formations.

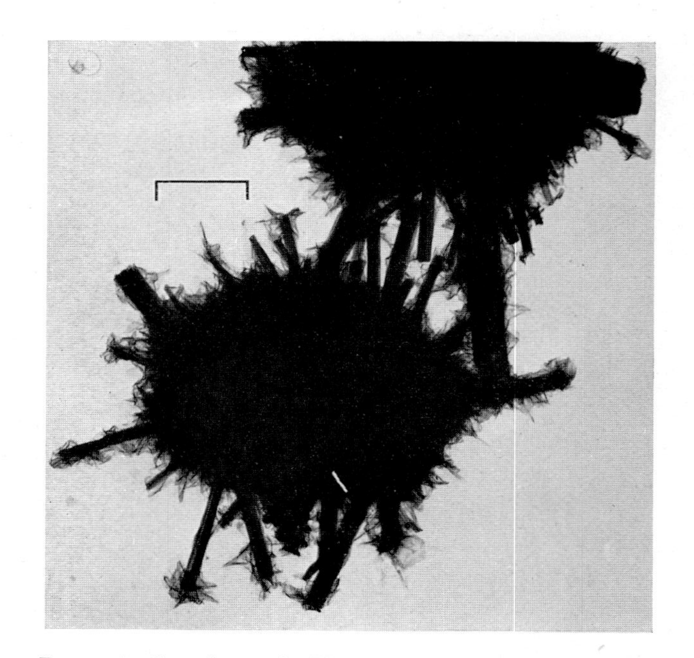

PLATE 7. Sample: as in Plate 5. EM: ettringite particles overgrown with foils of C–S–H (I).

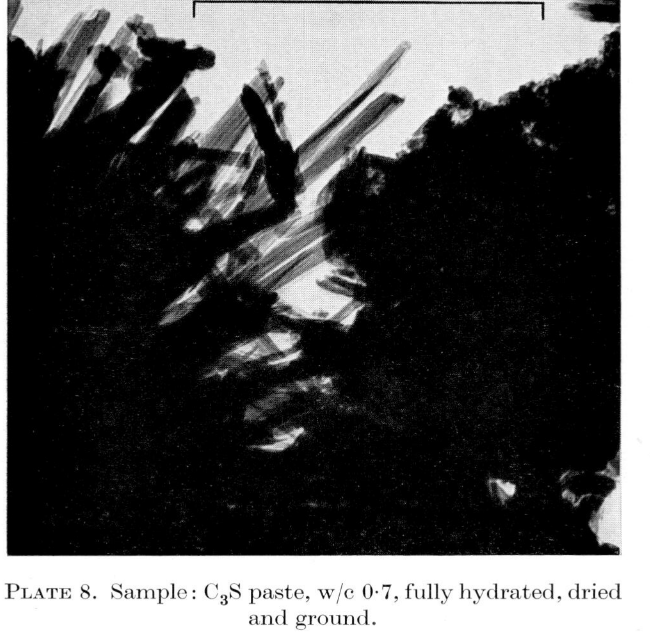

PLATE 8. Sample: C_3S paste, w/c 0·7, fully hydrated, dried and ground.

PLATE 9. Sample: same as in Plate 8, but with powdered paste dispersed in water for a few minutes.

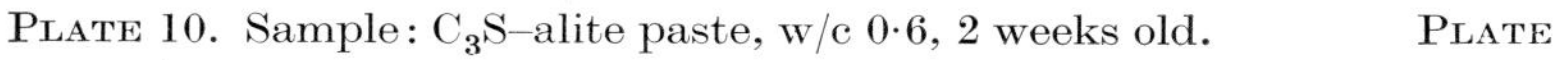

PLATE 10. Sample: C_3S–alite paste, w/c 0·6, 2 weeks old.

PLATE 11. Sample: β-C_2S paste, w/c 0·7, about 70% hydrated, dried and ground.

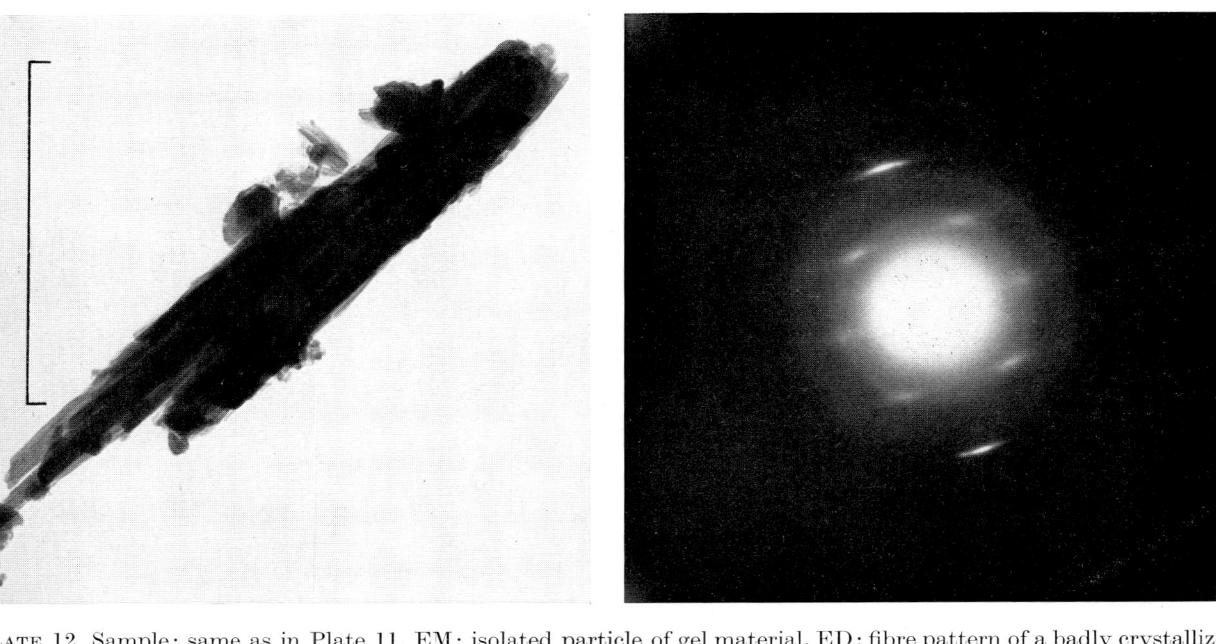

Plate 12. Sample: same as in Plate 11. EM: isolated particle of gel material. ED: fibre pattern of a badly crystallized high-lime tobermorite (six diffuse spots at about 2·9–3·0 Å, streak at 1·825 Å).

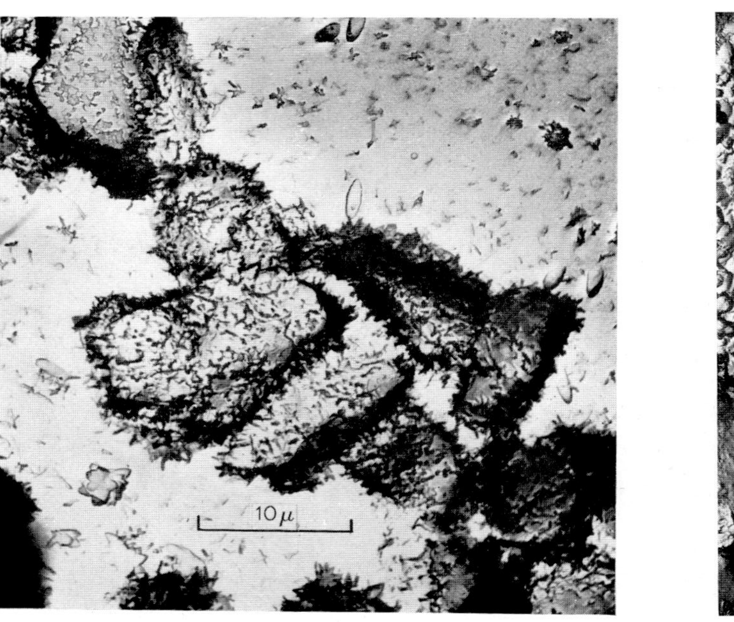

PLATE 13. Sample: C_3S paste, w/c 0·6, hydrated 1 h in thin layer on glass, carbon replica of surface. (*Courtesy of* Czernin)

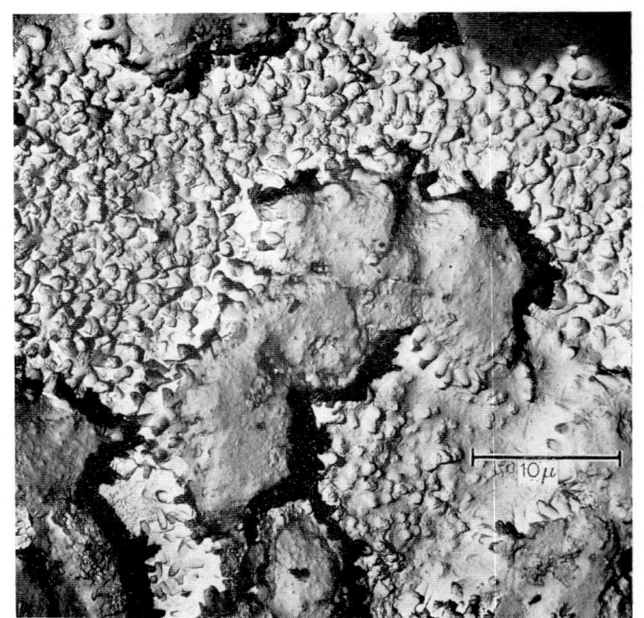

PLATE 14. Sample: Same as in Plate 13, hydrated 28 days. (*Courtesy of* Czernin)

Plate 15. Sample: cement paste, Type I cement, w/c 0·65, age $5\frac{1}{2}$ years, ground and dried. EM: particles of various characteristic types, plates, fibres and small, irregular flakes.

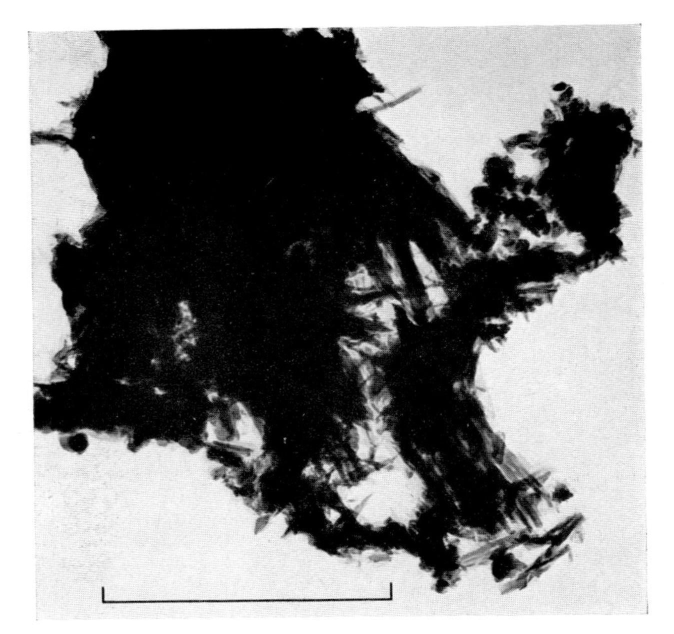

Plate 16. Sample: cement paste, Type I cement, w/c 0·55, age 5 months. EM: mainly fibrous aggregates of a high-lime tobermorite phase.

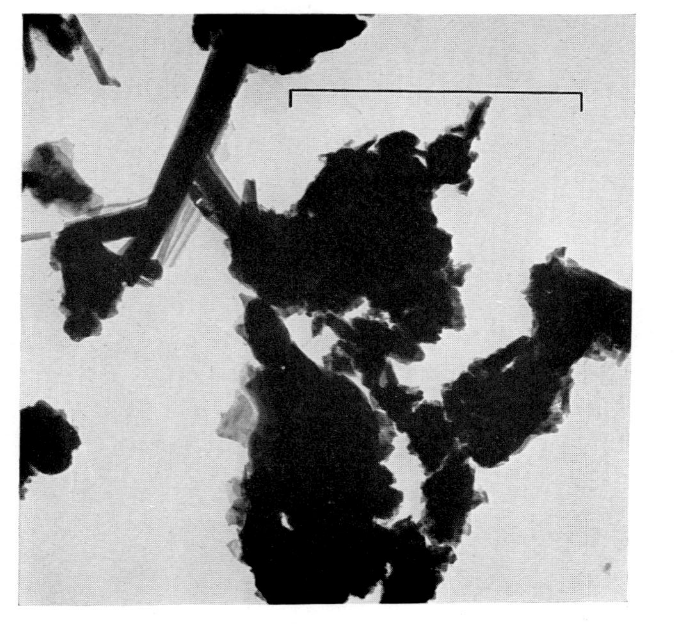

PLATE 17. Sample: cement paste, Type IV (low-heat) cement, w/c 0·6, age $7\frac{1}{2}$ years, ground and dried.

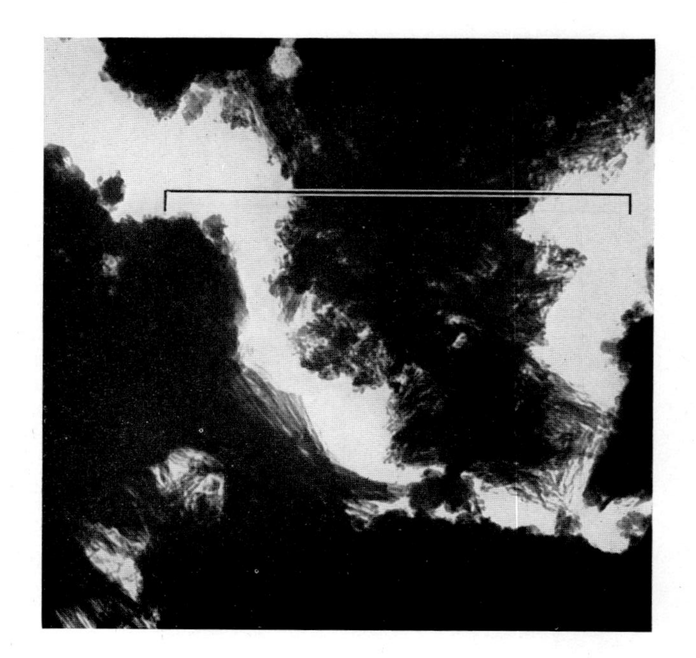

PLATE 18. Sample: cement paste, Type III (high early strength) cement, w/c 0·6, age $7\frac{1}{2}$ years, ground and dried.

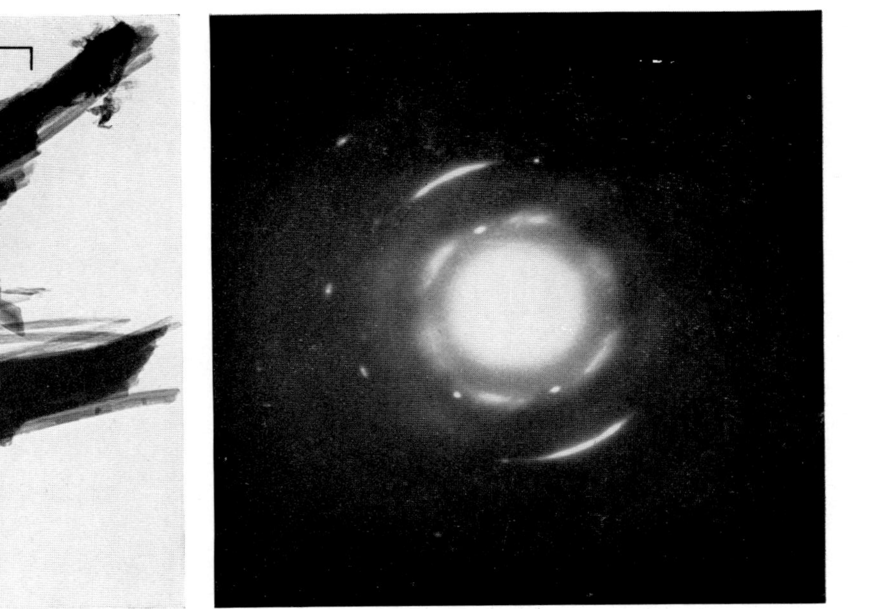

Plate 19. Sample same as in Plate 17. EM: bundle of parallel-oriented fibres or laths. ED: fibre pattern of a badly crystallized high-lime tobermorite (similar to Plate 12).

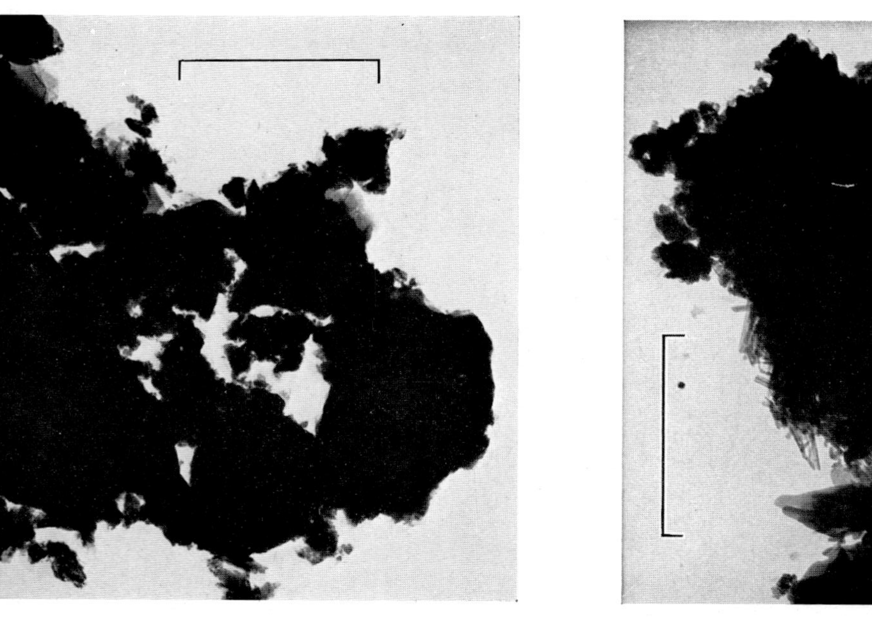

PLATE 20. Sample: cement paste, Type I cement, w/c 0·35, age $9\frac{1}{2}$ years.

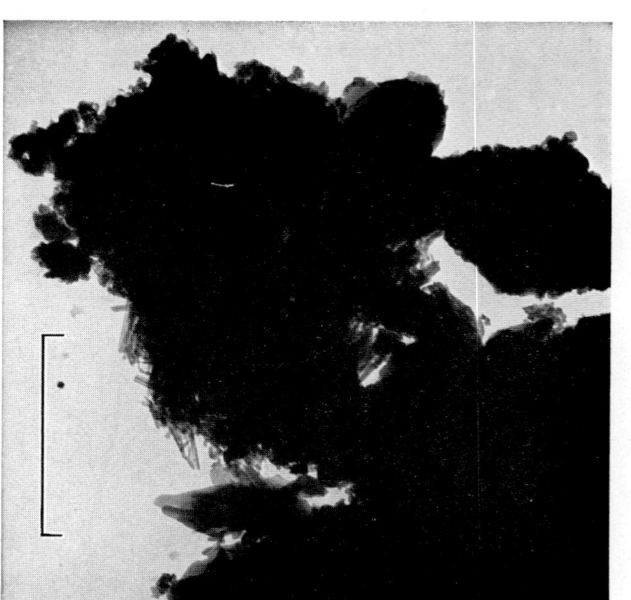

PLATE 21. Sample: cement paste, Type I cement, w/c 0·7, age $9\frac{1}{2}$ years.

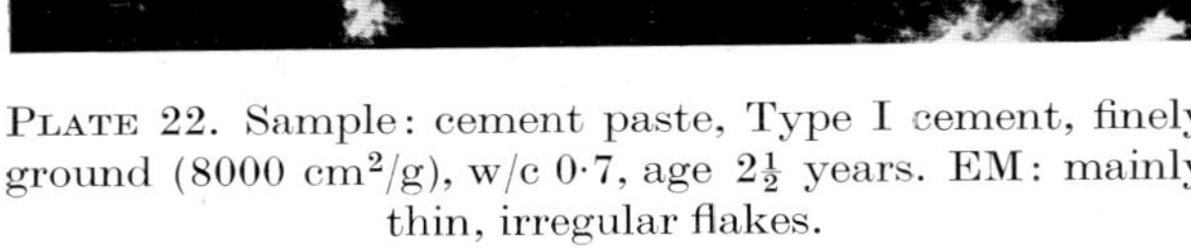

PLATE 22. Sample: cement paste, Type I cement, finely ground (8000 cm^2/g), w/c 0·7, age $2\frac{1}{2}$ years. EM: mainly thin, irregular flakes.

PLATE 23. Sample: cement paste, Type I cement, finely ground (8000 cm^2/g), w/c 1·5, age $3\frac{1}{2}$ years. EM: mass of intertwined fibres of a high-lime tobermorite phase.

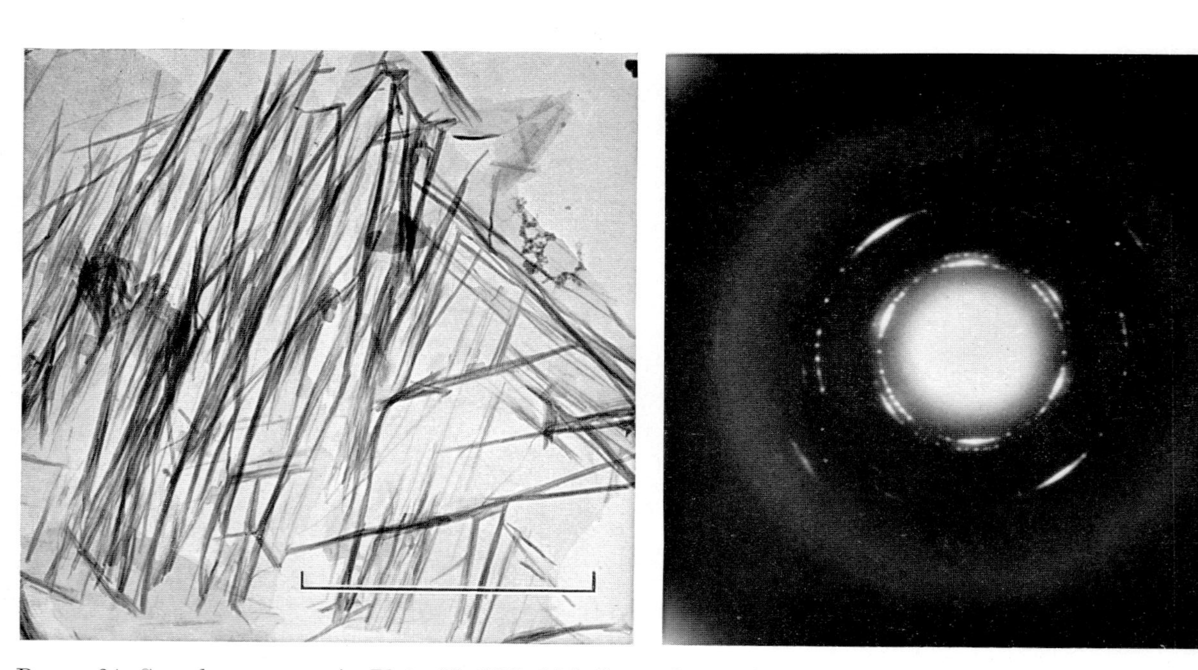

Plate 24. Sample: same as in Plate 23. EM: high-lime tobermorite fibres. ED: fibre pattern with tobermorite $hk.0$ reflections (3·05, 2·79, 1·82, 1·65, 1·53, 1·40 Å, etc.).

Plate 25. Sample: cement paste, standard cement, w/c 0·5, 6 h after mixing (during setting).

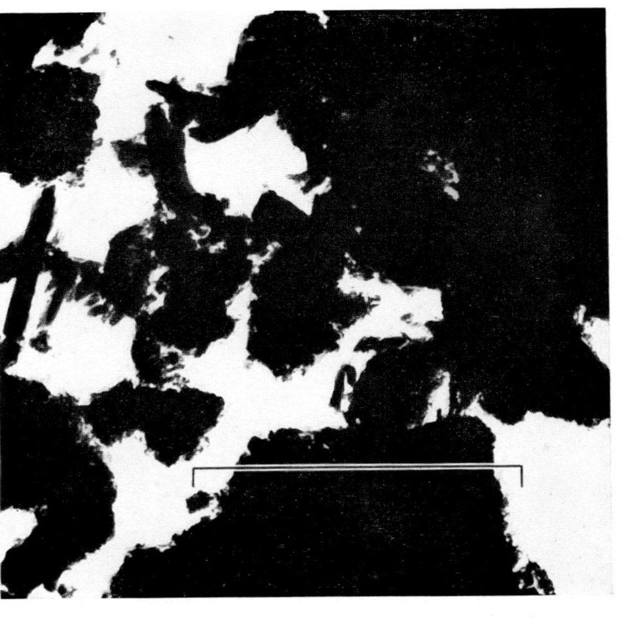

Plate 26. Sample: same as in Plate 25, after 4 days (paste hardened).

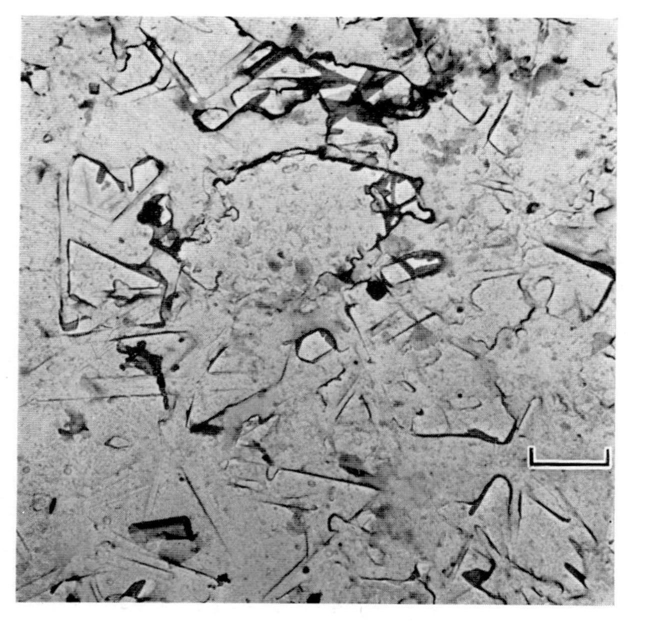

PLATE 27. Sample: cement paste, carbon replica of surface of hardened paste. EM: structureless gel mass with embedded crystals of fibrous or plate-like habits.
(*Courtesy of* Czernin)

PLATE 28. Sample: same as in Plate 27. EM: gel mass with protruding fibrous or rod-like particles, possibly ettringite-type formations.
(*Courtesy of* Czernin)

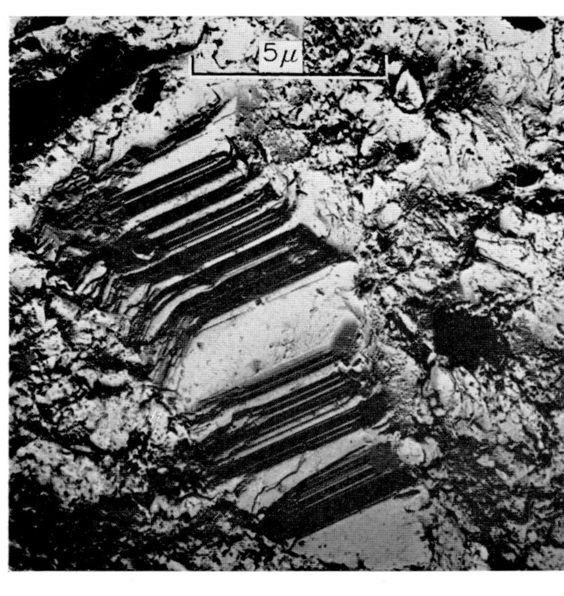

PLATE 29. Sample: cement–fly-ash paste, w/c 0·48, 20 weeks old, cellulose acetate–chromium–carbon replica of fracture surface. EM: gel mass with embedded stack of plate crystals, probably segregated CH phase.
(*Courtesy of* Saji)

PLATE 30. Sample: as in Plate 29, 6 weeks old (Plates 29 and 30: courtesy Saji). EM: impression of fly-ash filler particle, finely fibrous structure of fractured cement gel around filler grain.
(*Courtesy of* Saji)

Plate 31. Sample: cement paste, with normal content of alkali, hydrated 7 h, carbon replica of paste surface.
(*Courtesy of* zur Strassen)

Plate 32. Sample: same as in Plate 31, hydrated 56 days.
(*Courtesy of* zur Strassen)

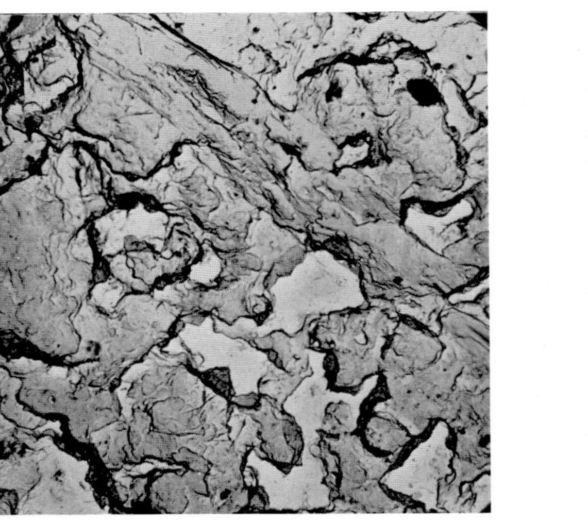

PLATE 33. Sample: cement paste, free from alkali, hydrated 7 days, carbon replica. (*Courtesy of* zur Strassen)

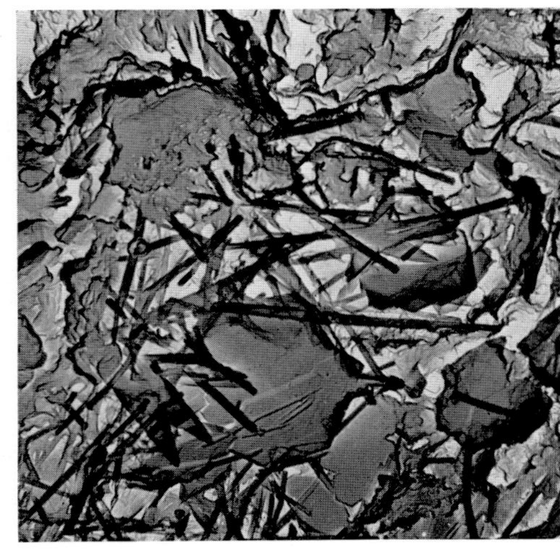

PLATE 34. Sample: same as in Plate 33, hydrated 56 days. (*Courtesy of* zur Strassen)

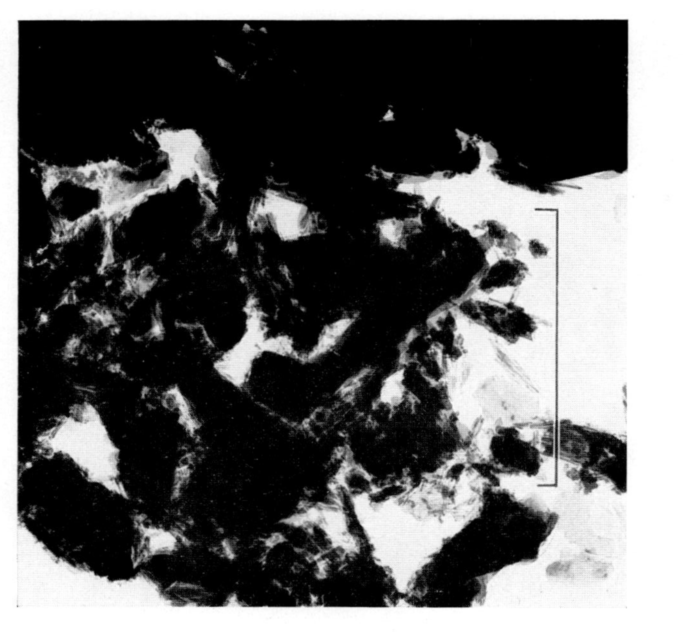

PLATE 35. Sample: cement paste, Type I cement, w/c 0·46, cured 24 h at 177°C in high-pressure steam (about 700 kg/cm^2), stored 7 years. EM: mainly a fibrous, badly crystallized tobermorite phase.

PLATE 36. Sample: cement paste, Type I cement, w/c 0·4, cured 24 h in saturated steam at 215°C, stored about 1 year. EM: mainly large particles of a recrystallized gel phase, probably α-C_2SH.

PLATE 37. Sample: same as in Plate 36. EM: cluster of lath-like or rod-like crystals of $C_6S_2H_3$. ED: pattern given by isolated single crystal, mainly the even-order reflections from a cell of dimensions $7{\cdot}5 \times 10{\cdot}0$ Å.

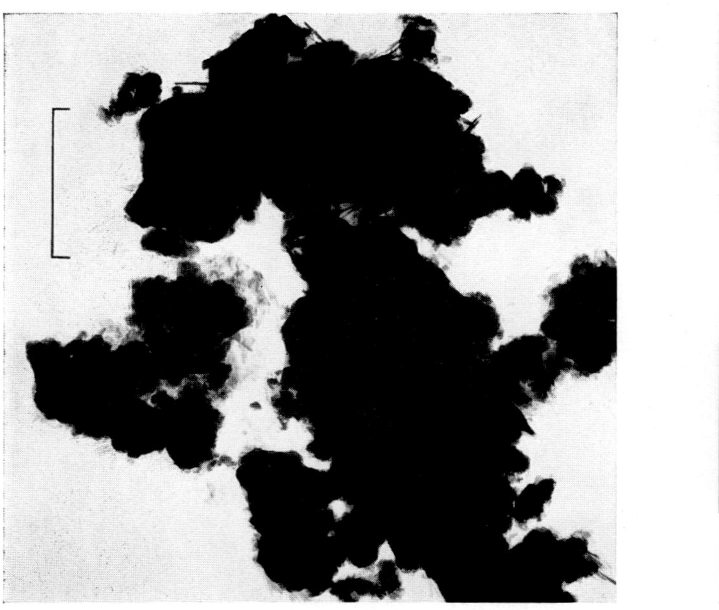

Plate 38. Sample: cement paste, high-alumina (Alcoa) cement, hydrated at room temperature.

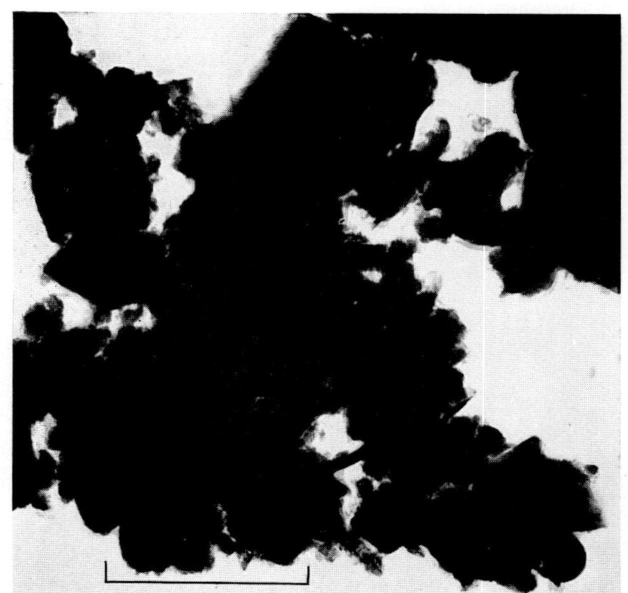

Plate 39. Sample: cement paste, high-alumina (Lumnite) cement, hydrated at 95°C for 72 h.

of lath-like or rolled-sheet particles, radiating from the surface. With addition of C_3A or gypsum or both, the development of this phase was virtually unchanged, the only visible difference being that the particles assumed slightly more acicular or fibrous habits when gypsum was present. However, in this case and particularly in the alite–C_3A–gypsum paste, ettringite-type particles were also formed and persisted for many weeks. In the alite–C_3A mixture the normal, rapid reaction of C_3A and the formation of a hexagonal plate phase seemed to be entirely suppressed, and such a phase could not be detected with certainty until after 2 weeks.

The early development of the major gel phases is likely to be influenced by uncontrolled variations in the small amount of alkali present in all normal cement clinkers. Because of the high rate at which the alkali ions go into solution, their concentration in the liquid phase during the first few hours may, under certain conditions, greatly exceed that of the calcium ions and will then depress the solubility of lime in the solution. This in turn will lead to variations in the composition and microstructure of the C-S-H phases formed.

The possible influence of alkali has been emphasized by, among others, zur Strassen [*60*], who also used EM methods to demonstrate its effect. The external surfaces of cement or clinker pastes of various compositions, although all with normal contents of alkali, were studied by carbon replicas. In all cases investigated, growth of networks of long, straw-like crystals was observed, starting at an age of a few hours and remaining even after many weeks (Plates 31 and 32). However, in a clinker paste from which all traces of alkali had been removed, the fibrous phase was absent even after 1 week's hydration but was observed in an 8-week sample (Plates 33 and 34). An isolated crystal of this phase, examined in transmission, gave an SED single-crystal fibre pattern, identified tentatively as that of a tobermorite crystal, although with an anomalous direction of the fibre axis. According to zur Strassen, these data imply that the formation of a well crystallized tobermorite phase is accelerated considerably in cement pastes containing appreciable amounts of alkali.

X. Observations on Products of Cement Hydration at Elevated Temperatures

Steam curing for sufficient lengths of time under autoclaving conditions is known to result in complex changes in the morphology of the paste hydration products of cement and cement constituents. EM data for some of the pure compounds formed at high temperatures have been reported, e.g. tobermorite [*25–27, 29–31, 34–37, 60*], xonotlite [*28, 35*], hillebrandite [*11, 35*], afwillite [*47, 59*] and tricalcium silicate hydrate

[*48*]. Only some EM observations of a general nature on hydration in pastes at 100–200° C [*22*] can be given here.

The rate of decomposition of the anhydrous phases increases with the temperature. Surface reactions and the formation of a coat of fine fibres or foils enveloping the cement crystals are greatly accelerated. Thus, even in β-C_2S paste cured at 100° C the crystals are rapidly covered with a felt-like hydration product, though most of the material remains unhydrated even after 3 days, which is perhaps the longest period of steam curing applicable in practical work. Even at temperatures as high as 200° C some β-C_2S may be left after corresponding periods of curing. Pastes of C_3S or of cement react faster at all temperatures, but require a few days for complete hydration even at about 150° C.

The finely fibrous or foil-like surface coat, which consists of a badly crystallized tobermorite phase, seems to be the first-formed product of hydration, together with crystalline plates of CH in cement and C_3S pastes, for all compositions of starting materials processed at 100–200° C. Although it is probably a metastable phase, it has been observed to persist for many days at 100° C. At somewhat higher temperatures, approaching 150° C, this cryptocrystalline material converts slowly to better crystallized phases. Typical secondary phases are large, rectangular, block-shaped particles of α-dicalcium silicate hydrate (α-C_2SH) embedded in a finely fibrous phase in β-C_2S paste, and similar formations together with long, narrow rods or needles of tricalcium silicate hydrate ($C_6S_2H_3$) in C_3S and cement pastes. The process of recrystallization becomes increasingly rapid at temperatures of about 200° C, and the primary, cryptocrystalline matrix seems to disappear during the first few hours of curing. The characteristic phases appearing at this temperature in β-C_2S paste are laths or striated sheets, probably of a high-lime tobermorite, together with α-C_2SH blocks. C_3S paste shows blocks or irregular particles of α-C_2SH mixed with bundled rods or needles of $C_6S_2H_3$, while cement paste shows a similar mixture of phases, but with a smaller proportion of needles (cf. Plates 35–37).

Upon addition of amorphous silica in varying proportions much of the primary, badly crystallized tobermorite is observed to form by reaction between the lime liberated from the anhydrous phases and the finely divided silica. The rate of decomposition of the anhydrous compounds may in certain cases be greatly increased by this process. Secondary phases are formed by recrystallization of the gel, mainly in accordance with the overall C/S ratio of the mix and the time–temperature relationship. Characteristic products of recrystallization are long, slender needles of xonotlite at 200° C and C/S 1, flat plates of well crystallized tobermorite at 150–200° C and C/S 1–1·5, block-shaped crystals of

α-C_2SH and fibres or needles of members of the foshagite–hillebrandite group at 150–200° C and C/S 1·5–2, and bundles of needles or slender rods of $C_6S_2H_3$ in mixes containing C_3S or alite at C/S ratios of about 2 or above.

It should be emphasized that all the products obtained under the conditions of composition and processing used in normal steam-curing practice appear as mixtures of several phases, although with one or two phases predominating. The attainment of true equilibrium, in the given range of temperatures and compositions, would certainly require many more days or weeks of curing.

It has not been possible to detect any signs of C-A-F-H phases in steam-cured cement preparations. The low-temperature phases described earlier are probably unstable above 100° C, and it has been suggested by Kalousek [*61*] that the aluminium and ferric ions are incorporated in the C-S-H phases. Alternatively, these ions may, at least at temperatures approaching 200° C, enter into reaction with silica, forming thin plates with a crystal symmetry resembling that of clays, as indicated by certain EM data [*16*].

XI. Reactions between Cement Paste and Aggregate Materials

The possible reactions at the interfaces between a cement paste and various rock materials are of great importance for the normal use of cement in mortar and concrete. Of such reactions, the so-called alkali–aggregate reaction is perhaps the best known, owing to the conspicuous damage which it may cause. However, hardly any silicate or carbonate mineral commonly used as aggregate is completely unreactive towards the liquid phase of cement paste. It has been demonstrated by mechanical tests that some kind of interaction occurs at the boundary between cement paste and various rocks. It would seem that EM techniques are very suitable for studying the microstructural aspects of these interaction phenomena, but practically no EM data on this problem have yet been reported.

The most likely cause of interaction would be the solution of silica from the surface of the aggregate and its subsequent effect on the microstructure of the paste in the vicinity. One case of interaction was observed by Gaze and Robertson [*39*] in a foamed asbestos–cement paste cured at room temperature. The asbestos fibres were enveloped in very thin, crinkly foils of a low-lime tobermorite, which otherwise is not readily formed in paste hydration products. Another example of aggregate reaction was reported by Saji [*57*] from replica studies of the fracture surfaces of cement pastes containing fly-ash as a filler material (Plate 30).

The rounded surfaces of broken-away fly-ash particles were mostly clean and smooth, whereas the corresponding impressions of the gel structure near the surfaces were covered with a layer of loosely aggregated, fibrous particles. The fracture seemed to occur preferentially in these surface layers.

REFERENCES

1. Radczewski, O. E., Müller, H. O. and Eitel, W. (1939). *Zement* **28**, 693.
2. Radczewski, O. E., Müller, H. O. and Eitel, W. (1939). *Naturwissenschaften* **27**, 807.
3. Eitel, W. (1942). *Zement* **31**, 489.
4. McMurdie, H. F., work reported in R. H. Bogue (1955). "The Chemistry of Portland Cement", 2nd Ed. Reinhold, New York.
5. Sliepcevich, C. M., Gildart, L. and Katz, D. L. (1943). *Industr. Engng Chem.* (*Industr. Ed.*) **35**, 1178.
6. Boutet, D. (1950). *Travaux* **183**, 1.
7. Bernal, J. D. (1954). *Proceedings of the Third International Symposium on the Chemistry of Cement, London 1952*, pp. 216, 257. Cement and Concrete Association, London.
8. Bogue, R. H. (1954). *Proceedings of the Third International Symposium on the Chemistry of Cement, London 1952*, p. 254. Cement and Concrete Association, London.
9. Swerdlow, M., McMurdie, H. F. and Heckman, F. A. *Proc. 3rd International Conference on Electron Microscopy, London, 1954*, p. 500, reprinted in PCAF, NBS, Paper No. 71 (1957).
10. Swerdlow, M. and Heckman, F. A., work reported in R. H. Bogue (1955). "The Chemistry of Portland Cement", 2nd Ed. Reinhold, New York.
11. Grudemo, Å. (1955). *Proc. Swed. Cement and Concrete Research Inst.* No. 26, 1.
12. Schimmel, G. (1957). *Zement-Kalk-Gips* **10**, 134.
13. Andrievskii, A. J., Tichonov, V. A., Shepinova, L. G. and Nabitovich, I. D. (1959). *Stroit. Mater., Lvov Polytech.*, No. 3, 33.
14. Danielsson, U. (1959). *Zement u. Beton*, No. 16, 22.
15. Grudemo, Å. (1959). *Gullkornet* **9**, 16.
16. Grudemo, Å. (1962). *Chemistry of Cement, Proceedings of the Fourth International Symposium, Washington 1960*, p. 615. National Bureau of Standards Monograph 43. U.S. Department of Commerce.
17. Uchikawa, H. and Takagi, S. (1961). *Zement-Kalk-Gips* **14**, 153.
18. Brunauer, S., Kantro, D. L. and Copeland, L. E. (1958). *J. Amer. chem. Soc.* **80**, 761.
19. Brunauer, S. and Greenberg, S. A. (1962). *Chemistry of Cement, Proceedings of the Fourth International Symposium, Washington 1960*, p. 135. National Bureau of Standards Monograph 43. U.S. Department of Commerce.
20. Buttler, F. G., Dent Glasser, L. S. and Taylor, H. F. W. (1959). *J. Amer. ceram. Soc.* **42**, 121.
21. Iwai, T. and Watanabe, K. (1957). *Proc. Regional Conference on Electron-Microscopy in Asia and Oceania, 1st Conference, Tokyo, 1956*, 288.
22. Chatterji, S. and Grudemo, Å. Unpublished work.
23. Stolnikov, V. V. (1950). *Zh. prik. Khim., Leningr.* **23**, 719.

24. Astreeva, O. M. and Lopatnikova, L. Y. (1957). *Tsement* **23**, 11.
25. Gard, J. A., Howison, J. W. and Taylor, H. F. W. (1959). *Mag. Concr. Res.* **11**, 151.
26. Kurczyk, H. G. (1960). Dissertation, T. H. Aachen.
27. Kurczyk, H. G. and Schwiete, H. E. (1962). *Chemistry of Cement, Proceedings of the Fourth International Symposium, Washington 1960*, p. 349. National Bureau of Standards Monograph 43. U.S. Department of Commerce.
28. Kalousek, G. L. (1954). *J. Amer. Concr. Inst.* **26**, 233.
29. Kalousek, G. L. (1955). *J. Amer. Concr. Inst.* **26**, 989.
30. Gaze, R. and Robertson, R. H. S. (1956). *Mag. Concr. Res.* **8**, 7.
31. Gard, J. A. and Taylor, H. F. W. (1956). *Miner. Mag.* **31**, 361.
32. Taylor, H. F. W. and Howison, J. W. (1956). *Clay Min. Bull.* **3**, 98.
33. Grudemo, Å. (1956). *Note on Research in Progress, Swedish Cement and Concr. Res. Inst.*, No. 5.
34. Grothe, H., Schimmel, G. and zur Strassen, H. (1962). *Chemistry of Cement, Proceedings of the Fourth International Symposium, Washington 1960*, p. 194. National Bureau of Standards Monograph 43. U.S. Department of Commerce.
35. Akaiwa, S. and Sudoh, G. (1956). *Semento Gijutsu Nenpo* **10**, 14.
36. Kalousek, G. L. and Prebus, E. F. (1958). *J. Amer. ceram. Soc.* **41**, 124.
37. Kalousek, G. L. (1957). *J. Amer. ceram. Soc.* **40**, 74.
38. Grudemo, Å. (1954). *Proceedings of the Third International Symposium on the Chemistry of Cement, London 1952*, p. 247. Cement and Concrete Association, London.
39. Gaze, R. and Robertson, R. H. S. (1957). *Mag. Concr. Res.* **9**, 25.
40. Funk, H. (1960). *Silikattech.* **11**, 375.
41. Bernard, P. (1957). *Rev. Matér. Constr.* No. 507, 351.
42. Stork, J. and Bystricky, V. (1959). *Rev. Matér. Constr.* Nos. 526/527, 173.
43. Stork, J. (1960). *Tsement* **1**, 25.
44. Funk, H. (1957). *Z. anorg. Chem.* **291**, 276.
45. Buckle, E. R. and Taylor, H. F. W. (1959). *J. appl. Chem.* **9**, 163.
46. Schwiete, H. E. and Müller-Hesse, H. (1959). *Zement u. Beton* No. 16, 25.
47. Copeland, L. E., Schulz, E. G. and Brunauer, S. (1960). *Silikattech.* **11**, 367.
48. Buckle, E. R., Gard, J. A. and Taylor, H. F. W. (1958). *J. chem. Soc.* 1351.
49. Grudemo, Å., quoted by Taylor, H. F. W. (1961). *Research Lond.* **14**, 154.
50. Czernin, W. (1959). *Betonsteinztg.*, No. 9, 1.
51. Funk, H. (1960). *Silikattech.* **11**, 373.
52. Gille, F. (1955). *Zement-Kalk-Gips* **8**, 128.
53. Gille, F. (1959). *Zement u. Beton* No. 16, 21.
54. Czernin, W. (1959). *Zement u. Beton* No. 16, 21.
55. Czernin, W. (1958). *Zement-Kalk-Gips* **11**, 381.
56. Saji, K. (1959). *Semento Gijutsu Nenpo* **13**, 1.
57. Saji, K. (1959). *Zement-Kalk-Gips* **12**, 418.
58. Copeland, L. E. and Schulz, E. G. (1962). *Chemistry of Cement, Proceedings of the Fourth International Symposium, Washington 1960*, p. 648. National Bureau of Standards Monograph 43. U.S. Department of Commerce.
59. Copeland, L. E. and Schulz, E. G. (1962). *J.P.C.A. Res. Dev. Labs* **4**, 2.
60. zur Strassen, H. *Tagungsber. Zementindustr.* **21**, 71.
61. Kalousek, G. L. (1954). *Proceedings of the Third International Symposium on the Chemistry of Cement, London 1952*, p. 334. Cement and Concrete Association, London.

CHAPTER 10

The Physical Structure of Portland Cement Paste

T. C. POWERS

Portland Cement Association Research and Development Laboratories, Skokie, Illinois, U.S.A.

CONTENTS

I. Introduction

A. Cement Paste

Portland cement paste is the essential ingredient of ordinary concrete. In the freshly mixed state it is composed mostly of grains of cement and an aqueous solution; in the hardened, mature state it is composed mostly of solid reaction products and spaces that are penetrable by water. Hardened paste is not simply a chemical product, however; it is a rigid structure formed by solid reaction products. Such a distinction is necessary because various structures can be formed, and in practice are formed, from identical chemical products.

B. Cement Gel and Pore Space

The hydration products that form within a body of cement paste occur as dense masses that have a characteristic porosity. This dense though porous substance is called *cement gel*, since it is composed of solid bodies having a specific surface area such as is characteristic of colloidal materials. The amount of cement gel is usually not sufficient to fill all the space within the visible boundaries of a specimen of cement paste. The space not filled with cement gel is called *capillary space*, or some variant of that term. Hence, hardened cement paste is composed of cement gel, capillary space, if any, and a residue of unreacted cement, if any. Total porosity comprises the capillary space and the pores in cement gel, the latter being called *gel pores*. Pores are generally submicroscopic, the gel pores being much smaller than capillary pores. Structural differences among pastes made from the same cement are first of all due to differences in capillary porosity.

C. Properties of Cement Gel

Cement gel is composed of the several chemical species that constitute hydrated Portland cement. Chemical constitution of this material is the subject of other chapters and therefore will be dealt with here only incidentally. Chemical constitution is not very much involved, because it turns out that the physical characteristics of hardened cement gel are influenced to only a minor degree by such differences in chemical composition as are found among the different types of Portland cement. So far as mechanical properties are concerned, the most significant characteristics are manifestations of the physical states of the solids composing cement gel, and of the porosity.

The specific volume of cement gel is about 0·567 cm^3/g dry weight; the porosity is about 28% of the total volume. Cement gel is permeable to fluids, but its coefficient of permeability is comparatively low, about 10^{-14} cm/sec. The solid matter of cement gel has a mean specific surface area of about 220 m^2/g dry weight, or about 550 m^2/cm^3 solid volume.

The main component of cement gel is believed to be tobermorite gel. Some components of cement gel are not colloidal, that is to say their specific surface area is relatively insignificant, but exactly what fraction is in this category is not yet known. The principal non-colloidal component is crystalline calcium hydroxide. It usually constitutes 20–30% of the weight of dry cement gel, or 1/6–1/5 of the overall volume.

D. General

Much of the information mentioned above has been obtained by indirect methods. Concepts of structure have been deduced from data on physical properties, aided by information gained by electron microscopy. In the following discussion we shall describe the evidence and the resulting concept of structure.

Formation of structure begins with the mixing of cement and water, which is to say with the formation of fresh paste. Fresh paste has structure, and its structure establishes a pattern for the subsequent gradual development of the final structure. Therefore, we begin by describing the physical properties of cement paste as they appear soon after the mixing process has ended. The primary units of this structure are the original cement grains, whose characteristics also must be considered.

II. Structure of Fresh Paste

A. Physical Properties of Portland Cement

The mean specific gravity of the particles composing Portland cement is about $3{\cdot}13 \pm 0{\cdot}05$, the indicated variation from the mean being a function of chemical composition. The fineness of grinding is controlled by the manufacturer and differs somewhat among cements prepared for different purposes. Also, the fineness of a given type of cement is not always the same in different countries. In American practice 90–99% of the product passes a sieve having square openings 44 μ wide (sieve No. 325) and 30–50% of it is represented by particles smaller than 10 μ. The specific surface area (American practice) ranges from about 1800 to 2600 cm^2/g as measured by the Wagner turbidimeter, or 3400 to 5300 cm^2/g as measured by the Lea and Nurse or Blaine apparatus.

B. Typical Proportions of Paste

In concrete mixtures the usual proportion of a mixture of cement and water is between 1 and 3 parts of cement to 1 part of water by weight, corresponding to weight ratios of water to cement of 0·33–1·00. Ratios between 0·45 and 0·80 are the most common.

C. Initial Reactions and Dormant Period

During a short period beginning when cement and water are first brought into contact (at room temperature) and during the time of mechanical stirring or kneading, relatively rapid chemical reactions occur. Within 5 min the rate of reaction subsides to a low level; there is then a period during which the paste normally remains plastic. This period, which has been called the *dormant period*, normally lasts 40–120 min, depending on the characteristics of the cement.

The fraction of cement used up in the initial reaction is small, perhaps 1%, and the insoluble part of the reaction product, which is most of it, adheres to the surfaces of the cement grains. The resultant alteration of the original cement grains is so slight that the Wagner specific surface area remains virtually unchanged and the grains appear to have retained their original shapes. The liquid phase of the mixture has become an aqueous solution having a pH in the neighbourhood of 13, the principal ions being Ca^{++}, Na^{+}, K^{+}, OH^{-} and SO_4^{2-}. (The sulphate ion virtually disappears, usually within about 24 h.)

Particles of cement in fresh paste are in a flocculent state, that is they tend to stick together. But cement paste does not consist of separate floccules of cement grains dispersed in water; on the contrary, it was found by Steinour [*1*] that the whole body of paste normally constitutes a single floc, the floc structure being a rather uniform reticulum of cement particles. The particle arrangement apparently tends to be that which best accommodates mutual attractions. Because cement particles, or rather the reaction products that coat them, are hydrophylic, and because of a positive zeta potential, the net interparticle forces of attraction are relatively weak and the cement particles remain appreciably separated from each other.

D. Rheological Characteristics

At ratios of water to cement between about 0·25 and 0·35 the consistencies of cement pastes range from that of putty to that of thick cream. At the other extreme, typically at ratios of water to cement exceeding 0·7, the mixture has quasi-fluid characteristics. In the intermediate range the flow characteristics of cement paste are like those of a Bingham body, although there are various complicating factors [*2*]. In all normal cases the pastes exhibit a yield value, and at shear stresses well above the yield value some (but not all) show a linear force–flow relationship.

E. Permeability

Having in general the properties of a solid, although a weak and plastic one, fresh cement paste behaves like a porous permeable solid. Its coefficient of permeability depends on the density of the structure (that is on its water content), on the specific surface area of the cement, on the shapes of the particles and on the amount of surface coating produced by the initial reaction. Typically the coefficient of permeability is of the order of 10^{-4} cm/sec.

F. Bleeding (Settlement)

The permeability of fresh paste is exhibited in the phenomenon known as bleeding. Because of the slight separation of adjacent particles at points of near contact, the particles in freshly mixed paste are subject to settlement through the fluid in which they are suspended, the settlement being activated by the force of gravity. Because of interparticle cohesive forces, large and small particles settle at the same rate, that is as a single structural unit. Such settlement occurs slowly, a typical rate being about 2 μ/sec. Since the particles near the bottom can settle in a shorter length of time than those near the top, the effect of settlement is to increase the density at the bottom relative to that at the top of a sample. However, this effect in small bodies of paste is not large, and as a good first approximation cement paste can be considered an isotropic material. The total amount of settlement for a small depth of paste ranges from about 5 to 25% of the original height, depending on water/cement ratio and cement characteristics, a typical amount being 10%.

The rate and amount of bleeding are smaller the finer the cement. Since, as given by the Kozeny–Carman equation for permeability [3], the coefficient of permeability is lower the higher the specific surface area, the rate of bleeding is also lower. The amount of bleeding per unit volume (termed bleeding capacity) is similarly affected; because of the slight separation at points of near contact, the larger the number of particles per gram the greater the bulking effect of particle separation at a given solid concentration. The greater cohesive strength of the floc structure could be another factor.

We have dealt with fresh paste as being composed of discrete cement particles suspended in a continuous fluid medium; we have also dealt with it as a network having a degree of mechanical continuity. When regarded in the latter way, the interstitial spaces of the paste can be thought of as capillaries and the water occupying those spaces can be called capillary water.

G. Influence of Fineness

All the properties of fresh paste, and some of the properties of mature paste, are influenced by the fineness of the cement. The effects of initial chemical reactions on the physical properties of fresh paste are greater the higher the specific surface area of the cement. At a given water content the cohesive strength and shear resistance are greater; effects on permeability and bleeding characteristics have already been mentioned.

Each cement particle is a source of supply for hydrated cement. Although it may become completely decomposed, a particle of cement marks a region in the cement gel it produces where the density is high, though not necessarily higher than in some other places. At a given cement concentration the number of such points of high density per unit volume is greater the finer the cement. Thus, one aspect of the physical structure of mature paste depends on the number of cement particles and hence on the fineness of the cement.

H. Typical Structure

Summing up, we may picture the structure of a typical fresh cement paste as that of a very weak, permeable solid formed of cement particles suspended in an aqueous solution. The network of particles is normally so dense that the average distance between the particles is considerably smaller than the diameter of the particles composing most of the mass. A unit volume typically might contain 40% solids and 60% fluid, although in practice the range of composition is considerable. At a given cement content the properties are different for cements having different specific surface area, and different early reaction characteristics.

III. Transition from the Fresh to Hardened State

A. Period of Setting

At the end of the dormant period a period of relatively rapid chemical reaction sets in which usually lasts about 3 h. During this time the paste loses its plasticity and, if it is a paste prepared according to standard test methods, passes through arbitrarily defined degrees of firmness known as the *initial set* and *final* set. Under standard test conditions final set commonly occurs at about the sixth hour after first contact with water. After final set, chemical reactions continue at a diminishing reaction rate until all the cement is consumed, or until one or more of the conditions necessary to the reaction is lacking. The solid product of these reactions, cement gel, tends to fill the capillary channels of the fresh paste, rapidly reducing the volume and size of the capillaries. At the same time cement gel develops in spaces made available as the cement particles are used up. If conditions are such as to allow the reactions to go to completion, the original grains of cement become completely consumed and are replaced by new solids that are different, both chemically and physically, from the source materials. The principal product is believed to belong to the class of materials called *tobermorite gel* [*4*]. Under other conditions some of the cement—residues of the largest particles—will remain as a constituent of hardened paste.

Of course, much of the water is transformed also; it loses its chemical identity and becomes a component of the new solid phases. The rest remains chemically free, but some of it is bound by surface forces.

B. Non-evaporable Water

Most of the hydrates are so imperfectly crystallized that they do not exhibit definite values for temperature and vapour pressure of dissociation. Because of this and because of adsorption and capillary effects, these hydrates cannot be separated from uncombined water simply by evaporating that water. Of necessity, studies of the physical structure of paste have been carried out in terms of the solid material that remains when all the water that will evaporate under arbitrarily selected conditions has been removed. For studies carried out by Powers and Brownyard before 1947 [*5*] the standard drying condition consisted of exposing granulated samples of standard granule size in a vacuum desiccator to a vapour pressure of 8 μ Hg at a temperature of 23° C. After 1947 standard conditions were substantially the same, with the important exception that the vapour pressure was maintained at 0·5 μ Hg [*6*]. Under either of these drying conditions the water that remains with the solid phase in the dry state is water of constitution and, at least at the higher of the two vapour pressures, some surface adsorbed water. But the water retained may not comprise all the water of constitution; Brunauer, Kantro and Copeland [*7*] estimated that a saturated specimen of pure tobermorite gel contains 1·5 moles water per mole SiO_2, but when dried to give a vapour pressure of 0·5 μ Hg at 23° C, it retained about 1 mole water. Once dried to this stage the material does not regain the $\frac{1}{2}$ mole when rewetted. It seems probable that this is true also of cement gel.

In view of the circumstance just described, the water retained by a sample under standard drying conditions is called *non-evaporable water*. That which is lost is of course called *evaporable water*.

As already mentioned, evaporable water comprises two categories: *gel water* and *capillary water*.

IV. Instability of Structure

From observations of such phenomena as irreversible stress–strain relationships and of irreversible shrinkage, it has long been known that the structure of hardened cement paste is not mechanically stable. More recently it has been learned that mechanical instability involves also instability of the specific surface area of the components of cement gel [*8*]. Structure becomes permanently altered when paste is dried for the first time, and the amount of alteration depends to a marked degree upon the rate at which the drying is carried out and on various other factors. It is imperative to follow a rigorously maintained routine of operations and of test conditions; reproducible experimental results are not readily obtainable. It is necessary to emphasize that knowledge of structure gained from the studies being reviewed is conditioned by methods of preparing the samples and controlling test conditions.

Since most of the information about structure was gained from once-dried samples, deductions from experimental data pertain to the once-dried state rather than to the original state of the material. However, there is as yet no reason to believe that the original state is much different from that observed.

V. The State of Evaporable Water

A. Adsorption Isotherms

Concepts of paste structure developed during the past two decades grew partly from studies of water vapour adsorption isotherms. These studies involved interpretations based on the assumption that the evaporable part of the total water in saturated hardened cement paste is restrained mainly by physical adsorption and by capillary forces. The foundation for this assumption has grown firmer as new information has accumulated.

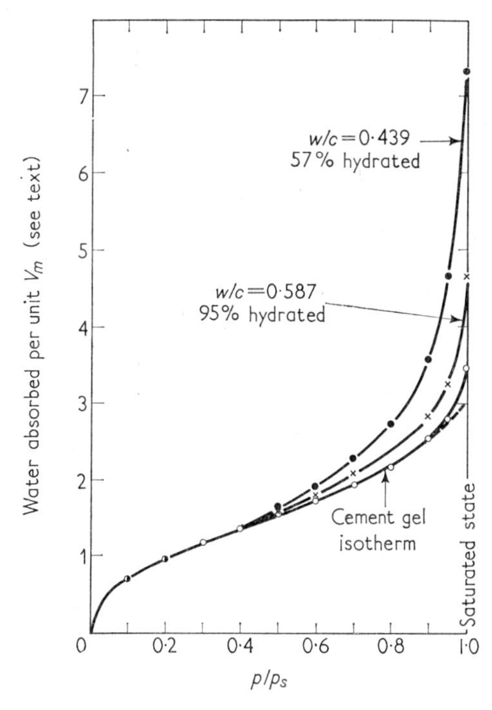

Fig. 1. Adsorption isotherms from cement pastes (25°).

The hydration products of Portland cement, in the form of specimens of hardened paste, give a smooth adsorption isotherm of Type II according to the Brunauer [*9*] classification. Examples are shown in Fig. 1. The scale of ordinates gives the amount of water adsorbed expressed as a multiple of the factor V_m, which is the Brunauer–Emmet–Teller (BET) surface area factor, but which may here be considered simply a quantity proportional to the amount of hydrated cement in the sample.

On the horizontal scale is the relative vapour pressure at 25° C, p being the existing pressure and p_s the vapour pressure at saturation. The lowest curve was found by Powers and Brownyard [*10*] to be the lowest possible position for any sample of hydrated Portland cement. (The broken line ending at 3 V_m indicates a later determination of the probable terminal point of this curve.) This limiting curve is approximated closely by Portland cements having different chemical compositions. It is therefore believed to be a characteristic of the structure of cement gel and is called the *cement gel isotherm.*

At relative vapour pressures below 0·45 the amount of water held at a given relative vapour pressure is proportional to the amount of cement gel and is independent of the porosity of the sample. This is evidence which, with other evidence, leads to the conclusion that the physical structure of cement gel is about the same in all pastes, but overall paste structure depends on the degree to which cement gel fills available space.

B. Hysteresis

Any adsorption curve such as one of those shown in Fig. 1 cannot be retraced by successively lowering the vapour pressure (desorption). When the desorption process is started from the saturated state, the amount of water retained at a point on the descending curve (not shown in Fig. 1) is always greater than at the corresponding point on the ascending curve, and this is true even in the range of low pressures, the closure of the loop not occurring until a relative vapour pressure of about 0·1 p_s is reached [*11*]. Thus, the adsorption and desorption curves for a given sample envelop an area; if at any stage of adsorption or desorption the process is reversed, the locus of the reverse curve falls inside the envelope. A given sample can produce innumerable adsorption–desorption relationships. This phenomenon, known as sorption hysteresis, is found among various chemically unrelated materials; the occurrence of adsorption hysteresis in cement paste is therefore regarded as one of the items of evidence leading to the conclusion that the evaporable water in cement paste is associated with the solid material predominantly by physical forces rather than by chemical combination.

C. Energy of Binding

The *mean* energy of binding of evaporable water, as determined by calorimetry [*12*], is greater the lower the porosity of the hardened paste; at the minimum porosity (28%) it is about 3600 cal/mole evaporable water. This amount is the excess over the normal heat of condensation, about 10,500 cal/mole. The energy of binding of evaporable water in the first adsorbed layer is estimated to be about 8000 cal/mole water. This is the order of magnitude characteristic of physical adsorption [*9*].

D. Mobility

Extensive experiments on the permeability of saturated hardened paste [*13*] to water indicated that under an externally produced hydraulic gradient all the evaporable water is mobile; that is, no appreciable amount of it was held by the solid phase in such a way that it did not respond to a pressure gradient, however small the gradient. The mean energy of activation for flow in cement paste is a function of porosity; it ranges in mature pastes from about 1·5 to 2·8 times the normal value for water at the same temperature [*14*].

E. Specific Volume

As already indicated in connection with the specific volume of the solids, the mean specific volume of evaporable water cannot be determined with certainty. However, it has been established that in all saturated cement pastes some of the evaporable water has the specific volume normal for the aqueous solution contained in the paste. This means that all the capillary water and some of the gel water has normal specific volume, but, as already indicated, the mean specific volume of gel water is not known with certainty. One estimate gives it a value of about 0·90 cm^3/g [*15*].

F. Conclusion

From evidence reviewed above it was concluded that nearly all the evaporable water in cement paste exists as a separate phase. Some of it is subject to the restraint of physical (van der Waals) adsorption and, if the relative humidity is sufficiently high, some of it may be restrained by hydrostatic tension induced by the capillary effect.

Some of the evaporable water in a saturated specimen may not be present as a separate phase. Tobermorite gel may lose $\frac{1}{2}$ mole water per mole silica during drying to a vapour pressure of 0·5 μ. This amount of evaporable water is apparently not regained during adsorption. Furthermore, there is evidence that some water becomes evaporable by partial decomposition of calcium sulphoaluminate and that some of this is regained in the middle range of relative vapour pressures. Possibly some of the other compounds show similar instability, although this has not yet been established as a matter of fact.

VI. Properties of Hydrated Cement

A. Mean Specific Surface Area

Since the evaporable water in hydrated paste is practically all in the condensed and physically adsorbed states, it was possible to estimate the internal specific surface area of the solid materials by means of the Brunauer–Emmett–Teller theory of multimolecular adsorption. As

TABLE I

Specific surface areas of hydrated cements by water vapour adsorption

Cement number	Computed composition of cement (%) C_3S	C_2S	C_3A	C_4AF	Specific surface (m^2/g) hydrated cement	
15754	45·0	27·7	13·4	6·7	219	267(1)
15756	48·5	27·9	4·6	12·9	200	253(1)
15763	28·3	57·5	2·2	6·0	227	265(1)
15758	60·6	11·6	10·3	7·8	193	249(1)
Average					210	258(1)
Ref. 7	100				210	293(1)
Ref. 7		100			279	299(1)

(1) Measured value for hydrated cement corrected for computed calcium hydroxide content.

already mentioned, the value for surface area determined by this method is somewhat dependent on the method of preparing the sample. The method used was such as to minimize and control the change due to the inherent structural instability.

The specific surface areas found for four different cements in the completely hydrated state are shown in Table I. The average of the four is 210 m^2/g and, although the range of chemical compositions is about as wide as is generally found among commercial cements of different types, the specific surface area of the hydration products differs from the average by less than 10%. This is an important part of the evidence that the physical characteristics of the hydration products are only slightly influenced by differences of chemical composition and that most of the products are in the colloidal state.

The average specific surface of hydrated cement is the same as that found by Brunauer *et al.* [*7*] for the specific surface area of hydrated tricalcium silicate, but it is less than that for a hydrated β-dicalcium silicate, the latter having a specific surface of 280 m^2/g. The difference between the areas of the two silicates is due to the difference in the amounts of calcium hydroxide in the product, the calcium hydroxide occurring as crystals having a negligible surface area. When the specific surface areas of the C_3S and C_2S pastes are corrected for their respective calcium hydroxide contents, the specific surface area of the tobermorite gel in those specimens turned out to be about 300 m^2/g. A similar correction applied to cement pastes gives an average specific surface of the gel material of 260 m^2/g. The difference between this figure and 300 is probably due to the presence of materials other than tobermorite gel and calcium hydroxide in Portland cement paste.

The specific surface area of hydrated cement as given by water vapour adsorption was confirmed by studies of permeability to liquid water. Powers, Mann and Copeland [*13*] obtained results in terms of "volume diameter" and "sphericity factor" quite compatible with the specific surface area determined by water vapour adsorption.

B. Specific Volume

One consequence of the high specific surface area of the solid material composing hardened cement paste and the concomitant smallness of the interstitial spaces, is that when different fluids are used to fill this space the capacity seems different for each fluid. Such an effect is believed to be due to differences in the size of molecules and in the energy of interaction between the molecules and the solid material. At present there is no unequivocal way to establish for hardened cement pastes the specific volume of the solids or its complement, the porosity. In Table II are shown the specific volumes of the hydration products of four different cements determined in two ways. In one case the specific volume is based on the assumption that the aqueous solution in cement paste has its normal specific volume, namely 0·99 cm^3/g [*16*]; in the other, the displacement of dry hydrated cement in helium is given [*15*].

Until it has been unequivocally established that the specific volume of adsorbed water is the same as that of physically free water, it seems preferable to assume that the best value for the specific volume of hydrated cement is that given by its displacement of helium.

TABLE II

Specific volumes and densities of hydrated cements

Cement number	Computed composition (%) C_3S	C_2S	C_3A	C_4AF	$CaSO_4$	Specific volume and density $v_{hc}^{(1)}$ (cm^3/g)	$\rho_{hc}^{(1)}$ (g/cm^3)	$v_{hc}^{(2)}$ (cm^3/g)	$\rho_{hc}^{(2)}$ (g/cm^3)
15754	45·0	27·7	13·4	6·7	4·0	0·397	2·52	0·411	2·43
15622	49·2	28·5	4·4	12·8	2·7	0·374	2·67	0·386	2·59
15669	33·0	54·2	2·3	5·8	3·1	0·374	2·67	0·386	2·59
15497	60·1	11·9	10·3	7·9	3·1	0·394	2·54	0·408	2·45

(1) Based on average value of hypothetical specific volume of non-evaporable water as determined from average displacements in helium of several dry hydrated cements.

(2) Based on pore volume equal to weight of evaporable water × 0·99 cm^3/g.

v_{hc} = Specific volume of hydrated cement, cm^3/g.
ρ_{hc} = Density of hydrated cement, g/cm^3.

C. MORPHOLOGY

Studies of hardened Portland cement paste with an electron microscope by Grudemo led him to the conclusion that "by far the largest part of the cement pastes examined consists of exceedingly ill formed colloidal products, in which it is sometimes difficult to discern any definite morphology" [*17*]. Copeland and Schulz [*18*], by observing pastes at early stages of hydration, arrived at slightly more definite impressions as to morphology, although they support Grudemo's conclusions. The colloidal material first appears as needle-like or lath-like protuberances. These bodies grow laterally and become irregular sheets. In mature samples the morphology is indefinite; attempts to describe it lead to using such terms as "thin plates", "flakes", "foil," "crumpled foil", "crepe paper"; in the denser parts of mature structures there appear to be "stacks of flat plates". Czernin reported similar observations [*19*].

Electron microscopists agree that the particles of colloidal material are very thin. Regardless of such irregularities as crumpling or rolling, a thin rectangular sheet serves as a suitable prototype. On that basis the average thickness of the colloidal particles in cement paste can be estimated as follows [*20*].

Consider the particle to be a rectangular prism having dimensions b, l, and t for width, length, and thickness. If σ is the specific surface area in cm^2/cm^3, then

$$\frac{\sigma}{2} = \frac{bt + lt + bl}{blt} = \frac{1}{b} + \frac{1}{l} + \frac{1}{t}$$

Multiplying by t and re-arranging, we obtain

$$t = \frac{2\left[1+\frac{t}{b}+\frac{t}{l}\right]}{\sigma}$$

The average specific surface area of cement gel, corrected for calcium hydroxide, is about 260 m^2/g (see Table I). Hence,

$$\sigma = 260\ m^2/g \times 2{\cdot}5\ g/cm^3 = 650\ m^2/cm^3$$

Using this figure and converting to Ångstrom units we obtain

$$t = 31\left(1+\frac{t}{b}+\frac{t}{l}\right)\ Å$$

On the basis of electron micrograph estimates, $b = 10t$ and $l = 300t$ seem reasonable minimum ratios. On this basis

$$t = 31 \times 1{\cdot}103 = 34\ Å$$

If the surface as determined by the permeability method had been used, the result would have been about 45 Å. Since the thickness of a unit-cell layer is about 10 Å, the indicated thickness is three or four layers.

VII. Porosity of Hardened Paste

A. Definition of Porosity

The porosity of cement paste is defined as the fraction of the volume of a saturated specimen occupied by evaporable water. Since the definition of evaporable water is somewhat arbitrary, as explained above, that of porosity is likewise arbitrary. Specifically, pore space is defined as the space in a saturated specimen occupied by that fraction of the total water content (including chemically bound water) which has a vapour pressure greater than 0·5 μ Hg at 23° C. Thus, if w_e is the weight of evaporable water and v_e its specific volume, the porosity, ϵ, is

$$\epsilon = \frac{w_e v_e}{V}$$

By definition

$$w_e = w_t - w_n = w_t - m w_n^0$$

where w_t is the total weight of water in the specimen, including the chemically combined part; w_n is the non-evaporable part; w_n^0 is the value of w_n for the completely hydrated state; m is the fraction of cement that has become hydrated. Taking v_e as approximately unity, we may use the relationship

$$\epsilon \sim \frac{w_e}{V} = (w_t - m w_n^0)\frac{1}{V} \quad (1)$$

Since data have usually been reported in terms of cement content, the following relationship, obtained by multiplying by V/c, is the most convenient one.

$$\frac{w_e}{c} = \frac{w_t}{c} - m\frac{w_n^0}{c} \tag{2}$$

Since w_n^0/c is a constant for a given cement, a plot of w_e/c vs. w_t/c for specimens having various porosities produces a straight line of unit slope for any fixed value of m; at $w_e/c = 0$, the intercept is mw_n^0/c.

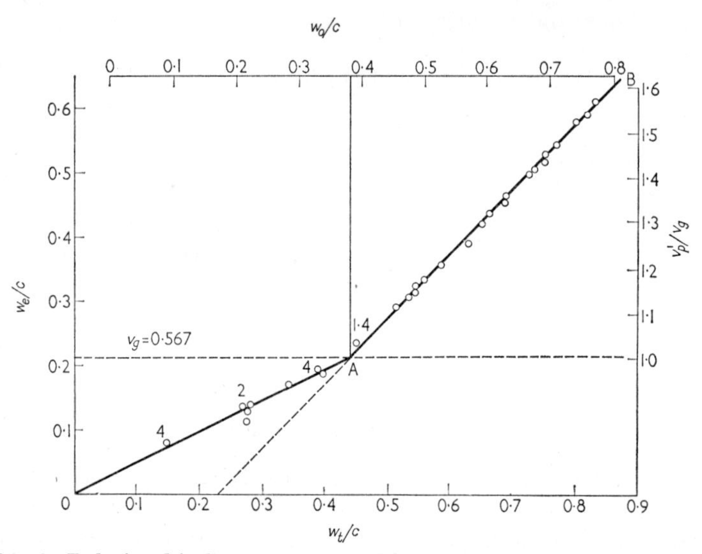

FIG. 2. Relationship between evaporable water content and total water content in specimens of mature cement paste. Numerals above points give period of wet curing in years.

Figure 2 shows a plot of experimental data in terms of equation (2) for specimens of paste made with a cement for which $w_n^0/c = 0{\cdot}227$. (Cement 15754, Table I.) The data divide themselves into two groups. One group, on line segment AB, conforms to equation 2 with $m = 1{\cdot}0$. Within this group porosity as indicated by w_e/c varies only with w_t/c, which in turn depends on w_o/c, the original net water/cement ratio given by the scale at the top; thus, for this group,

$$\frac{w_e}{w_t} = 1 - 1{\cdot}0\frac{w_n^0}{w_t} = 1 - \frac{0{\cdot}227}{w_t/c} \tag{3}$$

For the other group, on segment OA, m is less than 1·0 and the following ratios are constant with values as follows:

$$\left(\frac{w_e}{w_t}\right)^* = 0{\cdot}482; \quad \left(\frac{w_n}{w_t}\right)^* = 0{\cdot}518; \quad \left(\frac{w_e}{w_n}\right)^* = 0{\cdot}93$$

The asterisk denotes the minimum or maximum value attainable with the cement used.

Points along OA thus represent mixtures in which the porosity of the hydrated part of the paste is constant and which contain fractions of unhydrated cement. The fraction that has become hydrated is given by the following relationship:

$$m^* = \frac{w_o/c}{(w_o/c)^*} \tag{4}$$

where m^* is the ultimate value of m when a specimen contains an excess of cement, and $(w_o/c)^*$ is the lowest value of w_o/c that will permit complete hydration. For the cement represented in Fig. 2 $(w_o/c)^* = 0{\cdot}38$. For example, in a paste of "normal consistency" as prepared for a standard test, $w_o/c = 0{\cdot}23$ and $m^* = 0{\cdot}23/0{\cdot}38 = 0{\cdot}60$. This result shows that in such specimens 40% of the cement cannot become hydrated. Chemical equilibrium does not exist, but apparently the process stops because no spaces remain that are large enough to permit nucleation of new crystals and no existing crystals are able to grow into contiguous spaces [*22*].

B. Porosity of Cement Gel

From Fig. 2 and the above discussion it is apparent that the porosity of hydrated cement paste reaches its lowest value at point A, and that this same value applies at any possible point on segment OA. At point A the specific volume of the paste is 0·567 cm³ per g dry hydrated cement. The porosity at this point can be calculated from the following relationship:

$$\epsilon_g = 1 - \frac{\left(1+\frac{w_n^0}{c}\right) v_{hc}}{v_c' + v_w \left(\frac{w_o}{c}\right)^*} \tag{5}$$

where ϵ_g is the minimum porosity; v_{hc} is the specific volume of hydrated cement, in cm³ per g dry hydrated cement; v_c' is the apparent specific volume of cement in water [*21*]; v_w is the specific volume of the aqueous solution in fresh paste; $(w_o/c)^*$ is the lowest water/cement ratio that will permit complete hydration. For $w_n^0/c = 0{\cdot}227$, $v_{hc} = 0{\cdot}411$; $v_c' = 0{\cdot}315$; $v_w = 0{\cdot}99$ and $(w_o/c)^* = 0{\cdot}38$, equation (5) gives

$$\epsilon_g = 0{\cdot}28$$

This is considered to be the porosity of cement gel. No figure for cement gel produced by other cements has been established by data of the same kind. On the basis of similarities already discussed it is assumed that in any case the porosity of cement gel is that which corresponds to a capacity for evaporable water equal to $3V_m$. This means that none of the estimated figures for gel porosity would be substantially different from 28%.

C. Specific Volume of Cement Gel

As shown on Fig. 2, the specific volume of cement gel, defined as volume per unit weight of dry hydrated cement, is 0·567 cm³/g. (By definition the weight of dry hydrated cement is the same as that of cement gel.)

VIII. Increase of Solid Volume Due to Hydration

The factors by which the volume of solid material increases when cement is converted to hydrated cement and to cement gel are readily obtained from specific volume data given above. If the "specific volume" V/c of fresh paste (expressed as volume of paste per gram of dry cement) is smaller than a limiting value V^*/c, there will be an excess of cement that cannot become hydrated. Thus, the value V^*/c is defined as the least "specific volume" that will accommodate all the solid material that can be obtained from 1 g cement. The value of this limiting specific volume can be calculated as follows:

$$\frac{V^*}{c} = \left(1+\frac{w_n^0}{c}\right) v_g$$

where v_g is the specific volume of cement gel, in cm³ per g dry hydrated cement. Using the values given above, we obtain

$$\frac{V^*}{c} = 1{\cdot}227 \times 0{\cdot}567 = 0{\cdot}696 \text{ cm}^3 \text{ per g cement}$$

The least volume required by the *cement gel* produced by 1 cm³ cement is

$$\frac{V^*}{cv_c'} = \frac{0{\cdot}696}{0{\cdot}315} = 2{\cdot}2 \text{ cm}^3 \text{ per cm}^3 \text{ cement}$$

Similar computations for other cements give values close to 2·0. The principal conclusion is that the hydration products from 1 cm³ cement require at least 2 cm³ space.

IX. Size of Gel Pores

A. Specific Surface Area of Gel Pores

Since the specific surface of mature hydrated cement is independent of the porosity of the paste and since the pores are interstitial spaces among those particles, it is a simple matter to compute the boundary area of the voids per unit volume of void space, which is a specific surface area. Thus, the specific surface area of the void space may be expressed as

$$\sigma_v = \left[\frac{1-\epsilon}{\epsilon}\right] \sigma_{hc} \tag{7}$$

where σ_v is the specific surface area of the voids in cm^2/cm^3; ϵ is the porosity of the paste; σ_{hc} is the specific surface area of hydrated cement cm^2/cm^3. For example, for the same cement as above, the specific surface area of the hydrated cement is 219 m^2/g (adsorption method, Table I) and the specific volume of the hydrated cement is 0·41 cm^3/g (Table II); hence, the specific surface area of the solids is 530 m^2/cm^3. For a porosity of 28%, the minimum, the specific surface area of the void space is $(0 \cdot 72 \times 530)/0 \cdot 28 = 1360\ m^2/cm^3$. This result indicates that in cement gel the interstitial spaces are smaller than the solid bodies.

B. Average Width

The reciprocal of the specific surface area of void space is the hydraulic radius. That is, $1/\sigma_v$ is the hydraulic radius in cm when the specific surface area is given in cm^2/cm^3. The average distance from surface to surface is somewhere between two and four times the hydraulic radius, depending on the shape of the interstitial space [*10*]. Thus,

$$\text{Average distance between solid surfaces} = \frac{2 \text{ to } 4}{0 \cdot 136 \times 10^8}\ \text{cm} = 15 \text{ to } 30\ \text{Å}$$

If the pores in dense structure resembled thin slits, as seems likely, the shape factor is nearer to 2 than to 4. A reasonable estimate is that the *average* distance is about five times the diameter of a water molecule, or about 18 Å.

X. Continuity of Pores

The permeability of cement paste to water demonstrates that the pore system has continuity. In fresh paste the interstitial, water filled space consists of continuous interstitial capillaries, but in mature paste the capillaries may become discontinuous. Production of cement gel reduces capillary size, and, unless the original water/cement ratio exceeds about 0·7, more or less, depending on the fineness of the cement, cement gel eventually destroys the continuity of the capillaries [*23*]. When this stage of the process has developed, the capillary space consists of cavities scattered throughout the mass of cement gel. The term *capillary cavity* is used.

Destruction of capillary continuity does not destroy the continuity of the pore system as a whole. When there is no continuous capillary system, the flow cannot bypass the cement gel and, since the cement gel is intrinsically permeable, flow through cement gel does occur. This was proved by the data cited above showing that the solid surfaces over which flow takes place are the surfaces of the gel particles.

XI. Concept of Gel Structure

From the knowledge of the physical properties of the various constituents of Portland cement paste just reviewed, it is possible to deduce some of the processes involved in the maturing of cement paste and to develop concepts of submicroscopic structure such as have already been mentioned. We shall now refer to several observations and consider their bearing on structure.

A. Volume Constancy

During the processes of hydration after the time of final set a specimen of Portland cement paste remains relatively volume constant; in relation to the internal volume changes to be considered the external volume of paste may be considered invariant.† While the overall volume remains constant, the volume of solids within the boundaries of the specimen increases about 60% and, on account of gel pores, the effective increase in volume is at least 100%. This means that when the process of hydration is complete, about half the cement gel occupies the sites that were originally occupied by cement particles and the other half occupies space outside the original boundaries of grains. In effect the volume of each cement grain becomes doubled. Remembering that in fresh paste the cement particles form a network of cement grains almost touching each other, we see that the doubling of volume of those grains cannot have taken place by a symmetrical growth of cement gel outward from grain boundaries; if that should occur, a large expansion of the network of cement grains would be observed. Since the volume does remain relatively constant, the process by which anhydrous cement becomes cement gel must be such that cement gel is produced only where there is sufficient space to accommodate it; moreover, the cement gel evidently conforms to the shape of whatever space is available to it. Le Chatelier concluded that the process was one of solution followed by crystallization. This now seems to be an essentially correct explanation, but an incomplete one [*24*]. It seems that the process of dissolution is a special kind and that the subsequent crystallization involves also a complex

† Although it need not be considered here, expansion during curing, especially during the period of setting, is not always negligible. It varies widely among different Portland cements, and is not fully understood.

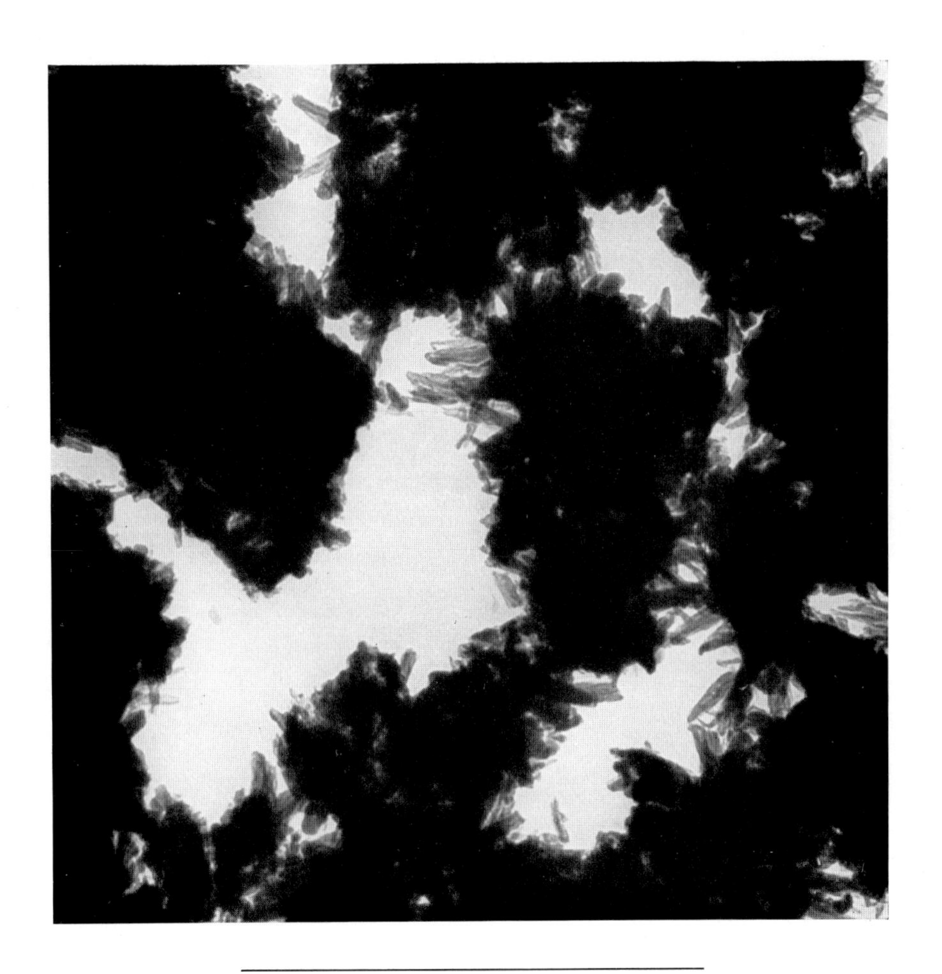

1μ

PLATE 1. Electronmicrograph of fragments of cements gel. (Photograph by Copeland and Schulz.)

diffusion process, with various periods of time between dissolution and crystallization. (The fact that crystallization is very imperfect is not especially significant in the present connection.) A diffusion process through pores too small to permit nucleation of a new solid phase enables half the dissolved hydration products to produce gel outside at the same time that an inward growth of cement gel follows the receding boundary of the anhydrous material. Presumably, a transition zone lies between the anhydrous phase and the gel, material in that zone and in gel pores being virtually in a state of solution.

B. Density

We have seen that most of the solid material of cement gel is composed of irregular sheets, foils, etc., only 3 or 4 mols thick. Such forms cannot possibly produce aggregates as dense as cement gel by a process of random aggregation, as for example by precipitation of crystallites from solution. To account for a density as high as the observed 0·72 cm^3 solid per cm^3 cement gel, it seems necessary to consider the process as one of gradual growth into available spaces. With a limit of particle thickness of 3 or 4 molecular layers growth must occur mostly at ends and edges, and one can imagine such growths invading the water filled interstitial region from all directions simultaneously. A density of about 0·72 cm^3 solid per cm^3 cement gel seems to be the highest average figure that can be attained by this process. It is to be kept in mind that this average figure includes not only the cement gel outside the original cement grain boundaries, but also that within those boundaries. It is possible, even probable, that the chemical composition of the products inside the grains is not identical with that outside; also, it is likely that there are some physical differences.

C. Uniformity

Although there are reasons to believe that cement gel is not chemically and physically homogeneous on a submicroscopic scale, there is also reason to believe that the inhomogeneity is of such kind and degree as to allow a considerable degree of uniformity in certain properties. In this connection, the significance of data on the permeability of cement pastes should be considered. As discussed above, in pastes having requisite density, resistance to the flow of water was found to be due to the same particles that, in dried pastes, adsorbed water vapour molecules. This means that the structure of dense cement paste is such that there is no way for water to flow through a specimen without encountering the individual gel particles and flowing over their surfaces. If the permeability of inside cement gel were very much different from that of outside cement gel, this experimental result could hardly have been attained. On this basis cement gel is regarded as having a density sufficiently uniform to account for the permeability results, and this may imply a rather high degree of physical uniformity.

Such uniformity is not especially difficult to account for. If the part of the hydration products that forms outside the original cement grain boundaries develops cement gel in the manner suggested, it would be natural for the growth from a given site to develop maximum possible density in the nearest spaces where there is opportunity for such growth. The process of outward growth is pictured as one of diffusion of the ions composing the solid material; the ions move through gel pores to a region where growth is possible and find permanent positions at ends and edges of the existing gel particles that border the region of growth. It is conceivable that a given particle has a growth limit and that all the particles bordering a given cavity may reach their growth limits before the cavity has become filled. In that case the solution concentration could rise sufficiently to allow the nucleation of a new particle. With or without the nucleation process one should expect the cavity to be filled to maximum density before the necessary concentration gradients for diffusion to a more remote cavity could develop.

XII. Paste Structure

A. Relation to Fresh Paste Structure

Although cement gel is believed to have a certain degree of physical homogeneity as just discussed, cement paste as a whole usually does not have such homogeneity. It is made up of cement gel, cavities not filled with gel, and usually some residue of cement. For reasons discussed above we may regard the cavities as residues of the originally water filled interstitial spaces and cement gel as being like a "sintered" mass of numerous porous bodies of hydration product, each such body being centred on the site of an original cement grain and sometimes enveloping a residue of the grain. Thus, we should expect the structure formed of gel and cavities to be built on a pattern established by the structure of the cement paste in its fresh state. This means that the structure of hardened paste, and thus its physical properties, depends first of all on the water/cement ratio and the maturity, and to a lesser degree on the number of cement particles per gram of cement.

The last mentioned aspect of paste structure was shown experimentally in terms of the lowest density of mature paste at which capillaries were found to be discontinuous, as indicated by permeability. (see Section X). Pastes having a wide range of densities and a corresponding range in water/cement ratio were made from one cement of normal specific surface area and from another having about double the normal specific surface. In terms of water/cement ratio the result was that pastes made with normal cement showed continuity of capillaries at all water/cement ratios above 0·7, whereas those made with superfine cement showed no capillary continuity until the water/cement ratio exceeded about 1·0, or, in terms of "specific volume" of fresh paste (cm^3 paste per g cement), the figures are about 1·0 and 1·3 cm^3 per g, respectively.

B. Effect of Rate of Hydration

Structure is probably influenced also by the temperature of the hydration reaction. Physical or chemical factors that accelerate early reactions generally reduce ultimate strength; presumably, the effect reflects some structural variant. Not much is known about such effects, except in the extreme case of high-pressure steam curing which causes a profound physical alteration and some chemical alteration in the structure; the principal effect is to lower the specific surface of the hydration products sufficiently to eliminate most of their colloidal attributes.

XIII. Visual Aids

Various attempts have been made to express concepts of paste structure by diagrams, each of them failing to depict more than a few aspects of the whole concept. Plate 1 serves perhaps as well as any that have been offered previously. It is an electron micrograph of fragments of cement gel. The dense bodies can be regarded as masses of cement gel, and the interstitial spaces as capillaries or capillary cavities. The fringe material is gel particles, and its relation to enclosed spaces illustrates the growth process discussed above.

This picture was selected mainly because it serves to illustrate certain aspects of the concept of structure deduced from various data. It may well be doubted that it is actually a picture of intact structure. Probably it is an accidental arrangement of fragments that happens to convey something of the author's concept. The gel particles seen here cannot be said to be typical, although they represent real cement gel. Such relatively regular shapes appear at early stages of hydration when only a little of the structure has reached maximum density. As congestion increases, particle form becomes less distinct, and finally may become so indistinct and various as to defy description. Nevertheless, the data here reviewed show that the specific surface area and density of gel particles remains nearly constant throughout the final 85 or 90% of the hydration process and that they will remain constant if the paste is not subjected to drying or other externally induced influences. Therefore, it seems that such features of structure as porosity and pore size are more distinctly revealed by measurements of related physical properties than they are by electron microscopy. Like the evidence from electron microscopy, the data from water vapour adsorption and permeability to water lead to the conclusion that the gel particles are far from being spherical and that they are so small that a particle of tobermorite gel, for example, contains perhaps as few as 40,000 formula units. In preparations in which particles can be seen distinctly, such as that shown in Plate 1, the visual information is quite compatible with such conclusions. Thus, there is reason to believe that even in preparations that fail to show distinct particles in a micrograph the indistinct particles are fundamentally like those that can be seen.

XIV. Mechanical Properties

A. Strength

Compressive strengths of specimens of cement paste of various water/cement ratios and at various states of maturity, cured at normal temperature, conform to the empirical relationship

$$f = AX^n$$

where f is compressive strength; X is the ratio of gel volume to the sum of gel volume and capillary space (it is called the gel/space ratio); A appears to represent the strength of the cement gel; n is a constant having a value usually between 2·6 and 3, depending on the characteristics of the cement. The above equation indicates that strength does not increase beyond the value of A; however, at $X = 1{\cdot}0$ strength increases as a function of increase in cement content, even though these increments of cement do not become hydrated. Using strength data that extended into the range where, presumably, $X = 1{\cdot}0$ for all *mature* specimens, Wischers [25] found that

$$f = 3100\,\rho_p^{2\cdot 7}\ \text{kg/cm}^2$$

or

$$f = 44{,}100\,\rho_p^{2\cdot 7}\ \text{p.s.i.}$$

where ρ_p is the volume of solid material, hydrated or not, in a unit volume of fully mature specimen.

All these relationships are empirical; they are valid only for results obtained under standardized conditions, and they involve constant factors whose physical significance, and hence limitations, are not yet clear.

B. Other Properties

1. *Elastic Modulus*

The modulus of the elasticity of specimens of cement paste is proportional to approximately the third power of the gel/space ratio. When the cement paste is a component of concrete or mortar, the modulus of elasticity is a complex function of the elastic properties of cement gel and of the aggregate (the aggregate including the residue of anhydrous cement, if any) and of the aggregate content of the mixture.

2. *Permeability*

The coefficient of permeability of mature paste to water is proportional to the following function of ϵ and T:

$$\frac{\epsilon^2}{1-\epsilon}\exp\left[\frac{-1250}{T}\left(\frac{1-\epsilon}{\epsilon}\right)\right]$$

The exponential term evaluates the increase above normal viscosity due to adsorption and other effects, ϵ being the total porosity of the paste and T the absolute temperature.

3. *Volume Changes*

Thermal volume changes, and volume changes in response to change in ambient humidity, are functions of paste structure. The thermal volume change for a given specimen is least when the specimen is either dry or saturated and greatest when the internal humidity of the specimen is about 70%. The lowest value of the thermal coefficient is about 11 millionths per degree C; the highest (at an intermediate state of dryness) may be three or four times as great.

Volume responses to variable ambient humidity (shrinking and swelling) are functions of the change in internal humidity of the material, and of the elastic and inelastic stress–strain characteristics of the material. When all the evaporable water is removed from a specimen for the first time, thereby reducing the internal humidity to about 0·02%, the resulting shrinkage is typically about 2% of the original volume, but it may be more or less than that, depending on the capillary porosity of the paste. After the first drying, different specimens swell on wetting and shrink on drying to extents that are nearly independent of differences in porosity.

The aggregate component of concrete restrains as much as 90% of the shrinkage of paste, the amount of restraint depending on the volume concentration of the aggregate material and the mechanical compressibility of that material.

REFERENCES

1. Steinour, H. H. (1944). *Industr. Engng Chem. (Industr. Ed.)* **36**, 840.
2. Ish-Shalom, Moshe and Greenberg, S. A. (1962). *Chemistry of Cement, Proceedings of the Fourth International Symposium, Washington 1960*, p. 731. National Bureau of Standards Monograph 43. U.S. Department of Commerce.
3. Carman, P. C. (1956). "Flow of Gases Through Porous Media". Academic Press, New York.
4. Copeland, L. E., Kantro, D. L. and Verbeck, G. (1962). *Chemistry of Cement, Proceedings of the Fourth International Symposium, Washington 1960*, p. 429. National Bureau of Standards Monograph 43. U.S. Department of Commerce.
5. Powers, T. C. and Brownyard, T. L. (1947). *Proc. Amer. Concr. Inst.* **43**, 249.
6. Copeland, L. E. and Hayes, John C. (1953). *ASTM Bull.* No. 194, 70.
7. Brunauer, S., Kantro, D. L. and Copeland, L. E. (1958). *J. Amer. chem. Soc.* **80**, 761.
8. Tomes, L. A., Hunt, C. M. and Blaine, R. L. (1957). *J. Res. nat. Bur. Stand.* **59**, 357.
9. Brunauer, S. (1945). "The Adsorption of Gases and Vapors". Princeton University Press.
10. Powers, T. C. and Brownyard, T. L. (1947). *Proc. Amer. Concr. Inst.* **43**, 469.
11. Powers, T. C. and Brownyard, T. L. (1947). *Proc. Amer. Concr. Inst.* **43**, 249.
12. Powers, T. C. and Brownyard, T. L. (1947). *Proc. Amer. Concr. Inst.* **43**, 549.
13. Powers, T. C., Mann, H. M. and Copeland, L. E. (1959). *Highw. Res. Bd, Wash., Special Report 40.*
14. Powers, T. C. (1962). *Chemistry of Cement, Proceedings of the Fourth International Symposium, Washington 1960*, p. 577. National Bureau of Standards Monograph 43. U.S. Department of Commerce.
15. Powers, T. C. and Brownyard, T. L. (1947). *Proc. Amer. Concr. Inst.* **43**, 669.
16. Copeland, L. E. (1955–1956). *Proc. Amer. Concr. Inst.* **52**, 863.
17. Grundemo, Å. (1962). *Chemistry of Cement, Proceedings of the Fourth International Symposium, Washington 1960*, p. 135. National Bureau of Standards Monograph 43. U.S. Department of Commerce.
18. Copeland, L. E. and Schulz, E. G. (1962). *J.P.C.A. Res. Dev. Labs* **4**, No. 2–12.
19. Czernin, W. (1959). Betonstein-Zeitung No. 9.
20. Brunauer, S. and Greenberg, S. A. (1962). *Chemistry of Cement, Proceedings of the Fourth International Symposium, Washington 1960*, p. 135. National Bureau of Standards Monograph 43. U.S. Department of Commerce.
21. Ford, C. L. (1958). *ASTM Bull.* No. 231, 81.
22. Powers, T. C. (1947). *Proc. Highw. Res. Bd, Wash.* **27**, 178.
23. Powers, T. C., Copeland, L. E. and Mann, H. M. (1959). *J.P.C.A. Res. Dev. Labs* **1**, 38.
24. Powers, T. C. (1961). *J.P.C.A. Res. Dev. Labs* **3**, 47.
25. Wischers, G. (1961). "Schriftenreihe der Zementindustrie", Vol. 28. Verein Deutscher Zement Werke, Düsseldorf.

CHAPTER 11

The Steam Curing of Portland Cement Products

H. F. W. TAYLOR

Department of Chemistry,
University of Aberdeen, Scotland

CONTENTS

I. Introduction

Steam curing of concrete has been developed for several reasons. Firstly, it greatly increases the rate of the hydration reactions; by curing in an autoclave at about 175° C, strengths can be obtained in about 15 h which are comparable to those obtained in 28 days at ordinary temperature. This permits the use of factory methods for the production of precast products; the demand for such methods seems likely to increase on account of the growth of prefabrication in building. Secondly, products can be made which are in some respects superior to those of normally cured concrete; the chemical resistance and the dimensional stability can be much improved. Lastly, there is a possibility of replacing the cement partly or wholly by waste materials, which are unreactive at ordinary temperatures but which possess cementing properties at higher temperatures.

Two essentially different types of high-temperature curing may be distinguished. Low-pressure steam curing is carried out at temperatures up to 100° C, under atmospheric pressure. High-pressure steam curing is carried out in autoclaves using saturated steam, usually at temperatures around 175° C.

II. Low-pressure Steam Curing

A. Methods of treatment and Properties of the Concrete

In low-pressure steam curing a normal Portland cement concrete mix is cured at 70–100° C in an atmosphere of saturated steam. In the most usual method the concrete is placed in heated chambers through which steam is passed. Other methods are: curing in hot water, and heating by passing an alternating electric current either through the concrete itself or, with reinforced concrete, through the reinforcement. The period of heating is usually 8–24 h.

The factors governing the choice of the curing cycle, and the physical and chemical properties of the product, have been discussed by Nurse [*1*]. There has been some disagreement as to the optimum heating rate and as to whether a period of initial curing is necessary. Nurse concluded that either an initial period of a few hours at room temperature or a slow rate of heating was desirable to avoid possible interference with the normal setting process, since there was some evidence that such interference could lower the strength ultimately reached. Figure 1 shows results obtained by Saul [*2*] for the maximum safe rate of heating, assuming a normal Portland cement to be used; different results were obtained with other types of cement.

Some investigators have postulated optimum curing temperatures, or ranges of temperature to be avoided; many of the early workers considered that lower strengths were obtained at 80–100° C than below or above this range. Saul [*2*] and Nurse [*1*] concluded that there was no basis for this view; satisfactory results could be obtained at all temperatures up to 100° C, provided the initial heating was not too rapid. Skramtaev [*3*] has recently expressed a similar view.

The strength of the concrete produced by low-pressure steam curing can be expressed as a percentage of that reached after 3-day normal curing. This percentage is, to a first approximation, a function of the product of curing time × curing temperature in °C [*1*]. Figure 2 shows the curve relating these parameters, which was derived by Nurse [*1*] on the basis of several independent studies. The experimental points showed a considerable spread, partly because of such factors as variations in the heating rate or the type of cement used, and Nurse emphasized the

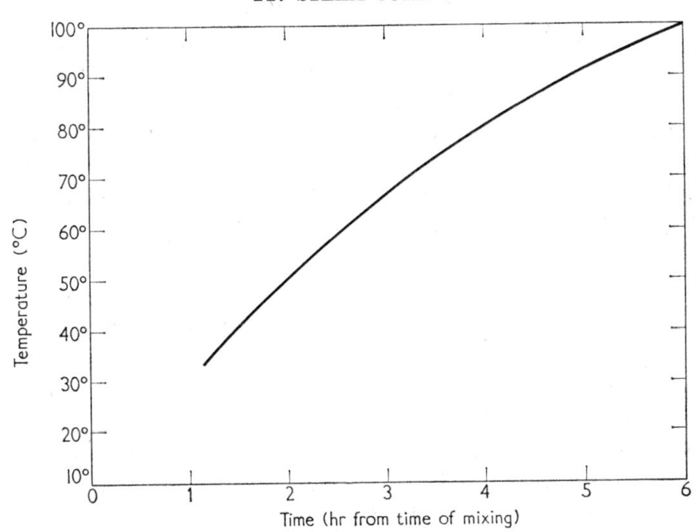

FIG. 1. Maximum rate of heating for steam-cured concrete [2].

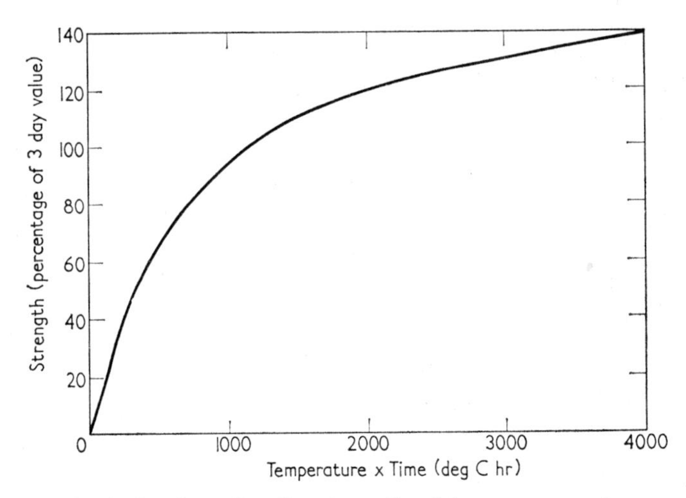

FIG. 2. Typical values for the strength of low-pressure steam-cured concrete expressed as a fraction of the 3-day strength obtained on normal curing and plotted as a function of the parameter (curing time × curing temperature in °C) [1].

approximate nature of the curve. It does not apply in the presence of reactive aggregates such as pozzolanas. More recently it has been recommended that the factor expressing the "maturity" of the concrete should be the product of the curing time and the temperature above $-10°$ C.

The soundness of these procedures has been discussed by Mironov and Ganin [*4*]. The converse procedure has also been advocated, namely, that curing in steam at atmospheric pressure or in hot water should be used to forecast the long-term strength of a normally cured cement or concrete [*5, 6*].

The effect on strength of variation in the type of cement has been studied recently by Malinovskiĭ [*7*] and by Keil and Narjes [*8*]. Malinovskiĭ recommended the use of a low-heat, slowly hardening cement. Like Nurse [*1*] he concluded that an initial period of normal curing was desirable; he considered it important that the period of maximum heat evolution by the cement should not coincide with steam curing. Keil and Narjes [*8*] also concluded that a cement low in C_3A should be used.

Low-pressure steam-cured concrete continues to gain in strength after the curing cycle has been completed, and the final strength, although slightly reduced, is comparable to that of normally cured concrete, provided the initial heating rate has not been excessive. In dimensional stability it does not differ significantly from normally cured concrete. The resistance to sulphate attack is not significantly altered by increase in curing temperature up to 90° C, but curing at 100° C has been reported to improve it [*9*]. The improvement is small compared with that obtained by high-pressure steam curing.

B. Phases Produced

1. Effect of Aggregate

Studies of pastes made from lime and finely divided quartz, discussed in Volume 2, Chapter 16, show that reaction is slow below about 120° C. Consequently, if Portland cement is used in conjunction with normal types of aggregate, reaction between cement and aggregate is not likely to be an important factor at temperatures up to 100° C.

2. C_3S and β-C_2S Pastes

The phases formed in β-C_2S pastes at temperatures around 100° C have been studied by Funk [*10*] and by Buckle and Taylor [*11*]. Both used X-ray and other methods. Buckle and Taylor also studied C_3S pastes. These studies showed that, broadly speaking, normal specimens of C_3S and β-C_2S behave in the same way as at room temperature; tobermorite gel and $Ca(OH)_2$ are formed, though reaction is more rapid.

Funk [*10*] showed that the behaviour of β-C_2S depends on the nature of the stabilizer. With some stabilizers, e.g. 0·5% B_2O_3, tobermorite gel and

$Ca(OH)_2$ are formed; the tobermorite gel formed at 100–120° C has a slightly higher Ca/Si ratio (1·88–1·89) than that formed at 50° C (1·76) but is closely similar in other respects. β-C_2S of this type was described as "completely stabilized". With other stabilizers, e.g. 0·2% Na_2O, a hydrated γ-C_2S-like phase was formed [*10, 12*]. β-C_2S of this type was described as "insufficiently stabilized". It seems likely that the β-C_2S in Portland cement is of the completely stabilized type.

Buckle and Taylor [*11*] confirmed that β-C_2S containing 0·5% B_2O_3 gave tobermorite gel and $Ca(OH)_2$. They showed that C_3S gave the same products; however, at temperatures above 50° C some tricalcium silicate hydrate was also formed. They obtained this compound below about 200° C only in freshly mixed pastes of C_3S; it did not form in aged pastes or in suspensions. Keevil and Thorvaldson [*13*] had much earlier obtained it by treating initially dry powders of C_3S with saturated steam at 110° C or above. Buckle and Taylor [*11*] considered that formation of tricalcium silicate hydrate occurred at relatively low temperatures only when the local and not merely the overall Ca/Si ratio was around 3; that is, it formed either by direct hydration of C_3S or else from a tobermorite gel from which the $Ca(OH)_2$ had barely started to segregate. The amount of tricalcium silicate hydrate formed within a few days at 100° C is not large; tobermorite gel and $Ca(OH)_2$ are the main products.

3. Neat Cement Pastes

The phases formed in neat cement pastes at temperatures around 100° C were first studied by Kalousek and co-workers [*14–18*], who used mainly DTA. Aitken and Taylor [*19*] later made a study using X-rays. These studies showed the main products to be tobermorite gel, probably in a modified form, and $Ca(OH)_2$. There are three main differences from the reaction at room temperature.

(a) Reaction is more rapid; Aitken and Taylor found that at 100–105° C the C_3A had mostly reacted within 1 day and that the C_3S had entirely reacted within 3 days, though the β-C_2S had not completely reacted within the latter period.

(b) Separate hydrated aluminate, ferrite or sulphoaluminate phases, if formed at all, are rapidly decomposed again; the Al^{3+}, Fe^{3+} and SO_4^{2-} ions apparently pass into the gel. Kalousek and co-workers [*14–16*] concluded from DTA evidence that, at all temperatures from 25° to 100° C, ettringite is formed at first, and that it is subsequently replaced by a C_4AH_{13}–$C_3A.CaSO_4.12H_2O$ solid solution, which in turn disappears, the Al^{3+} and $SO^2{}_4^-$ passing into the gel. At room temperature, in a typical case, the amount of ettringite reached a maximum in about

18 h and the amount of the solid solution in 7–28 days. The effect of temperature was to accelerate these changes; at 100° C only a little ettringite was detected after 30 min and the solid solution had disappeared within 4 h. Aitken and Taylor [*19*] could not detect any hydrated aluminate or sulphoaluminate phases at 100° C, even after only 3 h. These results do not necessarily conflict with Kalousek's, as the cement used by Aitken and Taylor may have reacted more rapidly; also, X-rays may be less sensitive than DTA for the detection of these phases. No indications were obtained, either by Kalousek and co-workers or by Aitken and Taylor, of the formation of C_3AH_6 or hydrogarnet type phases. The tobermorite gel formed in cement pastes at 100° C must differ from that formed in C_3S pastes at room temperature. Apart from the fact that it probably has a higher Ca/Si ratio [*10*], it contains Al^{3+}, Fe^{3+} and SO_4^{2-} ions.

(c) There are some indications that crystalline hydrated silicates are formed to a small extent, as well as tobermorite gel. Aitken and Taylor [*19*] obtained slender indications that some tricalcium silicate hydrate was formed at 50–75° C, while Kalousek and Prebus [*18*] reported that hillebrandite was formed at 50–120° C.

The results described here account in a general way for the observations on the strengths and other properties of low-pressure steam-cured concrete. The material is basically similar to that produced by normal curing; the principal effect is on the kinetics of the hydration reactions.

III. High-pressure Steam Curing

A. Introduction

In the high-pressure steam curing of Portland cement products typical procedure is as follows [*20*]. The mix contains Portland cement mixed with finely powdered silica, in addition to the aggregate; the silica/cement weight ratio is about 0·65. The product is allowed to stand for about 2 h at room temperature and is then placed in an autoclave. Saturated steam is admitted under pressure; the temperature and pressure are increased to about 175° C (9–10 kg/cm^2 or 125–140 lb/in^2) in not less than $1\frac{1}{2}$ h. These conditions are maintained for at least 8 h and the pressure is then allowed to fall to atmospheric in not less than 15 min.

The products thus obtained have compressive strengths which are normally at least as high as those obtained after 28-day curing at room temperature with comparable aggregates. They are superior to the latter in dimensional stability and in resistance to chemical attack.

Various types of concrete block and other precast products can be made in this way. The silica is generally added in the form of finely ground quartz. A particularly important application is in the production of asbestos cement products. The mix in this case consists of Portland cement, finely powdered silica and asbestos minerals.

The history of high-pressure steam curing goes back some 50 years; probably the earliest report is that of Wig [*21*] in 1912. Other early studies included those of Wig and Davis [*22*], Woodworth [*23*], Pearson and Brickett [*24*], Brown and Carlson [*25*] and Thorvaldson and co-

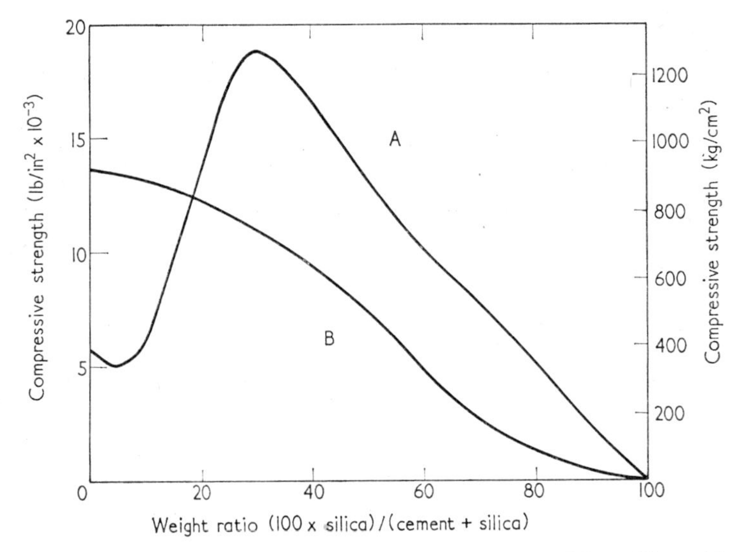

FIG. 3. Relationship between strength and amount of fine silica added to cement for mixes cured for 24 h at 177°C (curve A) and for 28 days at 21°C (curve B) [*29*].

workers [*13, 26, 27*]. These were reviewed by Thorvaldson in 1938 [*28*]. A very important advance was made in 1934–1936 by Menzel [*29*], who made a systematic study of the effects on the properties of the material of variations in the composition of the mix and the curing conditions. Typical results obtained by him are shown in Fig. 3. He showed that, if no finely ground silica or siliceous material was present, relatively poor strengths were obtained; they were lower than those obtained after 28-day curing at room temperature. Small additions of fine silica caused the strengths to drop even further, but with larger additions there was a marked increase in strength. The highest strengths were obtained at silica/cement weight ratios of 0·4–0·7. The proportion of silica required with a given cement depends on its particle size; if the silica is coarsely ground, more of it is needed. Menzel's results largely provide the basis for the typical procedure already described.

It is clear from the above that the development of strength in a high-pressure steam-cured product must depend, at least in part, on reaction between the cement and the fine silica. The laboratory studies described in the next sections confirm that this is the case. They show that the reactions are fundamentally different from those occurring during hydration at ordinary temperatures; in contrast, they have much in common with those taking place in the production of the autoclaved lime-silica, or calcium silicate products described in Volume 2, Chapter 16. There are, however, some important differences, particularly as regards the initial setting reactions and the consequent behaviour of the mix before the material is autoclaved.

B. Phases Produced in the Absence of a Reactive Aggregate

Studies of this type are only of background interest because, as already stated, the best products are not obtained from high-pressure steam curing unless a reactive aggregate is present.

1. C_3S and β-C_2S Pastes

The phases present in C_3S or β-C_2S pastes cured at 120–200° C have been studied by Butt and co-workers [*30, 31*], Malinina [*32*], Bozhenov and co-workers [*33, 34*] and Buckle and Taylor [*11*]. In some of these investigations [*30–34*] quantitative studies on strengths were also made. An electron microscope study has also been made (Chapter 9, p. 385). The principal results can be summarized as follows.

(a) With C_3S pastes autoclaved for times of 12–24 h higher strengths are obtained at about 175° C than at 200° C; with β-C_2S pastes the reverse is the case.

(b) With either C_3S or β-C_2S the initial products appear to be tobermorite gel and $Ca(OH)_2$. The tobermorite gel forms a matrix from which the crystalline phases, including $Ca(OH)_2$, gradually separate.

(c) α-C_2S hydrate forms at temperatures above about 120° C from either β-C_2S or C_3S. It is probably cryptocrystalline when first formed, but fairly quickly forms relatively large crystals. Tricalcium silicate hydrate forms only from C_3S. According to Buckle and Taylor [*11*], there are two distinct mechanisms for its formation: the low-temperature one, which has already been discussed and which is operative up to about 180°C, and a high-temperature one, which takes place by reaction of $Ca(OH)_2$ crystals with either α-C_2S hydrate or the tobermorite gel of the matrix. This occurs above about 180° C.

(d) The strength of the pastes depends on the continued existence of the matrix. With β-C_2S, reaction is sufficiently slow that a relatively strong product is obtained after 12–24 h at 200° C; although some α-C_2S hydrate is formed, much of the matrix remains. With C_3S, reaction is much more rapid, especially above 180° C, and at 200° C the matrix is largely destroyed after 12–24 h, giving α-C_2S hydrate, tricalcium silicate hydrate and $Ca(OH)_2$. The resulting product has little strength.

Butt, Rashkovich and Volkov [*35*] studied the behaviour in the autoclave at 200° C of C_3S, β-C_2S, C_3A, C_4AF and C_2F alone and in the presence of each other. Mixtures of C_3S or β-C_2S with C_3A were reported to yield free $Ca(OH)_2$ and hydrogarnets; C_3S and β-C_2S were found to be more reactive in the presence of C_3A. The reactivity of C_3S was also increased on addition of C_4AF; the reactivity of β-C_2S was increased by adding small amounts of C_4AF but decreased when larger amounts were added.

2. *Neat Cement Pastes*

The phases produced in neat cement pastes at 120–200° C were studied by Kalousek and Adams [*15*], Kalousek [*16*], Bozhenov and co-workers [*36–39*], Butt and Rashkovich [*31*], Sanders and Smothers [*40*] and Aitken and Taylor. [*41*] In the main, the results agree with what might be expected from the behaviour of C_3S and β-C_2S. No separate hydrated aluminate or sulphoaluminate phases appear to have been detected. Kalousek [*16*] studied five different cements at 125–175° C, and in no case were hydrogarnet-type phases observed. Most workers have observed the formation of tobermorite gel, $Ca(OH)_2$ and α-C_2S hydrate; Aitken and Taylor [*41*] found in addition that some tricalcium silicate hydrate was also formed at 170° C or above.

C. Phases Produced in the Presence of Finely Ground Quartz

The first systematic studies on the phases produced in autoclaved cement–silica pastes were made by Kalousek and co-workers [*14, 15, 42*]. They used DTA, soluble silica determinations and X-ray powder diffraction as the principal methods of identification. They showed that the high strengths obtained with optimum additions of fine quartz were associated with the formation of 11 Å tobermorite, and that the low strengths obtained with smaller additions of quartz were associated with the formation of α-C_2S hydrate.

Among later studies may be mentioned those of Butt and Rashkovich [*31*], Sanders and Smothers [*40*], Neese, Spangenberg and Weiskirchner [*45*], Bozhenov and Suvorova [*43*] and Aitken and Taylor [*41*] on cement–quartz pastes; of Butt and co-workers [*30, 31*], Malinina [*32*],

Bozhenov and co-workers [*33, 34*] and Berkovich *et al.* [*46*] on pastes of quartz with either β-C_2S or C_3S; and of Butt, Rashkovich and Tumarkina [*44*], Malinina [*32*] and Mironov, Astreeva and Malinina [*47*] on pastes of

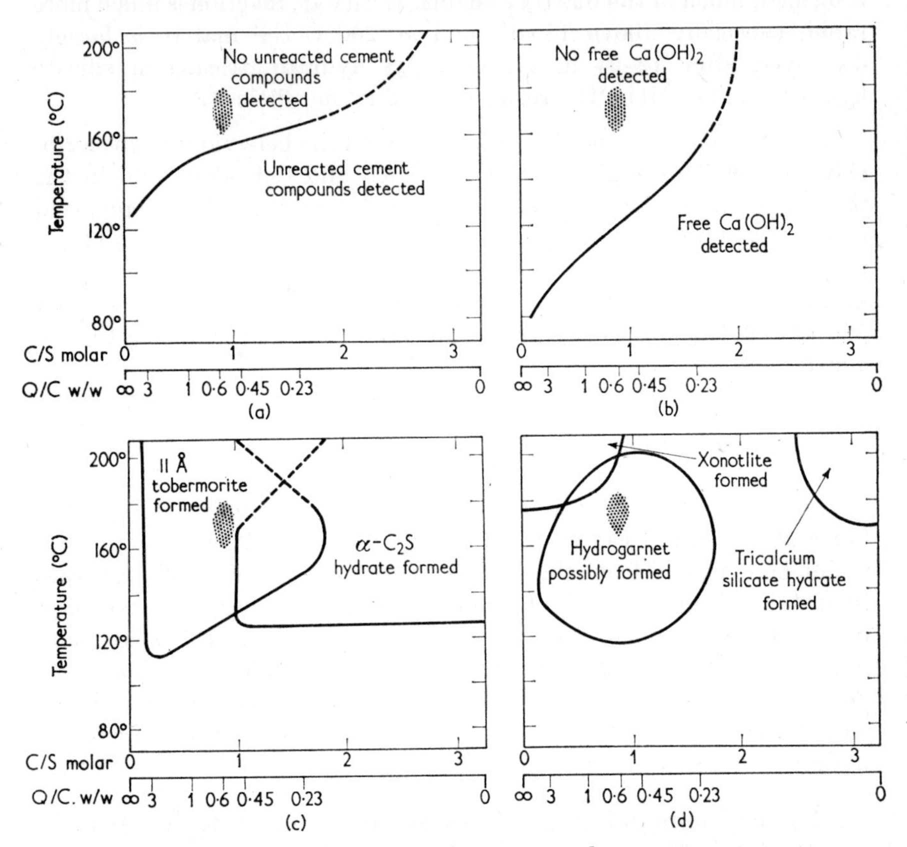

Fig. 4. Phases detected by X-rays in cement and cement–quartz pastes cured for 25 h [*41*]. (a) Occurrence of unreacted anhydrous compounds; (b) occurrence of free $Ca(OH)_2$; (c) occurrence of α-C_2S hydrate and of 11 Å tobermorite; (d) occurrence of xonotlite and tricalcium silicate hydrate, and possible occurrence of a hydrogarnet-type phase. Shaded area indicates conditions commonly used in practice.

quartz with C_3A or C_4AF. In many cases strength measurements were made [*30–33, 40, 46, 47*]. Pastes made from cement or calcium silicates together with amorphous silica have also been studied by electron microscopy (Chapter 9, p. 385). Broadly speaking, the results of these studies support and extend Kalousek's conclusions.

Figure 4 shows the principal results obtained by Aitken and Taylor, who used X-rays. The main conclusions resulting from this and the other studies are as follows.

The primary reaction at 85–200° C is probably always the hydration of the cement to give tobermorite gel and $Ca(OH)_2$. If any aluminate hydrate or sulphoaluminate hydrate phase is formed at all, it is quickly decomposed, though there are indications that within certain limits of quartz/cement ratio a hydrogarnet phase may separate out again at temperatures above about 120° C (Fig. 4(d)). In general, the Al_2O_3 seems to have only a minor effect on the subsequent behaviour of the tobermorite gel.

If no quartz is added, the tobermorite gel is slowly replaced by α-C_2S hydrate (above 125° C; Fig. 3(c)), and also by $C_6S_2H_3$ (above 165° C; Fig. 4(d)). Addition of small amounts of finely ground quartz (quartz/cement weight ratio = 0·1) suppresses the formation of $Ca(OH)_2$ and of $C_6S_2H_3$ (Figs. 4(b) and d)) and increases the production of α-C_2S hydrate (Fig. 4(c)). This corresponds to a minimum in strength. So long as unreacted quartz is present, there is probably a zoning within the tobermorite gel; material of low Ca/Si ratio, perhaps resembling C-S-H (I), is likely to form near the quartz grains, while material of higher Ca/Si ratio, perhaps resembling C-S-H (II), will tend to form further away.

At 120–180° C the C-S-H (II) tends to recrystallize to α-C_2S hydrate and the C-S-H (I) to 11 Å tobermorite. These processes can occur simultaneously over a considerable range of overall Ca/Si ratios (Fig. 4(c)). Formation of 11 Å tobermorite is favoured by having a high quartz/cement weight ratio; if this ratio rises above 0·7, however, unreacted quartz is liable to remain. Studies on lime–quartz pastes suggest that gyrolite might also form at high quartz/cement ratios, though this has not been observed in cement–quartz pastes. Xonotlite can also form within 2 h if the quartz/cement ratio is above 0·6 and the temperature above 175° C (Fig. 4(d)).

The addition of quartz seems to accelerate the reaction of the C_3S and β-C_2S (Fig. 4(a)). In contrast, there is some evidence that it retards the reaction of the C_3A [*41, 44*]. Butt *et al.* [*44*] and Mironov *et al.* [*47*] reported that hydrogarnets were formed in C_3A–quartz pastes; in C_4AF–quartz pastes various products were formed, including some free Fe_2O_3 and $Ca(OH)_2$.

To a fair approximation, the cement behaves as a source of all of the lime and part of the silica; the Al_2O_3 and Fe_2O_3 can be neglected. Provided the quartz is sufficiently finely ground to react completely, the phases formed are largely determined by the overall Ca/Si ratio for cement + quartz, as in an autoclaved lime–quartz paste; the cement can indeed be partly replaced by lime. 11 Å Tobermorite possibly forms more readily, and xonotlite less readily, when cement is used. This may be a consequence of the presence of Al^{3+}, which seems to stabilize 11 Å tobermorite.

No reliable quantitative data exist regarding the relative proportions of crystalline 11 Å tobermorite and of ill-crystallized tobermorite (C-S-H (I) and C-S-H (II)) which are needed to produce a product of maximum strength. Because of the correlation between the content of 11 Å tobermorite and strength, it has sometimes been assumed that the highest strength would be attained if the calcium silicate hydrate was entirely present in this form. A different view has been expressed by Bozhenov and co-workers [*33, 34*], who found that, if the curing time or temperature was increased beyond a certain point, the strength was reduced. They concluded that there exists a certain optimum ratio of crystalline to gelatinous products.

D. Influence of Type of Cement and Other Factors

High-pressure steam-cured products have until now nearly always been made using Portland cement designed to give the best possible results on normal curing. The chemical reactions that occur in the two cases are, however, so different that there is no reason to believe that such a cement is necessarily also the best when autoclave conditions are used. Very little work has yet been done on the design of Portland cement clinkers specifically for use in high-temperature steam curing.

Most manufacturers of autoclaved cement–silica products appear to have used normal types of Portland cement. Blair and Yang [*49*] reported that the sulphate resistance of autoclaved products made using a U.S. Type I (normal) cement was as high as that obtained if a Type V (sulphate resisting) cement was used, though for normally cured products the Type V cement gave better results. Butt, Rashkovich and Danilova [*30*] considered that clinkers of widely differing C_3S and β-C_2S contents could be used, provided the correct amounts of fine quartz were added; a clinker high in β-C_2S could, if properly used, give products at least equal in strength to those obtained using one high in C_3S. Butt [*48*] and Malinina [*32*] also concluded that a clinker low in C_3A and high in ferrite was desirable. Whether this last conclusion is true for any given case probably depends on the setting behaviour required.

The early studies on high-pressure steam curing [*21–29*] indicated optimum curing temperatures of 170–185° C, assuming curing times of 8–48 h. The two factors of time and temperature are not independent, because under saturated steam pressures 11 Å tobermorite is a metastable product at these temperatures; its formation from C-S-H (I) and subsequent conversion into xonotlite are governed by kinetic and not equilibrium considerations (Chapter 5). Higher temperatures can perhaps be used with advantage, provided that the technical difficulties of making suitable autoclaves can be overcome. Bozhenov *et al.* [*34*] reported the results of studies made on neat cements and cement–sand mortars at temperatures up to 364° C (200 kg/cm^2). They showed that the highest strengths could be obtained at 200–225° C (15–25 kg/cm^2). At these temperatures curing cycles could be used in which the temperature was brought up to the peak value in 2–4 h and brought down again in a similar period, without maintaining it at a steady value in between.

The optimum time and temperature have been reported to depend on the type of cement used [*31, 43*], cements high in C_3S and C_3A requiring less drastic treatment than those high in C_2S or ferrite. It has usually been considered desirable to limit the rates of heating and cooling or to allow a minimum period of preliminary curing at room temperature. The main stress has usually been placed on the rate of heating, but the cooling rate is perhaps no less important [*34*].

Butt [*50*] and Butt and Maĭer [*51*] have studied the possibility of seeding; the formation either of α-C_2S hydrate or of tobermorites can be assisted in this way. Chemodanov and Gavrilova [*52*] reported that 1—6% additions of NaCl or other chlorides accelerated the formation of crystalline products.

E. Properties of High-pressure Steam-cured Products

As already stated, the strengths of high-pressure steam-cured products are comparable with those of normally cured materials; typical results were given by Menzel [*29*]. Data on tensile strength have also been reported [*40, 53*]. Drying shrinkages are considerably lower than with normally cured concrete [*29, 53*]. This can be attributed to the high proportion of crystalline material.

The resistance to sulphate attack is markedly superior to that of normally cured Portland cement concrete [*9, 28, 50, 54*]. Blair and Yang [*49*] reported that autoclaved asbestos cement pipes which had been immersed for 24 years in a lake containing 12% of sulphates ($MgSO_4$ and Na_2SO_4) were physically sound; normally cured samples placed in the same environment were destroyed within a few years [*9*]. Blair and Yang attributed the superior stability of autoclaved products to four factors: (a) the elimination of free $Ca(OH)_2$, (b) the formation of a better crystallized calcium silicate hydrate, (c) the elimination of small amounts of sulphoaluminate hydrates and (d) the probable elimination of C_3AH_6.

F. Use of Materials Other than Cement and Quartz

The cement in a mix containing reactive silica can be partly or wholly replaced by lime [*42*]; to a first approximation, the properties of the product depend on the overall ratio of active lime to active silica. Products in which the cement is completely replaced by lime are discussed in Volume 2, Chapter 16.

It is also possible to use waste materials which are inert or possess only feeble hydraulic activity at ordinary temperatures. Examples are the use of γ-C_2S mixed with either lime or silica [*33*], and of nepheline slurry, a waste material from aluminium oxide extraction which contains about 80% of β-C_2S [*34*]. Many hydrated magnesium silicates can also be used as cements if they are first decomposed by heating in air to give reactive materials; thus, low-grade chrysotile asbestos wastes can be employed by heating to 700–900° C, followed by mixing into a paste with high-grade asbestos as an aggregate and autoclaving [*34, 55*].

Other siliceous materials may be used in place of finely ground quartz; Kalousek [*42*] investigated the use of pumice, expanded shale and expanded slag. With materials high in Al_2O_3 hydrogarnets are formed. Lime has been shown to react hydrothermally at 220° C or under with various silicates, including felspars [*56, 57*], muscovite [*57*], and fly ash, expanded shale, and granulated and foamed blastfurnace slag [*58*]. The most usual products are calcium silicate hydrates, often including 11 Å tobermorite under appropriate conditions, and hydrogarnets. Many of these substances, if sufficiently finely ground, could perhaps be used in conjunction with cement. The substitution of amorphous silica for quartz, however, appears to give unsatisfactory results in autoclaved products made with cement, little or no 11 Å tobermorite being formed under the conditions normally used [*40, 42*], though it can be used satisfactorily in products made with lime (Volume 2, Chapter 16). There is thus no certainty that a siliceous material that reacts hydrothermally with lime is also suitable for use in autoclaved cement products. Berkovich *et al.* [*46*] reported that reactions occur between the cement and the asbestos in asbestos–cement products. Ball and Taylor [*59*], in contrast, concluded that bulk reaction between lime and chrysotile is negligible at the temperatures used in industrial practice, though they considered that the chrysotile might undergo surface reactions which could affect the nature of the bond between it and the cement.

REFERENCES

1. Nurse, R. W. (1951). *Proc. Building Res. Cong.* 86.
2. Saul, A. C. A. (1951). *Mag. Concr. Res.* **1**, No. 6, 127.
3. Skramtaev, B. G. (1962). *Chemistry of Cement, Proceedings of the Fourth International Symposium, Washington 1960*, p. 1099. National Bureau of Standards Monograph 43. U.S. Department of Commerce.
4. Mironov, S. A. and Ganin, V. P. (1959). *Contract. Rec.*, December, p. 19.
5. King, J. W. H. (1957). *Civ. Engng, Lond.* **52**, 881.
6. Ackroyd, T. N. W. (1961). *Min. Proc. Inst. Civ. Engrs* **19**, 1.
7. Malinovskiĭ, R. K. (1956). *Trudy Soveshchaniya Khim. Tsementa, Moscow* 381 (*Chem. Abstr.* **52**, 6751).
8. Keil, F. and Narjes, A. (1959). *Zement-Kalk-Gips* **12**, 126.
9. Miller, D. G. and Manson, P. W. (1933). *U.S. Dept. Agric. Bull.* 358; (1940). *Proc. Amer. Soc. Test. Mater.* **40**, 988; (1951). *U. Minn. Tech. Bull.* 194.
10. Funk, H. (1957). *Z. anorg. Chem.* **291**, 276.
11. Buckle, E. R. and Taylor, H. F. W. (1959). *J. appl. Chem.* **9**, 163.
12. Funk, H. (1958). *Z. anorg. Chem.* **297**, 103.
13. Keevil, N. B. and Thorvaldson, T. (1936). *Canad. J. Res.* **B14**, 20.
14. Kalousek, G. L., Davis, C. W. and Schmertz, W. E. (1949). *J. Amer. Concr. Inst.* **45**, 693.
15. Kalousek, G. L. and Adams, M. (1951). *J. Amer. Concr. Inst.* **48**, 77.
16. Kalousek, G. L. (1954). *Proceedings of the Third International Symposium on the Chemistry of Cement, London 1952*, p. 334. Cement and Concrete Association, London.
17. Kalousek, G. L. (1955). *J. Amer. Concr. Inst.* **51**, 989.
18. Kalousek, G. L. and Prebus, A. F. (1958). *J. Amer. ceram. Soc.* **41**, 124.
19. Aitken, A. and Taylor, H. F. W. (1960). *J. appl. Chem.* **10**, 7.
20. Hansen, W. C. (1953). *J. Amer. Concr. Inst.* **49**, 841.
21. Wig, R. J. (1912). *Tech. Pap. nat. Bur. Stand.* No. 5.
22. Wig, R. J. and Davis, H. A. (1915). *Tech. Pap. nat. Bur. Stand.* No. 47.
23. Woodworth, P. M. (1930). *J. Amer. Concr. Inst.* **26**, 504.
24. Pearson, J. C. and Brickett, E. M. (1932). *J. Amer. Concr. Inst.* **3**, 537.
25. Brown, L. S. and Carlson, R. W. (1936). *Proc. Amer. Soc. Test. Mater.* **36**, 336.
26. Thorvaldson, T. and Shelton, G. R. (1929). *Canad. J. Res.* **1**, 148.
27. Thorvaldson, T. and Vigfusson, V. A. (1928). *Engng J., Montreal* **11**, 174.
28. Thorvaldson, T. (1939). *Proceedings of the Symposium on the Chemistry of Cements, Stockholm 1938*, p. 246. Ingeniörswetenskapsakademien, Stockholm.
29. Menzel, C. A. (1934). *J. Amer. Concr. Inst.* **31**, 125.
30. Butt, Yu. M., Rashkovich, L. N. and Danilova, S. G. (1956). *Dokl. Akad. Nauk S.S.S.R.* **107**, 571 (*Chem. Abstr.* **50**, 13399).
31. Butt, Yu. M. and Rashkovich, L. N. (1956). *Tsement, Moscow* **22**, 21 (*Chem. Abstr.* **51**, 14231).
32. Malinina, L. A. (1957). *Beton i Zhelezobeton*, No. 2, 65 (*Chem. Abstr.* **52**, 7647).
33. Bozhenov, P. I. and Kavalerova, V. I. (1959). 17-a Nauch. Konf. Professorsko-Prepodavat. Sostava Leningrad, p. 56 (*Chem. Abstr.* **55**, 15871).
34. Bozhenov, P. I., Kavalerova, V. I., Salnikova, V. S. and Suvorova, G. F. (1962). *Chemistry of Cement, Proceedings of the Fourth International Symposium, Washington 1960*, p. 327. National Bureau of Standards Monograph 43. U.S. Department of Commerce.

35. Butt, Yu. M., Rashkovich, L. N. and Volkov, V. V. (1958). *Izv. Vysshykh Uchebn. Zavedeniĭ, Khim. i Khim. Tekhnol.* **3**, 130 (*Chem. Abstr.* **53**, 2567).
36. Bozhenov, P. I. (1953). *Trud. Soveshch. Tsement. i Beton. Gidrotekh. Stroitel.* (*Lenizdat*) 53 (*Chem. Abstr.* **49**, 3503).
37. Bozhenov, P. I. and Suvorova, G. F. (1955). *Tsement, Moscow* **21**, 4 (*Chem. Abstr.* **50**, 5262).
38. Bozhenov, P. I. and Suvorova, G. F. (1956). *Trud. Soveshch. Khim. Tsementa, Moscow* 341 (*Chem. Abstr.* **52**, 6750).
39. Bozhenov, P. I. and Suvorova, G. F. (1957). *Tsement, Moscow* **23**, 8 (*Chem. Abstr.* **51**, 15917).
40. Sanders, L. D. and Smothers, W. J. (1957). *J. Amer. Concr. Inst.* **54**, 127.
41. Aitken, A. and Taylor, H. F. W. (1962). *Chemistry of Cement, Proceedings of the Fourth International Symposium, Washington 1960*, p. 285. National Bureau of Standards Monograph 43. U.S. Department of Commerce.
42. Kalousek, G. L. (1954). *J. Amer. Concr. Inst.* **50**, 365.
43. Bozhenov, P. I. and Suvorova, G. F. (1959). Doklady Mezhvuz. Konf. Izuchen. Avtoklavn. Materialov, Leningrad, p. 86 (*Chem. Abstr.* **56**, 9722).
44. Butt, Yu. M., Rashkovich, L. N. and Tumarkina, G. N. (1958). *Nauch. Dokl. Vyssheĭ Shkoly, Khim. i Khim. Tekhnol.* No. 3, 580 (*Chem. Abstr.* **53**, 2569).
45. Neese, H., Spangenberg, K. and Weiskirchner, W. (1957). *TonindustrZtg* **81**, 325.
46. Berkovich, T. M., Kheĭker, D. M., Gracheva, O. I. and Kupreyeva, N. I. (1958). *Dokl. Akad. Nauk S.S.S.R.* **120**, 372 (*Chem. Abstr.* **53**, 2950).
47. Mironov, S. A., Astreeva, O. M. and Malinina, L. A. (1957). *Tsement, Moscow* **23**, 9 (*Chem. Abstr.* **51**, 17135).
48. Butt, Yu. M. (1956). *Trud. Soveshch. Khim. Tsementa, Moscow*, 320 (*Chem. Abstr.* **52**, 6750).
49. Blair, L. R. and Yang, J. C. (1962). *Chemistry of Cement, Proceedings of the Fourth International Symposium, Washington 1960*, p. 849. National Bureau of Standards Monograph 43. U.S. Department of Commerce.
50. Butt, Yu. M. (1956). *Trud. mosk. khim.-tekhnol. Inst. Mendeleeva* No. 21, 144 (*Chem. Abstr.* **51**, 13349).
51. Butt, Yu. M. and Maĭer, A. A. (1957). *Trud. mosk. khim.-tekhnol. Inst. Mendeleeva* No. 24, 61 (*Chem. Abstr.* **52**, 13220).
52. Chemodanov, D. I. and Gavrilova, Z. Ya. (1959). *Sborn. Nauch. Trud. Tomsk. Inzh.-Stroitel. Inst.* **5**, 44 (*Chem. Abstr.* **56**, 8305).
53. Hansen, W. C. (1953). *J. Amer. Concr. Inst.* **49**, 745.
54. van Aardt, J. H. P. (1962). *Chemistry of Cement, Proceedings of the Fourth International Symposium, Washington 1960*, p. 835. National Bureau of Standards Monograph 43. U.S. Department of Commerce.
55. Mtschedlow-Petrossian, O. P. and Worobjow, J. L. (1960). *Silikattech.* **11**, 466.
56. Butt, Yu. M., Maĭer, A. A. and Manuĭlova, N. S. (1958). *Sborn. Trud. Resp. Nauch-issled. Inst. Mestnykh Stroitel. Materialov* No. 14, 20 (*Chem. Abstr.* **53**, 20747).
57. Assarsson, G. O. (1960). *J. phys. Chem.* **64**, 626.
58. Midgley, H. G. and Chopra, S. K. (1960). *Mag. Concr. Res.* **12**, 73.
59. Ball, M. C. and Taylor, H. F. W. (1963). *J. appl. Chem.* **13**, 145.

Author Index

Numbers in brackets are reference numbers and are included to assist in locating references in which the authors' names are not mentioned in the text. Numbers in italics indicate the page on which the reference is listed.

E

F

I

J

K

L

R

T

U

V

W

Y

Z

Subject Index

Formulae in conventional chemical notation are given at the end of the entry for the appropriate initial letter. Formulae in cement chemical notation are given immediately before those in conventional notation. Thus, for Ca...... see page 453; for C...... see pages 449–452.

C

D

E

I

J

K

L

M

N

O

P